AF773761

# FRONTIERS IN NUCLEAR STRUCTURE, ASTROPHYSICS, AND REACTIONS

## FINUSTAR 2

To learn more about AIP Conference Proceedings, including the Conference Proceedings Series, please visit the webpage
**http://proceedings.aip.org/proceedings**

# FRONTIERS IN NUCLEAR STRUCTURE, ASTROPHYSICS, AND REACTIONS

# FINUSTAR 2

*Crete, Greece 10 - 14 September 2007*

*EDITORS*

P. Demetriou
*Institute of Nuclear Physics, NCSR "Demokritos"*
*Athens, Greece*

R. Julin
*Department of Physics, University of Jyväskylä*
*Jyväskylä, Finland*

S. V. Harissopulos
*Institute of Nuclear Physics, NCSR "Demokritos"*
*Athens, Greece*

**SPONSORING ORGANIZATIONS**

NCSR "Demokritos"
University of Jyväskylä
National Technical University of Athens
Greek Ministry of Education

AMERICAN INSTITUTE OF PHYSICS

**Melville, New York, 2008**
**AIP CONFERENCE PROCEEDINGS ■ 1012**

**Editors**

Paraskevi Demetriou
Institute of Nuclear Physics
NCSR "Demokritos"
POB 60228
153.10 Aghia Paraskevi
Athens, Greece

E-mail: vivian@inp.demokritos.gr

Rauno Julin
Department of Physics
P.O. Box 35 (YFL)
40014 University of Jyväskylä
Finland

E-mail: julin@phys.jyu.fi

Sotirios V. Harissopulos
Tandem Accelerator Laboratory
Institute of Nuclear Physics
NCSR "Demokritos"
POB 60228
153.10 Aghia Paraskevi
Athens, Greece

E-mail: sharisop@inp.demokritos.gr

Authorization to photocopy items for internal or personal use, beyond the free copying permitted under the 1978 U.S. Copyright Law (see statement below), is granted by the American Institute of Physics for users registered with the Copyright Clearance Center (CCC) Transactional Reporting Service, provided that the base fee of $23.00 per copy is paid directly to CCC, 222 Rosewood Drive, Danvers, MA 01923, USA. For those organizations that have been granted a photocopy license by CCC, a separate system of payment has been arranged. The fee code for users of the Transactional Reporting Services is: 978-0-7354-0532-5/08/$23.00

© 2008 American Institute of Physics

Permission is granted to quote from the AIP Conference Proceedings with the customary acknowledgment of the source. Republication of an article or portions thereof (e.g., extensive excerpts, figures, tables, etc.) in original form or in translation, as well as other types of reuse (e.g., in course packs) require formal permission from AIP and may be subject to fees. As a courtesy, the author of the original proceedings article should be informed of any request for republication/reuse. Permission may be obtained online using Rightslink. Locate the article online at http://proceedings.aip.org, then simply click on the Rightslink icon/"Permission for Reuse" link found in the article abstract. You may also address requests to: AIP Office of Rights and Permissions, Suite 1NO1, 2 Huntington Quadrangle, Melville, NY 11747-4502, USA; Fax: 516-576-2450; Tel.: 516-576-2268; E-mail: rights@aip.org.

L.C. Catalog Card No. 2008926439

ISBN 978-0-7354-0532-5
ISSN 0094-243X

Printed in the United States of America

## CONTENTS

**Preface** .................................................... xv
**Committees** .................................................... xvii
**Conference Photo** .................................................... xix
**Scientific Program** .................................................... xxi


### ORAL CONTRIBUTIONS

**Quantum Phase Transitions in Atomic Nuclei** .................................... 3
J. Jolie and A. Dewald
**Recent Studies on Nuclear Phenomena at the Extremes Using $\gamma$-ray Spectroscopy** .................................... 11
S. Leoni, A. Bracco, F. Camera, G. Benzoni, N. Blasi, F. C. L. Crespi, B. Million, D. Montanari, and O. Wieland
**Shape-Transitional Nuclei: What Can We Learn from the Yrare States? Or Hello the Double Vacuum; Goodbye $\beta$-vibrations!** .................. 19
J. F. Sharpey-Schafer, S. M. Mullins, R. A. Bark, E. Gueorguieva, J. Kau, F. Komati, J. J. Lawrie, P. Maine, A. Minkova, S. H. T. Murray, N. J. Ncapayi, and P. Vymers
**Observation of a Double-humped $\gamma$-Ray Fold Distribution in $^{142}$Eu** ........... 26
R. M. Lieder, A. A. Pasternak, E. O. Lieder, W. Gast, G. de Angelis, and D. Bazzacco
**Neutrinos and Supernovae** .................................... 31
B. S. Meyer
**Spallation Nucleosynthesis by Accelerated Charged-Particles** .................. 38
S. Goriely
**Neutrino-nucleus Reaction Rates Based on the Relativistic Quasiparticle Random Phase Approximation** .................................... 46
N. Paar, D. Vretenar, T. Marketin, and P. Ring
**Spin-Dependent Transitions in Nuclei and Neutrino-Nucleus Reactions** ........ 51
T. Suzuki, S. Chiba, M. Honma, K. Higashiyama, T. Yoshida, H. Umeda, K. Nomoto, T. Kajino, and T. Otsuka
**Spectroscopy of Very Heavy Elements** .................................... 56
P. T. Greenlees, R.-D. Herzberg, S. Ketelhut, D. Ackermann, P. A. Butler, A. Chatillon, S. Eeckhaudt, B. J. P. Gall, T. Grahn, C. Gray-Jones, A. Görgen, F.-P. Heßberger, U. Jakobsson, G. D. Jones, P. Jones, R. Julin, S. Juutinen, T.-L. Khoo, W. Korten, M. Leino, S. Moon, M. Nyman, J. Pakarinen, P. Papadakis, P. Peura, P. Rahkila, M. Richer, D. Rostron, C. Santos, J. Sarén, C. Scholey, J. Sorri, S. K. Tandel, Ch. Theisen, J. Uusitalo, and M. Venhart

**Spectroscopy of Heavy Elements at Dubna** .................................... 64
O. Dorvaux, A. Lopez-Martens, K. Hauschild, A. V. Yeremin, A. Khouaja, A. V. Belozerov, Ch. Briançon, M. L. Chelnokov, V. I. Chepigin, D. Curien, P. Désesquelles, B. Gall, V. A. Gorshkov, M. Guttormsen, F. Hanappe, A. P. Kabachenko, F. Khalfallah, A. Korichi, A. C. Larsen, O. N. Malyshev, A. Minkova, Yu. Ts. Oganessian, A. G. Popeko, M. Rousseau, N. Rowley, R. N. Sagaidak, S. Sharo, A. V. Shutov, S. Siem, V. I. L. Stuttgé, A. I. Svirikhin, N. U. H. Syed, and Ch. Theisen

**Confirmation of Super Heavy Element Production in $^{48}$Ca-induced Fusion Reactions: A Handshake of Physics and Chemistry for Element 112** .................................... 69
S. Hofmann, D. Ackermann, S. Antalic, H. G. Burkhard, V. F. Comas, R. Dressler, Z. Gan, S. Heinz, J. A. Heredia, H. F. Heßberger, J. Khuyagbaatar, B. Kindler, I. Kojouharov, P. Kuusiniemi, M. Leino, B. Lommel, R. Mann, G. Müzenberg, K. Nishio, A. G. Popeko, S. Saro, H. J. Schött, B. Streicher, B. Sullignano, J. Uusitalo, M. Venhart, A. V. Yeremin, R. Eichler, N. V. Aksenov, A. V. Belozerov, G. A. Bozhikov, V. I. Chepigin, S. N. Dmitriev, H. W. Gäggeler, V. A. Gorshkov, F. Haenssler, M. G. Itkis, A. Laube, V. Ya. Lebedev, O. N. Malyshev, Yu. Ts. Oganessian, O. V. Petrushkin, D. Piguet, P. Rasmussen, S. V. Shishkin, A. V. Shutov, A. I. Svirikhin, E. E. Tereshatov, G. K. Vostokin, and M. Wegrzecki on behalf of the SHIP Collaboration and the 112 Chemistry Collaboration

**Properties of Heavy and Superheavy Nuclei in Supernova Environments** .................................... 74
T. J. Bürvenich, I. N. Mishustin, and W. Greiner

**Single-Neutron States in $^{101}$Sn** .................................... 79
D. Seweryniak, M. P. Carpenter, S. Gros, A. A. Hecht, N. Hoteling, R. V. F. Janssens, T. L. Khoo, T. Lauritsen, C. J. Lister, G. Lotay, D. Peterson, A. P. Robinson, W. B. Walters, X. Wang, P. J. Woods, and S. Zhu

**Quadrupole Collectivity of Neutron-Rich Nuclei around $^{132}$Sn** .................................... 84
Th. Kröll, T. Behrens, R. Krücken, V. Bildstein, T. Faestermann, R. Gernhäuser, M. Mahgoub, P. Maierbeck, M. Münch, W. Weinzierl, F. Ames, D. Habs, O. Kester, R. Lutter, T. Morgan, M. Pasini, K. Rudolph, P. Thirolf, J. Diriken, M. Huyse, O. Ivanov, P. Mayet, N. Patronis, I. Stefanescu, J. Van de Walle, P. Van Duppen, O. Niedermaier, H. Scheit, D. Schwalm, J. Eberth, F. Finke, D. Martin, P. Reiter, A. Scherillo, M. Seidlitz, N. Warr, D. Weisshaar, J. Iwanicki, P. Butler, J. Cederkäll, E. Clément, P. Delahaye, L. M. Fraile, G. Georgiev, U. Köster, T. Sieber, D. Voulot, F. Wenander, S. Franchoo, A. Hurst, A. Ekström, P. E. Kent, K.-H. Speidel, J. Leske, S. Scheilke, A. Jungclaus, V. Modamio, J. Walker, L. Coquard, M. Pantea, N. Pietralla, T. Davinson, S. Nardelli on behalf of the REX-ISOLDE and MINIBALL Collaborations

**Probing the Single-Particle Structure around $^{54}$Ca with One-Neutron Knockout** .......... 89
P. Maierbeck, R. Gernhäuser, R. Krücken, Th. Kröll, H. Alvarez-Pol, F. Aksouh, T. Aumann, K. Behr, E. A. Benjamin, J. Benlliure, V. Bildstein, M. Böhmer, K. Boretzky, M. J. G. Borge, A. Brünle, A. Bürger, M. Caamano, E. Casarejos, A. Chatillon, L. V. Chulkov, D. C. Gil, J. Enders, K. Eppinger, T. Faestermann, J. Friese, L. Fabbietti, M. Gascon, H. Geissel, J. Gerl, M. Gorska, P. G. Hansen, B. Jonson, R. Kanungo, O. Kiselev, I. Kojouharov, A. Klimkiewicz, T. Kurtukian, N. Kurz, K. Larsson, T. Le Bleis, K. Mahata, L. Maier, T. Nilsson, C. Nociforo, G. Nyman, C. Pascual-Izarra, A. Perea, D. Perez, A. Prochazka, C. Rodriguez-Tajes, D. Rossi, H. Schaffner, G. Schrieder, S. Schwertel, H. Simon, B. Sitar, M. Stanoiu, K. Sümmerer, O. Tengblad, H. Weick, and S. Winkler

**SPIRAL2 at GANIL: Status and Perspectives** .......... 94
S. Gales

**The HIE-ISOLDE Project** .......... 106
K. Riisager

**Neutron Capture Reactions at DANCE** .......... 111
T. Bredeweg on behalf of the DANCE Collaboration

**Proton-induced Fission on Actinide Nuclei at Energies between 20 and 70 MeV** .......... 118
S. Isaev, R. Prieels, Th. Keutgen, J. Van Mol, Y. El Masri, and P. Demetriou

**Towards Universal Predictions with the TALYS Code** .......... 123
S. Hilaire, S. Goriely, A. J. Koning, and E. Bauge

**Advancing Along the N=Z Line: A 30 Year Odyssey** .......... 128
C. J. (Kim) Lister

**High-Spin States of $^{84,85}$Br: Mapping the Proton Sub-Shells towards $^{78}$Ni** .......... 134
A. Astier, M.-G. Porquet, Ts. Venkova, I. Deloncle, F. Azaiez, A. Buta, D. Curien, O. Dorvaux, G. Duchêne, B. J. P. Gall, F. Khalfallah, I. Piqueras, M. Rousseau, M. Meyer, N. Redon, O. Stézowski, R. Lucas, and A. Bogachev

**Nuclear Structure Far from Stability at the N=50 Shell Closure** .......... 139
E. Sahin, G. de Angelis, A. Gadea, G. Duchene, T. Faul, D. R. Napoli, E. Farnea, J. J. Valiente-Dobón, R. Orlandi, D. Mengoni, F. Della Vedova, R. P. Singh, C. Ur, F. Recchia, M. N. Erduran, M. Bostan, S. Erturk, S. Aydin, K. T. Wiedemann, A. M. Stefanini, S. Lenzi, S. Lunardi, N. Marginean, L. Corradi, E. Fioretto, G. Montagnoli, F. Scarlassara, P. Mason, S. Szilner, D. Ackermann, G. Pollarolo, J. R. B. Oliveira, T. Byrski, D. Curien, B. Gall, F. Haas, O. Dorwaux, J. Robin, M. Nespolo, F. Azaiez, G. De France, A. Algora, X. Liang, F. Nowacki, Y. H. Zhang, and X. H. Zhou

**Recent Results on the $^{12}$C+ $^{12}$C Reactions** .......... 144
T. Spillane, F. Raiola, S. Zeng, C. Rolfs, D. Schürmann, F. Strieder, H.-W. Becker, C. Bordeanu, L. Gialanella, M. Romano, and J. Schweitzer

**$^{12}C(\alpha,\gamma)^{16}O$ Thermonuclear Reaction Rate: Present Status and Perspectives** .......... 150
A. Lefebvre-Schuhl

**AGB Fluorine Nucleosynthesis Studied by Means of Trojan-Horse Method: The Case of $^{15}N(p,\alpha)^{12}C$** .......... 155
R. G. Pizzone, C. Spitaleri, M. La Cognata, S. Cherubini, V. Crucillà, M. Gulino, L. Lamia, S. M. R. Puglia, G. G. Rapisarda, S. Romano, S. Sergi, A. Tumino, R. Tribble, L. Trache, C. Fu, V. Goldberg, A. Mukhamedzhanov, and G. Tabacaru

**Two Experimental Approaches in Nuclear Astrophysics Using Neutrons** .......... 160
Y. Parpottas, C. Iliadis, and H. Tsertos

**Recent Developments in Coulomb Breakup Calculations** .......... 165
P. Capel

**Enhanced Photoneutron Cross Sections for Zirconium Isotopes Near Threshold: Evidence for Extra M1 Strength?** .......... 173
H. Utsunomiya, S. Goriely, A. J. Koning, S. Hilaire, S. Goko, A. Makinaga, T. Kaihori, T. Kondo, T. Yamagata, H. Akimune, H. Toyokawa, T. Matsumoto, H. Harano, and Y.-W. Lui

**Cross-Section Measurements of $\alpha$-particle Capture Reactions Relevant to the P Process: The Case of $^{65}Cu(\alpha,\gamma)^{69}$ Ga** .......... 179
A. Lagoyannis, T. Konstantinopoulos, P. Demetriou, H.-W. Becker, and S. Harissopulos

**Microscopic Nature of $0^+$ States Populated in Large Abundance in Deformed Nuclei** .......... 183
N. Lo Iudice, A. V. Sushkov, and N. Yu. Shirikova

**Nature of the $K^{\pi}=4^+$ Bands in the Os Isotopes** .......... 188
P. E. Garrett, A. A. Phillips, L. Bettermann, N. Braun, D. G. Burke, G. A. Demand, T. Faestermann, P. Finlay, K. L. Green, R. Hertenberger, K. G. Leach, R. Krücken, M. A. Schmaker, C. E. Svensson, H.-F. Wirth, and J. Wong

**$\beta$-Delayed Particle Studies of the Halo Nucleus $^{11}Li$ and Its Core $^{9}Li$** .......... 193
M. Madurga, M. J. G. Borge, L. M. Fraile, H. O. U. Fynbo, B. Jonson, T. Nilsson, G. Nyman, Y. Prezado, K. Riisager, and O. Tengblad on behalf of the IS-417 and TONNERRE Collaborations

**Thermodynamic Properties of Atomic Nuclei with $T<1$ MeV** .......... 198
M. Guttormsen, A. C. Larsen, J. Rekstad, A. Schiller, S. Siem, N. U. H. Syed, and A. Voinov

**Coupled Channel Calculations for Reactions With The $^{16}O+{}^{64}Zn$ System** .......... 203
J. F. P. Huiza, E. Crema, D. S. Monteiro, J. M. B. Shorto, R. F. Simões, and P. R. S. Gomes

**First Evidence of Di-Proton Decay from $^{18}Ne$ Excited States** .......... 208
C. Sfienti, M. De Napoli, G. Raciti, E. Rapisarda, G. Cardella, F. Amorini, and F. Giacoppo

**Laser Spectroscopy at On-Line Isotope Separators** .......... 213
J. Billowes

**Isomer-Tagging with a Dual Multi-Wire Proportional Counter and a Differential Plunger** .......... 220
D. M. Cullen, P. J. R. Mason, S. V. Rigby, S. Khan, A. M. Kishada, B. J. Varley, C. Scholey, T. Grahn, P. Greenlees, P. Rahkila, P. M. Jones, R. Julin, S. Juutinen, M. Leino, A. P. Leppänen, P. Nieminen, M. Nyman, J. Pakarinen, and J. Uusitalo

**The Structure of $^{7}$He** .......... 225
A. H. Wuosmaa, K. E. Rehm, J. P. Greene, D. J. Henderson, R. V. F. Janssens, C. L. Liang, L. Jisonna, J. C. Lighthall, S. T. Marley, E. F. Moore, R. C. Pardo, N. Patel, M. Paul, D. Peterson, S. C. Pieper, G. Savard, J. P. Schiffer, R. E. Segel, R. H. Siemssen, X. Tang, and R. B. Wiringa

**Cluster Model for Near-barrier Fusion Induced by Weakly Bound and Halo Nuclei** .......... 233
C. Beck, N. Keeley, and A. Diaz-Torres

**Some Questions and Answers in Barrier Distributions** .......... 238
E. Piasecki, M. Kowalczyk, J. Jastrzębski, T. Krogulski, K. Piasecki, K. Rusek, Ł. Świderski, S. Khlebnikov, M. Mutterer, W. H. Trzaska, M. Sillanpää, S. Smirnov, G. Tiourin, S. Dmitriev, E. Kozulin, A. Ogloblin, and N. Rowley

**Spin-dipole Excitations Studied with Tensor Polarised Deuteron Beams** .......... 243
H. J. Wörtche

**Pygmy Dipole Strength and Neutron Skins in Exotic Nuclei** .......... 250
A. Klimkiewicz, N. Paar, P. Adrich, M. Fallot, K. Boretzky, T. Aumann, D. Cortina-Gil, U. D. Pramanik, Th. W. Elze, H. Emling, H. Geissel, M. Hellström, K. L. Jones, J. V. Kratz, R. Kulessa, C. Nociforo, R. Palit, H. Simon, G. Surówka, K. Sümmerer, D. Vretenar, and W. Waluś

**Neutron Skins in Spherical and Deformed Nuclei from Skyrme Hartree-Fock Calculations** .......... 255
P. Sarriguren, M. K. Gaidarov, E. Moya de Guerra, and A. N. Antonov

**Probing Collectivity in the Vicinity of Neutron Deficient Pb Nuclei** .......... 260
T. Grahn, A. Dewald, M. B. G. Hornillos, P. T. Greenlees, A. Görgen, K. Helariutta, J. Jolie, P. Jones, R. Julin, S. Juutinen, S. Ketelhut, R. Krücken, T. Kröll, M. Leino, J. Ljungvall, P. Maierbeck, B. Melon, M. Nyman, R. D. Page, A. Petts, Th. Pissulla, P. Rahkila, J. Sarén, C. Scholey, J. Sorri, J. Uusitalo, R. Wadsworth, and M. Zielińska

**Competing Shapes and Alignments in Neutron-Rich Hf Nuclei** .......... 265
P. Chowdhury, U. S. Tandel, S. K. Tandel, D. Cline, C.-Y. Wu, M. P. Carpenter, R. V. F. Janssens, T. L. Khoo, T. Lauritsen, C. J. Lister, D. Seweryniak, and S. Zhu

**Recent Progress in Theories of Reactions Off Stability** .......... 270
A. Bonaccorso

**Recent Developments in Mean-Field Theories** .......... 278
M. Bender

**Study of the Tensor Correlation in a Neutron-rich *sd*-Shell Region with the Charge and Parity-projected Hartree-Fock Method** .......... 286
S. Sugimoto, H. Toki, and K. Ikeda

**Coulomb Excitation of the N=50 Nucleus $^{80}$Zn** .............................. 291
J. Van de Walle, F. Aksouh, F. Ames, T. Behrens, V. Bildstein, A. Blazhev, J. Cederkäll, E. Clément, T. E. Cocolios, T. Davinson, P. Delahaye, J. Eberth, A. Ekström, D. V. Fedorov, V. N. Fedosseev, L. M. Fraile, S. Franchoo, R. Gernhauser, G. Georgiev, D. Habs, K. Heyde, G. Huber, M. Huyse, F. Ibrahim, O. Ivanov, J. Iwanicki, J. Jolie, O. Kester, U. Köster, T. Kröll, R. Krücken, M. Lauer, A. F. Lisetskiy, R. Lutter, B. A. Marsh, P. Mayet, O. Niedermaier, T. Nilsson, M. Pantea, O. Perru, R. Raabe, P. Reiter, M. Sawicka, H. Scheit, G. Schrieder, D. Schwalm, M. D. Seliverstov, T. Sieber, G. Sletten, N. Smirnova, M. Stanoiu, I. Stefanescu, J.-C. Thomas, J. J. Valiente-Dobón, P. Van Duppen, D. Verney, D. Voulot, N. Warr, D. Weisshaar, F. Wenander, B. H. Wolf, and M. Zielińska

**Sub-barrier Coulomb Excitation of $^{106,108,110}$Sn** .............................. 296
A. Ekström, J. Cederkäll, C. Fahlander, M. Hjorth-Jensen, F. Ames, P. A. Butler, T. Davinson, J. Eberth, G. Georgiev, A. Gorgen, M. Górska, D. Habs, M. Huyse, O. Ivanov, J. Iwanicki, O. Kester, U. Köster, B. A. Marsh, P. Reiter, H. Scheit, D. Schwalm, S. Siem, I. Stefanescu, G. M. Tveten, J. Van de Walle, P. Van Duppen, D. Voulot, N. Warr, D. Weisshaar, F. Wenander, and M. Zielińska

**Study of Multi-Nucleon Transfer Reactions with Light Nuclei** ............... 300
G. Benzoni, D. Montanari, A. Bracco, N. Blasi, F. Camera, F. C. L. Crespi, A. Corsi, S. Leoni, B. Million, R. Nicolini, O. Wieland, A. Zalite, F. Zocca, F. Azaiez, S. Franchoo, I. Stefan, F. Ibrahim, D. Verney, S. Battacharyya, G. De France, A. Navin, M. Rejmund, L. Corradi, G. de Angelis, A. Gadea, A. M. Stefanini, J. J. Valiente-Dobón, E. Farnea, S. Lunardi, P. Mason, G. Montagnoli, F. Scarlassara, S. Szilner, and G. Pollarolo

**Nuclear Astrophysics at LUNA: Status and Perspectives** .................... 305
P. Prati, D. Bemmerer, R. Bonetti, C. Broggini, P. Corvisiero, F. Confortola, E. Conti, H. Costantini, Z. Elekes, A. Formicola, Zs. Fülöp, G. Gervino, A. Guglielmetti, Gy. Gyürky, C. Gustavino, G. Imbriani, M. Junker, A. Lemut, B. Limata, M. Marta, C. Mazzocchi, R. Menegazzo, V. Roca, C. Rolfs, C. R. Alvarez, E. Somorjai, O. Straniero, F. Strieder, F. Terrasi, and H. P. Trautvetter on behalf of The LUNA Collaboration

**Nuclear Matter Equations of State and the Neutron Stars** .................. 313
M. Urbanec, E. Běták, and Z. Stuchlík

**Nucleosynthesis in Jets from Rotating Magnetized Stars during Core Collapse** ............................................................ 318
S.-I. Fujimoto, N. Nishimura, and M.-A. Hashimoto

**Asymptotic Normalization Coefficients from the $^{15}$N($^{3}$He,d)$^{16}$O Reactions and the Astrophysical Factor for the $^{15}$N(p,$\gamma$)$^{16}$O Capture** ......... 323
V. Burjan, P. Bém, S. Cherubini, C. A. Gagliardi, Z. Hons, M. La Cognata, V. Kroha, J. Mrázek, A. M. Mukhamedzhanov, J. Novák, Š. Piskoř, R. G. Pizzone, A. Plunkett, S. Romano, E. Šimečková, C. Spitaleri, L. Trache, R. E. Tribble, A. Tumino, and J. Vincour

**Charge Densities and the Electron Scattering Form Factors of Exotic Nuclei** ........................................................... 328
S. Karataglidis and K. Amos

**Near-barrier Scattering of $^{6}$He and $^{11}$Be** .................................... 333
L. Acosta, D. Escrig, D. Galaviz, A. M. Sánchez-Benítez, C. Angulo, M. A. G. Álvarez, M. V. Andrés, M. J. G. Borge, E. Casarejos, J. M. Espino, J. E. García-Ramos, J. Gómez-Camacho, I. Martel, A. M. Moro, I. Mukha, F. Pérez-Bernal, D. Rodríguez, K. Rusek, and O. Tengblad

**POSTER CONTRIBUTIONS**

**Triaxial Deformations in the Proton Emitters $^{161}$Re and $^{185}$Bi** ................ 341
P. Arumugam, E. Maglione, and L. S. Ferreira

**Systematic Studies of (n,$\alpha$) Reaction Cross Sections at 14.5 MeV Neutrons Energy** .................................... 344
M. Belgaid, F. Kadem, and A. Amokrane

**Mirrored Fragmentation Reactions—A New Technique for Probing Isospin Symmetry in Exotic Nuclei** .................................... 347
J. R. Brown, M. A. Bentley, M. J. Taylor, P. Aldrich, D. Bazin, J. M. Cook, C. A. Diget, A. Gade, T. Glasmacher, S. M. McDaniel, B. Pritychenko, A. Ratkiewicz, K. Siwek, and D. Weisshaar

**The Isovector Imaginary Neutron Potential: A Key Ingredient for the r-Process Nucleosynthesis** .................................... 350
S. Goriely and J.-P. Delaroche

**Latest Results of HFB Mass Formulas** .................................... 353
S. Goriely and J. M. Pearson

**Drip-line to Drip-line Microscopic Nuclear Level Densities** .................. 356
S. Hilaire and S. Goriely

**The AMEDEE Nuclear Structure Database** .................................... 359
S. Hilaire and M. Girod

**Quasi-elastic Barrier Distribution of the $^{18}$O + $^{64}$Zn System** ................. 362
J. F. P. Huiza, E. Crema, D. S. Monteiro, J. M. B. Shorto, R. F. Simões, and P. R. S. Gomes

**Systematic Analysis of Astrophysical S-factors and Thermonuclear Reactions Rates** .................................... 365
M. Katsuma

**Spectroscopy of Odd-proton Nuclei in the Region of $^{254}$No** .................. 368
S. Ketelhut, P. T. Greenlees, A. Chatillon, E. Bouchez, P. A. Butler, E. Clément, O. Dorvaux, S. Eeckhaudt, B. J. P. Gall, A. Görgen, T. Grahn, R.-D. Herzberg, F. Heßberger, A. Hürstel, G. D. Jones, P. Jones, R. Julin, S. Juutinen, H. Kettunen, F. Khalfallah, W. Korten, Y. Le Coz, M. Leino, A.-P. Leppänen, S. Moon, P. Nieminen, M. Nyman, J. Pakarinen, J. Perkowski, P. Rahkila, M. Rousseau, J. Sarén, C. Scholey, Ch. Theisen, J. Uusitalo, J. N. Wilson, P. Bonche, and P.-H. Heenen

**Giant Dipole Radiation and Evidence of Isospin Mixing Increase with Atomic Number in Hot Nuclei** .................................... 371
M. Kicińska-Habior, E. Wójcik-Gront, M. Kowalczyk, and M. Kisieliński

**Development of a Spectrometer for Multiple Prompt Gamma-Ray Measurement to Identify Nuclear Levels** .................................. 374
T. Kin, M. Oshima, K. Furutaka, M. Koizumi, Y. Toh, and A. Kimura
**$^{128}$Xe Lifetime Measurement Using the Coulex-Plunger Technique in Inverse Kinematics** .................................. 377
T. Konstantinopoulos, A. Lagoyannis, S. Harissopulos, A. Dewald, W. Rother, G. Ilie, P. Jones, P. Rakhila, P. Greenlees, T. Grahn, R. Julin, and D. Balabanski
**Puzzling $\gamma$-ray Strength Functions in $^{44,45}$Sc and $^{50,51}$V** .................................. 380
A. C. Larsen, M. Guttormsen, R. Chankova, F. Ingebretsen, T. Lönnroth, S. Messelt, J. Rekstad, A. Schiller, S. Siem, N. U. H. Syed, and A. Voinov
**DSAM Lifetime Studies fo Gd-Nd Nuclei with EUROBALL and AFRODITE** .................................. 383
E. O. Lieder, R. M. Lieder, A. A. Pasternak, B. G. Carlsson, I. Ragnarsson, R. A. Bark, E. A. Gueorguieva, J. J. Lawrie, S. M. Mullins, P. Papka, Y. Kheswa, J. F. Sharpey-Schafer, W. Gast, and G. Duchêne
**Fusion-Fission in the $^{86}$Kr + $^{238}$U Reaction** .................................. 386
M. Lipoglavšek, R. A. Bark, E. Gueorguieva, J. J. Lawrie, E. O. Lieder, R. M. Lieder, E. L. Hansen, S. M. Mullins, S. S. Ntshangase, P. Papka, T. Petrovič, and M. Vencelj
**Study of Dipole Excitations and the Single Particle Structure of Neutron Rich Ni Isotopes** .................................. 389
K. Mahata, S. Paschalis, P. Adrich, F. Aksouh, T. Aumann, M. Babilon, K.-H. Behr, J. Benlliure, T. Berg, M. Boehmer, K. Boretzky, A. Brünle, E. Casarejos, M. Chartier, A. Chatillon, D. Cortine-Gil, U. D. Pramanik, L. Deveaux, M. Elvers, H. Emling, B. Fernandez-Dominguez, M. Gorska, W. Hüller, G. Ickert, H. Johansson, A. Junghans, C. Karagiannis, L. Kern, O. Kiselev, A. Klimkiewicz, N. Kurz, M. Labiche, T. Le Bleis, R. Lemmon, K. Lindenberg, Y. Litvinov, P. Maierbeck, S. Müller, T. Nilsson, C. Nociforo, R. Palit, W. Prokopowicz, D. Rossi, H. Simon, K. Sümmerer, A. Wagner, W. Walus, H. Weick, and M. Winkler
**Neutron-Rich Hypernuclei Are Coming. What They Tell Us?** .................................. 392
L. Majling, J. Lukstins, A. N. Parfenov, M. Solar, and B. Sopko
**Decay of Relativistic Hypernuclei** .................................. 395
O. Majlingova
**U, Np, Pu & Am Prompt Fission Neutron Spectra** .................................. 398
V. M. Maslov
**Reaction Dynamics with Light Weakly Bound Radioactive Ion Beams at Near-barrier Energies** .................................. 401
M. Mazzocco, T. Glodariu, B. Martin, D. Pierroutsakou, C. Signorini, R. Bonetti, A. De Rosa, F. Farinon, A. Guglielmetti, G. Inglima, M. La Commara, C. Mazzocchi, M. Romoli, M. Sandoli, L. Stroe, F. Soramel, and E. Vardaci

**A DIAMANT Wedding for AFRODITE: Probing Structure and Characterizing Reaction Properties via Charged-Particle-$\gamma$ Correlations** .......... 404
S. M. Mullins, B. M. Nyakó, J. Timár, G. Berek, J. Gál, G. Kalinka, J. Molnár, S. H. T. Murray, R. A. Bark, E. Gueorguieva, K. Juhász, A. Krasznahorkay, J. J. Lawrie, E. O. Lieder, R. M. Lieder, M. Lipoglavšek, S. S. Ntshangase, P. Papka, J. N. Scheurer, J. F. Sharpey-Schafer, O. Shirinda, and L. Zolnai

**Alpha Particle Condensation in $^{12}C$ and Nuclear Rainbow Scattering** .......... 407
S. Ohkubo and Y. Hirabayashi

**Electron Screening and Stellar Rates in the $^{3}He$ ($^{3}He$, 2p) $^{4}He$ and $^{3}He$ (d,p) $^{4}He$ Fusion Reactions** .......... 410
L. Messahel, S. Ouichaoui, A. Belhout, M. Fouka, and A. Trabelsi

**Evidence for a Strongly-coupled Band in Very Neutron-deficient Odd-mass Nucleus $^{185}Pb$** .......... 413
J. Pakarinen, A. N. Andreyev, S. Antalic, L. Bianco, I. G. Darby, T. Grahn, P. T. Greenlees, R. Julin, S. Juutinen, H. Kettunen, M. Leino, A.-P. Leppänen, P. Nieminen, D. G. Jenkins, P. Jones, P. Joshi, M. Nyman, R. D. Page, J. Perkowski, P. M. Raddon, P. Rahkila, D. Rostron, J. Saren, C. Scholey, B. Streicher, J. Sorri, J. Uusitalo, K. Van de Vel, M. Venhart, D. Wiseman, and R. Wadsworth

**One Nucleon Transfer Reactions around $^{68}Ni$ at REX-ISOLDE** .......... 416
N. Patronis, R. Raabe, V. Bildstein, N. Bree, R. Gernhäuser, M. Huyse, Th. Kröll, R. Krücken, M. Mahgoub, P. Maierbeck, I. Stefanescu, J. Van de Walle, and P. Van Duppen

**On-line Separator for $\gamma$-spectroscopic Studies** .......... 419
A. G. Popeko, A. V. Belozerov, V. I. Chepigin, O. Dorvaux, K. Hauschild, A. P. Kabachenko, A. Korichi, A. Lopez-Martens, O. N. Malyshev, A. V. Shutov, A. I. Svirikhin, and A. V. Yeremin

**Production and Separation of Long-lived Radionuclides for Nuclear Physics Experiments** .......... 422
D. Schumann

**Level Densities and $\gamma$-Strength Functions in Sm Isotopes** .......... 425
S. Siem, M. Guttormsen, R. Chankova, A. C. Larsen, H. T. Nyhaus, J. Rekstad, N. U. H. Syed, A. Schiller, A. Voinov, and S. W. Ødegård

**The Oslo Method and its Application to Lead Isotopes** .......... 428
N. U. H. Syed, M. Guttormsen, F. Ingebretsen, A. C. Larsen, S. Siem, T. Lönnroth, A. Schiller, and A. Voinov

**Shape Coexistence in the Neutron-deficient Pt Isotopes in the Configuration-mixed IBM** .......... 431
C. E. Vargas, I. O. Morales, A. Frank, P. Van Isacker, and C. Campuzano

**List of Participants** .......... 435

**Author Index** .......... 445

# PREFACE

The 2nd International Conference "Frontiers In NUclear STructure, Astrophysics and Reactions (FINUSTAR 2)" was held in Aghios Nikolaos, the eastern capital city of the island of Crete, Greece, on September 10-14, 2007. The venue was the Convention Center at the Mirabello Beach Hotel and Village. It was organized by the Institute of Nuclear Physics of the National Center for Scientific Research "Demokritos", Athens, the Department of Physics of the University of Jyväskylä, Finland, and the Department of Physics of the National Technical University of Athens.

As a follow-up of FINUSTAR 2005, FINUSTAR 2 brought together once again, researchers from different nuclear physics communities, whose activities in recent years have shown a strong interplay and overlap, driven mainly by the utilization of common instrumentation and research facilities. The aim of FINUSTAR 2 was to create an open forum to discuss common problems, present recent results, get informed about the latest developments in theory as well as the most recent achievements in instrumentation. To meet these goals, FINUSTAR 2 covered a broad range of topics in nuclear structure, astrophysics and reactions as follows:

- Nuclear structure at the extremes
- Exotic excitations and decays
- Collective phenomena and symmetries
- Nuclear masses and ground-state properties
- Synthesis and structure of the heaviest elements
- Ab initio calculations and shell model
- Mean field theories, cluster models and molecular dynamics
- Nuclear scattering and dynamics at low and intermediate energies
- Nuclear reactions off stability and indirect methods
- Nucleosynthesis in the cosmos and nuclear physics aspects
- Weak-interaction processes – Neutrino physics
- Radioactive and exotic beams
- Facilities and instrumentation for the future

These topics were exhausted by 111 contributions, 77 of them given as oral and 34 as poster presentations. Also, 77 of them were based on experimental and 34 on theoretical results including the latest achievements from most of the leading nuclear physics laboratories in the world. The presentations were of excellent quality and invoked lively and instructive discussions. We are very proud of the high scientific level of the contributions and therefore we would like to express our sincere gratitude to all of the 134 participants for their efforts.

Our appreciation also goes to the members of the International Advisory Committee for their contribution to the scientific program through their valuable suggestions for invited speakers and oral presentations.

We are very grateful to the members of the Local Organizing Committee Rosa Vlastou, Dennis Bonatsos, Tassos Lagoyannis, Michalis Axiotis and Theo Konstantinopoulos for their hard work and excellent collaboration. Special thanks go to the rest of the local team, Sophia Karabaya, George Provatas and Varvara Foteinou for their tireless involvement during the conference.

Vivian Demetriou
Rauno Julin
Sotirios V. Harissopulos

**FINUSTAR 2**
Aghios Nikolaos, Crete, Greece, September 10 - 14, 2007

2nd International Conference on
Frontiers in Nuclear Structure, Astrophysics and Reactions
Aghios Nikolaos, Crete, Greece, September 10-14, 2007

## FINUSTAR Chairs

R. Julin (JYFL, University of Jyväskylä, Finland)

S. Harissopulos (NCSR Demokritos, Athens, Greece)

## Scientific Secretary

P. Demetriou (NCSR Demokritos, Athens, Greece)

## Local Organizing Committee

R. Vlastou (NTUA, Athens, GR)

D. Bonatsos (NCSR Demokritos, Athens, GR)

T. Konstantinopoulos (Demokritos, Athens, GR)

A. Lagoyannis (NCSR Demokritos, Athens, GR)

M. Axiotis (NCSR Demokritos, Athens, GR)

## International Advisory Committee

N. Alamanos (CEA, Saclay, France)

C. Angulo (UCL, Louvain-La-Neuve, Belgium)

M. Arnould (ULB, Brussels, Belgium)

C. Beck (IReS, Strasbourg, France)

M. Carpenter (ANL, Argonne, USA)

M. B. Chadwick (LANL, Los Alamos, USA)

G. De Angelis (LNL, Legnaro, Italy)

A. Dewald (IKP, Cologne, Germany)

B. Fulton (York, UK)

S. Gales (GANIL, Caen, France)

D. Guillemaud – Mueller (IPN, Orsay, France)

M. Harakeh (KVI, Groningen, The Netherlands)

P.-H. Heenen (ULB, Brussels, Belgium)

A. J. Koning (NRG, Petten, The Netherlands)

R. Kruecken (TUM, Munich, Germany)

S. Lenzi (INFN, Padova, Italy)

A. Maj (IFJ, Krakow, Poland)

A. Poves (UAM, Madrid, Spain)

K. Riisager (CERN, Geneva, Switzerland)

C. Rolfs (RUB, Bochum, Germany)

D. Rowe (Toronto, Canada)

J. Simpson (CCLRC, Daresbury, UK)

C. Spitaleri (LNS, Catania, Italy)

I. J. Thompson (LLNL, Livermore, USA)

R. Tribble (TAMU, Texas, USA)

H. Utsunomiya (Kobe, Japan)

P. Van Duppen (IKS, Leuven, Belgium)

D. Vretenar (Zagreb, Croatia)

R. Wyss (KTH, Stockholm, Sweden)

N. V. Zamfir (NIPNE, Bucharest, Romania)

Email: finustar2@inp.demokritos.gr

URL: http://www.inp.demokritos.gr/~finustar2

## MONDAY 10.09.2007

| | |
|---|---|
| 08:00 – 08:50 | **Registration** |
| 08:50 – 09:00 | **Opening – Welcome addresses** |
| | **SESSION I** |
| 09:00 – 09:30 | **J. Jolie** (Universität zu Köln, Germany) **Invited Talk**<br>*Quantum phase transitions in atomic nuclei* |
| 09:30 – 10:00 | **S. Leoni** (Università di Milano and INFN-Milano) **Invited Talk**<br>*Nuclear Structure at extreme conditions through ? spectroscopy measurements* |
| 10:00 – 10:20 | **J. F. Sharpey-Schafer** (Univ. Western Cape & Zululand, iThemba Labs, SA)<br>*Shape Transitional Nuclei: What can we learn from the Yrare States?* **Invited Contr.** |
| 10:20 – 10:40 | **R. M. Lieder** (iThemba Labs, South Africa and Fz Jülich, Germany)<br>*Observation of a Double-humped -Ray Fold Distribution in* $^{142}$*Eu* |
| 10:40 – 11:10 | **Coffee break** |
| | **SESSION II** |
| 11:10 – 11:40 | **B. S. Meyer** (Clemson University, SC, USA) **Invited Talk**<br>*Neutrinos and Supernovae* |
| 11:40 – 12:00 | **Ph. Chomaz** (GANIL, Caen, France) **Invited Contribution**<br>*Nuclear liquid-gas phase transition in asymmetric matter: from nuclei to compact stars* |
| 12:00 – 12:20 | **S. Goriely** (IAA, ULB, Belgium) **Invited Contribution**<br>*Spallation nucleosynthesis by accelerated charged-particles.* |
| 12:20 – 12:40 | **N. Paar** (University of Zagreb, Croatia)<br>*Exotic modes of excitation and charged-current neutrino-nucleus reactions* |
| 12:40 – 13:00 | **T. Suzuki** (Nihon University, Tokyo, Japan)<br>*Spin-Dependent Transitions in Nuclei and Neutrino-Nucleus Reactions* |
| 13:00 – 15:30 | **Lunch** |
| | **SESSION III** |
| 15:30 – 16:00 | **P.T.Greenlees** (JYFL, Jyväskylä, Finland) **Invited Talk**<br>*Spectroscopy of very heavy elements* |
| 16:00 – 16:20 | **O. Dorvaux** (IPHC-DRS/ULP, Strasbourg, France)<br>*Spectroscopy of heavy elements at Dubna* |
| 16:20 – 16:40 | **R. Dressler** (PSI, Villingen, Switzerland)<br>*Confirmation of super heavy element production in* $^{48}$*Ca induced fusion reactions a handshake of physics and chemistry for element 112* |
| 16:40 – 17:00 | **T. J. Bürvenich** (FIAS, J.W. Goethe University, Frankfurt, Germany)<br>*Properties of heavy and superheavy nuclei in supernova environments* |
| 17:00 – 17:30 | **Coffee break** |
| | **SESSION IV** |
| 17:30 – 18:00 | **O. Sorlin** (GANIL, Caen, France) **Invited Talk**<br>*The magic number N=28 and the spin-orbit interaction* |
| 18:00 – 18:20 | **D. Seweryniak** (ANL, Argonne, IL, USA)<br>*Single-neutron states in* $^{101}$*Sn* |
| 18:20 – 18:40 | **Th. Kröll** (TU München, Garching, Germany)<br>*Quadrupole collectivity of neutron-rich nuclei around* $^{132}$*Sn* |
| 18:40 – 19:00 | **P. Maierbeck** (TU München, Garching, Germany)<br>*Probing the single particle structure around 54Ca with one-neutron knockout* |
| 19:00 – 20:00 | **POSTER SESSION** |

## TUESDAY 11.09.2007

| | |
|---|---|
| | **SESSION V** |
| 09:00 – 09:30 | **S. Gales** (GANIL, Caen, France) **Invited Talk**<br>*SPIRAL2@GANIL: a world leading ISOL facility for the next decade* |
| 09:30 – 10:00 | **H. Geissel** (GSI, Darmstadt, Germany) **Invited Talk**<br>*The FAIR Project: Status and Perspectives for NUSTAR Studies* |
| 10:00 – 10:20 | **K. Riisager** (CERN, Geneva, Switzerland) **Invited Contribution**<br>*The HIE-ISOLDE Project* |
| 10:20 – 10:50 | **F. Azaiez** (INP, Orsay, France) **Invited Talk**<br>*The European Collaboration for Stable Ion Beams - ECOS* |
| 10:50 – 11:20 | **Coffee break** |
| | **SESSION VI** |
| 11:20 – 11:50 | **T. Bredeweg** (LANL, Los Alamos, USA) **Invited Talk**<br>*Neutron capture reactions at DANCE* |
| 11:50 – 12:10 | **M. Chadwick** (LANL, Los Alamos, USA)<br>*Off-stability nuclear reaction theory predictions for actinides* |
| 12:10 – 12:30 | **R. Vlastou** (NTUA, Athens, Greece) ------------ replaced by S. Harissopulos<br>*Isomeric cross section study of neutron induced reactions on Ge* |
| 12:30 – 12:50 | **Y. El Masri** (UCL, Louvain-la-Neuve, Belgium)<br>*Properties of proton-induced fission processes on actinide nuclei at energies between 20 and 80 MeV* |
| 12:50 – 13:10 | **S. Hilaire** (CEA,Bruyres-le-Chatel, France)<br>*Towards Universal Predictions with the TALYS code.* |
| 13:10 – 15:30 | **Lunch** |
| | **SESSION VII** |
| 15:30 – 16:00 | **K. Lister** (ANL, Argonne, USA) **Invited Talk**<br>*Shapes and correlations along the N=Z Line* |
| 16:00 – 16:20 | **S. Lunardi** (Univ. of Padova and INFN, Padova, Italy) **Invited Contribution**<br>*Spectroscopy of Neutron-rich Nuclei in the A≈60 Region* |
| 16:20 – 16:40 | **A. Astier** (CSNSM and Université Paris-Sud, Orsay, France)<br>*High-Spin States of $^{84,85}Br$: Mapping the Proton Sub-Shells towards $^{78}Ni$* |
| 16:40 – 17:00 | **E. Sahin** (INNF-LN Legnaro, Italy and Istanbul University, Turkey)<br>*Experimental study of neutron rich nuclei in the N=50 shell closure* |
| 17:00 – 17:30 | **Coffee break and CONFERENCE PHOTO** |
| | **SESSION VIII** |
| 17:30 – 18:00 | **T. Spillane** (Univ. of Connecticut, Storrs, USA) **Invited Talk**<br>*Recent results of the $^{12}C+^{12}C$ fusion reactions* |
| 18:00 – 18:20 | **A. Lefebvre-Schuhl** (CSNSM and Université Paris-Sud, Orsay, France)<br>*$^{12}C(\alpha,\gamma)^{16}O$ thermonuclear reaction rate: Present status and perspectives* |
| 18:20 – 18:40 | **R.G. Pizzone** (University of Catania and INFN Sez. Catania-LNS)<br>*AGB fluorine nucleosynthesis studied by means of Trojan-horse method: the case of $^{15}N(p,\alpha)^{12}C$* |
| 18:40 – 19:00 | **Y. Parpottas** (Univ. of Cyprus, Nicosia, Cyprus)<br>*Two experimental approaches in nuclear astrophysics using neutrons* |

## WEDNESDAY 12.09.2007

| | |
|---|---|
| | **SESSION IX** |
| 09:00 – 09:30 | **P. Capel** (ULB, Brussels, Belgium) **Invited Talk**<br>*Recent developments in Coulomb breakup calculations.* |
| 09:30 – 10:00 | **A. Zilges** (TU Darmstadt, Germany) **Invited Talk**<br>*Gamma-induced reactions in explosive nucleosynthesis.* |
| 10:00 – 10:20 | **H. Utsunomiya** (Konan Univ., Kobe, Japan) **Invited Contribution**<br>*Enhanced photoneutron cross sections for zirconium isotopes near threshold: evidence for an M1 strength?* |
| 10:20 – 10:40 | **A. Lagoyannis** (NCSR "Demokritos", Athens, Greece)<br>*Cross section measurements of α-particle capture reactions relevant to p-process nucleosynthesis* |
| 10:40 – 11:10 | **Coffee break** |
| | **SESSION X** |
| 11:10 – 11:30 | **N. Lo Iudice** (University of Naples "Federico II", Naples, Italy)<br>*Microscopic nature of $0^+$ states populated in large abundance in deformed nuclei.* |
| 11:30 – 11:50 | **P. E. Garett** (Univ. of Guelph, Guelph, Canada)<br>*Nature of the $K=4^+$ bands in the Os isotopes.* |
| 11:50 – 12:10 | **M. J. G. Borge** (CSIC, Madrid, Spain)<br>*β-delayed particle studies of the halo nucleus $^{11}Li$ and its core $^{9}Li$.* |
| 12:10 – 12:30 | **M. Guttormsen** (University of Oslo, Norway)<br>*Thermodynamic properties of atomic nuclei with T<1 MeV* |
| 12:30 – 12:50 | **C. Sfienti** (LNS-Catania, Italy)<br>*First evidence of di-proton decay from $^{18}Ne$ excited states.* |
| 12:50 – 13:10 | **J. F. P. Huiza** (Universidade de São Paulo, Brazil)<br>*Complete Experimental and Theoretical study of the $^{16}O+^{64}Zn$ System* |
| | **Lunch** |

## THURSDAY 13.09.2007

| | |
|---|---|
| | **SESSION XI** |
| 09:00 – 09:30 | **J. Billowes** (University of Manchester, Manchester, UK) **Invited Talk**<br>*Laser Spectroscopy at ISOL Facilities.* |
| 09:30 – 10:00 | **G. Neyens** (Katholieke Universiteit Leuven, Leuven, Belgium) **Invited Talk**<br>*g-RISING: Magnetic Moments Measurements on Relativistic Isomeric Beams.* |
| 10:00 – 10:20 | **A.M. Bruce** (University of Brighton, Brighton, UK)<br>*Isomeric states in neutron-rich A~110, Z~40 nuclei studied using RISING at GSI.* |
| 10:20 – 10:40 | **D.M. Cullen** (Univ.of Manchester, Manchester, UK)<br>*Recent Isomer-Tagging Results using a Dual Multi-Wire Proportional Counter setup and New Extensions to the Technique with a Differential Plunger.* |
| 10:40 – 11:10 | **Coffee break** |
| | **SESSION XII** |
| 11:10 – 11:30 | **A. H. Wuosmaa** (Western Michigan University, USA) **Invited Contribution**<br>*Structure of $^{7}He$.* |
| 11:30 – 11:50 | **S. Aoyama** (Niigata University, Niigata, Japan)<br>*Systematic Analyses of Super Neutron-Rich Nuclei, $^{4-7}H$.* |
| 11:50 – 12:10 | **C. Beck** (*IPHC, CNRS-IN2P3 et Université Louis Pasteur Strasbourg,, France)*<br>*Di-neutron cluster model for near-barrier fusion induced by halo nuclei.* |
| 12:10 – 12:30 | **E. Piasecki** (Warsaw Univ., Warsaw, Poland)<br>*Barrier Distributions from Fusion and Quasi-elastic Reactions: Some Puzzles, Questions and Answers.* |
| 13:00 – 15:30 | **Lunch** |
| | **SESSION XIII** |
| 15:30 – 16:00 | **H. J. Woertche** (KVI, Groningen, The Netherlands) **Invited Talk**<br>*Spin-dipole excitations studied with tensor polarised deuteron beams* |
| 16:00 – 16:20 | **K. Boretzky** (GSI, Darmstadt, Germany)<br>*Pygmy Dipole Strength and Neutron Skins in Exotic Nuclei.* |
| 16:20 – 16:40 | **P. Sarriguren** (CSIC, Madrid, Spain)<br>*Neutron skins in spherical and deformed nuclei from Skyrme Hartree-Fock calculations.* |
| 16:40 – 17:00 | **L. S. Ferreira** (Instituto Superior Tecnico, Lisbon, Portugal)<br>*Will neutron emission, from nuclei beyond the neutron drip-line be observed?* |
| 17:00 – 17:30 | **Coffee break** |
| | **SESSION XIV** |
| 17:30 – 17:50 | **A. Andreyev** (Katholieke Universiteit Leuven, Leuven, Belgium)<br>*Recent results in the Pb-Rn region in the vicinity of the neutron-midshell at N=104.* |
| 17:50 – 18:10 | **T. Grahn** (University of Jyväskylä, Jyväskylä, Finland)<br>*Probing collectivity in the vicinity of neutron deficient Pb nuclei.* |
| 18:10 – 18:30 | **H. Mach** (Uppsala University, Sweden)<br>*Systematic studies of B(E2) values in the exotic neutron-rich even-even nuclei.* |
| 18:30 – 18:50 | **P. Chowdhury** (Univ. of Massachusetts Lowell, Lowell, USA)<br>*Competing Shapes and Alignments in Neutron-Rich Hafnium Nuclei.* |
| | |
| 20:00 – 22:30 | **CONFERENCE BANQUET** |

# FRIDAY 14.09.2007

| | |
|---|---|
| | **SESSION XV** |
| 09:00 – 09:30 | **A. Bonaccorso** (University of Pisa and INFN, Pisa, Italy) **Invited Talk**<br>*Recent Progress in Theories of Reactions off Stability.* |
| 09:30 – 10:00 | **M. Bender** (CENBG, Gradignan, France) **Invited Talk**<br>*Recent developments in mean-field theories.* |
| 10:00 – 10:20 | **J. P. Draayer** (Louisiana State University, USA) **Invited Contribution**<br>*Symplectic No-Core Shell Model.* |
| 10:20 – 10:40 | **S. Sugimoto** (Osaka University, Osaka, Japan)<br>*Study of the tensor correlation in the neutron-rich sd-shell region with the charge and parity-projected Hartree-Fock method.* |
| 10:40 – 11:10 | **Coffee break** |
| | **SESSION XVI** |
| 11:10 – 11:40 | **G. Georgiev** (CSNSM and Université Paris-Sud, Orsay, France) **Invited Talk**<br>*Coulomb Excitation Studies at REX-ISOLDE.* |
| 11:40 – 12:00 | **G. C. Ball** (TRIUMF, Vancouver, BC, Canada) **Invited Contribution**<br>*Coulomb Excitation of Radioactive $^{20,21,28,29}Na$: First Results from TIGRESS at ISAC.* |
| 12:00 – 12:20 | **J. van de Walle** (Katholieke Universiteit Leuven, Leuven, Belgium)<br>*Coulomb excitation of the N=50 Nucleus $^{80}Zn$.* |
| 12:20 – 12:40 | **A. Ekström** (University of Lund, Sweden)<br>*Sub-Barrier Coulomb Excitation of 110Sn and First Results on $^{108,106}Sn$.* |
| 12:40 – 13:00 | **G. Benzoni** (Universitá degli Studi di Milano and INFN, Milano, Italy)<br>*Study of multi-nucleon transfer reactions with light nuclei.* |
| 13:00 – 15:30 | **Lunch** |
| | **SESSION XVII** |
| 15:30 – 16:00 | **P. Prati** (INFN-Genova, Genova, Italy) **Invited Talk**<br>*Nuclear astrophysics at LUNA: Status and perspectives.* |
| 16:00 – 16:20 | **E. Betak** (SAS, Bratislava, Slovakia)<br>*Nuclear Matter Equations of State and the Neutron Stars* |
| 16:20 – 16:40 | **S. Fujimoto** (Kumamoto National College of Technology, Kumamoto, Japan)<br>*Nucleosynthesis in Jets from Rotating Magnetized Stars during Core Collapse.* |
| 16:40 – 17:00 | **V. Burjan** (Institute of Nuclear Physics, CAS, Rez, Czech Republic)<br>*Asymptotic normalization coefficients from the $^{15}N(^3He,d)^{16}O$ reaction to determine the $^{15}N(p,\gamma)^{16}O$ direct capture S factor.* |
| 17:00 – 17:30 | **Coffee break** |
| | **SESSION XVIII** |
| 17:30 – 17:50 | **S. Karataglidis** (Rhodes University, South Africa)<br>*Charge densities and the electron scattering form factors of exotic nuclei.* |
| 17:50 – 18:10 | **T. Yamada** (Kanto Gakuin University, Yokohama, Japan)<br>*Single-α orbits and α condensation in $^{12}C$ and $^{11}B$.* |
| 18:10 – 18:30 | **I. Martel** (Universidad de Huelva, Huelva, Spain)<br>*Coulomb barrier scattering of the light halo nuclei $^6He$ and $^{11}Be$.* |
| 18:30 – 18:50 | **J. Lubian** (Universidade Federal Fluminense, Rio de Janeiro, Brazil)<br>*Searching for a polarization potential in the breakup of $^8B$.* |
| 18:50 – 19:00 | **CLOSING** |

# ORAL CONTRIBUTIONS

# Quantum Phase Transitions in Atomic Nuclei.

J. Jolie and A. Dewald

*Institute of Nuclear Physics, University of Cologne, Zülpicherstrasse 77, 50937 Cologne, Germany.*

**Abstract.** During the last few years there has been considerable interest in nuclei located at the onset of deformation. From the theoretical side this onset of deformation can be described as a quantum phase transition. Here detailed studies using the interacting boson and the collective model have been performed. In the latter atomic nuclei located at the phase transition between spherical and deformed can be described using the critical point symmetries X(5) and E(5). We review recent experimental studies in the oblate-prolate and sperical to deformed region which are looking for signatures of phase transitional behavior and/or critical point symmetries. We also discuss the difference between shape coexistence and these shape phase transitions.

**Keywords:** Quantum shape-phase transitions, critical point symmetries
**PACS:** 21.60Fw, 21.10Re

## INTRODUCTION

Phase transitions occur everywhere in physics. In nuclear physics classic phase transitions occur when heating up the nucleus by increasing the excitation energy (see ref [1]) or by smashing relativistic heavy ions so violently together that one reaches the quark-gluon phase transition [2]. On the otherhand there exist phase transitions in which pressure or temperature play no role as they take place at zero temperature. Such Quantum Phase Transitions (QPT) happen when the atomic nucleus changes his ground state shape when going from one (even-even) atomic nucleus to the next. In contrast with classical phase transitions, the control parameters are in this case given by the actual number of valence neutrons and protons.

We note that strictly speaking the QPT in atomic nuclei are precussors of the QPT which only occurs for nuclei with an infinite number of nucleons. Nevertheless we will use the term QPT also for the finite nucleus. The finite nucleon number itselves is of importance for three reasons. First, classically, phase transitions only occur in systems having an infinite number of constituents and therefore an extension of the concept to systems having a limited number of constituents is of importance. Secondly, since atomic nuclei contain only an integer number of nucleons, nature does not allow us to vary the control parameters continuously in the region where the QPT occurs. While the experimental limitation to integer nucleon number leads to discrete changes in the properties of atomic nuclei around the critical point, only theoretical calculations do allow one to continuously vary the appropriate control parameter. Finally, the limited number of N tends to wash out the QPT behavior due to quantum fluctuations making the search for this behavior more difficult.

This work is organised as follows. Using a very simple Hamiltonian we will in section 2 describe the shape phase transitions in the framework of the Interacting Boson Model (IBM) [3]. Then we will in section 3 go over to a collective model approach and discuss

CP1012, *Frontiers in Nuclear Structure, Astrophysics, and Reactions: FINUSTAR 2*
edited by P. Demetriou, R. Julin, and S. V. Harissopulos
© 2008 American Institute of Physics 0-7354-0532-5/08/$23.00

in detail experimental searches for examples of the spherical to deformed critical point symmetry X(5). In section 4 we will compare the IBM and X(5) descriptions. Finally, we will discuss in section 5 the difference between shape phase transitions and shape coexistence.

## SHAPE PHASE TRANSITIONS IN THE IBM

One of the best nuclear models to study shape phase transitions is the Interacting Boson Model [3] which is able to describe vibrational and deformed nuclei as well as the transitional regions in between them in an unified way. We will use the following very simple Hamiltonian:

$$\hat{H} = \alpha[\eta \hat{n}_d - \frac{(\eta - 1)}{N} \hat{Q}_\chi \hat{Q}_\chi] \quad (1)$$

with

$$\hat{Q}_\chi = (s^\dagger \tilde{d} + d^\dagger s)^{(2)} + \chi (d^\dagger \tilde{d})^{(2)}. \quad (2)$$

Here $\eta$ and $\chi$ play the role of the control parameters and $\alpha$ is an overall energy scaling factor. $N$ is the number of bosons and equals half the number of valence nucleons. The parameter $\eta$ governs the transition between spherical and deformed, and $\chi$ the transition between prolate and oblate deformation. The Hamiltonian (1) contains the three dynamical symmetries of the IBM: $U(5)$ when $\eta = 1$, $O(6)$ when $\eta = 0$ and $\chi = 0$ and $SU(3)$ when $\eta = 0$ and $\chi = -\sqrt{7}/2$ as well as the $\overline{SU(3)}$ limit when $\eta = 0$ and $\chi = +\sqrt{7}/2$. These correspond to vibrational nuclei with a spherical form ($U(5)$), an axially symmetric rotor which is prolate ($SU(3)$) or oblate ($\overline{SU(3)}$) , and an axially asymmetric rotor with a flat potential in $\gamma$ ($O(6)$).

The two parameter Hamiltonian (1) is thus able to exhibit four different dynamical symmetries, which each correspond to well established examples of different nuclear shapes. Furthermore is can generate the full range of intermediate structures of the IBM and is thus very well suited to study the shape phase transitions. The first term leads to a spherical solution while the second term, which is a quadrupole interaction between bosons, yields deformed solutions. The equilibrium shape is spherical for $\eta$ values ranging down from unity, crosses from spherical to deformed at some value of $\eta \equiv \eta_{crit}$, and is deformed for $\eta < \eta_{crit}$. The quadrupole operator introduces another parameter $\chi$ which is related to axial asymmetry, and which notably increases the richness of the structures possible with eq. 1. The parameter $\chi \in$ [-$\sqrt{7}/2$, $\sqrt{7}/2$]. Negative (positive) $\chi$ corresponds (if $\eta < \eta_{crit}$) to a prolate (oblate) axially symmetric rotor and $\chi$=0 to a completely $\gamma$–soft rotor. In order to visualise the Hamiltonian the Casten triangle [4] is mostly used. To use its full possibilities we use the so-called extended version [5]. The extended triangle maps also the phase transitions which occurs inside the triangle. They are the first order transition between oblate and prolate shapes occuring on the O(6)-U(5) axis to the point where the nucleus changes from spherical to deformed. From thereon the first order transition between spherical and deformed goes outwards to the legs. The central point itselfs represent the second order transition and can be seen as the triple point of nuclear deformation where spherical, prolate and oblate deformation coexist.

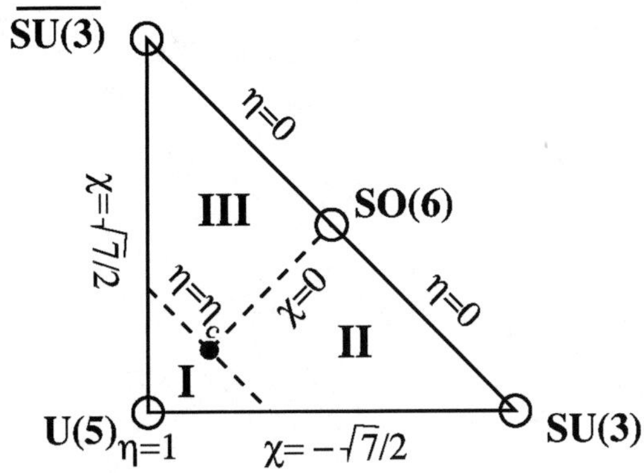


**FIGURE 1.** The extended casten triangle obtained using the Hamiltonian (1). Dashed lines indicate the first order phase transitions and the filled dot the isolated second order transition [5]. The open circles indicate the location of the different dynamical symmetries. The three different phases are spherical (I), prolate deformed (II) and oblate deformed (III).

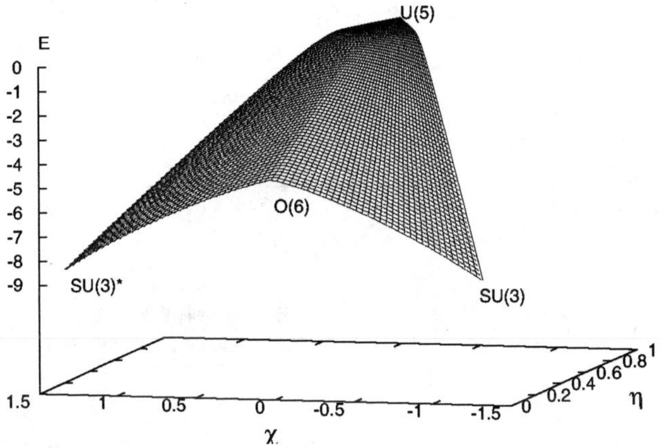


**FIGURE 2.** The groundstate energy as a function of $\eta$ and $\chi$ for N=40.

Figure 1 illustrates the such obtained phase diagram and indicates the position of the first and second order phase transitions. The phase transitions in the extended Casten triangle are exemplified in Figure 2 which shows the energy of the groundstate using (5) for N=40.

Up to now we discussed the results at large N values, which are not reached in atomic nuclei. Now the experimental evidence, hence at low N values, will be discussed. The question is whether there exist atomic nuclei whom exhibit the characteristics of the oblate-prolate phase transition. Although oblate nuclei or transitions from $\gamma$ soft to oblate shapes are rare they do occur in the platinum and mercury isotopes close to the well

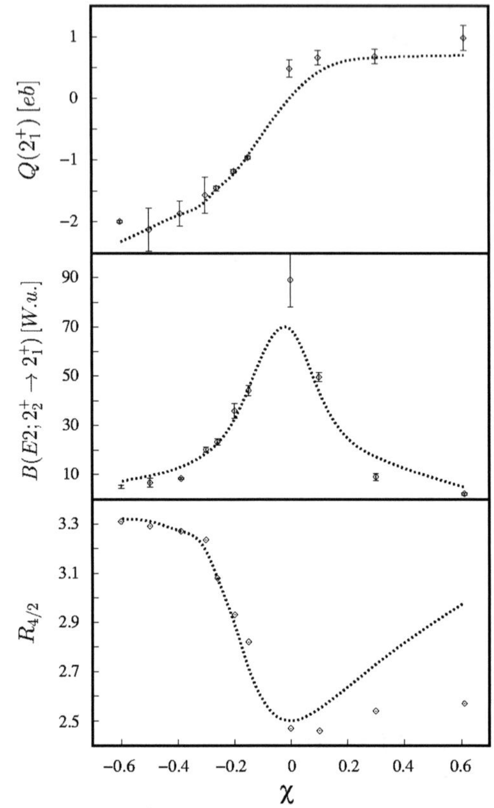


**FIGURE 3.** The experimental observables $R(4_1^+/2_1^+)$, $Q(2_1^+)$ and $B(E2;2_2^+ \rightarrow 2_1^+)$ for the nuclei: $^{180}Hf$, $^{182,184,186}W$, $^{188,190,192}Os$, $^{194,196}Pt$ and $^{198,200}Hg$. The dashed line gives the theoretical values (see text).

known O(6) nucleus $^{196}$Pt. In [5] three different signatures were indicated: $R(4_1^+/2_1^+)$, $Q(2_1^+)$ and $B(E2;2_2^+ \rightarrow 2_1^+)$. In an attempt to span a large part of the extended Casten triangle we plot these quantities in Figure 3 for nuclei ranging from the well deformed prolate rotor $^{180}Hf$ to $^{200}Hg$. For each nucleus the $\chi$ value was fitted with $\eta = 0$ and the appropriate N value. Clearly all experimental observables indicating the prolate-oblate phase transition [6] are observed. The very rapid change of the quadrupole moment might explain why $^{196}Pt$ has a non vanishing quadrupole moment while still being a good O(6) nucleus. Surprising is the structure in the mercury isotopes which does not resemble vibrational or shell model like structures whom have $R(4_1^+/2_1^+)$ around and below 2.

**TABLE 1.** P factors for all X(5) candidates.

| Nucleus | $^{150}Nd$ | $^{152}Sm$ | $^{154}Gd$ | $^{156}Dy$ | $^{176}Os$ | $^{178}Os$ |
|---|---|---|---|---|---|---|
| P | 4.4 | 4.8 | 5.1 | 5.3 | 4.5 | 4.6 |

## CRITICAL POINT SYMMETRIES

Recently, F. Iachello developed new critical symmetries which can describe atomic nuclei at the phase transition [7, 8]. The new symmetries, called X(5) and E(5), are obtained within the framework of the collective model [9] under some simplifying approximations. This new developments were first triggered by studies of $^{152}Sm$ by Casten and the Cologne group [10]. Remarkably, the parameter free predictions provided by these symmetries are closely realised in this atomic nucleus [11, 12]. The two new critical point symmetries focus on the $\beta$ degree of freedom and are based on a separation of the $\beta$ degree of freedom from the $\gamma$ degree of freedom. An infinite square well potential in the $\beta$ degree of freedom is imposed. Since these descriptions are embedded in the framework of the collective model, they do not incorporate finite N effects reflecting the finite number of nucleons in actual atomic nuclei. In the IBM the X(5) point resembles the phase transion on the U(5) to SU(3) leg of the triangle while the E(5) resembles the triple point. The first examples of X(5) nuclei were several other N=90 isotones: $^{150}Nd$ [13] and $^{154}Gd$ [14]. They follow the X(5) pattern in energies and in electromagnetic transition rates for the ground state band. A next candidate $^{156}Dy$ follows still the X(5) energy trend but is in its electromagnetic decay already behaving as a rotor [15]. This illustrates the need to measure electromagnetic decay in order to assess whether an atomic nucleus is described by a critical point symmetry.

A very useful observation was made by E. A. Mc Cutchan *et al.* [16]. If one uses the P-factor [17] which is:

$$P = \frac{N_n . N_p}{N_n + N_p} \tag{3}$$

then the spherical to well deformed phase transition takes place when P equals about 5. Strange enough there are only few regions near stability where this value is reached. They are the N=90 nuclei and the neutron deficient Os nuclei. Recent lifetime measurements using the Cologne plunger and Orange spectrometer indicate that indeed $^{176,178}Os$ exhibit the X(5) symmetry [18]. Figure 4 gives the results for the electromagnetic transitions in the ground state band of $^{176}Os$ obtained using the Cologne plunger and Orange spectrometer [19]. Table 1 lists the P factors of all X(5) candidates. Note that the P factor alone is not a sufficient condition for X(5) behavior. As an example $^{186}W$ has a P factor of 5.1 but lies on the O(6) to SU(3) leg of the triangle.

## COMPARISON OF X(5) AND THE IBM

The X(5) symmetry cannot be exactly realised in the IBM, but is situated in the neighborhood of the phase transition between a prolate rotor and a spherical vibrator. Dewald

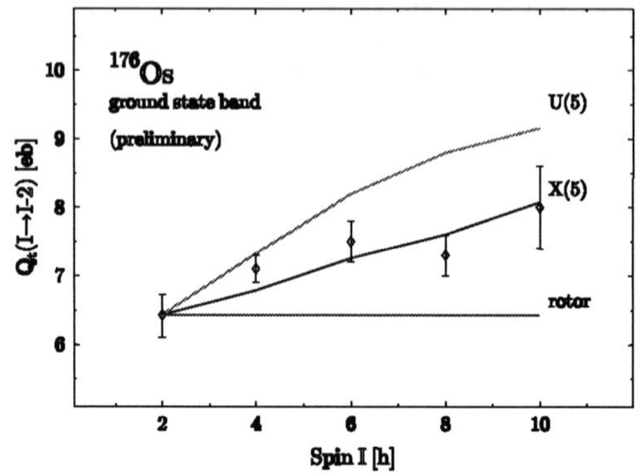


**FIGURE 4.** The ratios of transitional quadrupole elements $Q_t$ for the groundstate band transitions in $^{176}Os$ are compared with the predictions of X(5). Also given are the results for a rotor and a spherical vibrator.

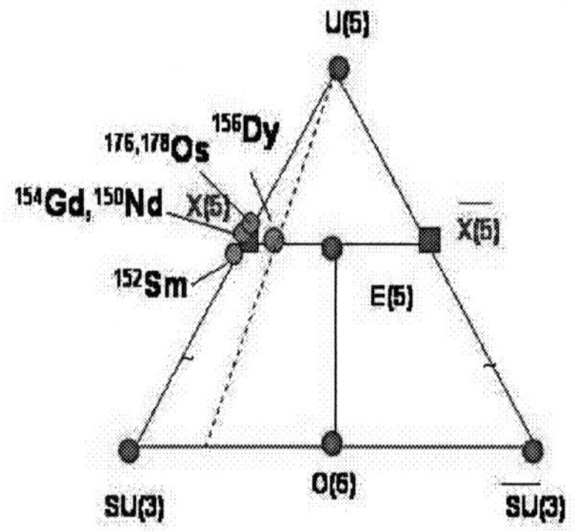


**FIGURE 5.** Location of the X(5) nuclei in the extended Casten triangle when fitted with Eq (1). Note that we have changed the $\eta$ scale so that the location is clearer.

*et al* have performed calculations using the simple hamiltonian (1) for all nuclei discussed before [18, 19, 20]. In general the description is improved compared to X(5) and values for $\eta$ very close to $\eta_{crit}$ are found. With the exception to $^{156}Dy$ with $\chi = -1.06$ the $\chi$ values equal $-\sqrt{7/4}$. Figure 5 shows the location of the fitted parameters in the extended triangle.

## SHAPE PHASE TRANSITIONS VERSUS SHAPE COEXISTENCE

Besides shape phase transitions atomic nuclei might also exhibit shape coexistence. Shape coexistence, invoking multiple shapes, was predicted and also observed in many spherical nuclei near magic shells and these particular phases could be linked to the occupation of very specific up- and /or downsloping orbitals, coined "intruder orbitals", which allowed for a simple understanding of the phenomenon of shape coexistence [21, 22]. The method put forward in those papers could be used to predict shape coexistence eg in the Sn nuclei, around mass number A=116 and in the Pb nuclei from mass number A=196 and below.

The point we like to stress here is that in a shape phase transition, the number of basis states is preserved through the transition and progressing from one limiting case into the other ( see Figure 6, left hand part). One notices that the eigenstates of one limit will be spread out and finally become the eigenstates of the other limit.

On the other hand, typical situations implying shape coexistence appear when a number of basis states, appearing at very high excitation energy (outside of the model space that is regularly considered as the space of low-lying configurations) as a function of neutron(or proton) number can profit from nucleon-nucleon correlations to drop considerably in excitation energy and thus a different set of states can occur at low excitation energy. Such classes of states are also quite often called 'intruder' excitations (see fig.1, right-hand part). In this situation, one essentially makes use of a single Hamiltonian $H$ but extends the model space by including (various) a different phase and its associated basis states. Therefore one has a coupling of two different Hilbertspace and $\mathscr{H} = \mathscr{H}_1 \otimes \mathscr{H}_2$. Mostly, this concerns in practical applications an $N$ sd boson space coupled to a $N+2$ sd bosonspace. They are used to describe shape coexistence between normal (valence) excitations and intruder excitations that include 2p-2h excitations. It should be mentioned that both sets of states can interact and mix with each other via a mixing hamiltonian that does not conserve the boson number. To distinguish between phase transition and shape coexistence the essential feature is that in the case of phase transitions the two different types of motion can only develop *incomplete* subsets of states resembling the motion generated by the changing phases. On the otherhand by phase coexistence complete structures can be generated.

## CONCLUSION

In conclusion, we have reviewed the occurence of QPT both in the interacting boson and the collective approach. Experimental examples for the prolate-oblate and spherical-rotor transitions (X(5)) are found. The difference between shape phase transitions and shape coexistence is stressed. The authors acknowledge the support of the BMBF under Project 06 K 203I.

## REFERENCES

1. Guttormsen, this proceeding.

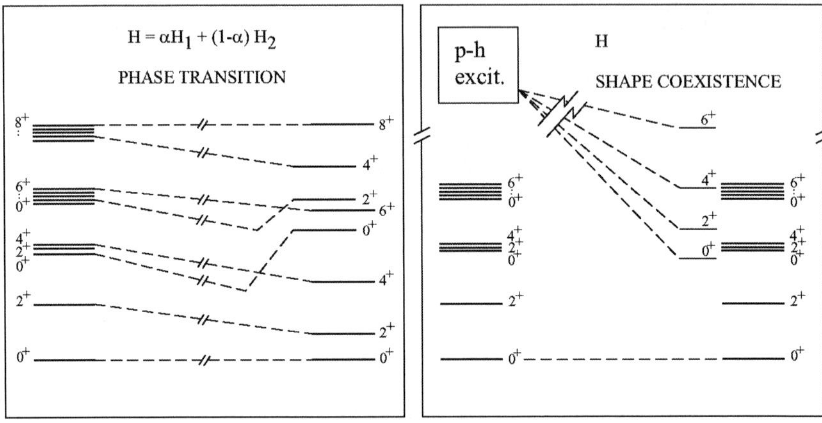


**FIGURE 6.** Schematic illustration for a possible shape phase transition (left-hand side) and for the situation where next to a family of regular states, another class of states can result an example of shape coexistence (right-hand).


2. Proc. of the 18th Int. Conf. on Ultra-Relativistic Nucleus-Nucleus Collisions, Nucl. Phys. **A774** (2006).
3. F. Iachello and A. Arima, The Interacting Boson Model (Cambridge University Press, Cambridge, 1987).
4. R.F. Casten in *Interacting Bose-Fermi Systems in Nuclei*, ed. F. Iachello Plenum, p 1 (1981).
5. J. Jolie, R.F. Casten, P. von Brentano, V. Werner Phys. Rev. Lett. **87**, 162501 (2001).
6. J. Jolie, A. Linnemann Phys. Rev. C **68**, 031301 (2003).
7. F. Iachello, Phys. Rev. Lett. **85**, 3580 (2000).
8. F. Iachello, Phys. Rev. Lett. **87**, 052502 (2001).
9. A. Bohr and B.R. Mottelson, Nuclear Structure (Benjamin, New York, 1975).
10. R.F. Casten, M. Wilhelm, E. Radermacher, N.V. Zamfir, and P. von Brentano, Phys. Rev. C **57**, R1553 (1998).
11. R.F. Casten and N.V. Zamfir, Phys. Rev. Lett. **85**, 3584 (2000).
12. R.F. Casten and N.V. Zamfir, Phys. Rev. Lett. **87**, 052503 (2001).
13. R. Kruecken *et al.*, Phys. Rev. Lett. **88**, 232501 (2002).
14. D. Tonev *et al.*, Phys. Rev. C **69**, 034334 (2004).
15. O. Moeller *et al.*, Phys. Rev. C **69**, 034334 (2004).
16. E.A. Mc Cutchan *et al.*, Phys. Rev. C **69**, 0244308 (2004).
17. R. F. Casten, D.S. Brenner, P.E. Haustein, Phys. Rev. Lett. **58** , 356 (1987).
18. A. Dewald *et al.*, J. Phys. G **31** (2005).
19. A. Dewald *et al.*, AIP Conf Ser **831**, 195 (2006).
20. B. Melon *et al.*, to be publ.
21. K. Heyde, P. Van Isacker, M. Waroquier, J.L. Wood and R.A.Meyer, Phys. Repts. **102**, 291 (1983).
22. J.L. Wood, K. Heyde, W. Nazarewicz, M. Huyse and P. Van Duppen, Phys. Repts. **215**, 101 (1992).
23. K.L.G. Heyde, J. Jolie, R. Fossion, S. De Baerdemacker and V. Hellemans, Phys. Rev. C **69**, 054304 (2004).

# Recent Studies On Nuclear Phenomena At The Extremes Using γ-ray Spectroscopy

S. Leoni, A. Bracco, F. Camera, G. Benzoni, N. Blasi, F.C.L. Crespi, B. Million, D. Montanari, O. Wieland

*University of Milano, Department of Physics and INFN sez. Milano, Via Celoria 16, 20133 Milano, Italy*

**Abstract.** The γ-decay of the excited nucleus provides experimental information on the nuclear structure properties at extreme conditions of temperature, angular momentum and isospin. First, the order-to-chaos transition is studied in the warm rotating $^{163}$Er nucleus, in terms of vanishing of selection rules on the K quantum number. Second, the properties of the excited superdeformed bands in $^{151}$Tb and $^{196}$Pb nuclei are investigated in connection with their decay-out into the low deformation states. Finally, the population of highly excited low-spin states in moderately neutron-rich systems is discussed in connection with deep-inelastic reactions.

**Keywords:** NUCLEAR REACTIONS. $^{150}$Nd($^{18}$O,5n), E=87,93 MeV; $^{130}$Te($^{27}$Al,6n), E=155 MeV; $^{170}$Er($^{30}$Si,4n), E=150 MeV; measured $E_\gamma$, $I_\gamma$, γγγ coin. $^{163}$Er, $^{151}$Tb, $^{196}$Pb rotational bands. Enriched targets, array of Compton suppressed Ge spectrometers.
**PACS:** 21.10.Re, 21.60.Ev, 21.60.Ka, 23.20.Lv, 27.70.+q

## INTRODUCTION

The study of the nucleus at the limits of angular momentum, temperature and neutron/proton number is one of the central topics currently addressed with selective γ-spectroscopy measurements.

In this contribution we present selected studies of the collective response of the atomic nucleus at extreme conditions, making use of nuclear reactions with stable ions. In particular, the temperature degree of freedom is used to investigate the transition between order and chaos in the deformed rotating nucleus $^{163}$Er, shape changes induced by the angular momentum at the highest values are studied in connection with the decay-out of the excited superdeformed (SD) bands in $^{151}$Tb and $^{196}$Pb. Finally, the population of highly excited low-spin states in moderately neutron-rich Ca isotopes via deep-inelastic reactions is discussed.

CP1012, *Frontiers in Nuclear Structure, Astrophysics, and Reactions: FINUSTAR 2*
edited by P. Demetriou, R. Julin, and S. V. Harissopulos
© 2008 American Institute of Physics 0-7354-0532-5/08/$23.00

## THE TRANSITION BETWEEN ORDER-TO-CHAOS

The transition between order and chaos in the atomic nucleus can be studied through the analysis of the γ-decay of warm rotating nuclei produced at high spins and moderate excitation energy by fusion reactions between heavy ions. After rapid evaporation of light particles, warm rotating nuclei de-excite emitting long sequences of γ transitions, ending up in discrete regular rotational bands when the nuclear temperature of the system is almost zero. Therefore, by detecting the largest number of γ-rays, nuclear structure properties can be investigated as a function of angular momentum and excitation energy, in the intermediate region between the ordered regime at T=0 and the chaotic region at the compound nucleus level [1,2].

By constructing γ–γ spectra from the measured $E_\gamma$ energies of high-fold γ cascades, one can distinguish between the contribution from the COLD region of regular decay at T=0 ("ridges") and from the WARM region above ≈ 1 MeV internal excitation energy ("valley"). This is where the nuclear rotation becomes damped and the rotational decay is fragmented over a large number of states with an energy spread $\Gamma_{rot}$ ≈ 200 keV, as a consequence of the rapidly increasing level density and of the presence of a residual two-body interaction mixing the nuclear states [1]. The study of such ridge-valley landscape can be used to investigate the transition between order and chaos in terms of a gradual vanishing (at T≠0) of the selection rules associated with the quantum numbers at T = 0, such as K (the projection of the angular momentum on the symmetry axis).

The deformed nucleus $^{163}$Er is an ideal case, being characterized by a number of rotational bands having low-K (K=5/2) and high-K (K=19/2) values [3-5]. The nucleus $^{163}$Er has been populated by the reaction $^{18}$O + $^{150}$Nd, at $E_{beam}$ = 87, 93 MeV, and its subsequent γ-decay has been detected using the EUROBALL array, consisting of more than 200 Ge crystals in 4π geometry around the reaction center [6]. The $^{150}$Nd target was made of a stack of two thin foils for a total thickness of 740 μg/cm$^2$ and the corresponding maximum angular momentum reached in the reaction has been calculated to be 40 and 45 , for the two different bombarding energies. Figure 1 shows examples of 60 keV wide projections, perpendicular to the $E_{\gamma 1} = E_{\gamma 2}$ diagonal of experimental matrices of $^{163}$Er, gated by transitions among states with low-K (K=5/2) and high-K (K=19/2) quantum numbers. The spectra, showing the typical ridge-valley structure resulting from the γ-decay of a deformed rotating nucleus, have been analyzed by statistical and line-shape analysis methods [3-5].

The analysis of the fluctuations of the events collected in the γ–γ spectra allows to estimate the number of bands ($N_{path}$) which populate the ridge and valley region [7]. As shown in panel a) of Fig. 2, a total number of ~40 discrete excited bands is found to populate the ridge structures of $^{163}$Er, half of which of high-K nature. On the contrary, many more bands, of the order of $10^3$-$10^5$, are found to populate the valley region, with large differences between low-K and high-K states, as shown in Fig. 2 b). This suggests that the K-quantum number is at least partially conserved up to moderate excitation energies, of the order of ~1.5 MeV above yrast, where the rotational motion is damped. At higher excitation energies (namely for $E_\gamma$ > 1.1 MeV in Fig. 2 b)) more similar number of bands are obtained for low-K and high-K gated spectra, pointing to a vanishing of selection rules on K and to the onset of a chaotic regime, in which quantum numbers and selection rules loose their meaning.

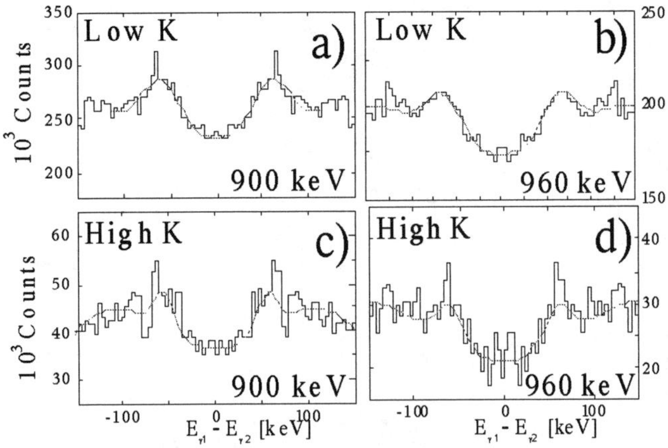


**FIGURE 1.** 60 keV wide projections on the $E_{\gamma 1}$-$E_{\gamma 2}$ axis of experimental matrices of $^{163}$Er, at the average transition energies $E_\gamma$ = 900 and 960 keV. Panels a) and b) (c) and d) show spectra obtained from $\gamma$–$\gamma$ matrices gated by low-K (high-K) configurations of $^{163}$Er. The smooth curves represent the interpolation of the data by a two-component spectral function [4].

The results of the fluctuation technique are supported by the analysis of the correlations in fluctuations between spectra gated by different K-states [3,5]. Figure 2 c) and d) give the correlation coefficient *r* obtained from the analysis of the ridge and valley region between low-K and high-K spectra. *r* assumes values between 0 (completely different event distributions) and 1 (identical spectra). The experimental data show that *r* is close to 0 in the case of the analysis of the ridges (which are populated by discrete excited bands at T=0), while in the case of the valley region *r* is found to gradually approach 1 at the highest transition energy values (corresponding to internal energy around 1.5 MeV), indicating strong similarities/correlation between the high-K and low-K gated distributions, as expected in a chaotic regime [3,5].

The experimental results shown in Fig. 2 are well reproduced by the shell model calculation of ref. [8] which combines a cranked mean-field and a residual two-body interaction, together with a terms taking into account the angular momentum carried by the K-quantum number. According to the model, K-mixing is induced by the interplay of the Coriolis and residual interaction, and it is found to gradually increase until a complete violation of the K-quantum number is reached above 2-2.5 MeV of internal energy, in good agreement with the experimental findings.

The spectral shape of the ridge-valley structure (see Fig. 1) can also be used to determine the rotational damping with $\Gamma_{rot}$ [4]. Figure 3 a) shows the values obtained for $\Gamma_{rot}$ by interpolating the ridge-valley event distribution of the total $\gamma$–$\gamma$ spectrum of $^{163}$Er, by a two component function (smooth curves in Fig. 1). The damping width is found to depend on the K value, being ~ 200 keV for low-K and ~ 150 keV for high-K states, in agreement with the calculation of the model [8]. This gives further support to the conservation of K up to moderate excitation energies.

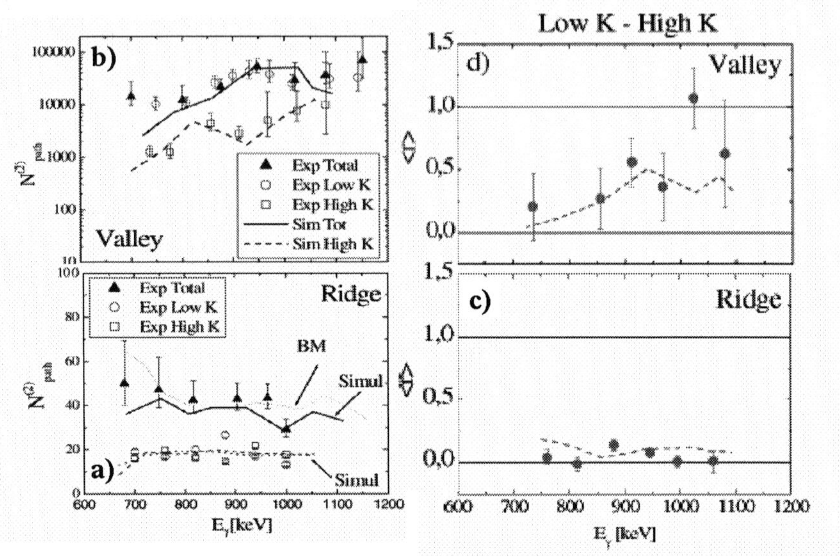


**FIGURE 2.** Results of the statistical analysis of the ridge and valley regions of the experimental γ–γ matrices of $^{163}$Er. Panel a) shows the number of discrete excited bands, obtained from the fluctuation analysis of the ridge structures of the total spectrum (triangles) and of spectra gated by Low-K and High-K structures (open symbols). The analysis of the corresponding valley regions is given in panel b). Panels c) and d) show the results of the covariance analysis between Low-K and High-K spectra, for the ridge and valley structures [3,5]. In all panel the lines represent prediction from the cranked-shell model calculation of ref. [8].

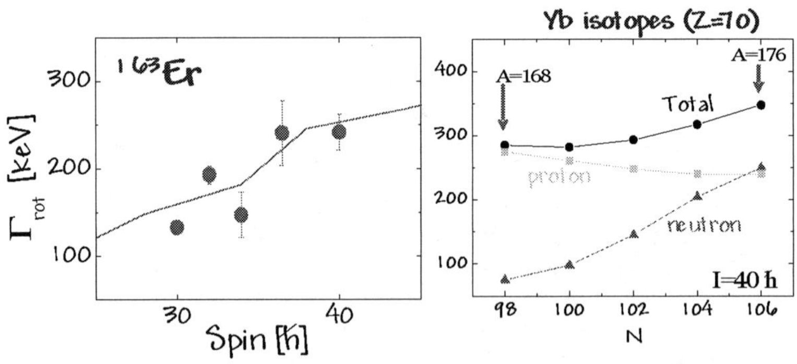


**FIGURE 3.** The left panel shows the experimental values of the rotational damping width $\Gamma_{rot}$,extracted from the spectral shape analysis of the total γ–γ spectrum of $^{163}$Er [4], together with predictions from cranked shell model calculations [8]. The right panel gives the prediction of the model for the rotational damping with in Yb isotopes, from A=168 and A=176, at spin I = 40 . The contributions due to protons and neutrons are also given separately.

It is important to remark that the experimental determination of the rotational damping width $\Gamma_{rot}$ can provide valuable information on the evolution of the shell structure at finite temperature (T≠0), as a function of the neutron number. As shown in Fig. 3 b), the cranked shell model of ref. [8] predicts, at spin I = 40 , an increase of almost 25% in the rotational damping width, moving from $^{168}$Yb (populated at present by stable heavy-ions reactions) to $^{176}$Yb (reachable in the future by means of radioactive beams, such as $^{132}$Sn).

## QUANTUM TUNNELING OF EXCITED SUPERDEFORMED BANDS

The study of the properties of the atomic nucleus as a function of angular momentum offers opportunities to learn about the phase transition between normal and superfluid systems [2]. This topic has been investigated in a number of nuclei, mostly in connection with the sudden decay-out from superdeformed (SD) to normal deformed (ND) structures, which are separated by a potential energy barrier. The decay between configurations occurs as a tunnelling process, which becomes suddenly possible over few units of spins when the nucleus undergoes an abrupt transition from normal to superfluid. Such a decay mechanism is well established for the discrete SD bands at T=0, and it can also be used to describe the decay of the excited states (T≠0). In this case the tunnelling model needs to be coupled to cranked shell model calculations at finite temperature [9]. Experimentally, only few cases have been studied in details [10-12], since selective data obtained by gating on transitions collecting only few percent of the decay flow are needed with high statistics.

In this paper we present results from two different EUROBALL experiments. The first one aimed at the population of $^{151}$Tb by means of the reaction $^{27}$Al, at 155 MeV, on a thin $^{130}$Te target [11], while the second experiment employed the reaction $^{30}$Si at 150 MeV on a thin target of $^{170}$Er, leading to $^{196}$Pb after the evaporation of 4 neutrons. The first experiment was performed using the EUROBALL array [6] in the standard configuration, while for the second experiment, the low efficiency Ge detectors in the forward hemisphere were substituted by 8 large volume $BaF_2$ scintillators to measure high energy γ-rays from the decay of the giant dipole resonance (GDR). In both experiments an InnerBall of BGO detectors was also used to determine the γ–multiplicity of each single event by measuring the number of fired detectors.

For both experiments the data have been sorted into a number of γ–γ matrices in coincidence with the isotope of interest ($^{196}$Pb or $^{151}$Tb), and in coincidence with the SD yrast band of each nucleus, with a condition on high-fold events (F≥20) to better focus on high-multiplicity cascades. In both sets of spectra cuts perpendicular to the main diagonal reveal the existence of ridge structures extending over a large range of spin (40 to 60 for $^{151}$Tb and 10 to 40 for $^{196}$Pb), with a spacing between the two most inner ridges equal to $4\times {}^2/J^{(2)}$, being $J^{(2)}$ the moment of inertia of the SD yrast band in each nucleus. Examples of such cuts are given in Fig. 4.

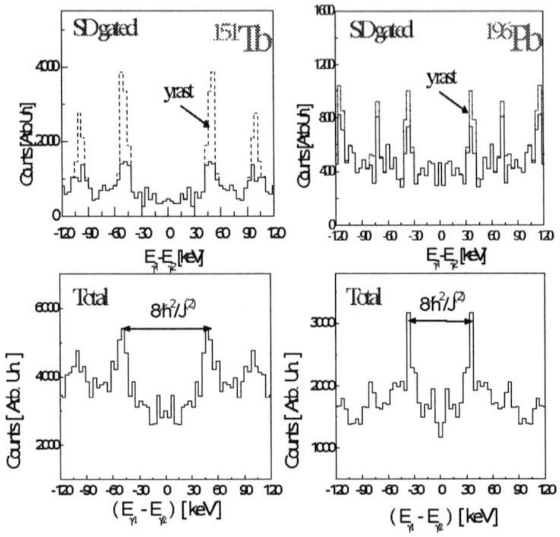


**FIGURE 4.** Projections on the $E_{\gamma 1}$-$E_{\gamma 2}$ axis of the Total (bottom) and SD gated (top) matrices of $^{151}$Tb (left) and $^{196}$Pb (right). In the bottom panels the arrow indicate the separation between the two most inner ridges, corresponding to ~ 100 keV and 70 keV, respectively. In the top panels the SD yrast intensity is shown by dotted lines. The cuts for $^{151}$Tb and $^{196}$Pb are performed at the average energy $<E_\gamma>$ = 1228 and 532 keV, corresponding to the spin values 52 and 26 , respectively.

In both nuclei, the ridge structures are found to be populated by several unresolved discrete transitions, as confirmed in first place by the evaluation of their intensities, up to 3 times grater than the intensity of the SD yrast. In addition, the number of bands populating the SD ridge structures have been evaluated by the fluctuation analysis technique [7]. The results are shown in Fig. 5 for $^{151}$Tb (left) and $^{196}$Pb (right). In both cases a large number of SD rotational bands (up to 40 in $^{151}$Tb and more than 50 in $^{196}$Pb) are found to populate the SD ridges (filled squares) [11].

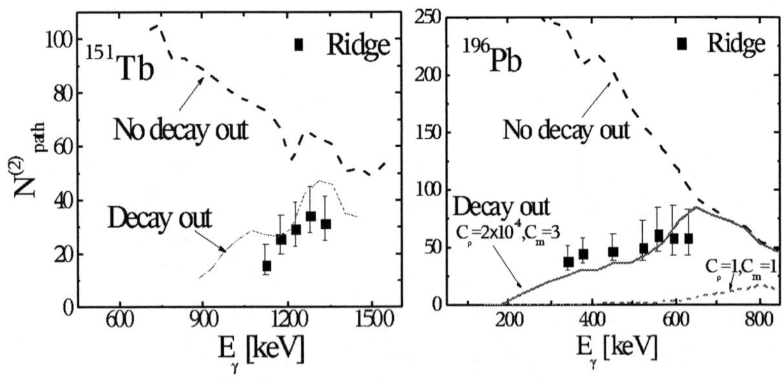


**FIGURE 5.** The fluctuation analysis of the ridge structures of $^{151}$Tb (left) and $^{196}$Pb (right). The number of discrete bands obtained from the Total matrix are shown by full symbols. The dotted lines are the predictions of the cranking model of ref. [9], not taking into account the decay-out towards the ND states. The full lines are obtained including also the probability to decay into the ND well, which in the case of $^{196}$Pb has been adjusted by the factor $C_p$ and $C_m$ (see text).

The results of the fluctuation analysis of the SD ridges in $^{151}$Tb and $^{196}$Pb have been interpreted in terms of the cranked shell model calculations of ref. [9]. In both cases the model largely overestimates the experimental data, especially at low transition energies, where the decay-out towards the ND structures occurs. Better agreement with the experiment is obtained once we take into account, for each SD excited band, a probability $P_{out}$ to decay towards ND states [9]. $P_{out}$ depends significantly on the level density $\rho_{ND}$ of the ND states and on the inertial mass $M_0$ of the nucleus, especially in the heaviest mass region A = 190. In this case, a renormalization of these two quantities by the constant $C_\rho = 2\times10^{-4}$ and $C_m = 3$ is needed to reproduce both the decay-out spin values and the number of excited bands, as shown by solid line in Fig. 5. The study of the decay-out of the excited SD bands, modeled as in the case of the SD yrast states, may therefore provide additional information on the key parameters governing the tunneling process.

## HIGHLY EXCITED STATES IN NEUTRON RICH NUCLEI

At present, one of most investigated issues in nuclear physics is the influence of the neutron/proton (N/Z) ratio on the structure of the atomic nucleus. For this reasons the fundamental modes of excitations (such as vibrations and rotations) are currently under investigation in different isotopic chains as a function of the N/Z ratio. Among the most interesting states are the low-lying electric dipole (pygmy) states which in the case of neutron-rich nuclei are expected to have a sizable strength. A large systematic of such states exists for stable isotopes in the mass regions A~40-50 and A~120-140, based on photon scattering experiments [13-15].

In a recent experiment at Legnaro National Laboratory (Italy) the Ge array CLARA coupled with the magnetic spectrometer PRISMA [16] has been used to investigate the possibility of populating highly excited states in neutron-rich Ca nuclei by means of a different reaction mechanism, namely deep-inelastic collisions with heavy ions (at ≤ 10 MeV/A). In the past, deep-inelastic scattering has been used at higher energies to study the decay of the Giant Dipole Resonance (GDR) in highly excited nuclei [17] and the emission of dipole strength in pre-equilibrium reactions [18]. It is therefore interesting to study with high resolution the excitation energy up to roughly 6-8 MeV, which is so far mainly unexplored. According to the calculations of Pollarolo et al. [19], a number of isotopes with sizable cross section (~ 100 mb) should be populated in the reaction $^{48}$Ca at 320 MeV on $^{64}$Ni, and this has been confirmed by the mass distribution of the Ca isotopes obtained in the experiment (left panel of Fig. 6). In addition, a sizable population of the nuclei of interest (mainly $^{48}$Ca and $^{50}$Ca) is expected at the highest excitation energy, around 10 MeV, as shown in the right panel of Fig. 6.

The analysis of the experiment is in its early stages and interesting results are expected also from the point of view of the dynamics of the reaction.

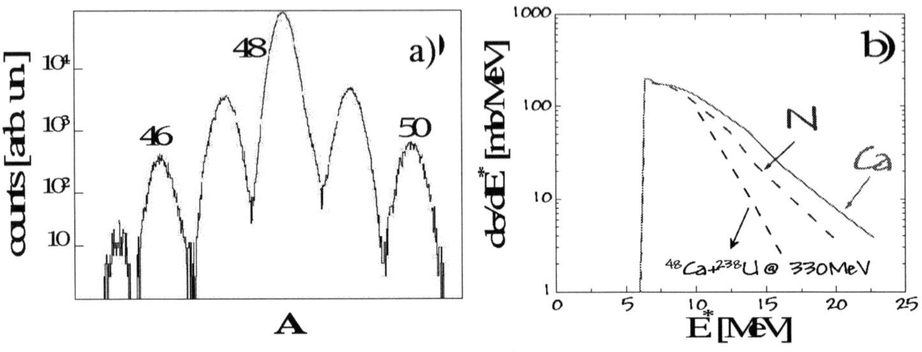


**FIGURE 6.** Panel a): mass distribution of the Ca isotopes measured in the deep-inelastic reaction $^{48}Ca$ + $^{64}Ni$ at 320 MeV. Panel b): cross section as a function of excitation energy for Ca and Ni fragments, in comparison with a similar reaction on an heavier U target [19].

## ACKNOWLEDGMENTS

The authors wish to thank the EUROBALL and PRISMA collaborations which made possible the realization of the experiments here discussed. This work has been partially supported by EU (contract No. EUROVIV: HPRI-CT-1999-00078) and by the Italian Institute of Nuclear Physics (INFN).

## REFERENCES

1. A. Bracco and S. Leoni, *Rep. Prog. Phys.* **65**, 299 (2002).
2. A. Bracco et al., *Cont. Phys.* **37**, 183 (1996).
3. G. B. Benzoni et al., *Phys. Lett.* **B615**,160 (2005).
4. S. Leoni et al., *Phys. Rev. Lett.* **93**, 022501 (2004).
5. S. Leoni et al., *Phys. Rev.* **C72**, 034307 (2005).
6. J. Simpson *Z. Phys.* **A358,** 139 (1997).
7. T. Døssing et al., *Phys. Rep.* **268**, 1 (1996).
8. M. Matsuo et al., *Nucl. Phys.* **A736**, 223 (2004).
9. K. Yoshida, M. Matsuo and Y.R. Shimizu, *Nuc. Phys.* **A696**, 85 (2001).
10. S. Leoni et al., *Phys. Lett.* **B498**,137 (2001).
11. G. Benzoni et al., *Phys. Rev.* **C75**,047301 (2007).
12. T. Lauritzen et al., *Phys. Rev.* **C75**, 064309 (2007).
13. T. Hartmann et al., Phys. Rev. Lett. **85**, 271 (2000).
14. T. Hartmann et al., Phys. Rev. Lett. **93**, 192501-1 (2004).
15. J. Enders et al., Nuc. Phys. A741, 3 (2004).
16. A. Gadea et al., Eur. Phys. J. **A20**, 193 (2004).
17. G. Enders et al., Phys. Rev. Lett. **69**, 249 (1992).
18. M. Trotta et al., RIKEN review 23 (1999): *Focused on selected Topics in Nuclear Collective Excitations.*
19. A. Winther, Nucl. Phys, **A549**, 203 (1995).

# Shape Transitional Nuclei: What can we learn from the Yrare States? or Hello the Double Vacuum; Goodbye β-vibrations!

J. F. Sharpey-Schafer[§¶†], S. M. Mullins[†], R. A. Bark[†], E. Gueorguieva[†], J. Kau[†], F. Komati[†], J. J. Lawrie[†], P. Maine[§†], A. Minkova[#], S. H. T. Murray[†], N. J. Ncapayi[†], P. Vymers[§]

[§] *University of Western Cape, Department of Physics, P/B X17, Bellville,ZA-7535 South Africa*
[¶] *University of Zululand, Department of Physics and Engineering, P/B X1001, Kwa Dlangezwa, ZA-3886 South Africa*
[†] *iThemba Laboratory for Accelerator Based Sciences, PO Box 722, Somerset-West, ZA-7129 South Africa.*
[#] *Faculty of Physics, St. Kliment Ohridski University of Sofia, Sofia 1164, Bulgaria*

**Abstract.** The results of our measurements on the yrare states up to spin 20ħ in $^{152,154,155}Gd$, using (α,xn) reactions and the AFRODITE γ-ray spectrometer, are presented. We find that in $^{155}Gd$ the decay scheme is divided into levels feeding the [505]11/2$^-$ band, that is extruded by the prolate deformation from the $h_{11/2}$ orbital, and levels feeding the $i_{13/2}$ [651]3/2$^+$ intruder orbital and the $h_{9/2}$ [521]3/2$^-$ orbital. The decay scheme of $^{154}Gd$ is very complex. We find no evidence for the existence of β-vibrational levels below 1.5MeV. We discover that the level scheme can be best understood as a set of collective states built on the ground state configuration $|0_1^+>$ plus a "congruent" set of collective states based on the $|0_2^+>$ state at 681keV. The data suggest that this second vacuum has reduced pairing. Our data do not support IBA and phonon interpretations of these transitional nuclei.

**PACS:** 21.10.Re, 23.20.Lv, 25.55.-e, 27.70.+q

## INTRODUCTION

The rapid shape changes over small ranges of nucleon number in transitional nuclei bring different single particle orbitals close to the Fermi Surface. For these nuclei, structures with different shapes can coexist and the softness of the nucleus to β and γ vibrations can also be affected. Exotic shapes have also been predicted to occur for some nuclei, for example, tetrahedral shapes have been predicted to occur for nuclei near N = 90 and Z = 64, 70. Hence experimental data is required on as many levels as possible and as far from the yrast line as can be reached: "complete spectroscopy".

The N = 90 nuclei have for a long time been a very fruitful source of new and complex phenomena, especially at high spins. This has been recently emphasized by the discovery [1] of triaxial superdeformed bands in $^{158}Er_{90}$ . At very low spins there have been a series of attempts to explain the properties of the ground state band and

CP1012, *Frontiers in Nuclear Structure, Astrophysics, and Reactions: FINUSTAR 2*
edited by P. Demetriou, R. Julin, and S. V. Harissopulos
© 2008 American Institute of Physics 0-7354-0532-5/08/$23.00

the lowest $0^+$ "β-band" in terms of the IBA model and critical point symmetries, X(5) in particular [2-8]. However these IBA models only address the first few positive parity states and their decay transition strengths and contain nothing about negative parity states and the alignments beyond spin $10\hbar$ which are generally well described, at least qualitatively, by the Cranked Shell Model (CSM) [8]. It has been pointed out by Clark et al [9] that a better description of these $K^\pi = 0^+$ bands can be achieved with the traditional collective pairing-plus-quadrupole model with $\Delta K = 0$ mixing. Also it is well known [10-12] that octupole vibrations and even permanent octupole deformations play a significant role in the structure of these nuclei. In addition it has been proposed by Dudek and his collaborators [13,14] that nuclei with N = 90 and Z = 64 or 70 might contain bands that correspond to tetrahedral nuclear shapes.

Inspired by the "complete spectroscopy" that Davidson et al [15] achieved at ILL for the nucleus $^{168}$Er, we decided to look at the uncomplex (α,xn) reactions to the transitional Gd nuclei using our escape suppressed γ-ray spectrometer array, AFRODITE. This array consists of 8 clover detectors and 8 four-fold segmented LEPS detectors. Using 5 mg cm$^{-2}$ targets of $^{152}$Sm and $^{154}$Sm we obtained about $5.10^8$ γγ coincidences for each of the reactions $^{152}$Sm(α,4n)$^{152}$Gd at 45MeV, $^{152}$Sm(α,2n)$^{154}$Gd at 25MeV, $^{154}$Sm(α,4n)$^{154}$Gd at 45MeV and $^{154}$Sm(α,3n)$^{155}$Gd at 35MeV.

## EXPERIMENTAL RESULTS

We reported on our data on the $^{152}$Sm(α,4n)$^{152}$Gd reaction at a previous conference [16] and the results have not changed substantially since.

The current state of our analysis of our $^{155}$Gd data is shown in the decay scheme in Fig. 1. The backbones of the decay scheme are;

(a) the ground state band based on the $h_{9/2}$ [521]3/2$^-$ orbital.

(b) the yrast band based on the $i_{13/2}$ [651]3/2$^+$ intruder equatorial orbital.

(c) the high spin negative parity band based on the [505]11/2$^-$ extruded polar orbital.

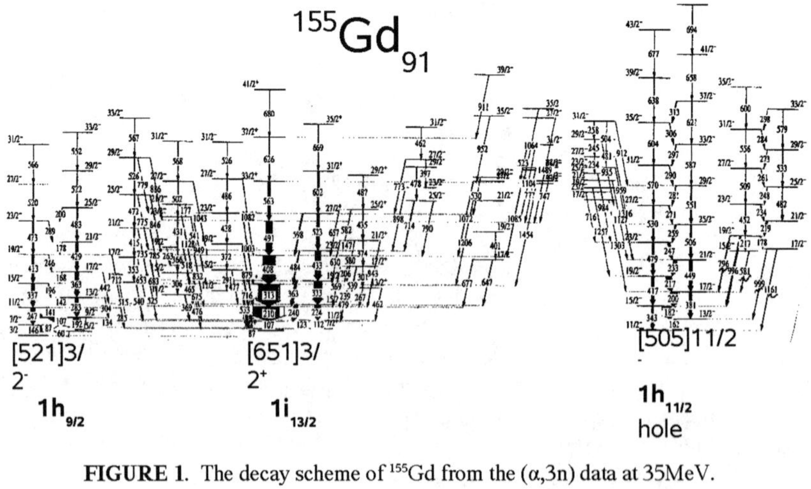


**FIGURE 1.** The decay scheme of $^{155}$Gd from the (α,3n) data at 35MeV.

It can be seen in Fig. 1 that the decay scheme is sharply divided between decays to the [505]11/2$^-$ polar orbital and decays to the more equatorial (a) and (b) structures. Also there are many E1 transitions which emphasize the importance of octupole collectivity.

The decay scheme of $^{154}$Gd is even more complex, so in Fig. 2 we break the scheme up into separate sections. The ($\alpha$,2n) data was most useful in determining the existence of levels between 1.5 and 3.0 MeV excitation energy. The ($\alpha$,4n) data extended several structures to higher angular momentum. The extensive and precise work of Kulp et al [17] was most helpful as a starting point and many transitions at lower spins are in agreement with the work of Spitz et al [18]. The higher spin states generally agree with the work of Morrison et al [19].

In Fig. 2(a) we show the decays of both signatures of the negative parity octupole vibrational bands. We have found all even spin members up to 16$^-$. There is also a series of new levels decaying to these states some of which could be candidates for a $\gamma$+octupole vibrational structure. In Fig. 2(b) we show levels that decay through the known [20] 7$^-$ 68ns isomer. We see a band parallel to that observed by Morrison et al [19] which seems to have the 7$^-$ isomer as its band head. We therefore label this $K^\pi$ = 7$^-$. The Morrison band we label as $K^\pi$ = 9$^-$. In Fig. 2(d) we show states decaying to the band based on the 0$^+$ level at 681keV, previously though to be a $\beta$-vibrational band but which we will now label $|0_2^+\rangle$ as the second vacuum state. On the left of Fig. 2(d) we find a series of levels decaying to $|0_2^+\rangle$ at and around its back-bend. The "Pairing Isomer" band of Kulp et al [21] is very tentatively extended and there are also levels decaying in a manner similar to the 2$^+$ and 3$^+$ levels observed in ref. [17] at 1531 and 1661keV. This 2$^+$ level has the characteristics of a $\gamma$-vibrational $K^\pi$ = 2$^+$ band head based on the $|0_2^+\rangle$ configuration. In Fig. 2(c) the $\gamma$-band has been extended to 17$^+$ and the AF band of Morrison et al [19] has an additional 8$^-$ level at the bottom. We have found levels decaying in a similar manner to the decay of the 2$^-$ and 3$^-$ levels established by Kulp et al [17]. These levels are at an energy where they could be the octupole vibration based on the second vacuum $|0_2^+\rangle$. In Fig. 2(e) we show the decays of the $K^\pi$ = 4$^+$ band which we have extended by an additional 4 levels. Our data is consistent with the conclusion of Burke et al [22] that this has mainly the proton particle-hole configuration [413]5/2$^+$☼[411]3/2$^+$ . There are several new levels which decay strongly to the $K^\pi$ = 4$^+$ band.

The main conclusion of our experiment is that there is no evidence that the 681keV level is the $\beta$-vibration based on the ground state. A much more consistent interpretation of the physics of this nucleus is achieved if the 681keV level is regarded as a second vacuum $|0_2^+\rangle$ which then has its own $\gamma$- and octupole vibrations built upon it. If it is a second vacuum it must also have its own CSM alignments characteristic of its own deformation and pairing. As $|0_2^+\rangle$ is not lowered in energy as much as $|0_1^+\rangle$, we would expect it to have less pairing strength. Hence the $A_2B_2$ band crossing should occur at a lower frequency than that for $A_1B_1$. Indeed the back-bend in $|0_2^+\rangle$ occurs at $\hbar\omega$ = 0.22MeV [23] rather than the usual $\hbar\omega$ = 0.28MeV for yrast bands in the rare earth region, consistent with reduced pairing. It has been clearly established in lifetime measurements [24,25] that the B(E2)s, and therefore the deformation, are slightly smaller in $|0_2^+\rangle$ than in $|0_1^+\rangle$. However the moment-of-inertia is larger for $|0_2^+\rangle$ than for $|0_1^+\rangle$, characteristic of a more rigid rotation due to reduced pairing.

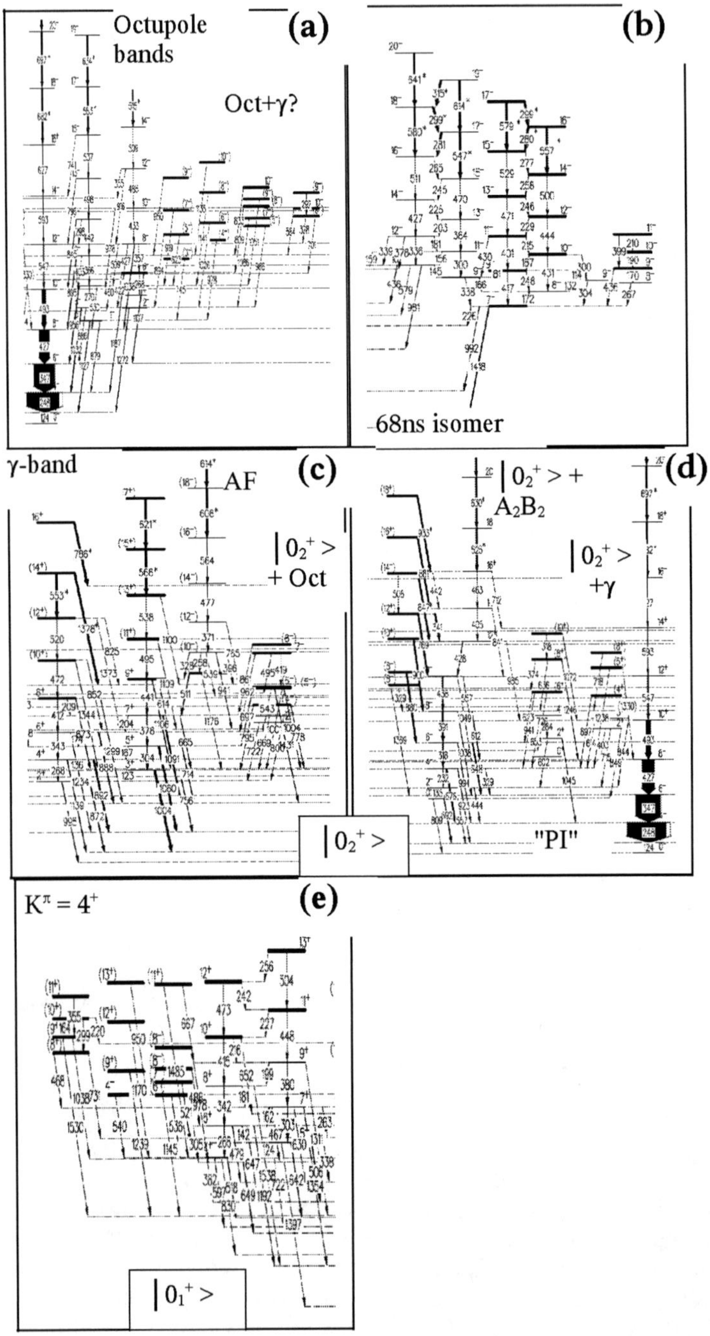


**FIGURE 2.** The decay scheme of $^{154}$Gd. New levels are bold and transitions only seen in the (α,4n) reaction are marked *. Spins in brackets are not established and are consistent with the observed decays.

In Fig. 3 we show the excitation energies, minus a rigid rotor energy, of the collective and particle-hole structures in $^{154}$Gd. It can be seen that the γ-band tracks $|0_1^+>$, as do the $K^\pi = 4^+$ and $7^-$ bands, with no signs of band crossings but a steady alignment characteristic of the very strong interaction for $A_1B_1$. The octupole bands, interpreted as $A_1E_1$ and $A_1G_1$, and the high-K band ($9^-$) of ref [19] all become more yrast at higher spins. Only the AF band has an upbend, the BC crossing, at $\hbar\omega$ = 0.31MeV [19].

Assuming our interpretation of the structure of $^{154}$Gd is correct, the question then arises; what is the difference in the underlying configurations of $|0_1^+>$ and $|0_2^+>$? The proton stripping [22] from $^{153}$Eu, ground state configuration [413]5/2$^+$ from the $1g_{9/2}$ orbital, populates both the $|0_1^+>$ band and the γ-band but there is very little cross-section to the $|0_2^+>$ band. Surprisingly the only data in the literature on one neutron pickup from $^{155}$Gd is a reference in [26] to a thesis [27]. Again in the pickup [27] the $|0_1^+>$ and the γ-band are seen strongly but the $|0_2^+>$ band very weakly. The $|0_2^+>$ state is strongly populated in the (t,p) two neutron stripping reaction but not in the (p,t) pick up reaction [26,28,29 and references therein]. In none of these references is the sanctity of the β-vibration interpretation of the 681keV $0_2^+$ questioned!

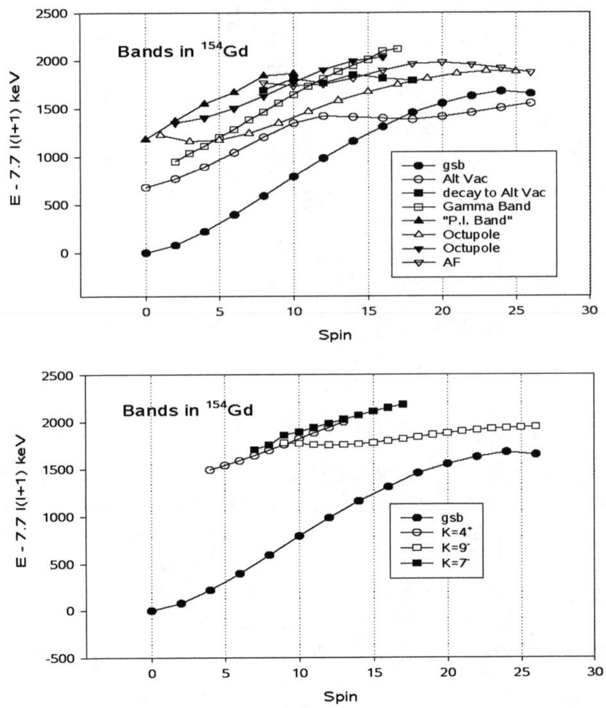


**FIGURE 3.** Bands in $^{154}$Gd observed in this work. "gsb" = $|0_1^+>$ and "Alt Vac" = $|0_2^+>$.

In the actinides $0_2^+$ states had been observed by Maher et al [30] in the (p,t) reaction that could not be described as a pairing or β vibration. The suggestion of Griffin et al [31] that these states could arise from the weakness of the oblate-prolate pairing force was taken up by van Rij and Kahana [32] and studied in detail by Ragnarsson and Broglia [33] in the framework of BCS and RPA. The core idea is that the oblate-prolate pairing force $G_{op}$ is significantly weaker than the oblate-oblate $G_{oo}$ and prolate-prolate $G_{pp}$ pairing forces. Here oblate refers to Nilsson configurations that have a polar orbit (usually high-K) and prolate to equatorial orbitals (usually low-K). Central to this model [32] is the paucity of oblate Nilsson levels near the Fermi surface in the actinides, a situation that also occurs in the transitional nuclei with N ~ 90 at the start of the rare earths. This decoupling of the polar and equatorial orbitals leads to the oblate pairing energy $\Delta_o$ , and hence the oblate quasi-particle energy, being reduced drastically permitting the existence of a low-lying $0^+$ state. Also the two neutron transfer cross-section to this state is no longer reduced by the normal pairing effects. Reference [33] labels such $0^+$ levels "Pairing Isomers".

Peterson and Garrett [34] have studied the single neutron stripping and pickup cross-sections to the [505]11/2$^-$ band heads in odd Gd, Dy and Er nuclei. They find that their results can be explained assuming a small pairing parameter for this oblate orbit. In Fig.2 we see that in $^{155}$Gd the [505] structure is completely separated from the [521] and [651] prolate orbitals. This suggests that the separation we observe in $^{154}$Gd between $|0_1^+>$ and $|0_2^+>$ is due to the two vacua having very different populations of the [505] orbital.

If this decoupling has a significant part of the physical processes leading to the "congruent" structures we observe in $^{154}$Gd, then it is clearly important to trace the band structures to higher spin where pairing energy dependent alignments take place.

## CONCLUSIONS

We have shown that our data gives no support to the interpretation of the $0_2^+$ state in $^{154}$Gd being a β-vibration of the ground state configuration. Rather there are two competing vacua, $|0_1^+>$ and $|0_2^+>$ , with "congruent" rotational, γ- and octupole vibrations built upon them. Very beautiful and comprehensive data, obtained by Kulp et al [35], support this conclusion in the neighbouring even-even N=90 nucleus $^{152}$Sm. Our data also suggest that each vacuum has its own CSM alignments, whose properties are determined by the pairing appropriate to that vacuum and that the pairing for $|0_2^+>$ is less than for $|0_1^+>$ .

These data lend support to Garrett, who has questioned [36] the trite assumption that most first excited $0_2^+$ states in prolate deformed nuclei are β-vibrations. These data also cast doubt on the usefulness of interpreting the structure discussed here in terms of the IBA and X(5) symmetries.

## ACKNOWLEDGEMENTS

We would like to thank David Kulp, John Wood, Paul Garrett et al for very constructive discussions and for sending us their unpublished data. One of us (JFS-S) would like to thank John Singer Sargent and the Joyce Frances Adlard Cultural Fund for financial support.


## REFERENCES


1. E. S. Paul et al, Phys. Rev. Lett. **98**, 012501 (2007).
2. F. Iachello, Phys. Rev. Lett. **87**, 052502 (2001).
3. R. F. Casten and N. V. Zamfir, Phys. Rev. Lett. **87**, 052503 (2001).
4. A. Leviatan and J. N. Ginocchio, Phys. Rev. Lett. **90**, 212501 (2003).
5. R. Krücken et al, Phys. Rev. Lett. **88**, 232501 (2002).
6. J. Jolie et al, Phys. Rev. Lett. **89**, 182502 (2002).
7. D. Bonatsos et al, Phys. Rev. **C 69**, 014302 (2004).
8. R. Bengtson and S. Frauendorf, Nucl. Phys. **A 314**, 27 (1979).
9. R. M. Clark et al, Phys. Rev. **C 67**, 041302R (2003).
10. M. Sugawara et al, Nucl. Phys. **A 557**, 653c (1993).
11. N. Minkov et al, Phys. Rev. **C 73**, 044315 (2006).
12. M. Babilon et al, Phys. Rev. **C 72**, 064302 (2005).
13. J. Dudek et al, Phys. Rev. Lett. **88**, 252502 (2002).
14. J. Dudek et al, Phys. Rev. Lett. **97**, 072501 (2006).
15. W. F. Davidson et al, J. Phys. **G 7**, 455 (1981).
16. J. F. Sharpey-Schafer et al, Proc. XLIV Int. Winter Meeting on Nuc. Phys.; Bormio, Italy (2006); Univ. degli Studi di Milano No. 125, Eds. I Iori, A. Tarantola, pp295-303
17. W. D. Kulp et al, Phys. Rev. **C 69**, 064309 (2004).
18. A. Spitz et al, http://bib120.sckcen.be/blg-703.pdf
19. J. D. Morrison et al, J. Phys. **G 15**, 1871 (1989).
20. R. L. West, E. G. Funk and J. W. Mihelich, Phys. Rev. **C 18**, 679 (1978).
21. W. D. Kulp et al, Phys. Rev. Lett. **91**, 102501 (2003).
22. D. G. Burke et al, Nucl. Phys. **A 688**, 716 (2001).
23. D. Ward, R. L. Graham, J. S. Geiger and H. R. Andrews, Phys. Lett. **44 B**, 39 (1973).
24. N. Rud et al, Nucl. Phys. **A 191**, (1972); S.H. Sie et al, Nucl. Phys. **A 291**, 443 (1977).
25. D. Tonev et al, Phys. Rev. **C 69**, 034334 (2004).
26. M. A. M. Shahabuddin et al, Nucl. Phys. **A 340**, 109 (1980).
27. Om Prakah Jolly, PhD thesis McMaster Univ. (1976);Jim Waddington, private communication.
28. D. G. Fleming et al, Phys. Rev. Lett. **27**, 1235 (1971); Phys. Rev. **C 8**, 806 (1973).
29. Th. W Elze et al, Nucl.. Phys. **A 187**, 473 (1972).
30. J. V. Maher et al, Phys. Rev. Lett. **25**, 302 (1970).
31. R. E. Griffin, A. D. Jackson and A. B. Volkov, Phys. Lett. **36B**, 281 (1971).
32. W. I. van Rij and S. H. Kahana, Phys. Rev. Lett. **28**, 50 (1972).
33. I. Ragnarsson and R. A. Broglia, Nucl. Phys. **A 263**, 315 (1976).
34. R. J. Peterson and J. D. Garrett, Nucl. Phys. **A 414**, 59 (1984).
35. W. D. Kulp, J. L. Wood, P. E. Garrett, C. Y. Wu, D. Cline, J. M. Allmond, D. Bandyopadhyay, D. Dashdorj, S. N. Choudry, A. B. Hayes, H. Hua, S. R. Lesher, M. Mynk, M. T. McEllistrem, C. J. McKay, J. N. Orce, R. Teng, S. W. Yates, private communication and to be published.
36. P. E. Garrett, J. Phys. **G 27**, R1 (2001).

# Observation of a Double-humped $\gamma$-Ray Fold Distribution in $^{142}$Eu

R.M. Lieder*,†, A.A. Pasternak**,†, E.O. Lieder*, W. Gast†, G. de Angelis‡ and D. Bazzacco§

*iThemba Laboratory for Accelerator Based Sciences, Somerset West 7129, South Africa
†Institut für Kernphysik, Forschungszentrum Jülich, D-52425 Jülich, Germany
**A.F. Ioffe Physical Technical Institute RAS, RU-194021 St. Petersburg, Russia
‡Istituto Nazionale di Fisica Nucleare, Laboratori Nazionali di Legnaro, I-35020 Legnaro, Italy
§Dipartimento di Fisica dell'Università and INFN Sezione di Padova, I-35131 Padova, Italy

**Abstract.** An investigation of $\gamma$-ray fold distributions has been carried out for the $^{97}$Mo($^{51}$V,$z\alpha\, yp\, xn$) reaction. For the $\alpha$2n channel leading to $^{142}$Eu a double-humped fold distribution was found. An explanation in terms of an incomplete-fusion reaction mechanism is proposed.

**Keywords:** $\gamma$-ray fold distributions, complete and incomplete fusion reactions
**PACS:** 29.30.Kv, 25.70.-z, 25.70.Gh

## INTRODUCTION

In heavy-ion induced reactions high excitation energies and large angular momenta are transfered to the produced nuclei. The highly-excited nuclei deexcite by long cascades of $\gamma$-ray transitions to the ground state. The number of $\gamma$-rays in each cascade is refered to as $\gamma$-ray multiplicity $M_\gamma$. The multiplicity distribution can be measured with $\gamma$-detector arrays equipped with an inner scintillator ball. The number of responding detectors to a $\gamma$-ray cascade of multiplicity $M_\gamma$ is called fold $K$. A study of fold distributions allows to obtain information on the reaction mechanism.

## EXPERIMENTAL RESULTS

Fold distributions for the population of specific final nuclei around $^{142}$Eu have been deduced from high-spin $\gamma$-spectroscopy data obtained with the $\gamma$-detector array GASP of the Laboratori Nazionali di Legnaro, Italy. In this investigation a $^{97}$Mo target has been bombarded with 238 MeV $^{51}$V projectiles. The target consisted of two $^{97}$Mo foils (enrichment 94.2 %) with a total thickness of $\approx$ 1.0 mg/cm$^2$. The GASP array, consisting of 40 bismuth-germanate (BGO) suppressed Ge detectors and a 80 element BGO ball, was equipped with the charged-particle detector array ISIS [1], consisting of 40 Si particle telescopes with thicknesses of $\approx$140 $\mu$m and $\approx$1 mm for the $\Delta$E and E detectors, respectively. Coincidence events were recorded when $\geq$ 3 escape-suppressed Ge detectors and $\geq$ 3 BGO scintillators detected $\gamma$-rays in coincidence. In total $2 \cdot 10^9$ coincidence events have been collected.

CP1012, *Frontiers in Nuclear Structure, Astrophysics, and Reactions: FINUSTAR 2*
edited by P. Demetriou, R. Julin, and S. V. Harissopulos
© 2008 American Institute of Physics 0-7354-0532-5/08/$23.00

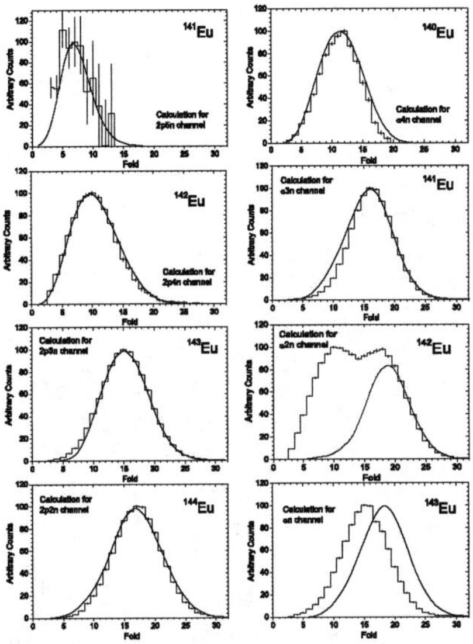


**FIGURE 1.** Experimental fold distributions for $^{140-144}$Eu obtained for the $\alpha$xn and 2pxn channels in the $^{97}$Mo + $^{51}$V reaction at 238 MeV are shown as histograms. Monte-Carlo simulations of the fold distributions calculated assuming a complete fusion reaction meachanism are shown as lines.

Fold distributions, measured in the $^{97}$Mo($^{51}$V,$z\alpha\, yp\, xn$) reaction with the inner BGO ball of GASP, have been obtained for the final nuclei $^{142-144}$Gd, $^{140-144}$Eu and $^{140,141}$Sm by gating on discrete single $\gamma$-lines in fold – $E_\gamma$ matrices. In the analysis p- and $\alpha$-gated matrices have been used, utilizing the charged particle information obtained with the ISIS array. Here, the results for Eu nuclei will be reported as populated in $\alpha$xn and 2pxn reaction channels.

Fold distributions for the $^{97}$Mo($^{51}$V,2pxn) and $\alpha$xn reaction channels leading to $^{141-144}$Eu are shown in fig. 1. The centroids of the 2pxn fold distributions are gradually shifting to lower folds with an increasing number of emitted neutrons as expected for complete-fusion reactions. However, a completely different picture is found for the $\alpha$xn channels leading to the nuclei $^{140-143}$Eu, as can be seen in fig. 1. The centroids of the fold distributions change irregularly with the number of emitted neutrons ($\alpha$4n: $< F >= 11$; $\alpha$3n: $< F >= 16$; $\alpha$n: $< F >= 15$) and the fold distribution of the $\alpha$2n channel leading to $^{142}$Eu shows a double-humped structure with maxima at $< F >= 11$ and 17. Such a fold distribution has to our knowledge not been reported so far. The fold distributions of $^{142}$Eu has been obtained by gating on four transitions of 283, 303, 420 and 1032 keV. They all show the same double-humped fold distribution within statistical uncertainties. Several checks for contaminations and possible leakage between p-gated and $\alpha$-gated spectra have been made for $^{142}$Eu. A contribution of $\approx 9\%$ may be expected from contaminations in the high-fold peak and $\leq 2\%$ due to leakage in the low-fold peak of the $^{142}$Eu fold distribution, which does not provide any explanation of the double-humped fold distribution as due to admixtures.

## DISCUSSION

For the interpretation of the observed fold distributions Monte-Carlo simulation calculations have been carried out.

### Complete fusion

To calculate fold distributions a new approach has been developed, based on Monte-Carlo simulations of entry-state population distributions as well as of the depopulation to known levels. Details of the formalism can be found in ref. [2]. In the first step the input angular momentum is calculated taking into account the dependence of the fission barrier on the angular momentum $L$. In the second step, for complete fusion reactions, the light-particle evaporation is simulated with Monte-Carlo methods based on a statistical theory of nuclear reactions [3]. The code COMPA has been used [4] for these calculations. The competition between particle evaporation and fission mainly defines the locations of the entry-state population distributions of the various final nuclei in a plot of the excitation energy $E^*$ vs. angular momentum $I$. In the third step the deexcitation of the entry states is simulated with the code GAMMA [4, 5] taking into account statistical $E1$, $M1$ and $E2$ transitions, stretched $E2$ bands (including the rotational damping effect with a width of $\Gamma = 0.3$ MeV [6]), superdeformed (SD) bands in the continuum and the existence of a large amount of particle-hole excitations which generate cascades of stretched magnetic dipole transitions. The latter two components are particular important in the light Gd nuclei, since SD and magnetic rotational bands are known to exist in these nuclei. The contributions of the $M1$ cascades to the $\gamma$-ray multiplicities depend on the relative probability $S_{MR}$ that the depopulation proceeds through them. This parameter is determined by a global fit of the experimental fold distributions for Gd – Eu – Sm nuclei obtained in three different reactions [2]. Parameters for the calculation of statistical transitions have been chosen in accordance with systematics of $\gamma$-ray strength functions for $E1$, $M1$ and $E2$ multipolarities, respectively [7]. In this way for each cascade the number of emitted $\gamma$-rays $M_\gamma$ is determined giving the multiplicity distribution considering a large number of cascades. A comparison of the experimental fold distributions with the simulated multiplicity distributions is not carried out by an unfolding procedure applied to the fold distributions. Instead, the simulated $\gamma$-ray multiplicity distributions are transformed into fold distributions by Monte-Carlo methods. The response of the BGO ball is simulated for each $\gamma$-ray transition separately, using detection probabilities for $\gamma$-quanta of given energy, which have been extracted from fold-multiplicity calibrations of the BGO ball.

In fig. 1 the experimental fold distributions are compared with simulation calculations assuming a complete fusion reaction mechanism. It can be seen that for the 2pxn, $\alpha$3n and $\alpha$4n reaction channels the experimental fold distributions are described rather well, indicating that these reactions proceed through a complete fusion reaction mechanism. However, the fold distributions for $^{142,143}$Eu nuclei populated in the $\alpha$2n and $\alpha$n reaction channels, respectively, cannot be described. Furthermore, the complete-fusion calculations predict very small cross sections for these channels contradictory to the experimental observation. Both facts indicate that an incomplete-fusion reaction mechanisms may play a role.

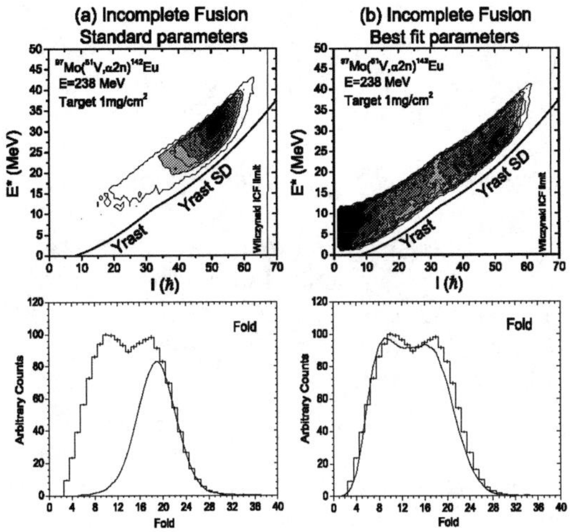


**FIGURE 2.** Calculation of fold distributions for $^{142}$Eu, as populated through the $\alpha$2n channel, assuming an incomplete-fusion reaction mechanism.

## Incomplete fusion

Incomplete fusion reactions can be described as a heavy cluster stripping from the incident ion in the reaction. They are also called "massive transfer reactions" since the remaining part of the projectile is transfered to the target nucleus and fuses with it. The direct transfer of a heavy cluster is accompanied with a considerable kinetic energy dissipation in the entrance and exit channels, defining the angular and energy distributions of the emitted particles. Semi-classical methods can be applied for the calculations of the relative motion of the particles because of the small wavelength and the high level density of excited states leading to an exhibition of the classical properties of relative motion. The transfer process, however, is not classical. Such a multi-step process provides the dominant contribution to the cross section of $\alpha$-particle formation almost over the whole ranges of their energies and angles.

The mechanism of the formation of light particles in heavy-ion collision is not yet fully understood. There are several mechanisms leading to the light particle emission, the contributions of which depend on the masses and energies of the colliding nuclei and also on the energies and emission angles of the light particles. The multi-step massive transfer process seems to contribute to the formation of light particles over the whole range of their energies. The quasidirect massive transfer process makes the dominant contribution to light-particle formation at the high-energy tail of their spectrum especially at large angles. At small angles the break-up process seems to contribute significantly. The particles are formed in the initial stage of the reaction before thermodynamical equilibrium of the system is reached.

In order to carry out Monte-Carlo simulation calculations of fold distributions for the $\alpha$2n channel assuming an incomplete fusion reaction mechanism it is considered that:

• Incomplete fusion occurs above a transferred angular momentum $L_{crit}$ before a compound nucleus is formed as confirmed by experimental data of Siwek-Wilczynska et al. [8]. In the present case $L_{crit} = 68$. The complete fusion reaction mechanism is assumed to be negligible in this angular momentum region, because for complete fusion the fission barrier is lower than $L_{crit}$.
• The $\alpha$-particles escaping from the projectiles experience an effective barrier $V_{eff}$ which can differ from the Coulomb barrier by a factor $f_b = V_{eff}/V_C$.
• The effective barrier $V_{eff}$ is assumed to have a Gaussian distribution of width $\sigma_V$.
• The initial energies of the particles emitted in the incomplete fusion process are defined at the moment of escape from the projectile. The initial energies of the $\alpha$-particles $E_{in}$ can have a maximum value $E_{max}$ corresponding to the projectile velocity if the breakup occurs far from the target nucleus. A mimimum value $E_{min}$ is obtained for a projectile breakup near the target nucleus, when the projectile is slowed down in the Coulomb field of the target nucleus. The distribution of energies $E_{in}$ is calculated as $E_{in} = (E_{max} - E_{min})N_r^{\delta} + E_{min}$, $N_r$ being a random number between 0 and 1.

The angular, energy and angular momentum distributions of the $\alpha$-particles after the interaction with the fields of both the projectile and target nuclei are calculated using the four parameters $L_{crit}$, $f_b$, $\sigma_v$ and $\delta$.

With parameters used to reproduce calculations of Zagrebaev [9] for a $^{181}$Ta($^{22}$Ne,$\alpha$) reaction at 178 MeV, shown in figs. 19 and 20 of his work, only the high-fold part of the experimental fold distribution for $^{142}$Eu can be explained as can be seen in fig. 2a. However, if the parameters are chosen such that the $\alpha$-particles carry away large angular momenta then a second minimum occurs in the entry-state population distribution at low spins and the double-humped fold distribution of $^{142}$Eu can be described as shown in fig. 2b.

To prove this assumption an experiment to measure energy spectra and angular distributions of $\alpha$-particles for the $^{97}$Mo + $^{51}$V reaction at 238 MeV with the $\gamma$-detector array AFRODITE and the charged-particle ball DIAMANT is planned.

## ACKNOWLEDGMENTS

The work was partly supported by IB of BMBF at DLR, Germany under the WTZ contract RUS 99/191 and by the Heinrich Hertz Foundation, Germany under the contract Nr. 16/03.

## REFERENCES

1. E. Farnea et al., Nucl. Instrum. Methods **A400**, 87 (1997).
2. R.M. Lieder et al., to be published in Eur. Phys. J. A (2007).
3. J. Domscheit et al., Nucl. Phys. **A689**, 655 (2001).
4. R.M. Lieder et al., Eur. Phys. J. **A21**, 37 (2004).
5. E. Grodner et al., Eur. Phys. J. **A**, (2006).
6. F.S. Stephens et al., Nucl. Phys. A **587**, 513 (1995).
7. Handbook for Calc. of Nuclear Reaction Data, RIPL-2, IAEA-TECDOC-1506 IAEA, Vienna (2006).
8. K. Siwek-Wilczynska et al., Nucl. Phys. A **330**, 150 (1979).
9. V.I. Zagrebaev, Ann. Phys. **197**, 33 (1990).

# Neutrinos and Supernovae

Bradley S. Meyer

*Department of Physics and Astronomy, Clemson University, Clemson, SC 29634-0978*

**Abstract.** Core-collapse supernovae are one of the few astrophysical environments in which neutrinos play a dominant role. Neutrinos emission is the means by which a newly-born neutron star formed in a core-collapse event cools. Neutrinos may play a significant role in causing the supernova explosion. Finally neutrinos may significantly affect the nucleosynthesis occurring in the layers of the exploding star that are eventually ejected into interstellar space. This paper reviews some interesting neutrino-nucleus processes that may occur in the cores of exploding massive stars and then discusses some effects neutrinos may have on explosive nucleosynthesis in supernovae.

**Keywords:** supernovae,nucleosynthesis,neutrinos,r-process
**PACS:** 23.40.Bw,26.30.-k,26.30.Hj,26.50.+x>

## INTRODUCTION

Supernovae are one of the few events in the Universe in which neutrinos play a dominant, if not the dominant, role. Neutrino emission is the means by which nascent neutron stars cool, and neutrino interactions may be the principal mechanism for reviving the stalled supernova shock and causing the supernova explosion. Neutrinos also interact with material undergoing nucleosynthesis during the supernova event and modify the resulting yields compared to nucleosynthesis in the absence of the neutrinos. This paper presents a brief review of supernovae and neutrinos, a brief discussion of neutrino-nucleus processes in supernova cores, some considerations of the effects of neutrinos on explosive nucleosynthesis, and finally some comments on how to include neutrinos in nucleosynthesis calculations using the libnucnet library.

## A BRIEF REVIEW OF STELLAR EVOLUTION AND CORE-COLLAPSE SUPERNOVAE

Throughout most of their lives, stars are stable configurations in which the force that tends to cause contraction, namely, gravity, is balanced by the force that tends to push the layers of the star out, namely, pressure. This balance is known as hydrostatic equilibrium. As the star shines, however, it loses energy and, consequently, pressure. To counter this loss, nuclear reactions occur in the stellar interior and release energy locally. This replenishes energy lost to radiation and maintains the star's pressure.

The first reactions to occur are those of hydrogen burning, that is, those reactions that convert the abundant hydrogen in the stellar core into helium. Once the hydrogen is exhausted, the stellar core contracts and heats until it begins helium burning in which it converts helium into carbon and oxygen. A shell of hydrogen burning may ignite above the hydrogen exhausted core and burn contemporaneously with the core helium burning.

CP1012, *Frontiers in Nuclear Structure, Astrophysics, and Reactions: FINUSTAR 2*
edited by P. Demetriou, R. Julin, and S. V. Harissopulos
© 2008 American Institute of Physics 0-7354-0532-5/08/$23.00

Stars with mass less than about eight times the mass of our Sun develop electron-degenerate carbon/oxygen cores. The electron degeneracy is largely independent of temperature, which thereby halts contraction. Such stars have successive stages of hydrogen and helium shell burning. They eventually drive off their hydrogen-rich envelopes and leave behind C/O white dwarf stars.

In contrast to these stars, stars with mass larger than about eight times the mass of our Sun develop non-degenerate C/O cores. Such cores are able to contract after helium exhaustion and begin carbon burning in which carbon nuclei from helium burning fuse to form neon and magnesium. Stars with mass less than about 11 solar masses form degenerate O/Ne/Mg cores (where the oxygen is left over from the helium burning) and, hence, leave O/Ne/Mg white dwarf stars once the envelope is ejected.

For stars with mass greater than about 11 solar masses, subsequent burning phases can occur. In particular, neon burning produces oxygen and magnesium from neon. Oxygen burning produces silicon from oxygen and magnesium. And silicon burning predominantly produces iron-group isotopes, predominantly $^{56}$Fe, from $^{28}$Si. Iron-group isotopes are at the top of the binding energy per nucleon curve; hence, any fusion of these nuclei absorbs energy rather than releasing it. Silicon burning is thus the endpoint in mainline stellar evolution.

Once a massive star develops an iron core, electron captures, disintegrations, and neutrino losses lead to a dramatic drop in the pressure. The stellar core collapses on a timescale of $\sim$1 second. Collapse proceeds until the material reaches supranuclear densities. At this point, the pressure rises dramatically as nucleons now push against each other. This halts the collapse. Nevertheless, material outside the inner part of the iron core continues to rain in supersonically. As this material crashes onto the inner core, a shock wave develops that then works its way out through the outer stellar layers and drives them out into the interstellar medium. This event is the death of the star and is known as a core-collapse supernova. In fact the supernova would dim fairly rapidly were it not for the radioactive decay of $^{56,57}$Co, produced in the explosion as $^{56,57}$Ni. The energy resulting from these decays is deposited in the stellar layers and keeps the supernova shining for months after the original core collapse.

The bulk energetics of the supernova are easy to work out. In effect, the original Fe core, an object with the mass of about 1.4 times the mass of our Sun and the size of our Earth, collapses to a neutron star, which has the same mass but a radius of about 10 km. If we simply imagine the stellar core as a homogeneous sphere, this is a change of the gravitational binding of the stellar core

$$\Delta E = -\frac{3}{5}\frac{GM^2}{R_f} - \left(-\frac{3}{5}\frac{GM^2}{R_i}\right) \tag{1}$$

where $M$ is the 1.5 times the mass of our Sun (i.e., $1.4 \times 2 \times 10^{33}$ g), $R_i$ is the radius of our Earth ($\sim$ 6000 km), and $R_f$ is 10 km, the final radius of the neutron star. From this equation, we find $\Delta E = -3 \times 10^{53}$ ergs $= -2 \times 10^{59}$ MeV. The negative sign indicates that this gravitational binding energy must be radiated away, which occurs dominantly by neutrino emission.

Since the gravitational energy initially goes into internal energy in the core, we may estimate the initial temperature. We imagine the instant after the initial collapse. The radius of the hot neutron star is $\sim 50$ km, which is a volume of about $5 \times 10^{59}$ fm$^3$. The energy density in the stellar core is $U \approx 0.4$ MeV/fm$^3$. If we imagine that all this energy goes into a blackbody spectrum of neutrinos, we find

$$U = aT^4 = \frac{\pi}{30} g_\nu \frac{(kT)^4}{(\hbar c)^3} \tag{2}$$

where $g_\nu = \frac{7}{8} \times 6$ for the six neutrino flavors and the fact that the neutrinos are leptons. The result is that $kT \approx 40$ MeV! This simple estimate is in reasonable agreement with detailed computations.

Because of the high density in the stellar core, the neutrinos must diffuse out. Detailed models show that this occurs over a period of several seconds over which time the central temperature drops from 10's of MeVs to less than an MeV. This neutrino emission represents the Kelvin-Helmholtz cooling of the neutron star, which, in turn, allows the star to contract to its final size of $\sim$ 10km in radius. For more details, the interested reader may wish to consult ref. [1].

## HIGH-TEMPERATURE NEUTRINO-NUCLEUS INTERACTIONS

Because neutrinos are so abundant in the central regions of a core-collapse supernova, a number of processes that might seem exotic occur there. In the central regions, where nuclei no longer exist, neutrinos can be produced by exotic processes such as nucleon-pair bremsstrahlung. In regions where nuclei are present, other interesting processes can occur. Inelastic scattering of neutrinos by nuclei can increase or decrease the scattering neutrino's energy. Neutrino-pair annihilation can occur in the presence of a nucleus (to conserve momentum). Another interesting process is de-excitation of a nucleus into a neutrino-antineutrino pair. This process is likely one of the dominant producers of low energy neutrinos in the supernova core (e.g., [2]).

A good general rule of thumb for neutrino processes in supernovae is Feynman's dictum that "Anything that is not forbidden (by a conservation or selection rule) must occur." The question for supernova dynamics is how dominantly a given process is amongst all the other processes. For example, while one may not habitually think of neutrino/anti-neutrino pair annihilation, it occurs with frequency in supernovae due to the almost unimaginably high abundance of neutrinos.

Another important rule is that it is necessary to include all processes. For example, neutrino spectra in dense media can develop low-energy windows. This is because electron degeneracy prevents low-energy neutrinos from scattering off electrons due to the fact that the final electron state is filled. The mean-free path for the low-energy neutrinos becomes large and they can escape from the core more easily than higher-energy neutrinos. If a low-energy window develops, electron capture can occur more easily. This could dramatically lower $Y_e$, the electron-to-baryon ratio, in the core. The low-energy window, however, is rapidly filled in part by neutrino-pair de-excitation. This hinders electron capture and limits the decrease in $Y_e$ [3]. The lesson, then, is that many neutrino processes are occuring during a supernova and they are often counterbalancing each other. If one process is left out of a calculations, the balance may incorrectly shift.

# NEUTRINOS AND EXPLOSIVE NUCLEOSYNTHESIS IN SUPERNOVAE

Neutrinos can set the initial electron-to-baryon ratio $Y_e$ in matter undergoing explosive nucleosynthesis in supernovae. Because the processes $\bar{\nu}_e + p \rightarrow n + e^+$ and $\nu_e + n \rightarrow p + e^-$ are keys to $\bar{\nu}_e$ and $\nu_e$ opacities, respectively, the increasing neutron richness of the neutron star causes the $\nu_e$'s to have a higher opacity than the $\bar{\nu}_e$'s and the former decouple at larger stellar radius than the latter. Since the temperature decreases with increasing radius in the star, this means that the $\bar{\nu}_e$'s emerging from the neutron star will typically be hotter than the $\nu_e$'s. These higher energy $\bar{\nu}_e$'s will then interact more readily with material outside the neutron star and tend to drive it neutron rich (that is, drive $Y_e < 0.5$).

Once neutrinos have set the initial $Y_e$, however, we generally neglect further neutrino interactions with the ongoing nucleosynthesis. Because of the slowness of the interactions, this is generally a good approximation. Nevertheless, depending on the flux, these neutrino interactions can have interesting consequences.

## The r-Process of Nucleosynthesis

One area where the neutrinos might play a role is in the high-entropy r-process, which might occur in the high-entropy region just outside a newly-born neutron star [4, 5, 6]. While these models showed early promise, it soon became apparent that the entropies or neutron richness required for successful production of the heaviest r-process elements were not attained in realistic supernova models (e.g., [7]).

A possible aid to the models was the neutrinos streaming from the neutron star. The idea was that neutrinos could spall neutrons from the abundant nuclei which would then enable more r-processing. In fact the reverse turned out to be true. When neutrino spallation reactions were included in the r-process calculations, neutrons were indeed spalled from the alpha particles. These, however, quickly captured back onto the alpha particles with no consequence for the nucleosynthesis. The problem was that neutrinos also spalled protons. Because of the neutron richness of the environment, the resulting $^3$H from a spallation reaction captured one of the abundant $^4$He nuclei to become $^7$Li. Subsequent $^4$He captures carried the $^3$H up to higher mass. Since the spalled proton simply captured two neutrons to become another $^3$H, the result is that one proton spallation event converted a $^4$He into two heavy nuclei. This reduced the number of neutrons per nucleus and hindered the r-process [8].

An even more detrimental effect is the so-called "alpha effect". This occurs even earlier than the process mentioned above in the expansion and cooling of a fluid element while neutrons and protons are assembling into $^4$He. As discussed above, the higher temperature of the electron-type anti-neutrinos tends to drive protons into neutrons. As the free nucleons assemble into $^4$He, however, the abundances are dominated by neutrons and $^4$He. The $^4$He are essentially inert to neutrino capture. Neutrino capture still occurs on the neutrons, however. This produces protons, which capture neutrons to become $^4$He. As long as the neutrino flux continues, this tends to drive all nucleons into

$^4He$, eliminates any neutron richness, and prevents an r-process [9, 10].

It seems likely, then, that neutrino interactions in general tend to hinder the r-process rather than help it. A recently proposed site for the r-process, shocked surface layers of an O/Ne/Mg white dwarf star, has material expanding so rapidly that neutrino interact very little with the matter undergoing the nucleosynthesis [11]. While this model may or may not be correct, the feature that once the nucleosynthesis starts, matter gets driven away rapidly from the damaging neutrino flux is likely to be preserved in any successful r-process model in the future.

## Production of Light p-Process Nuclei

While astrophysical models do a good job of explaining the production of most p-process nuclei, the light p-isotopes, and particularly $^{92,94}Mo$ and $^{96,98}Ru$ are generally underproduced [12]. This has suggested to many that there is another site for the production of these isotopes. One such site might be the high-entropy alpha-rich freezeouts just outside nascent neutron stars in core-collapse supernovae [13, 14]. While standard alpha-rich freezeouts could not lead to cases in which the light p-process isotopes were the most overproduced (a requirement that the site be a dominant producer of those species), we found that intense neutrino fluxes could lead to neutrino capture on stable species [15, 16] which greatly enhanced the production of 92, 94Mo and $^{96,98}Ru$. For example, neutrino capture on a species like $^{90}Zr$, normally stable, could produce $^{90}Nb$. This would likely be followed by a neutron capture. Another neutrino capture plus neutron capture would produce $^{92}Mo$.

The principal difficulty with the above alpha-rich freezeout models was that it required a high flux of neutrinos during the nucleosynthesis and considerably higher electron-type anti-neutrino temperature than electron-type neutrino temperature. Supernova models could not deliver the required conditions. In the meantime, it has become increasingly clear that matter just outside the newly-born neutron star is likely to be only slightly neutron rich or perhaps even slightly proton rich (e.g., [17]) early in the explosion. In this case, the nucleosynthesis differs from that studied in [14]. In particular, in the late stages of the expansion, the high abundance of protons leads to a sequence of proton captures. Once a nucleus becomes proton rich enough, it beta decays and continues to proton capture. This is an example of the rp-process of nucleosynthesis (e.g., [18]).

Such nucleosynthesis can lead to production of light p-nuclei, but not in the requisite amounts. The reason is that the proton-capture flow reaches nuclei whose beta decay times are too long before producing those isotopes. If neutrinos impinge on the matter during nucleosynthesis, however, the reaction $p(\bar{\nu}_e, e^+)n$ can occur on the abundant protons. The neutrons produced in this way can drive $(n,p)$ reactions on the waiting point nuclei, and the reaction flow can proceed to higher masses in what has been termed the "nu rp-process" [19]. Models of this process show good production of the light p-process isotopes [17], although the models produce somewhat less $^{92}Mo$ than is needed to explain its Solar System's abundance. Errors in the input nuclear physics may be to blame here.

Despite a potential difficulty with $^{92}Mo$, the nu rp process shows great promise for explaining the Solar System's abundance of light p-process isotopes. It is an instance in which neutrinos may play a key role in the synthesis of particular nuclear species.

## LIBNUCNET AND NEUTRINO INTERACTIONS

Inclusion of neutrino interactions in nucleosynthesis calculations presents an interesting computational twist to nuclear network calculations. The neutrinos are, in a sense, external to the parcel of matter whose evolution one is following with the network code. For normal nuclear reactions, one needs the temperature and density of the parcel of matter to compute the reaction rates. For the neutrino interactions, the rates are determined by the neutrino interaction cross section and the neutrino flux. The former are properties of the local nuclei (including, for example, the distribution of the nuclei among their excited states according to the local temperature). The latter, the neutrino flux, however, requires knowledge of the spectrum of the neutrinos, which is set by the details of the neutrino transport in the supernova, and the location of the parcel of matter. For simplicity, one often uses a blackbody spectrum of neutrinos, which is generally not a bad approximation. Also, one can parameterize the trajectory of the parcel of matter to know how far it is from the neutrino source. More detailed calculations in the context of a realistic supernova model are, of course, also possible (e.g., [17]).

We have developed a library of C codes, called libnucnet, for managing and evolving nuclear reactions networks [20]. The codes are freely available at:

`http://www.webnucleo.org/home/modules/libnucnet`

and are released under the GNU General Public License. Tutorials at the above site demonstrate how to download, compile, and run codes based on libnucnet. A variety of example codes demonstrate the API (Applications Programming Interface). The API is straightforward and should allow users to incorporate libnucnet into their own codes. Among the example codes is one that demonstrates how to use the libnucnet API to include neutrino interactions into a nuclear network calculation. The interested reader is referred to the above web site for more information.

## CONCLUSION

Two themes emerge from the above discussion. First, because of the high densities and temperatures in supernovae, nearly every conceivable neutrino process occurs at some level. The question for a modeler is, how much does that process contribute and are the counterbalancing processes accounted for? A second theme emerges especially from the study of explosive nucleosynthesis. In order to understand the effect of neutrinos on explosive nucleosynthesis, it is necessary to follow the full "ecology". While the particular neutrino reaction under consideration may be straightforward, for example, spallation of nucleons from alpha particles, its effect on the nucleosynthesis may be less clear and generally can only be determined by following detailed nuclear network calculations. This may be a frustrating finding for those calculating cross sections of neutrino processes in astrophysics because the consequences of their work for astrophysics may not be immediately clear. Nevertheless it is simply a reminder of the complexity of supernovae and the complicated role neutrinos play in them.

## ACKNOWLEDGMENTS

This work has been supported in part by DOE grant DE-FG02-87ER40328 and NASA Cosmochemistry Program grant NNX07AJ04G.

## REFERENCES

1. S. E. Woosley, A. Heger, and T. A. Weaver, *Rev. Mod. Phys.* **74**, 1015–1071 (2002).
2. E. Kolb, and T. Mazurek, *Astrophys. J.* **234**, 1085 (1979).
3. G. M. Fuller, and B. S. Meyer, *Astrophys. J.* **376**, 701–716 (1991).
4. B. S. Meyer, G. J. Mathews, W. M. Howard, S. E. Woosley, and R. D. Hoffman, *Astrophys. J.* **399**, 656–664 (1992).
5. K. Takahashi, J. Witti, and H. T. Janka, *Astron. Astrophys.* **286**, 857–869 (1994).
6. S. E. Woosley, J. R. Wilson, G. J. Mathews, R. D. Hoffman, and B. S. Meyer, *Astrophys. J.* **433**, 229–246 (1994).
7. Y.-Z. Qian, and S. E. Woosley, *Astrophys. J.* **471**, 331–351 (1996).
8. B. S. Meyer, *Astrophys. J. Lett.* **449**, L55–L58 (1995).
9. G. M. Fuller, and B. S. Meyer, *Astrophys. J.* **464**, 521 (1996).
10. B. S. Meyer, G. C. McLaughlin, and G. M. Fuller, *Phys. Rev. C* **58**, 3696–3710 (1998).
11. H. Ning, Y. . Qian, and B. S. Meyer, *ArXiv e-prints* **708** (2007), 0708.1748.
12. M. Arnould, and S. Goriely, *Phys. Rep.* **384**, 1–2 (2003).
13. S. E. Woosley, and R. D. Hoffman, *Astrophys. J.* **395**, 202–239 (1992).
14. R. D. Hoffman, S. E. Woosley, G. M. Fuller, and B. S. Meyer, *Astrophys. J.* **460**, 478–488 (1996).
15. G. M. Fuller, and B. S. Meyer, *Astrophys. J.* **453**, 792–809 (1995).
16. B. S. Meyer, *Nuclear Physics A* **719**, 13–20 (2003).
17. J. Pruet, R. D. Hoffman, S. E. Woosley, H.-T. Janka, and R. Buras, *Astrophys. J.* **644**, 1028–1039 (2006), arXiv:astro-ph/0511194.
18. L. van Wormer, J. Goerres, C. Iliadis, M. Wiescher, and F. . Thielemann, *Astrophys. J.* **432**, 326–350 (1994).
19. C. Fröhlich, G. Martínez-Pinedo, M. Liebendörfer, F.-K. Thielemann, E. Bravo, W. R. Hix, K. Langanke, and N. T. Zinner, *Physical Review Letters* **96**, 142502 (2006).
20. B. S. Meyer, and D. C. Adams, *70th Annual Meteoritical Society Meeting, held in August 13-17, 2007, Tucson, Arizona. Meteoritics and Planetary Science Supplement, Vol. 42, p.5215* **42**, 5215 (2007).

# Spallation nucleosynthesis by accelerated charged-particles

S. Goriely

*Institut d'Astronomie et d'Astrophysique, Université Libre de Bruxelles, Campus de la Plaine, CP 226, 1050 Brussels, Belgium*

**Abstract.** Recent observations have suggested the presence of radioactive elements, such as Pm and $84 \leq Z \leq 99$ elements) at the surface of the magnetic star HD101065, also known as Przybylski's star. This star is know to be a chemically peculiar star and its anomalous $38 < Z \leq 92$ abundance pattern has been so far explained by diffusion processes in the stellar envelope. However, such processes cannot be called for to explain the origin of short-lived radioelements. The large magnetic field observed in Ap stars could be at the origin of a significant acceleration of charged-particles, mainly protons and $\alpha$-particles, that in turn can by interaction with the stellar material modify the surface content.

The present contribution explores to what extent the spallation processes resulting from the interaction of the stellar material with stellar energetic particle can by themselves only explain the abundances determined by observation at the surface of HD101065. We show that specific parametric simulations can explain many different observational aspects, and in particular that a significant production of $Z > 30$ heavy elements can be achieved. In this nucleosynthesis process, the secondary-neutron captures play a crucial role. The most attractive feature of the spallation process is the systematic production of Pm and Tc and the possible synthesis of actinides and sub-actinides.

Based on such a parametric model, it is also shown that intense fluences of accelerated charged-particles interacting with surrounding material can efficiently produce elements heavier than iron. Different regimes are investigated and shown to be at the origin of p- and s-nuclei in the case of high-fluence low-flux events and r-nuclei for high-fluence high-flux irradiations. The possible existence of such irradiation events need to be confirmed by hydrodynamics simulations, but most of all by spectroscopic observations through the detection of short-lived radio-elements.

**Keywords:** stellar nucleosynthesis – nuclear reactions – CP stars
**PACS:** 26.30.+k,24.60.Dr,21.10.Dr, 21.10.Ma

## INTRODUCTION

Chemically peculiar (CP) stars are spectral type A stars that have chemical abundances differing significantly from that of other stars of similar classification. They exhibit a remarkable variety of elemental enhancements and depletions in their photospheres, sometime 5 to 6 orders of magnitudes different than found in the sun. The two major types of CP stars are Am stars that have no detectable magnetic field and Ap stars that have magnetic fields with large-scale structure.

Rapidly oscillating Ap (roAp) stars are a subgroup of magnetic Ap stars which oscillate with non-radial, low order acoustic $p$-modes with the axis of oscillation aligned with the axis of the magnetic field [1]. The mechanisms responsible for the excitation of roAp stars and other physical parameters that distinguish them from non-pulsating CP stars remains an open question. The comparison between roAp and non-roAp star

CP1012, *Frontiers in Nuclear Structure, Astrophysics, and Reactions: FINUSTAR 2*
edited by P. Demetriou, R. Julin, and S. V. Harissopulos
© 2008 American Institute of Physics 0-7354-0532-5/08/$23.00

surface composition do not reveal large abundance differences. So far, the CP star abundances have been explained almost uniquely on the basis of diffusion processes, i.e diffusive segregation of ionic and elemental species resulting from a balance between radiative and gravitational forces within the atmosphere and sub-atmospheric levels of warm stars.

Recent observations have suggested the presence of short-lived radioactive elements, such as Tc, Pm and $84 \leq Z \leq 99$ elements at the surface of the CP roAp star HD 101065, also known as Przybylski's star [2, 3]. Such an observation can in no way be explained by diffusion processes. Observations of CP-stars in the X-ray regime show evidence that high energy phenomena are associated with, if not produced by, some CP stars, so that high-energy events can be expected in the vicinity of (if not on) many Ap stars. The large magnetic field observed in Ap stars can be at the origin of a significant acceleration of charged-particles, mainly protons and $\alpha$-particles, that in turn can by interaction with the stellar material modify the surface content. Due to the success of diffusion, only a few studies back in the 60s have been devoted to explaining the CP stars' abundance anomalies by nuclear reaction in the stellar surface layers (see in particular, [4]).

The purpose of the present work is to show that the irradiation process resulting from the interaction of the stellar material with energetic particles can by itself only explain the abundances determined by observation at the surface of HD 101065. This includes the particularly large abundances of the heavy elements with $38 \leq Z \leq 92$ by 3-4 dex [5], so far explained by diffusion processes [6], but also the possible presence of radioactive elements, such as Tc, Pm and $84 \leq Z \leq 99$ elements that have been identified by [2, 3]. No effort is done at the present stage to explain the possible mechanisms responsible for the particle acceleration. As already discussed, such particles could be locally accelerated, but they could also come from an external source.

## MODELLING

To describe the changes in abundance of the nuclei as a result of the irradiation process, a nuclear reaction network including all reactions of relevance is used. All nuclei with $0 \leq Z \leq 102$ and located between the proton and the neutron drip lines are included in the network. The chosen set of nuclear species are then coupled by a system of differential equations corresponding to all the reactions affecting each nucleus, i.e. mainly proton, $\alpha$, $^3$He and neutron captures, $\beta$- and $\alpha$-decays, as well as spontaneous and induced fissions. The rate of change of the molar fraction $Y_{(Z,A)}$ of a nucleus $(Z,A)$ with charge number Z and mass number $A$ can be written as

$$\begin{aligned}
\frac{dY_{(Z,A)}}{dt} = & - Y_{(Z,A)}\Phi_p\langle\sigma\rangle_p + \sum_k Y_k\Phi_p\langle\sigma\rangle_{p,k}^{(Z,A)} \\
& - Y_{(Z,A)}\Phi_\alpha\langle\sigma\rangle_\alpha + \sum_k Y_k\Phi_\alpha\langle\sigma\rangle_{\alpha,k}^{(Z,A)} \\
& - Y_{(Z,A)}\Phi_{^3\mathrm{He}}\langle\sigma\rangle_{^3\mathrm{He}} + \sum_k Y_k\Phi_{^3\mathrm{He}}\langle\sigma\rangle_{^3\mathrm{He},k}^{(Z,A)} \\
& - Y_n\rho N_a\langle\sigma v\rangle_n^{(Z,A)} Y_{(Z,A)} + \sum_k Y_n\rho N_a\langle\sigma v\rangle_{n,k}^{(Z,A)} Y_k
\end{aligned}$$

$$
\begin{aligned}
&- \quad Y_p \rho N_a \langle\sigma v\rangle_p^{(Z,A)} Y_{(Z,A)} + \sum_k Y_p \rho N_a \langle\sigma v\rangle_{p,k}^{(Z,A)} Y_k \\
&- \quad Y_\alpha \rho N_a \langle\sigma v\rangle_\alpha^{(Z,A)} Y_{(Z,A)} + \sum_k Y_\alpha \rho N_a \langle\sigma v\rangle_{\alpha,k}^{(Z,A)} Y_k \\
&- \quad \lambda_\gamma^{(Z,A)} Y_{(Z,A)} + \sum_k \lambda_{\gamma,k}^{(Z,A)} Y_k \\
&- \quad \lambda_{\beta-}^{(Z,A)} Y_{(Z,A)} - \lambda_{\beta+}^{(Z,A)} Y_{(Z,A)} + \lambda_{\beta-}^{(Z-1,A)} Y_{(Z-1,A)} + \lambda_{\beta+}^{(Z+1,A)} Y_{(Z+1,A)} \\
&- \quad \lambda_\alpha^{(Z,A)} Y_{(Z,A)} + \lambda_\alpha^{(Z+2,A+4)} Y_{(Z+2,A+4)} - \lambda_{sf}^{(Z,A)} Y_{(Z,A)} + \sum_k \lambda_{sf,k}^{(Z,A)} Y_k \quad (1)
\end{aligned}
$$

where $\lambda_{\beta^\pm}$, $\lambda_\alpha$ and $\lambda_{sf}$ are the $\beta^\pm$, $\alpha$ and spontaneous fission decay rate, respectively; $N_a$ is the Avogadro number and $\rho$ the local density. Note that in Eq. 1, the recycling by n-, p- or $\alpha$-induced fission is also taken into account with a full fission fragment distribution. $\langle\sigma\rangle_p$, $\langle\sigma\rangle_\alpha$ and $\langle\sigma\rangle_{^3\mathrm{He}}$ are the total effective cross sections for proton, $\alpha$ and $^3$He-captures averaged over the energy distribution $\Phi(E)$, i.e for the proton case

$$
\langle\sigma\rangle_p = \frac{1}{\Phi_p}\int_0^\infty \Phi(E)\,\sigma_p(E)dE \quad (2)
$$

where $\sigma_p(E)$ is the $E$-dependent total proton capture cross section and $\Phi_p = \int_0^\infty \Phi(E)dE$ the proton flux amplitude (see below). Similarly, $\langle\sigma\rangle_{p,k}^{(Z,A)}$ is the partial effective cross sections for proton capture on a nucleus $k$ leading to the residual $(Z,A)$ nucleus (including fission modes).

Thermonuclear charged-particles capture reactions as well as photoreactions are also included in Eq. 1, since in hot environments they may come into play (see some of the simulations below). A special attention is paid to the abundance of the neutrons emitted by the spallation process. We assume that the so-produced neutrons are thermalized on timescales much shorter than the typical capture timescales, as discussed in [8]. In Eq. 1, $\langle\sigma v\rangle_{x,k}^{(Z,A)}$ (where $x$ can be neutrons, protons or $\alpha$-particles) corresponds to either the $(x,\gamma)$, $(x,n)$, $(x,p)$, $(x,\alpha)$ or $(x,f)$ reaction rate leading to the formation of the $(Z,A)$ nucleus. Similarly $\lambda_\gamma$ corresponds to the photoemission rate. The rates $\langle\sigma v\rangle$ and $\lambda_\gamma$ are estimated by the reaction code Talys [10, 11, 12].

The proton,$\alpha$ and $^3$He fluxes are assumed to remain constant in time during the irradiation period $\tau_{irr}$. Many different energy distributions could be studied. We will restrict ourselves to two types of energy spectra, namely a constant distribution in a restricted $E_{min} \le E \le E_{max}$ energy range and the widely-used power law introduced by [4] and defined by

$$
\Phi(E) = K\,E^{-\gamma} \quad (3)
$$

over an energy range between $E_{min}$ and $E_{max}$ ($E$ is the kinetic energy per nucleon always expressed in the laboratory system). The constant $K$ is related to the flux amplitudes $\Phi_p$, $\Phi_\alpha$ or $\Phi_{^3\mathrm{He}}$ after normalization of Eq. 3, i.e $\int_{E_{min}}^{E_{max}} \Phi(E)dE = \Phi_p$ for the proton case.

The proton, $\alpha$ and $^3$He capture cross sections $\sigma(E)$ are either taken from experiments (essentially for light species [9]) or estimated with the Talys nuclear reaction code [10, 12]. Due to the large number of open channels at high energies, typically about

30 different (p,$xn$ $yp$ $z\alpha$) reaction types corresponding to the emission of $x$ neutrons, $y$ protons, and $z$ $\alpha$-particles, need to be taken into account per target nucleus. This implies that the reaction network includes for the 5600 different species with $0 \leq Z \leq 102$ some 300000 proton, $\alpha$ and $^3$He-induced reactions. Finally, note that the initial abundance distribution of the irradiated material is assumed to be solar in most of the cases studied here.

No effort is done in the present work to explain the possible mechanisms responsible for the particle acceleration. The energy spectrum, the proton, $\alpha$ and $^3$He fluxes, as well as the irradiation time are taken as free parameters. It can however be kept in mind that in the specific case of solar flares typical proton fluences (i.e. time-integrated fluxes) are of the order of $10^{17-18}$cm$^{-2}$. Early solar system models call for fluences of about $10^{23}$cm$^{-2}$ to explain the origin of now-extinct short-lived radionuclides known to have been present in the early solar system [13]. If HD 101065 is not a T-Tauri star, its putative companion could possibly be one. In addition, the proton fluence can also be expected to be increased by a factor as high as $10^6$ if the protons are guided along the magnetic field lines [4]. On the other hand, it can also be of interest to apply such spallative model to the jet particles emitted, for example, in the case of bipolar supernova explosions [14]. In such a situation, the jet can reach fluences by far larger than some $10^{30}$ cm$^{-2}$ at a speed of about $c/6$, i.e some 10 MeV per nucleon.

## RESULTS

In the above-described model, the parameter space spanning the different possible characteristics of the accelerated particles is extremely vast. We leave here the braod discussion to Ref. [8] and restrict ourselves here to a specific exemplary event characterized by a proton flux $\Phi_p = 10^{-4}$ mb$^{-1}$ s$^{-1}$ and a $\alpha$-to-proton flux ratio $\Phi_\alpha/\Phi_p = 0.1$ typical of solar flares ($\Phi_{^3\mathrm{He}} = 0$ is assumed here). The adopted energy spectrum is a constant distribution with $E_{min} = 0$ and $E_{max} = 10$ MeV.

Quite generally, in the case of small fluxes, the proton and $\alpha$ captures are slower than the $\beta^+$ decays, so that the nuclear flow follows a path along the valley of $\beta$-stability on the neutron-deficient side. For larger values of $\Phi_p$ and $\Phi_\alpha$, neutron-deficient nuclei are produced significantly and many unstable nuclei are produced up to the proton-drip line.

For the adopted energy distribution, the low energy component of the projectile flux is responsible for radiative captures, while the high energy one favours a large number of transfer reactions. In particular, the (p or $\alpha$,$xn$ $yp$ $z\alpha$) reactions are responsible, at least at large fluences for a significant production of neutrons (especially on the initially abundant light species) which through their radiative neutron captures tend to bring the flow back to the valley of $\beta$-stability. Such secondary neutrons may play a crucial role, especially if we are concerned by a possible production of actinides. No p- or $\alpha$-captures could possibly lead to the synthesis of actinides or sub-actinides due to the low $\mu$-second $\alpha$-unstable island on the $85 \leq Z \leq 89$ p-rich side. Only a nuclear path on the neutron-rich side could be taken to produce transuranium significantly. The neutron abundance is followed with the same time-dependent equation as Eq. 1.

In Fig. 1, we show the evolution of the two radio-elements Tc and Pm as well as their neighbour elements Mo and Nd, respectively, as a function of time, or equivalently the

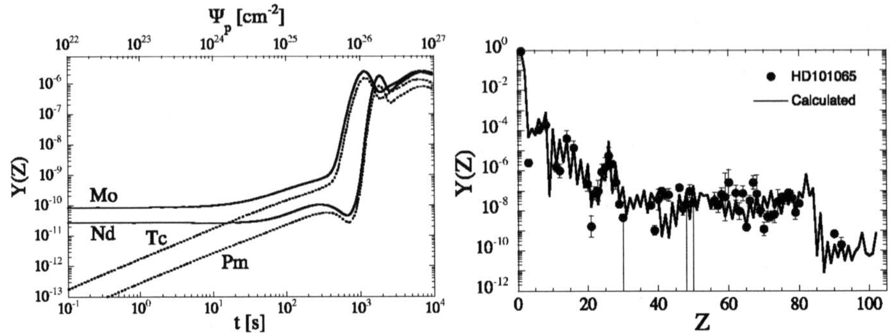


**FIGURE 1.** Abundances resulting from the irradiation by energetic particles corrresponding to $\Phi_p = 10^{-4}\,\mathrm{mb}^{-1}\,\mathrm{s}^{-1}$, $\Phi_\alpha/\Phi_p = 0.1$, and a constant $E$-distribution with $E_{max} = 10$ MeV. Left panel: Evolution with time t, or equivalently proton fluence $\Psi_p$, of the Tc and Pm molar fractions as well as their neighbour element Mo and Nd, respectively. Right panel: final abundance distribution resulting from the mixing of solar abundances with material irradiated during a time $\tau_{irr} = 5000$ s. The dilution factor is set to $f_{mix} = 0.02$. The full dots correspond to the HD 101065 abundance determination [5].

proton fluence $\Psi_p = \int_0^t \Phi_p \, dt$. Minimum values of $10^{23}$ cm$^{-2}$ are required to produce Tc and Pm significantly. But for such fluences, heavy actinides away from the already existing Th and U cannot be formed. If fluences of the order $10^{26-27}$ cm$^{-2}$ can be achieved, not only globally the abundances of the elements heavier than iron can be increased by 5 orders of magnitude, but also the neutron flux becomes large enough to bridge the $N > 126$ $\alpha$-unstable region between Bi and Ra and produce actinides as heavy as $Z \simeq 100$ in very large amount. Obviously, such high fluences can hardly be expected in normal stellar flares and the relevance of such a scenario must be confirmed by more realistic astrophysics models.

The abundance pattern resulting from spallation nucleosynthesis is usually rather flat, and indeed with the dominant capture of charged-particles with energies up to 10 MeV per nucleon, all nuclear structure effects are smoothed out. The high-fluence abundance distribution is consequently quite different from the observed one. Some structure effects are, however, recovered if, at the end of the irradiation, neutrons are still largely present and keep on participating to the nucleosynthesis.

If we now assume that the material irradiated for a given time $\tau_{irr}$ is mixed with unaffected surounding layers of solar-type composition, or equivalently that only part of the observed material has been irradiated, the final abundances can be estimated by

$$Y^{mix}_{(Z,A)} = f_{mix} Y^{irr}_{(Z,A)} + (1 - f_{mix}) Y^{\odot}_{(Z,A)} \tag{4}$$

where $f_{mix}$ is an arbitrary dilution factor. For the specific parameter sets considered in the case of Fig. 1, we see that globally the observed abundances can be remarkably well explained, including the presence of Tc, Pm and transuranium. Even the isotopic ratio of $^6$Li$/^7$Li=0.27 is close to the value of 0.3 derived from observation [15], though the Li abundance is here overestimated by about 1 dex. More details can be found in [8].

## SPALLATION NUCLEOSYNTHESIS OF HEAVY ELEMENTS

Assuming that high-energy particle irradiations, as discussed above, have polluted the envelope of CP stars, similar spallation events could also possibly be at the origin of additional nucleosynthesis processes. We now investigate the possibility for high-fluence irradiation events to produce elements heavier than iron, as observed in the sun and not anymore in HD 101065. We first consider high-fluence low-flux events. In this case, the lack of neutrons as well as the fast $\beta$- and $\alpha$-decays constrain the nuclear flow to the bottom of the valley of stability, so that at high fluences, essentially p- and s-elements are produced. We show in Fig. 2 (left panel) the overabundances (i.e. the mass fractions relative to the solar value) of all nuclei produced in the high-fluence low-flux irradiation corresponding to the conditions called for to explain the origin of now-extinct short-lived radionuclides known to have been present in the early solar system [13]. These are $\Phi_p = 10^{-12}\ \mathrm{mb}^{-1}\,\mathrm{s}^{-1}$, $\Phi_\alpha/\Phi_p = 0.1$, $\Phi_{^3\mathrm{He}}/\Phi_\alpha = 1$ and $\gamma = 2.7$. While early solar system model requires a total fluence of about $10^{23}\ \mathrm{cm}^{-2}$, the total fluence considered here reaches $10^{27}\ \mathrm{cm}^{-2}$. The corresponding irradiation timescale is a few hundred thousands years. In this case, a significant nucleosynthesis takes place with a large production of p-nuclei (see [16] for a review on the p-process nucleosynthesis). The nucleosynthesis of p-nuclei by low-fluence spallation was introduced decades ago [17] and shown to form only the heaviest p-nuclei $^{184}$Os, $^{190}$Pt, $^{196}$Hg, but abandoned in favour of the thermonuclear p-process [16]. In our high-fluence irradiation, a smooth solar-like p-abundance distribution is obtained, and interestingly, the $^{92,94}$Mo and $^{96,98}$Ru the origin of which remains a mystery (see [16]) are significantly synthesized. $^{138}$La and $^{180}$Ta are underproduced in this specific example. Additionally, the s-only nuclei, i.e the nuclei produced uniquely by the s-process, are seen to be synthesized in non-negligible amount, but still a factor of 10 less than the p-nuclei.

In Fig. 2 (right panel), a second low-flux irradiation event is shown to synthesize s-nuclei. Its properties are $\Phi_\alpha = 10^{-12}\ \mathrm{mb}^{-1}\,\mathrm{s}^{-1}$, $\Phi_p = \Phi_{^3\mathrm{He}} = 0$, and a constant energy distribution between 2 and 3 MeV. The initial composition of the irradiated matter is taken as pure C. The irradiation timescale is 1000 y, corresponding to $\Psi = 3\ 10^{25}\ \mathrm{cm}^{-2}$. The accelerated $\alpha$ particles could for example originate from the wind of a Wolf-Rayet star and be irradiating an amorphous graphite meteorite. In this specific case, s-only nuclei are seen to be produced in solar distribution. All the p-nuclei, except $^{92,94}$Mo, are underproduced in this case. Neutron densities between $10^8$ and $10^{10}\ \mathrm{cm}^{-3}$ are achieved during the irradiation. The light Be and B isotopes are also strongly produced.

Finally, we also explore here the high-fluence high-flux regime by considering an irradiation event with the following characteristics: $\Phi_\alpha = 10^4\ \mathrm{mb}^{-1}\,\mathrm{s}^{-1}$, $\Phi_\alpha = \Phi_{^3\mathrm{He}} = 0$, $\gamma = 2$ and $E_{max} = 25$ MeV. Only $\alpha$-particles are accelerated, as could be expected for example in supernovae jets where at decreasing temperatures neutrons and protons have recombined. A high-fluence is again simulated up to $\Psi = 3\ 10^{26}\ \mathrm{cm}^{-2}$, i.e $\tau_{irr} = 3\ 10^{-5}$ s. The temperature is set to $T = 5\ 10^8$ K, although ideally the temperature profile need to be estimated by hydrodynamical simulations. During this very short irradiation, a large amount of neutrons (with neutron densities up to some $10^{23}\ \mathrm{cm}^{-3}$) is produced bringing the nuclear flow into the neutron-rich region. The $\alpha$ captures are therefore followed by neutron captures, $\beta$-decays having no time to take place on such short

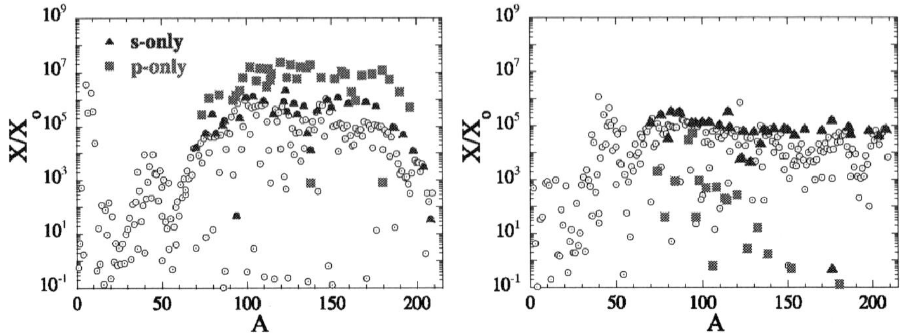


**FIGURE 2.** *Left panel*: abundances relative to solar resulting from a low-flux irradiation by energetic particles corrresponding to $\Phi_p = 10^{-12}$ mb$^{-1}$ s$^{-1}$, $\Phi_\alpha/\Phi_p = 0.1$, $\Phi_{^3\mathrm{He}}/\Phi_\alpha = 1$ and $E_{max} = 25$ MeV; $\gamma = 2.7$ and a final total fluence $\Psi = 10^{27}$ cm$^{-2}$. The triangles represent s-only nuclei and the squares p-nuclei. *Right panel*: same for the event $\Phi_\alpha = 10^{-12}$ mb$^{-1}$ s$^{-1}$, $\Phi_p = \Phi_{^3\mathrm{He}} = 0$; $2 \leq E$ [MeV] $\leq 3$ and $\Psi = 3\ 10^{25}$ cm$^{-2}$. The initial composition of the irradiated material in this case is taken as pure C.

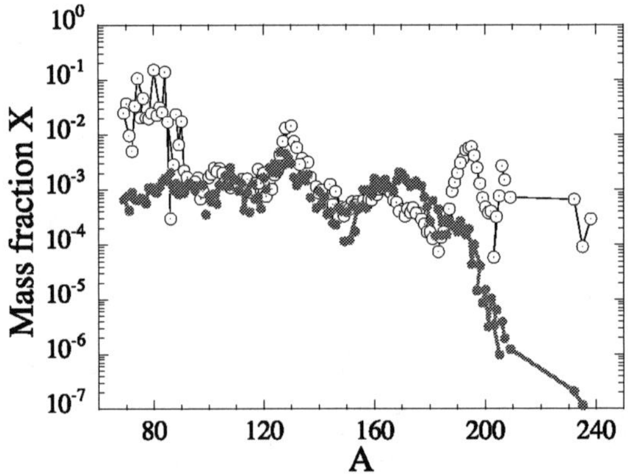


**FIGURE 3.** Abundances of r-nuclei (full circles) resulting from the irradiation with $\Phi_\alpha = 10^4$ mb$^{-1}$ s$^{-1}$, $\Phi_\alpha = \Phi_{^3\mathrm{He}} = 0$ and $E_{max} = 25$ MeV; $\gamma = 2$ and $\Psi = 3\ 10^{26}$ cm$^{-2}$ (corresponding to $\tau_{irr} = 3\ 10^{-5}$ s). The solar system r-abundance distribution (open circles) arbitrarily normalized is also shown to guide the eye.

timescales. The resulting abundance distribution for the r-nuclei are compared with the solar r-abundance in Fig. 3. Although this case does not match perfectly the solar r-abundance distribution, it can be seen that r-elements are clearly produced and that spallation event can be at the origin of a significant nucleosynthesis even in the exotic neutron-rich region of the nuclear chart. In other conditions, the heaviest r-elements up to the actinides can also be formed.

## CONCLUSION

The present study aims at exploring spallation nucleosynthesis as a possible explanation of the peculiar abundances observed at the surface of HD 101065. We have shown that at least the specific simulation shown here with large fluences tend to be compatible with many different observational aspects. In particular, a significant production of $Z > 30$ heavy elements by radiative and spallation reactions can be achieved, and most of all a significant synthesis of radio-elements, such as Tc, Pm but also actinides up to at least $Z \simeq 100$. This requires, however, fluences by far larger than what would be expected in normal or even extreme stellar flares. The possible existence of such irradiation events need to be confirmed by hydrodynamics simulations, but most of all by spectroscopic observations through the detection of short-lived radio-elements. Finally, if such high-fluence irradiation events are confirmed, they could also be at the origin of a new rich nucleosynthesis including the production of p-, s- and r-nuclei.

## ACKNOWLEDGMENTS

S.G. is FNRS research associate.

## REFERENCES

1. D.W. Kurtz, Mon. Not. R. Astron. Soc. **200**, 807 (1982).
2. C.R. Cowley, W.P. Bidelman, S. Hubrig, G. Mathys, D.J. Bord, Astron. Astrophys. **419**, 1087 (2004).
3. V.F. Gopka, A.V. Yushchenko , A.V. Shavrina, D.E. Mkrtichian, A.P. Hatzes, S.M. Andrievsky and L.V. Chernysheva, in Proc. IAU symposium Nr 224, (eds. Zverko et al.), 734 (2004).
4. P.J. Brancazio and A.G.W. Cameron, Can. J Phys. **45**, 3297 (1967).
5. C.R. Cowley, T. Ryabchikova, F. Kupka, D.J. Bord, G. Mathys and W.P. Bidelman, Mon. Not. R. Astron. Soc. **317**, 299 (2000).
6. G. Michaud, Astroph. J. **160**, 641 (1970).
7. M. Arnould and S. Goriely, Nucl. Phys. **A777**, 157 (2006).
8. S. Goriely, Astron. Astrophys., **466**, 619 (2007).
9. J.P. Meyer, Astroph. J. Suppl. **7**, 417 (1972).
10. A.J. Koning, S. Hilaire and M.C. Duijvestijn, in Nuclear Data for Science and Technology (eds. C. Haight et al.) AIP Conference Vol. 769, p. 1154, 2005.
11. S. Goriely, S. Hilaire and A.J. Koning, in Nuclear Data for Science and Technology, in press (2007).
12. S. Hilaire , S. Goriely and A.J. Koning, this conference (2007).
13. I. Leya, A.N. Halliday and R. Wieler, Astroph. J. **594**, 605 (2003).
14. K. Maeda and K. Nomoto, Astroph. J. **598**, 1163 (2003).
15. A.V. Shavrina, N.S. Polosukhina, Ya. V. Pavlenko *et al.*, Astron. Astrophys. **409**, 707 (2003).
16. M. Arnould and S. Goriely, Phys. Rep. **384**, 1 (2003).
17. J. Audouze, Astron. Astrophys., **8**, 436 (1970).
18. M. Arnould, S. Goriely and K. Takahashi, Phys. Rep. **450**, 97 (2007).

# Neutrino-nucleus reaction rates based on the relativistic quasiparticle random phase approximation

N. Paar*, D. Vretenar*, T. Marketin* and P. Ring†

*Physics Department, Faculty of Science, University of Zagreb, Croatia
†Physik-Department der Technischen Universität München, D-85748 Garching, Germany

**Abstract.** Neutrino-nucleus cross sections are described in a novel theoretical framework where the weak interaction of leptons with hadrons is expressed in the standard current-current form, the nuclear ground state is described in the relativistic Hartree-Bogoliubov model, and the relevant transitions to excited states are calculated in the relativistic quasiparticle random phase approximation. The model is employed in studies of neutrino-nucleus reactions in several test cases.

**Keywords:** neutrino-nucleus cross sections, relativistic QRPA, relativistic mean field, charge-exchange excitations, supernova neutrinos, neutrino detectors
**PACS:** 21.30.Fe, 21.60.Jz, 23.40.Hc, 26.30.+k

## INTRODUCTION

Neutrino-nucleus reactions in the low-energy range up to 100 MeV play an important role in many astrophysical processes, including stellar nucleosynthesis. Understanding the production of heavy elements during the r-process necessitates accurate knowledge on the neutrino-nucleus cross sections not only in stable nuclei but also in nuclei away from the valley of $\beta$-stability, which cannot be studied within presently available models such as the shell model or the continuum random phase approximation [1], because the ground-state input for these models is adjusted to experimental data.

Since nuclei are used as neutrino detectors in solar experiments, supernova observatories, and in oscillation measurements, it is of outmost importance to achieve a reliable and quantitative description of neutrino detector response in a consistent microscopic theory. Studies of neutrino-nucleus reactions will also provide the relevant information for feasibility studies and simulations of a low-energy beta beam facility with perspective to produce neutrino beams of interest for particle physics, nuclear physics and astrophysics [2]. The general expressions for relevant transition matrix elements have been known since many years [3], however, only very recently first consistent calculations involving open shell nuclei have been performed by systematic implementation of the same effective interaction both in description of the nuclear ground and excited states [4].

The principal objective of our approach is to treat transitions relevant for neutrino-nucleus cross sections in a fully microscopic framework based on a relativistic energy density functional and to include pairing correlations in a consistent way. The nuclear ground state is described in the relativistic Hartree-Bogoliubov (RHB) model, and the

CP1012, *Frontiers in Nuclear Structure, Astrophysics, and Reactions: FINUSTAR 2*
edited by P. Demetriou, R. Julin, and S. V. Harissopulos
© 2008 American Institute of Physics 0-7354-0532-5/08/$23.00

relevant excitations are treated within the self-consistent relativistic quasiparticle random phase approximation (RQRPA). Several test cases for neutrino-nucleus cross sections will be presented, in comparison with previous theoretical studies and experimental data.

## NEUTRINO-NUCLEUS CROSS SECTIONS BASED ON THE RHB+RQRPA MODEL

We consider the neutrino-nucleus charged current reaction,

$$\nu_l + _Z X_N \rightarrow _{Z+1} X^*_{N-1} + l^-, \tag{1}$$

where $l$ denotes different types of leptons, i.e. electron and muon. The detailed expressions for weak interaction rates are given in Ref. [3]. By employing the multipole expansion of the weak nuclear current, and assuming the extreme relativistic limit (ERL) where the energy of the outgoing lepton is higher than its rest mass, the differential neutrino-nucleus cross section reads,

$$\begin{aligned}
\left(\frac{d\sigma_\nu}{d\Omega}\right)_{ERL} &= \frac{2G_F^2 cos^2\theta_c}{\pi}\frac{E_l^2}{2J_i+1} \\
&\times\left\{\left(\frac{q_\lambda^2}{2q^2}cos^2\frac{\theta}{2}+sin^2\frac{\theta}{2}\right)\sum_{J\geq 1}\left[|\langle J_f||\hat{\mathscr{T}}_J^{MAG}||J_i\rangle|^2+|\langle J_f||\hat{\mathscr{T}}_J^{EL}||J_i\rangle|^2\right]\right. \\
&-sin\frac{\theta}{2}\sqrt{\frac{q_\lambda^2}{q^2}cos^2\frac{\theta}{2}+sin^2\frac{\theta}{2}}\sum_{J\geq 0}2Re\langle J_f||\hat{\mathscr{T}}_J^{MAG}||J_i\rangle\langle J_f||\hat{\mathscr{T}}_J^{EL}||J_i\rangle^* \\
&\left.+cos^2\frac{\theta}{2}\sum_{J\geq 0}|\langle J_f||\hat{\mathscr{M}}_J-\frac{q_0}{|q|}\hat{\mathscr{L}}_J||J_i\rangle|^2\right\}.
\end{aligned} \tag{2}$$

It is a function of the angle $\theta$ between the incoming and outgoing leptons, the energy of the final lepton $E_l$, momentum transfer $q$, and 4-momentum transfer $q_\lambda$. The transition matrix elements contain the multipole operators $\hat{\mathscr{M}}_J$, $\hat{\mathscr{L}}_J$, $\hat{\mathscr{T}}_J^{EL}$, $\hat{\mathscr{T}}_J^{MAG}$, which correspond to the charge, longitudinal, transverse electric, and transverse magnetic operators, respectively. The operators involved in weak or electromagnetic processes can be expressed in terms of 7 fundamental single-particle operators, which contain basic multipole operators with spherical Bessel functions, spherical harmonics, and vector spherical harmonics [3].

The nuclear ground state is calculated with the RHB model, and the corresponding RQRPA is used to determine the excited states [5, 6]. The proton-neutron(PN)-RQRPA [7] is formulated in the canonical single-nucleon basis of the RHB model. For the interaction in the particle-hole channel effective Lagrangians with density-dependent meson-nucleon couplings are used [8], and pairing correlations are described by the pairing part of the finite-range Gogny interaction [9]. Both in the particle-hole and pairing channels, the same interactions are used in the RHB equations that determine

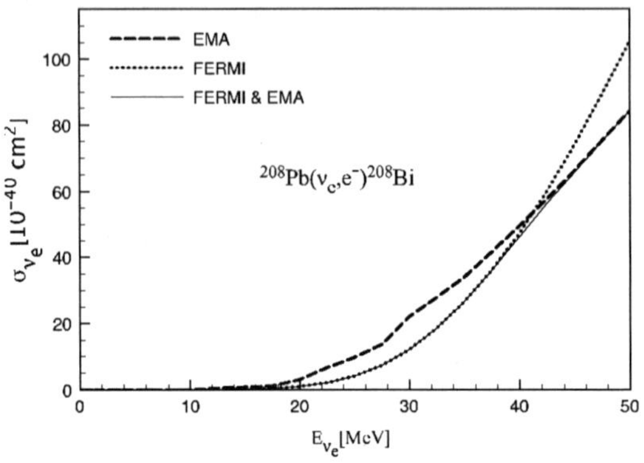


**FIGURE 1.** Inclusive neutrino-nucleus cross sections for $^{208}$Pb$(\nu_e, e^-)^{208}$Bi reaction evaluated using effective momentum approximation (dashed line), Fermi function (dotted line), and their combination (solid line).

the canonical quasiparticle basis, and in the matrix equations of the RQRPA. The spin-isospin dependent interaction terms are generated by the $\rho$- and $\pi$-meson exchange. Because of parity conservation, the one-pion direct contribution vanishes in the mean-field calculation of a nuclear ground state. It is however, essential to include the pion contributions in description of excitations that involve spin and isospin degrees of freedom. In addition to configurations built from two-quasiparticle states of positive energy, the RQRPA configuration space contains quasiparticle excitations formed from the ground-state configurations of fully or partially occupied states of positive energy and the empty negative-energy states from the Dirac sea.

By using Eq. (2) and the RHB+RQRPA framework to determine the reduced matrix elements for the 7 key transition operators, we evaluate the electron and muon neutrino-nucleus cross sections. The Coulomb effect of a nucleus on outgoing leptons is taken into account by multiplying the cross section with the Fermi function obtained from numerical solution of the Dirac equation for an extended nuclear charge [1]. For neutrinos at higher energies, the Coulomb effect is better described by the effective momentum approximation (EMA), where the lepton momentum $p_l$ and energy $E_l$ are modified by a constant electrostatic potential evaluated at the center of the nucleus [10]. These approximations are illustrated in Fig. 1 for the reaction $^{208}$Pb$(\nu_e, e^-)^{208}$Bi, which involves a high-Z target nucleus. The RHB+RQRPA calculations are based on the DD-ME2 effective interaction [11], and we display the corresponding cross sections corrected either with the Fermi function, or obtained with EMA. At higher neutrino energies the values of cross sections obtained with the Fermi function are above the EMA results. Following Ref. [4], we will use the Fermi function approximation in the range of neutrino energies for which the cross section is below the EMA result, whereas at higher energies EMA will be employed (solid line in the figure).

Probably one of the most extensively studied neutrino-nucleus reactions is $^{12}$C$(\nu_e, e^-)^{12}$N, because of its relevance for neutrino detectors and available data.

**TABLE 1.** The PN-RQRPA flux averaged neutrino-nucleus cross sections for $^{12}$C target nucleus compared with previous studies and experimental data.

| | $(\nu_\mu,\mu^-)$ $\langle\sigma\rangle(10^{-40}cm^2)$ | $(\nu_e,e^-)$ $\langle\sigma\rangle(10^{-42}cm^2)$ |
|---|---|---|
| RPA [4] | 19.23 | 55.10 |
| QRPA [4] | 20.29 | 52.0 |
| CRPA [14] | 18.18 | 19.28 |
| PN-RQRPA | 19.59 | 12.14 |
| Exp. | 12.4±0.3±1.8 [13] | 14.1±1.6 [12] |

For this reaction we evaluate the flux averaged cross section

$$\langle\sigma_\nu\rangle = \frac{\int dE_\nu \sigma_\nu(E_\nu) f(E_\nu)}{\int dE'_\nu f(E'_\nu)}, \tag{3}$$

where in the case of $\nu_e$ we employ the Michel flux from decay at rest of $\mu^+$ [12], whereas for $\nu_\mu$ we use the experimental flux obtained by the LSND collaboration from the decay in flight of $\pi^+$ [13]. In Table 1 we list the PN-RQRPA muon and electron neutrino-nucleus cross sections for the reactions $^{12}$C$(\nu_e,e^-)^{12}$N and $^{12}$C$(\nu_\mu,\mu^-)^{12}$N, averaged over experimental neutrino fluxes, in comparison with several theoretical studies and experimental data. In the case of $\nu_\mu$ absorption, the calculated cross section is in excellent agreement with previous theoretical studies which are, however, systematically larger than experimental value. For the reaction $^{12}$C$(\nu_e,e^-)^{12}$N the agreement with experiment is significantly improved when compared to results of non-relativistic (Q)RPA calculations.

Next we evaluate the averaged cross sections over a Supernova neutrino flux, which is given by the Fermi-Dirac spectrum,

$$f(E_{\nu_e}) = \frac{1}{T^3}\frac{E^2_{\nu_e}}{exp[(E_{\nu_e}/T)-\alpha]+1}. \tag{4}$$

The corresponding PN-RQRPA cross sections for the reaction $^{16}$O$(\nu_e,e^-)^{16}$F and $\alpha = 0$ are shown in Fig. 2, calculated for a set of temperatures in the interval T=2-10 MeV. The PN-RQRPA values are compared with the shell model results of Ref. [15]. In this particular case, the two metods are in excellent agreement. However, an important advantage of the present approach is the possibility to model nuclei which are beyond the reach of the shell model approach: medium-heavy and heavy nuclei, as well as systems far from stability which play an important role in neutrino nucleosynthesis.

## CONCLUSION

We have introduced a novel theoretical framework for the description of neutrino-nucleus cross sections: weak interactions of leptons and hadrons are expressed in the standard current-current form, the nuclear ground state is described in the RHB model,

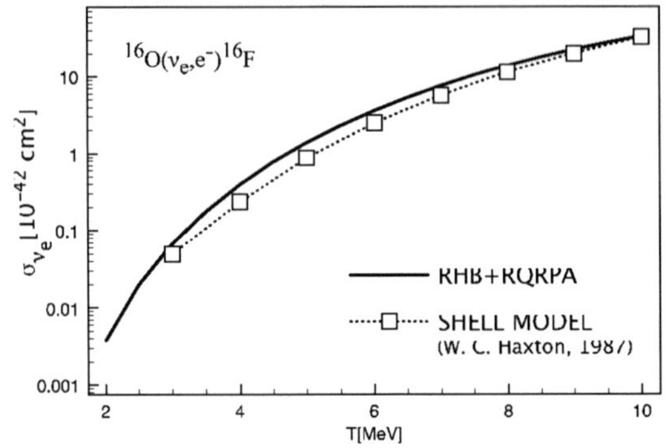


**FIGURE 2.** Inclusive neutrino-nucleus cross sections for $^{16}O(\nu_e, e^-)^{16}F$ reaction including all transitions with multipolarities up to $J = 5$, and averaged over the neutrino flux described by the Fermi distribution for temperatures T=2-10 MeV. Shell model cross sections are shown in the same range of temperatures [15].

and the transition matrix elements to excited states are evaluated using the RQRPA. It is important to note that a reliable theoretical description of neutrino-nucleus reaction rates, not only in stable nuclei but also far from stability, necessitates input from consistent models of nuclear structure that can provide accurate ground state properties and relevant transitions to excited state. The present approach fully meets these conditions, and illustrative examples of calculated neutrino-nucleus cross sections exhibit very good agreement with previous theoretical studies and data.

## REFERENCES

1. E. Kolbe, K. Langanke, G. Martinez-Pínedo, and P. Vogel, *J. Phys. G* **29**, 2569 (2003).
2. C. Volpe, *J. Phys. G* **30**, L1 (2004).
3. J. S. O'Connel, T. W. Donnelly, and J. D. Walecka, *Phys. Rev.* C **6** 719, (1972).
4. C. Volpe, N. Auerbach, G. Colò, T. Suzuki, and N. Van Giai, *Phys. Rev. C* **62**, 015501 (2000).
5. D. Vretenar, A. V. Afanasjev, G. A. Lalazissis, and P. Ring, *Phys. Rep.* **409**, 101 (2005).
6. N. Paar, D. Vretenar, E. Khan, and G. Colò, *Rep. Prog. Phys.* **70**, 691 (2007).
7. N. Paar, T. Nikšić, D. Vretenar, and P. Ring *Phys. Rev. C* **69**, 054303 (2004).
8. T. Nikšić, D. Vretenar, P. Finelli, and P. Ring, *Phys. Rev. C* **66**, 024306 (2002).
9. J. F. Berger, M. Girod, and D. Gogny, *Comp. Phys. Comm.* **63**, 365 (1991).
10. J. Engel, *Phys. Rev. C* **57**, 2004 (1998).
11. G. A. Lalazissis, T. Nikšić, D. Vretenar, and P. Ring, *Phys. Rev. C* **71**, 024312 (2005).
12. D. A. Krauker et al. *Phys. Rev. C* **45**, 2450 (1992).
13. C. Athanassopoulos et al., *Phys. Rev. Lett.* **81**, 1774 (1998).
14. E. Kolbe, K. Langanke, and P. Vogel, *Nucl. Phys. A* **652**, 91 (1999).
15. W. C. Haxton, *Phys. Rev. D* **36**, 2283 (1987).

# Spin-Dependent Transitions in Nuclei and Neutrino-Nucleus Reactions

Toshio Suzuki*, Satoshi Chiba†, Michio Honma**, Koji Higashiyama‡, Takashi Yoshida§, Hideyuki Umeda, Ken'ichi Nomoto¶, Toshitaka Kajino‖ and Takaharu Otsuka††

*Department of Physics, College of Humanities and Sciences, Nihon University, Sakurajosui 3-25-40, Setagaya-ku, Tokyo 156-8550, Japan,
Center for Nuclear Study, University of Tokyo, Hirosawa, Wako-shi, Saitama, 351-0198, Japan
†Advanced Science Research Center, Japan Atomic Energy Agency, 2-4 Shirakata-Shirane, Tokai, Naka-gun, Ibaraki 319-1195, Japan
**Center for Mathematical Sciences, University of Aizu, Aizu-Wakamatsu, Fukushima, Japan
‡Department of Physics, Chiba Institute of Technology, Narashino, Chiba 275-0023, Japan
§National Astronomical Observatory of Japan, Mitaka, Tokyo 181-8588, Japan
¶Department of Astronomy, School of Science, University of Tokyo, Tokyo 113-0033, Japan
‖Department of Astronomy, Graduate School of Science, University of Tokyo, Hongo, Bunkyo-ku, Tokyo 113-0033, Japan
National Astronomical Observatory of Japan, Mitaka, Tokyo 181-8588, Japan
††Department of Physics and Center for Nuclear Study, University of Tokyo, Hongo, Bunkyo-ku, Tokyo 113-0033, Japan
RIKEN, Hirosawa, Wako-shi, Saitama 351-0198, Japan

**Abstract.** Neutrino-induced reactions on $^{12}$C, $^{4}$He as well as Fe and Ni isotopes are studied based on recent advances in shell model calculations in $p$-shell and $fp$-shell. The reaction cross sections are found to be enhanced compared with conventional Hamiltonians. The production yields of $^{7}$Li and $^{11}$B during supernova explosions are found to be enhanced and the effects of neutrino oscillations are discussed. A possible implication of neutrino-induced reactions on the production yields of heavy elements such as Mn and Co is also discussed.

**Keywords:** shell model, neutrino-nucleus reactions, nucleosynthesis
**PACS:** 21.60.Cs, 25.30.Pt, 24.60.Dr, 26.30.Jk

## NEUTRINO-INDUCED REACTIONS ON $^{12}$C AND $^{4}$HE

Neutrino induced reactions on $^{12}$C have been studied with the use of a new shell model Hamiltonian, SFO[1], which properly takes into account important roles of spin-isospin interactions that leads to significant improvements in magnetic and spin properties of $p$-shell nuclei as well as a shell evolution toward drip-lines. In particular, a general important role played by the tensor interaction is pointed out[2].

The exclusive charge-exchange and neutral-current reactions on $^{12}$C induced by the Gamow-Teller transition are well described by the SFO Hamiltonian for the DAR neutrinos[3]. The inclusive charge-exchange reaction to excited states is also well reproduced with a quenching for the axial-vector coupling constant, $g_A^{eff}/g_A^{free}$ = 0.70 (0.90) for the $2^-$ ($0^-$, $1^-$ etc.) states, which is consistent with the electron scattering data[4]. Reactions induced by supernova neutrinos have been investigated, and the cross

CP1012, *Frontiers in Nuclear Structure, Astrophysics, and Reactions: FINUSTAR 2*
edited by P. Demetriou, R. Julin, and S. V. Harissopulos
© 2008 American Institute of Physics 0-7354-0532-5/08/$23.00

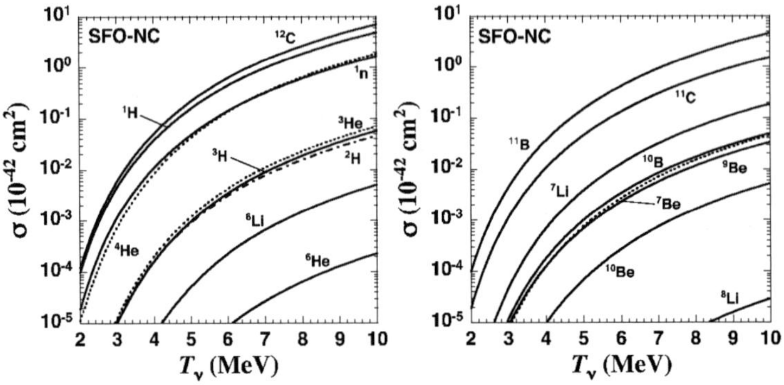


**FIGURE 1.** Calculated neutral current reaction cross sections for $^{12}$C induced by supernova neutrinos with temperature $T_\nu$ obtained by shell model calculations with the use of the SFO Hamiltonian. Knocked-out particles and residual nuclei are denoted in the left and right figures, respectively.

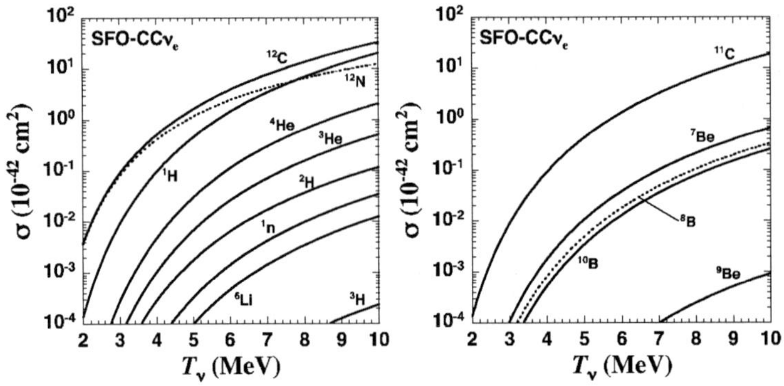


**FIGURE 2.** Calculated reaction cross sections for $^{12}$C $(\nu, e^- \ X)$ B induced by supernova neutrinos with temperature $T_\nu$ obtained by shell model calculations with the use of the SFO Hamiltonian. Knocked-out particles X and residual nuclei B are denoted in the left and right figures, respectively.

sections are found to be enhanced compared with those by the conventional shell model Hamiltonians (e. g. PSDMK2[5]) as well as previous calculations[6]. Reaction cross sections for the $\gamma$ decay, proton, neutron, deuteron, $^3$He, t and $\alpha$ knock-out processes are shown in Fig. 1 and 2 for supernova with temperature $T_\nu$. The branching ratios for the processes have been calculated by the Hauser-Feshbach theory. Here, the production of $^{10}$Be by direct $\nu$-processes is included for the first time.

Cross sections for neutrino-induced reactions on $^4$He are shown in Fig. 3 for the WBP[7] and the SPSDMK[5, 8] Hamiltonians. Calculated results are found to be enhanced compared with the previous calculations[6]. The enhancement of the $\nu$-$^4$He and $\nu$-$^{12}$C reaction cross sections is found to lead to the increase of the production yields of $^7$Li and $^{11}$B during supernova explosions [3] about by 30% and 20%, respectively, in case of WBP+SFO compared with those for the previous case[6]. When there are neutrino oscillations, the yield ratio Y($^7$Li)/ Y($^{11}$B) becomes sensitive to the mixing angle

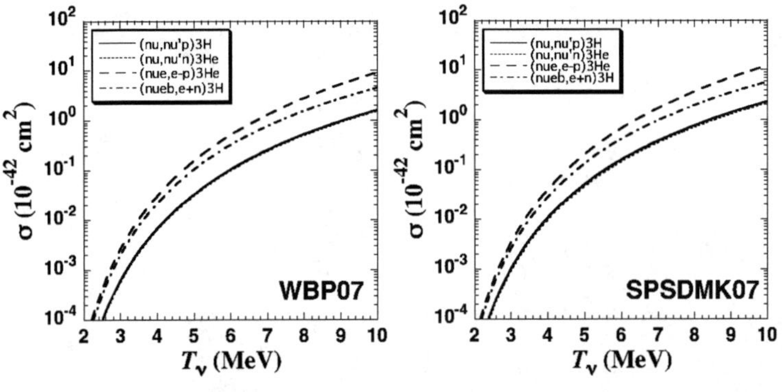


**FIGURE 3.** Calculated reaction cross sections for $^4$He induced by supernova neutrinos with temperature $T_\nu$ obtained by shell model calculations with the use of the WBP (left) and SPSDMK (right) Hamiltonians.

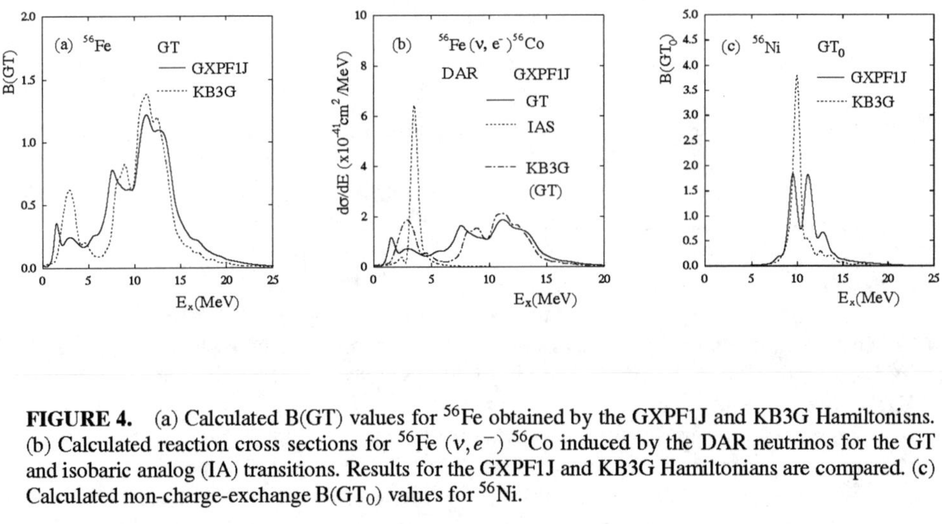


**FIGURE 4.** (a) Calculated B(GT) values for $^{56}$Fe obtained by the GXPF1J and KB3G Hamiltonisns. (b) Calculated reaction cross sections for $^{56}$Fe $(\nu, e^-)$ $^{56}$Co induced by the DAR neutrinos for the GT and isobaric analog (IA) transitions. Results for the GXPF1J and KB3G Hamiltonians are compared. (c) Calculated non-charge-exchange B(GT$_0$) values for $^{56}$Ni.

$\theta_{13}$ for the normal neutrino mass hierarchy, which can be used to determine the lower limit of $\theta_{13}$ as well as if the neutrino mass hierarchy is normal or inverted[9].

## NEUTRINO-INDUCED REACTIONS ON FE AND NI ISOTOPES

Neutrino-induced reaction on Fe and Ni isotopes have been studied with the use of a new shell-model Hamiltonian for the $fp$-shell, GXPF1J [10]. The Gamow-Teller strengths in $^{54,56}$Fe and $^{56-64}$Ni are found to be more fragmented than those obtained by the KB3G Hamiltonian[11]. The summed B(GT) value with a universal quenching factor of 0.74

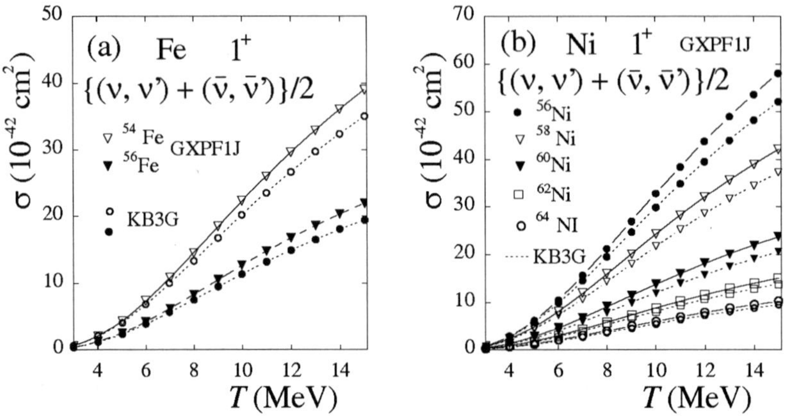


**FIGURE 5.** Neutral current reaction cross sections for (a) Fe and (b) Ni isotopes induced by supernova neutrinos with temperature $T$. Calculated results obtained by the GXPF1J and KB3G Hamiltonians are compared.

is 9.47 and 8.86 for the GXPF1J and KB3G Hamiltonian, respectively, which are close to the observed value, B(GT) = 9.9±2.4[12]. Calculated reaction cross sections for $^{56}$Fe $(\nu, e^-)$ $^{56}$Co induced by the DAR neutrinos are shown in Fig. 4 for the Gamow-Teller (GT) and isobaric analog (IA) transitions. Cross sections for the GT transitios are rather close; $\sigma = 1.60\ (1.56) \times 10^{-40}$cm$^2$ for the GXPF1J (KB3G) case, which gives about half of the observed value of $\sigma = 2.56 \pm 1.08 \pm 0.43 \times 10^{-40}$ cm$^2$.

Cross sections induced by the supernova neutrinos are also studied for the Gamow-Teller transitions. In case of neutral-current reactions, the fragmentation of GT$_0$ (T$_f$ = T, T+1) strengths for the GXPF1J Hamiltonian gets more pronounced compared with the KB3G one than the charge-exchange cases (see Fig. 4(c) for $^{56}$Ni). The neutral-current reaction cross sections are, thus, found to be enhanced for the GXPF1J compared with the KB3G, especially for high $\nu$ temperatures as shown in Fig. 5.

Now, we study particle knock-out processes in neutral current reactions on $^{56}$Ni, and discuss production of $^{55}$Mn through a path, $^{56}$Ni $(\nu, \nu' p)$ $^{55}$Co $(e^-, \nu)$ $^{55}$Fe $(e^-, \nu)$ $^{55}$Mn in supernova explosions. Calculated reaction cross sections are shown in Fig. 6 for both GXPF1J and KB3G Hamiltonians and compared with previous nucleon knock-out cross sections[6]. Proton knock-out cross sections are found to be much enhanced for the GXPF1J case. This is due to the fact that the strength is more fragmented and a large fraction of the strength is found in the proton emission channel in case of the GXPF1J Hamiltonian (see Fig. 4(c)). Production yields of elements during supernova explosions for a population III star with mass 15 M$_\odot$[13] are investigated. Total $\nu$ energy of $E_\nu$ = $3 \times 10^{53}$ ergs and $\nu$ temperatures of $(T_{\nu_e}, T_{\bar{\nu}_e}, T_{\nu_{\mu,\tau}})$ = (4, 4, 6) MeV are assumed. Cross

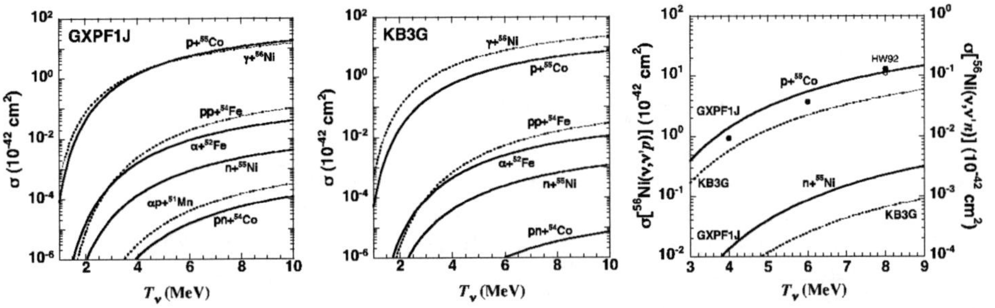


**FIGURE 6.** Neutral current reaction cross sections for $^{56}$Ni induced by supernova neutrinos with temperature $T_\nu$ obtained by the GXPF1J (left) and KB3G (middle) Hamiltonisnas. Knocked-out particles and residual nuclei are denoted. Nucleon knock-out cross sections are compared with those of ref.[6] which are denoted as HW92 (right).

sections of ref. [6] are used except for neutral current reactions on $^{56}$Ni. Enhancement of the production yield of $^{55}$Mn is obtained for the GXPF1J Hamiltonian as shown in Table 1. Enhancement of the production yield of $^{59}$Co is also obtained.

**TABLE 1.** Production yields of Mn in a supernova explosion model of a population III star with 15 $M_\odot$

| Model | GXPF1J | GXPF1J × 2 | KB3G | KB3G × 2 | HW02 | No $\nu$ |
|---|---|---|---|---|---|---|
| M($^{55}$Co)/$M_\odot$ ($10^{-4}$) | 4.16 | 5.00 | 3.56 | 4.00 | 3.80 | 2.29 |
| M(Mn)/$M_\odot$ ($10^{-4}$) | 4.31 | 5.16 | 3.72 | 4.16 | 3.96 | 2.30 |
| Log (Mn/Fe) | -0.25 | -0.17 | -0.32 | -0.27 | -0.29 | -0.53 |

# REFERENCES


1. T. Suzuki, R. Fujimoto and T. Otsuka, *Phys. Rev.* C **67**, 044302 (2003).
2. T. Otsuka, T. Suzuki, R. Fujimoto, R. Grawe and Y Akaishi, *Phys. Rev. Lett* **95**, 232502 (2005).
3. T. Suzuki, S. Chiba, T. Yoshida, T. Kajino and T. Otsuka, *Phys. Rev.* C **74**, 034307 (2006).
4. T. E. Drake, E. L. Tomusiak and H. S. Caplan, *Nucl. Phys.* **A118**, 138 (1968);
   A. Yamaguchi, T. Terasawa, K. Nakahara and Y. Torizuka, *Phys. Rec.* C **3**, 1750 (1971).
5. D.J. Millener and D. Kurath, *Nucl. Phys.* **A255**, 315 (1975).
6. S. E. Woosley, D. H. Hartmann, R. D. Foffman and W. C. Haxton, *Astrophys. J.* **356**, 272 (1990).
7. S. E. Warburton and B. A. Brown, *Phys. Rev.* C **46**, 923 (1992).
8. OXBASH, B. A. Brown, A. Etchegoyen and W. D. M. Rae, *MSU Cyclotron Laboratory Report* No. 524 (1986).
9. T. Yoshida et al., *Phys. Rev. Lett* **96**, 091101 (2006); ibid. **94**, 231101 (2005); *Astrophys. J.* **649**, 319 (2006).
10. M. Honma et al., *Phys. Rev.* C **65**, 061301 (2002); ibid. C **69**, 034335 (2004); *Journal of Physics: Conference Series* **20**, 7 (2005).
11. A. Poves, J. Sánchez-Solano, E. Caurier and F. Nowacki, *Nucl. Phys.* **A694**, 157 (2001).
12. J. Rapaport et al., *Phys. Rev.* C **24**, 335 (1981).
13. T. Yoshida, H. Umeda and K. Nomoto, submitted to *Astrophys. J.*

# Spectroscopy of Very Heavy Elements

P.T. Greenlees*, R.-D. Herzberg†, S. Ketelhut*, D. Ackermann**, P.A. Butler†, A. Chatillon‡,**, S. Eeckhaudt*, B.J.P. Gall§, T. Grahn*,†, C. Gray-Jones†, A. Görgen‡, F.-P. Heßberger**, U. Jakobsson*, G.D. Jones†, P. Jones*, R. Julin*, S. Juutinen*, T.-L. Khoo¶, W. Korten‡, M. Leino*, S. Moon†, M. Nyman*, J. Pakarinen*,†, P. Papadakis†, P. Peura*, P. Rahkila*, M. Richer§, D. Rostron†, C. Santos§, J. Sarén*, C. Scholey*, J. Sorri*, S.K. Tandel‖, Ch. Theisen‡, J. Uusitalo* and M. Venhart††

*Department of Physics, University of Jyväskylä, FIN-40014 Jyväskylä, Finland
†Department of Physics, University of Liverpool, Oxford Street, Liverpool, L69 7ZE, U.K.
**GSI, D-64291 Darmstadt, Germany
‡DAPNIA/SPhN, CEA-Saclay, F-91191 Gif-sur-Yvette Cedex, France
§Institut Pluridisciplinaire Herbert Curien, F-67037 Strasbourg, France
¶Argonne National Laboratory, Argonne, Illinois, IL 60439, U.S.A.
‖Department of Physics, University of Massachusetts Lowell, Lowell, MA 01854, U.S.A.
††Comenius University, Bratislava, Slovakia

**Abstract.** Advances in spectrosopic techniques have meant that heavy nuclei in the upper right-hand corner of the nuclear chart have become amenable to detailed study. This detailed spectroscopic data can provide a stringent test for current nuclear structure theories. Experiments to investigate the structure of nuclei in the region of $^{254}$No can yield information concerning moments of inertia, stability against fission with rotation, single-particle properties, excitation energies of two quasi-particle states, and so on. A brief overview of the techniques used and recent results from studies in the region of $^{254}$No are presented, along with a summary of future developments which will allow further advances to be made.

**Keywords:** Superheavy Elements, Recoil-Decay Tagging, Nuclear Structure
**PACS:** 27.90.+b, 21.10.-k, 29.30.-h, 23.35.+g

## INTRODUCTION

The pursuit of new nuclei at the upper end of the known nuclear chart has always been one of the major goals of nuclear physics. Coulomb repulsion between the large number of protons in these nuclei means that they should not exist, and their stability is solely due to the shell-correction energy. A detailed understanding of this shell correction (arising from gaps in the single-particle spectrum) is therefore desirable in order to test current nuclear structure models. Constraint of such models may also contribute to discussion on the theoretical location of the next closed proton- and neutron- shells beyond $^{208}$Pb (see e.g. ref. [1] for a comparison of theoretical approaches). Experiments which aim to produce superheavy elements directly may only produce a small number of atoms for study, which precludes the extraction of detailed spectroscopic data [2, 3, 4], though information concerning decay modes and half-lives can be extracted. Favored $\alpha$ decays connect states of similar spin and parity, and in the case of odd-mass nuclei,

CP1012, *Frontiers in Nuclear Structure, Astrophysics, and Reactions: FINUSTAR 2*
edited by P. Demetriou, R. Julin, and S. V. Harissopulos
© 2008 American Institute of Physics 0-7354-0532-5/08/$23.00

the presence of spin isomers may "mask" the decay path to the ground state. In recent years an alternative approach to study the heaviest elements has been to produce nuclei with lower $Z$, which can be produced in significant quantities. This then means that improved spectroscopic data can be obtained. This contribution focuses on recent in-beam measurements of even-even nuclei in the region of $^{254}$No, mainly in experiments carried out at the Department of Physics of the University of Jyväskylä (JYFL). To date, there are relatively few in-beam measurements of odd-mass nuclei in the region. A contribution concerning experiments to study the odd-proton nuclei $^{251}$Md and $^{255}$Lr can be found elsewhere in these proceedings [5, 6]. The only other case so far studied is $^{253}$No [7, 8]. For a recent review of detailed decay spectroscopic measurements using $\alpha$-$\gamma$ and $\alpha$-electron coincidence techniques, see ref. [9].

## GROUND-STATE BANDS OF EVEN-EVEN NUCLEI

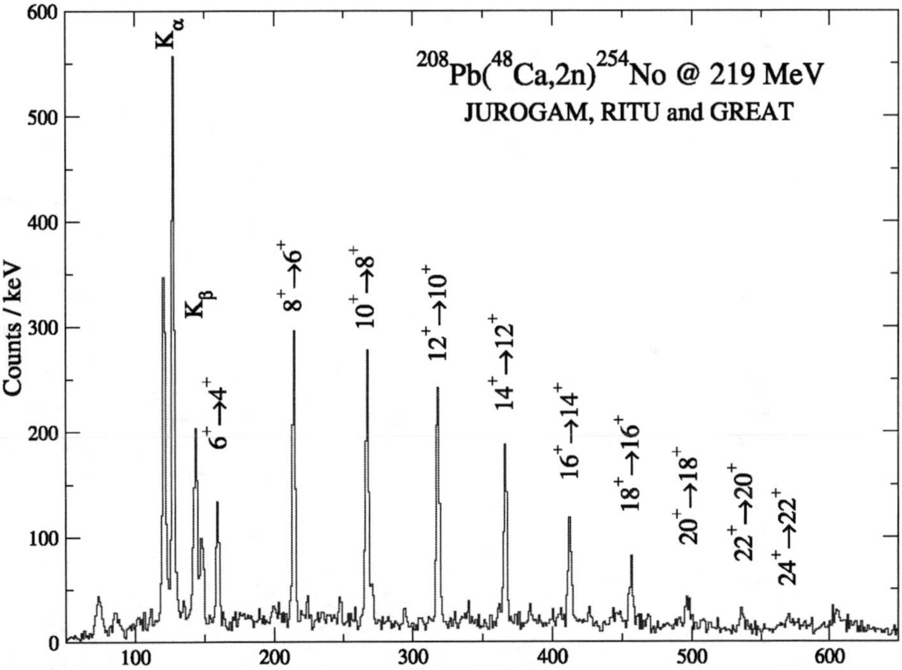


**FIGURE 1.** Recoil-gated $\gamma$ ray singles spectrum of $^{254}$No. The reaction used is shown in the figure. Data from two experiments have been combined to create the spectrum.

The ground-state band of $^{254}$No was first measured in an experiment carried out at the Argonne National Laboratory (ANL) using the GAMMASPHERE array of germanium detectors coupled to the Fragment Mass Analyzer (FMA) (see ref. [10] and references therein). The experiment employed the recoil-decay tagging technique, whereby only those $\gamma$ rays associated with fusion-evaporation residues detected at the focal plane of the separator are accepted [11, 12]. These first measurements allowed the ground-state band to be established up to a spin of $14\hbar$, and showed that $^{254}$No was indeed deformed. Subsequent measurements carried out at JYFL using the RITU gas-filled separator [13] coupled to the JUROGAM array of germanium detectors and the GREAT focal plane spectrometer [14] extended the ground-state band and provided the first evidence for the observation of non-yrast states [15, 16]. The most recent spectrum to be produced for $^{254}$No is shown in figure 1. The spectrum shows recoil-gated $\gamma$ ray singles produced in two experiments carried out using the reaction shown in the figure. The beautiful rotational band structure can be clearly seen up to a spin of $24\hbar$, along with other weaker peaks which have not yet been placed in the level scheme. In addition to $^{254}$No, studies have been made in JYFL of the ground-state bands of $^{252}$No and $^{250}$Fm [17, 18]. These studies showed that the dynamical moment of inertia for the $N$=150 isotones increases more rapidly than that for $^{254}$No, with $N$=152. This difference can be related to the alignment of proton $i_{13/2}$ and neutron $j_{15/2}$ orbitals. An experimental plot of the systematic behavior of the moment of inertia for several nuclei in the region can be found in ref. [19]. A number of theoretical papers dealing with the moments of inertia of these ground-state bands can also be found [20, 21, 22, 23].

## Ground-state band of $^{248}$Fm

The latest in this series of experiments was dedicated to the study of $^{248}$Fm. The experiment was carried out at JYFL and again employed the JUROGAM, RITU and GREAT devices. Data were readout using the Total Data Readout (TDR) system and analyzed using the GRAIN analysis package [24, 25]. A recoil-decay tagged spectrum of $\gamma$ rays obtained through the $^{202}$HgS($^{48}$Ca,2n)$^{248}$Fm reaction at a bombarding energy of 213 MeV is shown in figure 2.

Due to technical problems, the level of statistics obtained was much lower than that expected. The spectrum shown corresponds to the observation of approximately 700 $\alpha$ decays. Despite these problems, the ground-state band can be observed up to a spin of $16\hbar$. Further experiments are planned to expand this study and to confirm the observation of a new isomer in $^{248}$Fm for which evidence was also found [26].

## STUDIES OF K-ISOMERISM

Over the past few years, a number of experiments have been carried out at JYFL, ANL and at GSI to investigate the decay of isomeric states in $^{252,254}$No and in $^{250}$Fm [27, 28, 29, 30, 31]. The existence of isomers in $^{254}$No and $^{250}$Fm was postulated in the 1970's, though the experiments carried out did not allow a determination of the spins or of the de-excitation path to the ground state [32]. The nuclei in question are deformed and have a high proton number, Z. These facts lead to the possibility that the isomeric states decay via a cascade of low-energy transitions which are strongly converted, resulting in the emission of a number of low-energy electrons. It was suggested by Jones that as the nuclei decay whilst being implanted in a silicon detector, the energies of these electrons should sum, giving a calorimetric signal which can be measured above

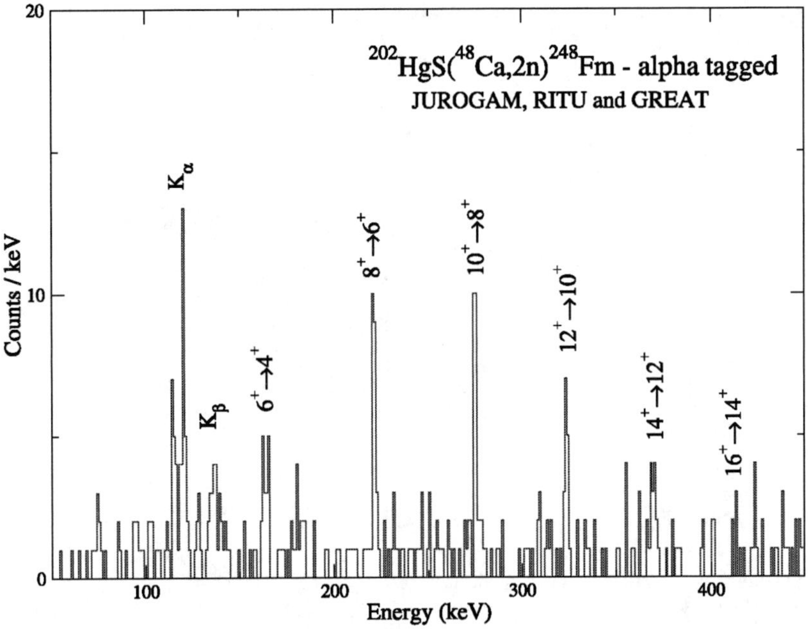


**FIGURE 2.** Spectrum of $\gamma$ rays detected in the JUROGAM array when a fusion-evaporation product is detected at the focal plane of RITU and followed by a $^{248}$Fm $\alpha$ decay at the same position within 120 seconds.

threshold [33]. The signal can then be used as a "tag" to extract the half-life of the isomer and for any coincident $\gamma$ rays detected at the focal plane of the separator. After more than thirty years, the new technique allowed the isomers in $^{254}$No and $^{250}$Fm to be confirmed. The half-life measurements are in remarkable agreement with the values from the pioneering experiment of Ghiorso. In $^{254}$No, the $K^\pi$=8$^-$ isomer decays via a $K^\pi$=3$^+$ band to the ground-state rotational band. By extracting $M1/E2$ transition intensity ratios, it was possible to unambiguously assign the $K^\pi$=3$^+$ band head to the $\pi$[521]1/2$^-\otimes\pi$[514]7/2$^-$ proton two-quasiparticle configuration. This in turn allowed the $K^\pi$=8$^-$ isomeric state to be assigned the configuration $\pi$[514]7/2$^-\otimes\pi$[624]9/2$^+$ [29, 30]. In $^{250}$Fm and $^{252}$No, the $K^\pi$=8$^-$ isomers are deduced to be neutron two-quasiparticle states, with the configuration $\nu$[624]7/2$^+\otimes\nu$[734]9/2$^-$. In both cases, the isomer decays via a low-lying $K^\pi$=2$^-$ band [31, 27, 28]. Studies using various other techniques ($\beta$ and EC decays, $\alpha$ decay and spontaneous fission studies) have allowed a number of $K$-isomers to be identified over a rather large range of $Z$ and $N$. The properties of these known isomers are summarized in Table 1. In many cases the assignments are tentative and are not based on unambiguous experimental data (see relevant references for details).

**TABLE 1.** Table of known $K$-isomers in the heavy and superheavy elements. In many cases the $K^{\pi}$ or configuration assignments are tentative and have not been made on the basis of unambiguous experimental data. See relevant references for details.

| Nucleus | $K^{\pi}$ | $T_{1/2}$ | $E_x$ | Decay Mode | Configuration | Reference |
|---|---|---|---|---|---|---|
| $^{270}$Ds | $9^-,10^-$ | 6 ms | $\simeq$1.13 MeV | $\alpha$ | $\nu[725]11/2^- \otimes \nu[613]7/2^+$<br>$\nu[725]11/2^- \otimes \nu[615]9/2^+$ | [34] |
| $^{254}$No | $8^-$ | 266 ms | 1.293 MeV | $\gamma$ | $\pi[514]7/2^- \otimes \pi[624]9/2^+$ | [29, 30] |
| $^{252}$No | $8^-$ | 110 ms | 1.254 MeV | $\gamma$ | $\nu[624]7/2^+ \otimes \nu[734]9/2^-$ | [27, 28] |
| $^{250}$No | $6^+$ | 42 $\mu$s | - | SF,$\gamma$? | $\nu[622]5/2^+ \otimes \nu[624]7/2^+$ | [35] |
| $^{256}$Fm | $7^-$ | 70 ns | 1.425 MeV | $\gamma$,SF | $\pi[633]7/2^+ \otimes \pi[514]7/2^-$ | [36] |
| $^{250}$Fm | $8^-$ | 1.92 s | 1.195 MeV | $\gamma$ | $\nu[624]7/2^+ \otimes \nu[734]9/2^-$ | [31] |
| $^{248}$Fm | - | $\simeq$8 ms | - | $\gamma$ | - | [26] |
| $^{246}$Cm | $8^-$ | - | 1.179 MeV | $\gamma$ | $\nu[624]7/2^+ \otimes \nu[734]9/2^-$ | [37] |
| $^{244}$Cm | $6^+$ | 34 ms | 1.040 MeV | $\gamma$ | $\nu[622]5/2^+ \otimes \nu[624]7/2^+$ | [38] |

## SYSTEMATICS OF TWO-QUASIPARTICLE STATES

The data shown in Table 1, along with further data on non-yrast states for Cf isotopes taken from ref. [39] allow limited systematics for the $N$=150 and $N$=152 isotones to be built up. A plot of the level energies of the $K^{\pi}=8^-$ states and low-lying non-yrast states is shown in figure 3.

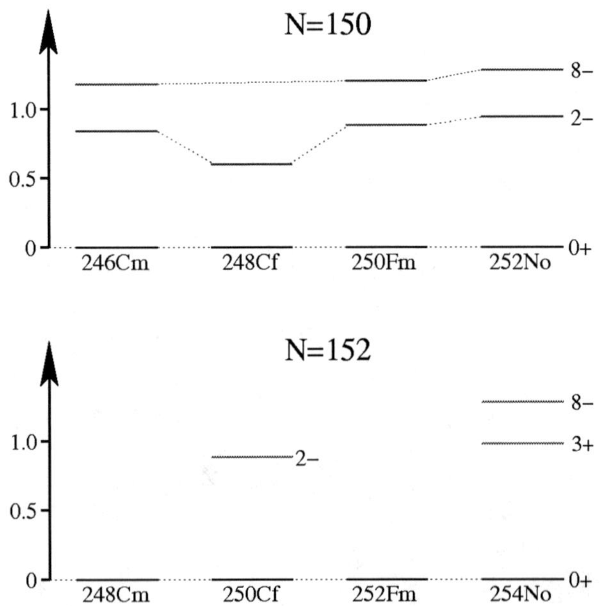


**FIGURE 3.** Systematics of known two-quasiparticle and non-yrast states in the $N$=150 and $N$=152 isotones.

As can be seen, the systematics for the $N$=152 isotones are somewhat lacking compared to those for $N$=150. It is unfortunate that the $K^{\pi}$=$8^-$ state is not known in $^{248}$Cf. For the $N$=150 isotones, the $8^-$ states are at around 1.2 MeV for all isotones, indicating that they are rather pure, whilst the energies of the $K^{\pi}$=$2^-$ states vary somewhat. This is especially pronounced in $^{248}$Cf where the energy of the $2^-$ state is very low. It is difficult to reproduce the energies of the $2^-$ states in terms of pure two-quasiparticle configurations and the bands must be brought to lower energies by collective effects. In $^{254}$No, the energies of the $3^+$ and $8^-$ states have been reasonably well reproduced by calculations based on a Woods-Saxon potential [30]. As of yet, there are few self-consistent calculations of the energies of these two-quasiparticle states. A comprehensive set of Hartree-Fock-Bogoliubov calculations with the Gogny interaction can be found in ref. [40], and further calculations based on Skyrme interactions are expected in the near future [41]. One general remark is that so far the calculations do not reproduce particularly well the energies of the $8^-$ states, a discrepancy which can in part be traced back to the level ordering in the single-particle spectrum [41]. It is hoped that detailed analysis and discussion of these discrepancies will be forthcoming in the near future.

## FUTURE DEVELOPMENTS

One of the major problems in $\gamma$-ray spectroscopic measurements of high-$Z$ nuclei is the competition from internal conversion in the decay of low-energy transitions. In the nobelium isotopes, the internal conversion coefficient for a 100 keV $E2$ transition is 31.6, meaning that the transition is about 97% converted. This can be seen from the $\gamma$-ray spectra shown in figures 1 and 2, where the transitions below the $6^+$ to $4^+$ transition cannot be observed. The problem is exasperated in odd-mass nuclei where a considerable fraction of the decay intensity may go through $M1$ transitions, which are even more strongly converted. In order to try and overcome these problems, a new spectrometer known as SAGE will be constructed at JYFL. The device will be built as a collaborative effort between the University of Liverpool, Daresbury Laboratory and JYFL, with the majority of funding coming from the U.K. EPSRC. The spectrometer (shown schematically in figure 4) is a combination of the JUROGAM2 germanium detector array and a conversion electron spectrometer centered around a magnetic solenoid. The JUROGAM2 array consists of 24 Clover detectors and up to 15 Phase-I type detectors, and will increase the efficiency for $\gamma$-ray detection from the 4.2% of JUROGAM to around 6%. The magnetic solenoid design is based on experience gained with the SACRED electron spectrometer, which was coupled successfully to RITU and used to perform conversion electron spectroscopy of $^{254}$No and $^{250}$Fm [42, 43, 18]. A further major change in the spectrometer will be to instrument all germanium and silicon detector channels with digital electronics. It is hoped that this will allow higher count rates to be used, which in turn means that higher beam intensities can be tolerated, giving higher yield of the nuclei of interest in a given experiment time.

In a broader perspective, the development of the highly efficient $\gamma$-ray tracking arrays AGATA and GRETA coupled to recoil-separators, should allow a new regime of production cross-section to be reached. Present devices are capable of performing measurements at the level of tens of nanobarns, so it is not too great a leap to predict that in-beam

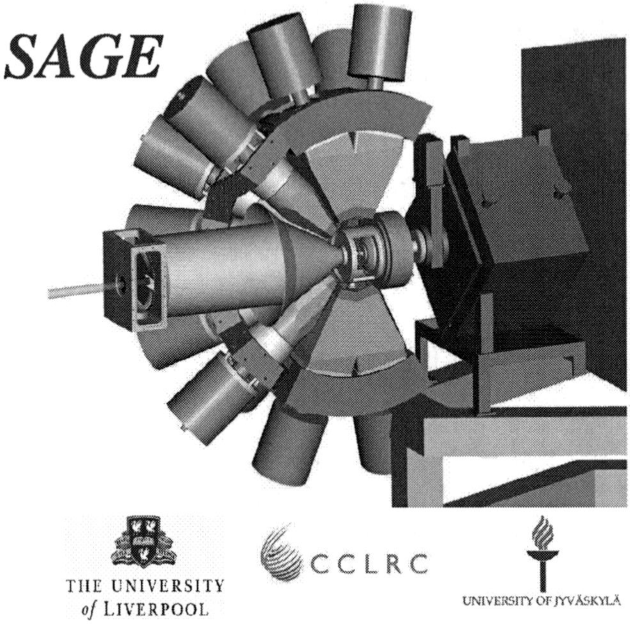


**FIGURE 4.** Schematic drawing of the SAGE spectrometer. SAGE is a collaborative effort between the University of Liverpool, Daresbury Laboratory and JYFL. The device will be constructed in JYFL in 2008. Drawing courtesy of D.Seddon, University of Liverpool.

spectroscopic measurements at the nanobarn level will soon be achievable. The advent of next generation radioactive beam facilities may also allow an extension of studies in this region, for example by populating nuclei over a wider range of neutron number which are otherwise inaccessible. The required intensities are not too far removed from those predicted for the most intense beams at SPIRAL2 and EURISOL.

## ACKNOWLEDGMENTS

This work was supported by the Academy of Finland under the Finnish Center of Excellence Programme 2006-2011, the UK Engineering and Physical Sciences Research Council (EPSRC) and the European Union 6th Framework Programme "Integrating Infrastructure Initiative - Transnational Access", No. 506065 (EURONS). We also thank the U.K./France (EPSRC/IN2P3) detector Loan Pool and EUROBALL owners committee ($\gamma$-pool network) for the EUROGAM detectors of JUROGAM. We are indebted to GSI for the loan of two large volume segmented Clover detectors for several experiments.

# REFERENCES


1. M. Bender *et al.*, *Phys. Rev. C* **60**, 034304 (1999).
2. Yu. Ts. Oganessian *et al.*, *Eur. Phys. J. A* **25**, Supplement 1, 589 (2005).
3. K. Morita *et al.*, *J. Phys. Soc. Jpn.* **73**, 2593 (2004).
4. S. Hofmann *et al.*, *Nucl. Phys. A* **734**, 93 (2004).
5. S. Ketelhut *et al.*, *These proceedings.*
6. A. Chatillon *et al.*, *Phys. Rev. Lett.* **98**, 132503 (2007).
7. P. Reiter *et al.*, *Phys. Rev. Lett.* **95**, 032501 (2005).
8. R.-D. Herzberg *et al.*, *To be published.*
9. M. Leino and F.-P. Heßberger, *Ann. Rev. Nucl. Part. Sci.* **54**, 175 (2004).
10. P. Reiter *et al.*, *Phys. Rev. Lett.* **82**, 509 (1999).
11. R. S. Simon *et al.*, *Z. Phys. A* **354**, 229 (1986).
12. E. S. Paul *et al.*, *Phys. Rev. C* **51**, 78 (1995).
13. M. Leino *et al.*, *Nucl. Instr. and Meth. B* **99**, 653 (1995).
14. R.D. Page *et al.*, *Nucl. Instr. and Meth. B* **204**, 634 (2003).
15. M. Leino *et al.*, *Eur. Phys. J. A* **6**, 63 (1999).
16. S. Eeckhaudt *et al.*, *Eur. Phys. J. A* **26**, 227 (2005).
17. R.-D. Herzberg *et al.*, *Phys. Rev. C* **65**, 014303 (2002).
18. J. E. Bastin *et al.*, *Phys. Rev. C* **73**, 024308 (2006).
19. P. T. Greenlees, *AIP Conf. Proc.* **853**, 416 (2006).
20. T. Duguet, P. Bonche and P.-H. Heenen, *Nucl. Phys. A* **679**, 427 (2001).
21. M. Bender, P. Bonche, T. Duguet and P.-H. Heenen, *Nucl. Phys. A* **723**, 354 (2003).
22. A. V. Afanasjev *et al.*, *Phys. Rev. C* **67**, 024309 (2003).
23. J. L. Egido and L. M. Robledo, *Phys. Rev. Lett.* **85**, 1198 (2000).
24. I. H. Lazarus *et al.*, *IEEE Trans. Nucl. Sci.* **48**, 567 (2001).
25. P. Rahkila, *To be published.*
26. R.-D. Herzberg *et al.*, *To be published.*
27. B. Sulignano *et al.*, *Eur. Phys. J. A* **33**, 327 (2007).
28. T. L. Khoo *et al.*, *To be published.*
29. R.-D. Herzberg *et al.*, *Nature* **442**, 896 (2006).
30. S. K. Tandel *et al.*, *Phys. Rev. Lett.* **97**, 082502 (2006).
31. P. T. Greenlees *et al.*, *Nucl. Phys. A* **787**, 507c (2007) and *To be published.*
32. A. Ghiorso *et al.*, *Phys. Rev. C.* **7**, 2032 (1973).
33. G. D. Jones, *Nucl. Instr. and Meth. A* **488**, 471 (2002).
34. S. Hofmann *et al.*, *Eur. Phys. J. A* **10**, 5 (2001).
35. D. Peterson *et al.*, *Phys. Rev. C.* **74**, 014316 (2006).
36. H. L. Hall *et al.*, *Phys. Rev. C.* **39**, 1866 (1989).
37. L. G. Multhauf *et al.*, *Phys. Rev. C.* **13**, 771 (1976).
38. R. W. Hoff *et al.*, *Phys. Rev. C.* **29**, 618 (1984).
39. R. B. Firestone and V. S. Shirley, eds., *Table of Isotopes* 8th edition, Wiley, New York, 1996.
40. J.-P. Delaroche, M. Girod, H. Goutte and J. Libert, *Nucl. Phys. A* **771**, 103 (2006)
41. P.-H. Heenen, *Private Communication*
42. H. Kankaanpää *et al.*, *Nucl. Instr. and Meth. A* **534**, 503 (2004).
43. P. A. Butler *et al.*, *Phys. Rev. Lett.* **89**, 202501 (2002).

# Spectroscopy of heavy elements at Dubna

O. Dorvaux[1], A. Lopez-Martens[2], K. Hauschild[2], A.V. Yeremin[3], A. Khouaja[1], A.V. Belozerov[3], Ch. Briançon[2], M.L. Chelnokov[3], V.I. Chepigin[3], D. Curien[1], P. Désesquelles[2], B. Gall[1], V.A. Gorshkov[3], M. Guttormsen[4], F. Hanappe[5], A.P. Kabachenko[3], F. Khalfallah[1], A. Korichi[2], A.C. Larsen[4], O.N. Malyshev[3], A. Minkova[6], Yu.Ts. Oganessian[3], A.G. Popeko[3], M. Rousseau[1], N. Rowley[1], R.N. Sagaidak[3], S. Sharo[7], A.V. Shutov[3], S. Siem[4], V.I. L. Stuttgé[1], A.I.Svirikhin,[3], N.U.H. Syed[4], and Ch. Theisen[8]

[1]*IPHC-DRS/ULP, $IN_2P_3$-CNRS, UMR 7178, F-67037 Strasbourg, France.*
[2]*CSNSM, $IN_2P_3$-CNRS, UMR 8609, F-91405 Orsay, France.*
[3]*FLNR, JINR, Dubna, Russia.*
[4]*Department of Physics, Oslo University, 0316 Oslo, Norway.*
[5]*Université Libre de Bruxelles, 1650 Bruxelles, Belgium.*
[6]*Faculty of Physics, University of Sofia, 1164 Sofia and INRNE, Bas, Bulgaria.*
[7]*Department of Physics, Comenius Unversity, SK-84215, Bratislava, Slovakia.*
[8]*CEA_Saclay, DSM/DAPNIA/SPhN, 91191 gif-sur-Yvette, France.*

**Abstract..** In 2004, the GABRIELA (Gamma Alpha Beta Recoil Investigation with the Electromagnetic Analyzer VASSILISSA) collaboration started a twofold scientific program at the FLNR (Dubna) laboratory : (a) the systematic study of the behaviour of single particle states within isotopic and isotonic chains with Z=100-104 as N varies from 152 to 162 and (b) the study of isotopes for which very little spectroscopic information is known. After discussing the Physics motivations and the experimental setup, some new results concerning the structure of $^{249}$Fm and $^{253}$No nuclei, and some preliminary results on the decay of $^{217}$Pa decay are presented.

**PACS:** 21.10.-k, 23-20.Lv, 23.20.Nx, 23.60.+e , 23.90.+W, 21.10.Hw , 27.90.+b

## SCIENTIFIC MOTIVATIONS

It is well established now that nuclei beyond Fm (Z=100) are stabilized only by shell effects and therefore predictions of their shape, lifetime, and structure strongly depend on the parameters of model calculations. Indeed, the prediction of the position of the next double magic gap after $^{208}$Pb, the Very Heavy Element (VHE) region, remains uncertain and controversial as the Nilsson-Strutinsky, the Hartree-Fock and the Relativistic Mean Field calculations predict different position for this new island of stability (Z=114, 120 or 126 for the next proton shell, and N=172 or 184 for the next neutron shell) [1-4]. These divergences mainly come from the way the spin-orbit interaction is treated. Moreover, the various parameterisations of the nuclear effective force or nuclear potential are obtained from the "known" region of stability. Therefore, calculations in the VHE region represent a big extrapolation. Since very

CP1012, *Frontiers in Nuclear Structure, Astrophysics, and Reactions: FINUSTAR 2*
edited by P. Demetriou, R. Julin, and S. V. Harissopulos
© 2008 American Institute of Physics 0-7354-0532-5/08/$23.00

deformed nuclei around $^{254}$No (Z=102) sample single particle orbitals, which lie close the shell gap for the VHE nuclei, it becomes crucial to determine the ordering and relative excitation energies of these single-particle. This is what motivated the GABRIELA (Gamma Alpha Beta Recoil Investigation with the Electromagnetic Analyzer VASSILISSA) collaboration to launch in 2004 a scientific program at the FLNR (Dubna) laboratory based on the systematic study of the behaviour of single particle states within isotopic and isotonic chains with Z=100-104 as N varies from 152 to 162, the FLNR laboratory being a unique place for nuclear reactions with high intensity beams and radioactive targets favourable for this kind of spectroscopic studies.

## THE GABRIELA EXPERIMENTAL SETUP

The GABRIELA experimental setup and its performances are fully described in ref. [5]. It is installed at the focal plane of the electrostatic analyzer VASSILISSA [6-7] and is composed of a new Aluminium vacuum chamber, a 60x60 mm$^2$ 16-strip position sensitive 300 μm thick Si implantation detector for recoil detection and alpha decay measurements surrounded in the backward direction by four 50x50 mm$^2$ segmented 500 μm thick Si detectors disposed in a tunnel configuration for conversion electron detection. Around the implantation detector, six Ge phase I detectors from the French-UK loan pool are disposed in a ring with their anti-Compton shields. A seventh detector, improved by shortening the distance between the Ge crystal and the Aluminium end cap, is placed upstream from the implantation detector. The total γ-ray detection efficiency of the Ge array (visible in fig. 1) is about 17% at 200 keV.

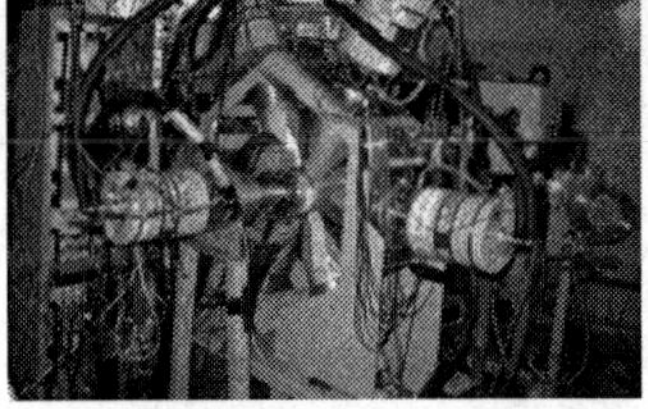

**FIGURE 1.** Photograph of the GABRIELA setup at the focal plane of VASSILISSA

## DATA ANALYSIS

### The $^{249}$Fm nucleus

Excited states in $^{249}$Fm were populated via the α decay of $^{253}$No, which was produced in the reaction $^{207}$Pb($^{48}$Ca, 2n). A total of 10,500 α particles from the decay of $^{253,254}$No were detected during the experiment. A real effort has been carried out on the γ-ray and conversion electron spectra analyses (see fig. 2) and has permitted to identify the first members of the ground-state rotational band as well as a new excited $5/2^-$ state at 211 keV excitation energy. Internal conversion coefficients were measured for most of the observed transitions and they suggest that the 279-keV transition de-exciting the $9/2^-$ state has anomalous E1 conversion coefficients [8].

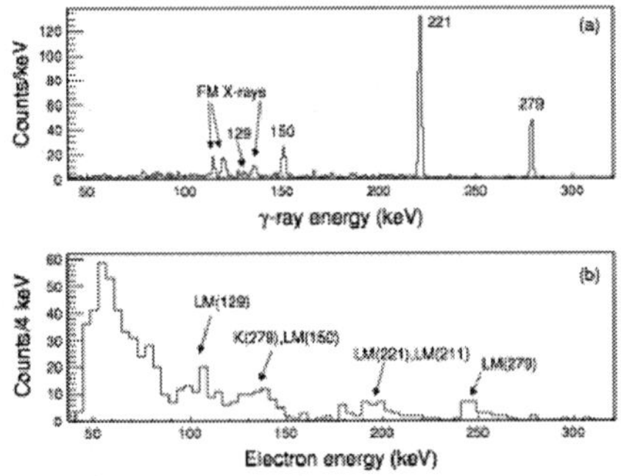


**FIGURE 2.** (a)Compton-suppressed energy spectrum of γ rays detected in prompt coincidence with the $^{253}$No α decay. (b) Energy spectrum of conversion electrons following the α decay of $^{253}$No.

## The $^{253}$No nucleus

Isomeric states are also very revealing of the nuclear structure in the Z~100 region since their properties depend on the nature and ordering of single-particle levels, on the shape of the nucleus and the quasi-particle configuration. Recently, a $K=8^-$ isomeric state has been established in the $^{254}$No nucleus as a two-proton state [9]. Other long-lived isomeric states have been highlighted in ref. [10-13].

In the same data set from the $^{207}$Pb($^{48}$Ca, 2n) reaction, previously mentioned, isomeric states study of $^{253}$No has been undertaken [14]. It has revealed, in addition of the confirmation of a 31 μs isomer reported more than 30 years ago decaying through a M2 transition to the ground state of $^{253}$No, a new long-lived isomeric state of about 0.9 ms (see fig. 3). Unfortunately because of a lack of statistics, no spectroscopic information could have been obtained.

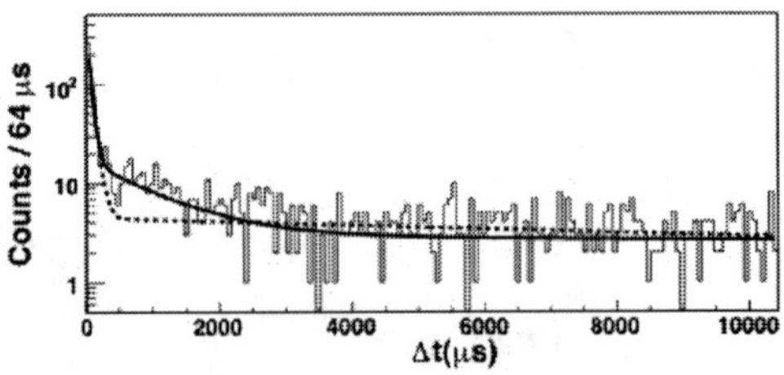


**FIGURE 3.** Difference in time between the detection of recoils and electrons. The dashed-line represents a fit when only one isomeric state is considered; the solid line represents the case when a second isomeric state is taken into account.

## The $^{217}$Pa nucleus

For isotopes of elements at the limit of β stability, α spectroscopy is one of the most important tools to obtain information on the nuclear structure of the daughter nuclei. In the neutron deficient Pa isotope region, very few information is known and in particular for the $^{217}$Pa nucleus.

The reaction $^{181}$Ta($^{40}$Ar, 4n)$^{217}$Pa has been studied with the GABRIELA setup. The preliminary coincidence spectra between alpha-decay and prompt γ-ray or electron emission [15] are shown in figure 4. All α transitions are clearly identified with some improved lifetime and internal conversion electron spectra may help clarify and improve the level scheme of ref. [16]. The analysis is in progress.

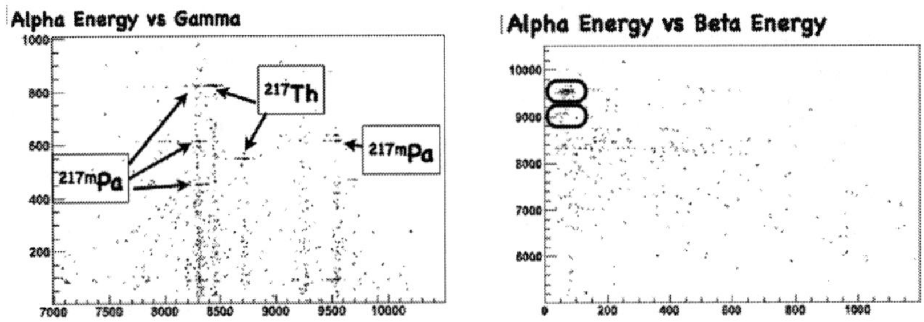


**FIGURE 4.** α-γ energy correlation (left panel) and α-conversion electron correlation (right panel).

## CONCLUSIONS AND PERSPECTIVES

Some new results in the field of spectroscopic studies of very heavy elements in the Fermium region have been presented and have shown the sensitivity of the GABRIELA setup placed at the focal plane of the recoil separator VASSILISSA at the FLNR laboratory. This setup will be improved mainly by the modernization of VASSILISSA (cf. A. Popeko poster contribution) in a way to get a dedicated experimental setup to Transfermium spectroscopy elements using radioactive targets. Some efforts will be focused on the optimization of the transmission efficiency for asymmetric reactions using actinides targets such as $^{22}$Ne+$^{242}$Pu as well as for symmetric reactions.

## ACKNOWLEDGMENTS


The GABRIELA project is jointly funded by JINR (Russia) and $IN_2P_3$/CNRS (France). Work at FLNR was performed partially under the financial support of the Russian Foundation for Basic Research, contract N 05-02-16198 and JINR - BMBF (Germany), JINR - Polish, JINR - Slovak Cooperation Programmes. Financial support from the Norwegian Research Council and from the Bulgarian National Foundation, contract MON-204/06, is also acknowledged.

## REFERENCES


1. S. Cwiok et al.., Nucl. Phys. **A611,** 211 (1996).
2. M. Bender et al.., Phys. Rev. **C60,** 034304 (1999).
3. A.T. Kruppa et al.., Phys. Rev. **C61**, 034313 (2000).
4. M. Bender et al., Phys. Lett. **B515,** 42 (2001).
5. K. Hauschild et al., Nucl. Instr. Meth. **A560**, 388 (2006).
6. A.V. Yeremin et al., Nucl. Instr. Meth. **B126,** 329 (1997).
7. O.N. Malyshev et al., Nucl. Instr. Meth. **A440,** 86 (2000).
8. A. Lopez-Martens et al., Phys. Rev. **C74**, 044303 (2006).
9. R.-D. Herzberg et al.., Nature **442**, 896 (2006).
10. F.P. Hessberger et al.., Z. Phys. **A359**, 415 (1997).
11. F.P. Hessberger et al.., E. Phys. J **A12**, 57 (2001).
12. F.P. Hessberger et al.., E. Phys. J **A22**, 417 (2004).
13. A. Chatillon et al.., E. Phys. J **A30**, 397 (2006).
14. A. Lopez-Martens et al., E. Phys. J **A32,** 245 (2007).
15. A. Khouaja (private communication).
16. F.P. Hessberger et al.., E. Phys. J **A15,** 335 (2002).

# Confirmation Of Super Heavy Element Production In $^{48}$Ca Induced Fusion Reactions A Handshake Of Physics And Chemistry For Element 112


**SHIP Collaboration**

S. Hofmann[1,2], D. Ackermann[1], S. Antalic[3], H.G. Burkhard[1], V.F. Comas[4], R. Dressler[5], Z. Gan[6], S. Heinz[1], J.A. Heredia[4], F.P. Heßberger[1], J. Khuyagbaatar[1], B. Kindler[1], I. Kojouharov[1], P. Kuusiniemi[7], M. Leino[8], B. Lommel[1], R. Mann[1], G. Münzenberg[1], K. Nishio[9], A.G. Popeko[10], S. Saro[3], H.J. Schött[1], B. Streicher[3], B. Sulignano[1], J. Uusitalo[8], M. Venhart[3], A.V. Yeremin[10]

*[1] Gesellschaft für Schwerionenforschung (GSI), D-64291 Darmstadt, Germany*
*[2] Institut für Kernphysik, Johann Wolfgang Goethe-Universität, D-60438 Frankfurt, Germany*
*[3] Department of Nuclear Physics and Biophysics, Comenius University, SK-84248 Bratislava, Slovakia*
*[4] InSTEC, Habana 10400, Cuba*
*[5] Paul Scherrer Institute, CH-5232 Villigen, Switzerland*
*[6] Institute of Modern Physics, Chinese Academy of Sciences, Lanzhou 730000, China*
*[7] CUPP, University of Oulu, FIN-90014 Oulu, Finland*
*[8] Department of Physics, University of Jyväskylä, FIN-40351, Finland*
*[9] Japan Atomic Energy Agency, Tokai, Ibaraki 319-1195, Japan*
*[10] Flerov Laboratory of Nuclear Reactions, JINR, RU-141 980 Dubna, Russia*

**112 Chemistry Collaboration**

R. Eichler[1,2], N.V. Aksenov[3], A.V. Belozerov[3], G.A. Bozhikov[3], V.I. Chepigin[3], S.N. Dmitriev[3], R. Dressler[1], H.W. Gäggeler[1,2], V.A. Gorshkov[3], F. Haenssler[2], M.G. Itkis[3], A. Laube[1], V.Ya. Lebedev[3], O.N. Malyshev[3], Yu.Ts. Oganessian[3], O.V. Petrushkin[3], D. Piguet[1], A.G. Popeko[3], P. Rasmussen[1], S.V. Shishkin[3], A.V. Shutov[3], A.I. Svirikhin[3], E.E. Tereshatov[3], G.K. Vostokin[3], M. Wegrzecki[4], A.V. Yeremin[3]

*[1] Paul Scherrer Institute, CH-5232 Villigen, Switzerland*
*[2] Department of Chemistry and Biochemistry, University of Bern, CH-3012 Bern, Switzerland*
*[3] Flerov Laboratory of Nuclear Reactions, JINR, RU-141 980 Dubna, Russia*
*[4] Institute of Electron Technology, PL-02-668 Warsaw, Poland*



**Abstract.** The production of $^{283}$112 in $^{48}$Ca induced nuclear reactions was investigated using physical and chemical separation techniques. In the reaction $^{48}$Ca on $^{238}$U, four events were registered at the SHIP velocity filter. The mean atomic mass of the evaporation residues (EVR)



CP1012, *Frontiers in Nuclear Structure, Astrophysics, and Reactions: FINUSTAR 2*
edited by P. Demetriou, R. Julin, and S. V. Harissopulos


© 2008 American Institute of Physics 0-7354-0532-5/08/$23.00

was determined to be 286 ± 10. In the $^{48}Ca$ on $^{242}Pu$ reaction, five events were found using the IVO thermochromatographic system. The observed chemical properties are typical for group 12 elements. Therefore, the atomic number is deduced to be 112. The measured decay properties of all events obtained in both experiments are in agreement with the results published from the DGFRS group. Hence, the findings bolster the production and observation of $^{283}112$, $^{287}114$, and $^{291}116$.

**Keywords:** $^{283}112$, $^{48}Ca$, nuclear reactions, chemical separation, SHIP, COLD
**PACS:** 07.75.+h, 23.60.+e, 25.70.-z, 25.85.Ca, 82.80.Bg, 82.80.Jp, 82.80.Rt

## INTRODUCTION

The discovery of relative long-lived isotopes in so-called "warm" fusion reactions using $^{48}Ca$ beams and actinide targets was one of the most important steps into the direction of the "Island of Stability" within the last 20 years. Up to now 35 new isotopes of the elements 104 to 116 and 118 are discovered at the Flerov Laboratory of Nuclear Reactions (FLNR) in Dubna (for review see [1]). They show an enhanced stability approaching Z=114 N=184, as predicted almost 40 years ago [2-4]. These results offer new possibilities not only for the physics of super heavy elements but also for investigations of their chemical properties.

In 1999 the VASSILISSA separator group announced the discovery of $^{283}112$ in the $^{48}Ca$ on $^{238}U$ reaction [5]. This isotope should undergo spontaneous fission (SF) with a half-life of about 5 min. In subsequent experiments using targets of $^{238}U$ [6] and $^{242}Pu$ [7] consistent decay properties of $^{283}112$ were observed. 2004 the Dubna gas-filled recoil separator (DGFRS) became operational for investigating "warm" fusion reactions. Experiments carried out with this device using the same nuclear reactions revealed decay properties for $^{283}112$ [8, 9] contradicting the previous claims. According to the now accepted properties $^{283}112$ decays mainly (>90%) by emission of a (9.54 ± 0.06) MeV α-particle with a half-life of $(3.8^{+1.2}_{-0.7})$ s.

The lack of critical discussions from the Dubna groups about these inconsistencies of both observations started a world-wide controversy [10-12] about the reproducibility and meaning of the results obtained at the FLNR. Later on, due to new findings at the DGFRS their results win acceptance. In a recently published review of all experiments carried out with $^{48}Ca$ Yu.Ts. Oganessian [1] declares the need for clarifying the reasons for these discrepancies. The very high importance of the discoveries of the DGFRS triggered several attempts to reproduce these findings [11, 12]. Here we present results of two recently performed experiments dealing with the confirmation of the production and of the decay properties of $^{283}112$.

## SHIP-EXPERIMENTS

The velocity filter SHIP and its detection system [13,14] is one of the most successful devices to investigate the heaviest achievable elements up to $^{277}112$.

In 2005 an experiment series was started to confirm the formation of $^{283}112$ in the 3n channel of the reaction $^{48}Ca$ on $^{238}U$. The course of irradiations was split into three parts. During the first part targets made of natural $UF_4$ were used, whereas in both other parts the targets were prepared as metallic U layers via magnetron sputtering. In the case of the $UF_4$ targets an additional carbon layer (10-20 μg/cm$^2$) was added to reduce sputtering losses. The metallic U targets show almost no material loss during the irradiations, but they are much more difficult to produce, though. All U targets were prepared on carbon backing foils (42 μg/cm$^2$).

The separation efficiency $\varepsilon_{SHIP}$ for fusion products depends on the field settings of the two quadrupole magnet triples. A Monte Carlo code was used to calculate optimized field configurations, which were tested in the first part of the experiment. It was shown that a transmission of 17% can be reached using a symmetrical quadrupole field setting, whereas an asymmetrical field setting yields up to 24%. In the case of metallic U targets, for the asymmetrical field configuration a value of 34% was determined. For a detailed description of the irradiation parameters, the experimental setup, and data evaluation see [15].

In total four events were registered, which are assigned to evaporation residues (EVR) of the reaction $^{48}$Ca on $^{238}$U. All of these events were observed at excitation energies of 34-35 MeV. A summary of the observed events is given in TABLE 1.

The total kinetic energy (TKE) of all recorded SF signals is exceptionally high, as expected for the decay of super heavy elements. The α-decay of event 4 as well as the SF fragments of the events 1, 2, and 4 were registered during the beam off period of the accelerator, which is a strong argument to assign these events to a decay of implanted nuclei. In addition, the consecutive signals of event 2 and 4 were detected in the same detector strip and the vertical positions of the EVR and decay signal vary less then 1.5 mm for all observed events as is shown in [15].

**TABLE 1.** SHIP events assigned to $^{283}$112 ($E^*_{cot}{}^a$ = 34.6 MeV)

| Event no | Date Time | $E_{EVR}{}^b/E_\alpha/E_{TKE}$ [MeV] | A$^c$ | $n_\gamma$ | strip/box | Life-time [s] |
|---|---|---|---|---|---|---|
| EVR | 08.05.200 | 31.5 | 28 | 0 | 9 | |
| 1 | 5 | $241^{+15}_{-15}$ | 6 | 2 | 9/13 | 7.6 |
| SF | 22:45 | | | | | |
| EVR | 23.01.200 | 34.3 | 29 | 0 | 14 | |
| 2 | 7 | 9.519 ± 0.0015 | 3 | 0 | 14 | 16.3 |
| α | 22:45 | $224^{+45}_{-15}$ | | 2 | 14 | 0.397 |
| SF | | | | | | |
| EVR | 26.01.200 | 34.3 | 28 | 0 | 9 | |
| 3 | 7 | $213^{+45}_{-15}$ | 3 | 0 | 9 | 10.9 |
| SF | 01:52 | | | | | |
| EVR | 30.01.200 | 34.3 | 28 | 0 | 6 | |
| 4 | 7 | 9.525 ± 0.027 | 1 | 0 | 6/17 | 5.1 |
| α | 13:05 | $195^{+45}_{-15}$ | | 0 | 6 | 0.118 |
| SF | | | | | | |

[a] excitation energy at the center of the target (cot)
[b] calculated kinetic energy of EVRs before implantation into the detector
[c] atomic mass number determined from energy-TOF measurement (uncertainty ± 21)

Scattered $^{48}$Ca as well as sputtered $^{238}$U ions are measured in the focal plane detector in addition to the wanted EVR. A rough mass determination of the EVR can be achieved from the energy versus TOF spectra of all registered ions. These numbers are given in TABLE 1. The mean mass of the EVR was deduced to be A = 286 ± 10. The following decay properties of $^{283}$112 using all four observed events are: half-life ($6.9^{+6.9}_{-2.3}$) s, α-SF branching ratio 50% ± 20%, TKE (238 ± 14) MeV. A cross section of $(0.72^{+0.58}_{-0.35})$ pb can be calculated at the mean excitation energy of 34.6 MeV.

## 112 CHEMISTRY EXPERIMENTS

The Insitu-Volatilization and On-line detection (IVO) technique [16] together with the Cryo On-Line Detector (COLD) allows thermochromatographic studies of volatile chemical elements or compounds. This technique was successful used to determine the chemical properties of Hs in the form of its gaseous $HsO_4$ [17].

The experiments to investigate chemical properties of element 112 were carried out at the Flerov Laboratory of Nuclear Reactions (FNLR) in Dubna in two parts. In contrary to the SHIP experiments the nuclear reaction $^{48}$Ca on $^{242}$Pu was used which exhibits a significant higher peak cross section of 3-4 pb (see [1]). To monitor the chemical separation of the setup, an admixture of 30-60 μg $^{nat}$Nd was added to the target material producing short-lived Hg isotopes. The irradiation parameters of the chemistry experiments are given in TABLE 2.

**TABLE 2.** Irradiation parameters used at 112 chemistry in the reaction $^{48}$Ca on $^{242}$Pu.

| year | $E_{cot}$[a] [MeV] | dose [$10^{18}$] | $\rho_{target}$[b] [$\mu g/cm^2$] | gas flow [ml/min] | Temp. range [°C] | n[c] | event type |
|---|---|---|---|---|---|---|---|
| 2006 | 236 ± 3 | 0.7 | 1400 | 900 | -24 to -182 | 1 | EVR-α-SF |
| | 236 ± 3 | 2.8 | 1400 | 900 | 34 to -182 | 1 | EVR-α-SF |
| 2007 | 236 ± 3 | 2.5 | 1480 | 1300 | 34 to -166 | 3 | EVR-α-SF |

[a] beam energy at the center of the target (cot)
[b] content of Pu inside the target material
[c] number of observed events

The nuclear reaction products are stopped in the carrier gas (He-Ar 70%:30%) after leaving the target material and are transported through 8 m long PFA capillary (within about 2-4 s) to the thermochromatographic COLD detector. The primary reaction product $^{287}$114 decays during this transport to $^{283}$112. Along the detector array a negative temperature gradient of about -6°C/cm was established. The surface temperatures of the gold covered detectors ranged from 34°C down to -184°C. For a detailed description of the experimental setup and data evaluation see [18].

Chemical species which are volatile enough to reach the COLD were deposited along the temperature gradient according to their interaction with the chromatographic gold surface. The deposition place and hence the deposition temperature depends on the gas composition, the linear flow velocity, the half-life of the investigated isotopes, and the adsorption enthalpy of the chemical species on the substrate material. Since, the latter property follows characteristic trends among elements of the same group in the periodic table, the assignment of the observed events to an atomic number is feasible. However, relativistic effects have to be taken into account [19, 20].

In addition to the wanted element 112, isotopes of Hg, At, Rn as well as their decay products were found in the recorded spectra. All other reaction by-products are well separated and do not reach the COLD. In side the COLD only the life-times of subsequent decaying nuclei but not of the primary deposited isotopes can be measured. This is a principal drawback of the IVO technique. Therefore, the assignment of the events to $^{283}$112 is based on the decay pattern (α-SF) and the life-time of the SF decaying $^{279}$Ds.

In the course of the experiment, 5 event chains attributed to the decay of $^{283}112$ were observed. The deposition temperatures are very different, according to the applied temperature gradients and flow rates. A summary of the decay properties of all observed events is given in TABLE 3.

The adsorption enthalpy of -52 kJ/mol for the adsorbed element with Au surfaces can be deduced using a Monte Carlo simulation of the gas-adsorption chromatography [18]. This value points to the formation of a metal-metal bond between element 112 and the gold surface, which is typical for group 12 elements. Hence, the assignment of the events to the decay of an isotope of element 112 is based on the deduced chemical properties as well as the observed decay properties.

**TABLE 3.** 112 chemistry events assigned to $^{283}112$ ($E^{*}_{cot}{}^{a} \approx 37$ MeV)

| **Event no** | **Date Time** | **$E_\alpha/E_{TKE}$ [MeV]** | **detector** | **deposition temp. [°C]** | **life-time [s]** |
|---|---|---|---|---|---|
| α 1 | 11.05.2006 02:40 | 9.38 ± 0.10 | 2 bot | -24 | |
| SF | | 231 ± 20 | 2 bot + 2 top | | 0.592 |
| α 2 | 25.05.2006 08:37 | 9.48 ± 0.10 | 7 bot | -5 | |
| SF | | 232 ± 20 | 7 bot + 7 top | | 0.536 |
| α 3 | 10.04.2007 20:24 | 9.52 ± 0.10 | 14 top | -44 | |
| SF | | 145 ± 20 | 14 bot + 14 top | | 0.072 |
| α 4 | 11.04.2007 07:15 | 9.35 ± 0.10 | 25 top | -127 | |
| SF | | 97 ± 20 | 26 bot + 26 top | | 0.773 |
| α 4 | 13.04.2007 23:45 | 9.52 ± 0.10 | 11 top | -26 | |
| SF | | 145 ± 20 | 10 bot + 11 top | | 0.088 |

[a] excitation energy at the center of the target (cot)

# REFERENCES


1 Yu. Oganessian, J. Phys. G: Nucl. Part. Phys. **34**, R165-R242 (2007).
2 W.D. Myers, W.J. Swiatecki, Nucl. Phys. **81**, 1 (1966).
3 A. Sobiczewski,et al., Physics Letters **22**, 500 (1966).
4 H. Meldner, Arkiv Fysik **36**, 593 (1967).
5 Yu.Ts. Oganessian, et al., Eur. Phys. J. **A 5**, 63-68 (1999).
6 Yu.Ts. Oganessian, et al., Eur. Phys. J. **A 19**, 3-6 (2004).
7 Yu.Ts. Oganessian, et al., Nature, **400**, 242-245 (1999).
8 Yu.Ts. Oganessian, et al., Phys. Rev. **C 69,** 054607 (2004).
9 Yu.Ts. Oganessian, et al., Phys. Rev. **C 70**, 064609 (2004).
10 P. Armbruster, Eur.Phys.J **A 7**, 23-33 (2000).
11 W. Loveland, et al., Phys. Rev. **C 66,** 044617 (2002).
12 K.E. Gregorich, et al., Phys. Rev. **C 72**, 014605 (2005).
13 G. Münzenberg, et al., Nucl. Instr. Methods **161**, 65-82 (1979).
14 S.Hofmann, G.Münzenberg: Rev. Mod. Phys. **72**, 733 (2000).
15 S. Hofmann, et al., Eur. Phys. J. A **32**, 251-260 (2007).
16 Ch. Düllmann, et al., Nucl. Instr. Methods **A 479**, 631-639 (2002).
17 Ch. Düllmann, et al., Nature **418**, 859-862 (2002).
18 R. Eichler, et al., Nature **447**, 72-75 (2007).
19 C. Sarpe-Tudoran, et al., Eur. Phys. J. **D 24**, 65-67 (2003).
20 V. Pershina, T. Bastug, Chem. Phys. **311**, 139-150 (2005).

# Properties of heavy and superheavy nuclei in supernova environments

T. J. Bürvenich, I. N. Mishustin and W. Greiner

*Frankfurt Institute for Advanced Studies Johann Wolfgang Goethe University, Ruth-Moufang-Str. 1, 60438 Frankfurt am Main, Germany*

**Abstract.** The properties of nuclei embedded in an electron gas are studied within the relativistic mean-field approach. These studies are relevant for nuclear properties in astrophysical environments such as neutron-star crusts and supernova explosions. The electron gas is treated as a constant background in the Wigner-Seitz cell approximation. We investigate the stability of nuclei with respect to $\alpha$ and $\beta$ decay. We find that the presence of the electrons leads to stabilizing effects for $\alpha$ decay at high electron densities. Furthermore, the screening effect shifts the proton dripline to more proton-rich nuclei, and the stability line with respect to $\beta$-decay is shifted to more neutron-rich nuclei. Implications for the creation and survival of very heavy nuclear systems are discussed.

**Keywords:** superheavy nuclei, supernova explosions, electron gas
**PACS:** 21.10.Dr, 21.10.Ft, 21.10.Tg, 23.60.+e, 26.50.+x

## INTRODUCTION

The properties of nuclei in astrophysical environments have enjoyed continuous interest for over more than 30 years. The seminal work by Negele and Vautherin [1] layed out the path for microscopic studies of nuclei in stellar environments, see e.g. Refs. [2, 3, 4, 5, 6, 7].

In this contribution we focus on the physics of nuclei embedded in a dense electron gas [8] at zero temperature. We consider nuclei across the periodic chart up to superheavy nuclei embedded in a Wigner-Seitz cell with constant electron density. The presence of the electrons will effect the location of the $\beta$-stability line and the proton dripline as well as decay modes such as $\alpha$-decay. The calculations are performed within the Wigner-Seitz (WS) approximation by dividing the system into (spherical) WS cells, each containing only one nucleus and the number of electrons equal to the nuclear charge Z. The construction of the WS cell is aimed at an approximate and convenient segmentation of an ensemble of nuclei surrounded by the uniform background of electrons. Its construction should ensure that the physics contained in the Wigner-Seitz cell is to an optimally large degree isolated from the surrounding. This requires, for example, charge neutrality of the cell.

For the description of nuclei we employ the relativistic mean-field (RMF) model [9, 10]. The effective in-medium nucleon-nucleon interaction is parametrized via the exchange of several meson fields: scalar-isoscalar ($\sigma$), vector-isovector ($\omega_\mu$) and vector-isovector ($\vec{\rho}_\mu$). The meson fields are treated as classical potentials and the nucleons are represented by Dirac spinors. The isoscalar-scalar $\sigma$ meson delivers the intermediate-range attraction, while the isoscalar-vector $\omega$ meson is responsible for the short-range

CP1012, *Frontiers in Nuclear Structure, Astrophysics, and Reactions: FINUSTAR 2*
edited by P. Demetriou, R. Julin, and S. V. Harissopulos
© 2008 American Institute of Physics 0-7354-0532-5/08/$23.00

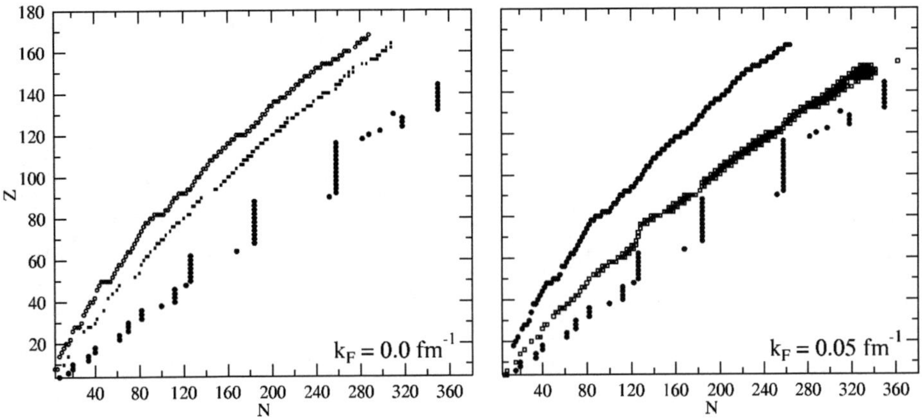


**FIGURE 1.** The line of $\beta$-stability (line in the middle), the proton drip-line (left-most line) and the two-neutron drip-line (right-most line) for various electron fermi momenta

repulsion. The isovector-vector $\rho$ meson couples to the isovector nucleon density and thus parametrizes the isovector properties of the model. The photon is coupled in standard fashion. No (explicit) exchange terms are taken into account.

In order to account for the electron background, we modify the source term of the photon field by adding the electron density, i.e. $\Delta\phi = -e(\rho_p - \rho_e) = -e\rho_{ch}$, which is the Poisson equation for the electrostatic potential $\phi \equiv A_0$. All other equations remain the same as in vacuum. We employ the mean-field parametrization NL3 in these studies [11] which delivers accurate values for nuclear ground-state properties.

## PERIODIC CHART

In this section we analyze how the nuclear $\beta$-stability valley and the proton and neutron drip lines change with increasing electron Fermi momentum. The proton single-particle potential experiences a constant downward shift. This downward shift of the proton potential and hence the gap between proton and neutron chemical potentials is the reason for the new $\beta$-stability line. In the presence of electrons the equilibrium condition with respect to weak decays ($n \to p + e^- + \nu_e,\ \ p \to n + e^+ + \bar{\nu}_e$) reads

$$\mu_n = \mu_p + \mu_e, \qquad (1)$$

where $\mu_n$ and $\mu_p$ are neutron and proton chemical potentials calculated within the framework of the RMF model, i.e.

$$\mu_N = \sqrt{(m_N + g_\sigma\sigma)^2 + p_{FN}^2} + g_\omega\omega^0 + g_\rho\rho^0\tau_3, \quad N = n, p \qquad (2)$$

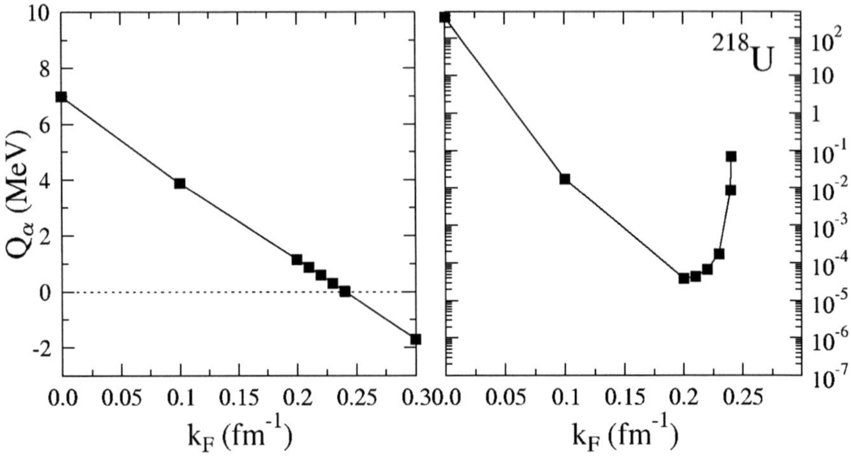


**FIGURE 2.** The evolution of $Q_\alpha$ values (left) and $\alpha$ decay half-lifes (right) for the nucleus $^{218}$U as a function of the electron Fermi momentum. In the right panel, the full squares correspond to calculations within the model of Ref. [12], multiplied with a prefactor taking into account tunneling near the threshold $Q_\alpha = 0$. The lines are drawn to guide the eye.

The electron chemical potential is simply given by $\mu_e = \sqrt{k_F^2 + m_e^2}$ where the contribution of the electrostatic potential has been neglected since it cancels out with the corresponding contribution from the proton in Eq. (1).

If $\mu_n > \mu_p + \mu_e$ holds, neutrons can decay into available proton states with the electron going on the top of the Fermi distribution. On the other hand, if $\mu_n < \mu_p + \mu_e$, protons can capture electrons from the Fermi distribution and occupy neutron states. Only when the condition in Eq. (1) is fulfilled, nuclear systems are stable with respect to these decay modes. Note that this relation becomes the well-known condition for $\beta$-stability in vacuum, $\mu_n \approx \mu_p$, for $k_F = 0$.

For zero temperature, the proton and neutron chemical potentials coincide with their Fermi energies. We note that this correspondence is not uniquely defined anymore for systems with pairing, where the effective Fermi energy is determined to yield $\langle \hat{N} \rangle = A$, but can lie somewhere between the last level contributing with nonzero occupancy and the next one. Thus, in order to obtain the (N,Z) dependence of the stability line, instead of Eq. (1) we employ the criterion $|\mu_e - (\mu_n - \mu_p)| < \Delta$, where $\Delta = 1$ MeV.

In Fig. 1, the $\beta$-stability lines are plotted for various values of $k_F$. In the same plot we show also the proton and neutron drip lines (see below). As $k_F$ increases, the line of $\beta$-stability is shifted more and more to the neutron-rich side, i.e. the presence of electrons stabilizes neutron-rich nuclei. This shift is so strong already for $k_F = 0.1 fm^{-1}$ that $\beta$-stability is reached only in the region of the two-neutron drip line. This means that the $\beta$-equilibrium is reached only when free neutrons appear in the system.

## $\alpha$ DECAY

In this section we investigate the influence of the electron background on $\alpha$-decay of nuclei. The $Q_\alpha$ value of the reaction is defined as

$$Q_\alpha(N,Z) = B(N,Z) - B(N-2,Z-2) - B(2,2) \tag{3}$$

and corresponds approximately, neglecting nuclear recoil, to the kinetic energy of the $\alpha$ particle leaving the nucleus. In the considered environment, the binding energies of mother and daughter nuclei as well as the binding energy of the $\alpha$-particle increase due to the attractive electromagnetic interaction of protons with the electron background.

The calculation of the $\alpha$-decay lifetimes in the presence of electrons is not completely trivial. There are two competing effects. The $Q_\alpha$ value is lowered due to the increased nuclear binding caused by electrons, which alone (for the same barrier) could increase the lifetime. The presence of the electrons leads to screening and modification of the Coulomb potential. Therefore, we need also to take into account the change of the barrier through which the $\alpha$ particle has to penetrate. To quantify this effect we use a simple one-parameter model of Ref. [12] which expresses the $\alpha$-particle potential $V_\alpha$ as $V_\alpha = 2\,V_p + 2\,V_n$, where $V_p$ and $V_n$ are, respectively, proton and neutron single-particle potentials.

The half-life is written as $\tau_{1/2} = \ln 2/\lambda$, where the decay constant $\lambda$ is parametrized as $\lambda = c\,P$. Here the pre-formation factor and the knocking frequency in the Gamov picture are absorbed into one parameter $c$, which is adjusted to known data. The probability for transmission through the barrier, $P = e^S$, is calculated within the WKB approximation

$$S = -2\int_{R_1}^{R_2} dr \sqrt{\frac{2\mu[V_\alpha(r) - Q_\alpha]}{\hbar^2}}, \tag{4}$$

where $R_1$ and $R_2$ are the turning points of the barrier. The $\alpha$-particle potential is lowered as a function of increasing electron Fermi momentum, making it easier for the $\alpha$ particle to escape the nucleus. Together with the $Q_\alpha$ value this leads to an overall decreasing of $\alpha$ half lives, see Fig. 2. This holds true at low electron densities. However, this trend changes in the vicinity of the point where the $Q_\alpha$ value turns negative at a certain electron Fermi momentum $k_F^{crit}$. At $k_F > k_F^{crit}$, $\alpha$-decay becomes forbidden and the nucleus becomes stable with respect to this decay mode. In the Uranium isotope discussed here, $k_F^{crit} = 0.24\ fm^{-1}$. However, the potential barrier experienced by the $\alpha$ particle at $k_F^{crit} = 0.24\ fm^{-1}$ is still $\approx 13$ MeV high. It goes to zero, however, at the WS cell radius.

## CONCLUSIONS AND OUTLOOK

We have studied the properties of atomic nuclei embedded in an electron gas as occuring in, e.g., neutron star crusts and supernova explosions. Nuclear structure calculations have been performed within the relativistic mean-field approach employing the force NL3. The calculations have been performed within the Wigner Seitz cell approximation in coordinate-space.

The presence of electrons leads to effects of rather macroscopic character. Since the electron background is constant over the nuclear volume – by construction – it leads predominantly to a constant downward shift of the proton potential.

The electron gas alters the stability condition for $\beta$ decay and shifts the stability line to more neutron-rich nuclei. Electron capture becomes a relevant process, the transformation of protons into neutrons favors large isospin in nuclei. Furthermore, the two-proton dripline is shifted to more proton-rich nuclei since the protons gain additional binding due to the attractive interaction with the electrons. The neutron drip line is not altered by the presence of the electrons since the neutron single-particle potential remains (to a very good approximation) unaffected.

We have found that in a dense electron background $\alpha$-decay is suppressed. Our calculations show that the $\alpha$ decay half lives decrease as a function of the electron Fermi momentum $k_F$ until they increase again for larger $k_F$. A general trend is that the $Q$ values of those decay modes decrease with increasing electron density. In extreme astrophysical environments, the production of very exotic and superheavy nuclei could become possible. This might happen during the r-process (rapid neutron capture) and the rp-process (rapid proton capture) when the electron background prevents the heavy (and superheavy) nuclei from fast decay by spontaneous fission or alpha decay. As a consequence, the nuclei, which otherwise would be unstable, can provide a bridge to the island of superheavy elements. If long-lived or even stable superheavy nuclei exist, they could be created in such an environment and later ejected into space.

## ACKNOWLEDGMENTS

We gratefully acknowledge fruitful discussions with J. A. Maruhn, S. Reddy, P.–G. Reinhard, N. Sandulescu, L. Satarov, J. Schaffner-Bielich and S. Schramm. This work was supported in part by the Gesellschaft für Schwerionenforschung (Germany) and by grants RFFR-02-04013 and NS-8756.2006.2 (Russia).

## REFERENCES

1. J. W. Negele, and D. Vautherin, Nucl. Phys. **A207** (1973) 298
2. A.S. Botvina and I.N. Mishustin, *Phys. Lett.* **B584**, 233 (2004).
3. P. Magierksi, A. Bulgac, and P.–H. Heenen, nucl-th/0112003 v1
4. P. Magierski and P.–H. Heenen, nucl-th/112018 v1
5. N. Sandulescu, nucl-th/0403019 v2
6. K. S. Cheng, C. C. Yao, and Z. G. Dai, Phys. Rev. C **55** (1997) 2092
7. T. Maruyama, T. Tatsumi, D. N. Voskresenky, T. Tanigawa, S. Chiba, and T. Maruyama, nucl-th/0402002 v1
8. T. J. Bürvenich, I. N. Mishustin, and W. Greiner, accepted for publication in Phys. Rev. C
9. P.–G. Reinhard, Rep. Prog. Phys. **52**, 439 (1989).
10. M. Bender, P.–H. Heenen, P.–G. Reinhard, Rev. Mod. Phys. **75**, 121 (2003).
11. G. Lalazissis, J. König, and P. Ring, Phys. Rev. C **55** (1997) 540.
12. Z. A Dupre and T. J. Bürvenich, Nucl. Phys. A**767**, 81 (2006).

# Single-Neutron States in $^{101}Sn$

D. Seweryniak[a], M.P. Carpenter[a], S. Gros[a], A.A. Hecht[b], N. Hoteling[b], R.V.F. Janssens[a], T.L. Khoo[a], T. Lauritsen[a], C.J. Lister[a], G. Lotay[c], D. Peterson[a], A.P. Robinson[a], W.B. Walters[b], X. Wang[a,d], P.J. Woods[c], and S. Zhu[a]

[a]*Argonne National Laboratory, Argonne, Illinois 60439 USA.*
[b]*University of Maryland, College Park, Maryland 20742 USA.*
[c]*University of Edinburgh, Edinburgh, EH9 3JZ United Kingdom.*
[d]*University of Notre Dame, Notre Dame, IN 46556 USA.*

**Abstract.** A search for γ-ray transitions in $^{101}$Sn, which contains only one neutron outside the $^{100}$Sn core, was carried out at the Argonne Tandem-Linac Accelerator System. $^{101}$Sn nuclei were produced using the $^{46}$Ti($^{58}$Ni,3n)$^{101}$Sn reaction. In-beam γ rays were detected in the Gammasphere Ge-detector array and were correlated with $^{101}$Sn β-delayed protons using the Recoil-Decay Tagging method. A γ-ray line at 172 keV was assigned to $^{101}$Sn and was interpreted as the transition between the single-neutron $g_{7/2}$ and $d_{5/2}$ states. The measured $\nu g_{7/2}$-$\nu d_{5/2}$ energy splitting is compared with predictions of various mean-field potentials. It is also used to calculate multi-neutron configurations in light Sn isotopes. Prospects of using β-delayed protons for tagging other exotic nuclei in the $^{100}$Sn region are discussed.

**Keywords:** Gamma-ray spectroscopy, β-delayed proton emission, shell model
**PACS:** 23.20.Lv, 21.10.Pc, 21.60.Cs, 27.60.+j

## THE $^{100}$SN REGION

Doubly-magic nuclei are the cornerstones of the nuclear landscape. Properties of nuclei such as $^{48}$Ni, $^{78}$Ni, $^{100}$Sn and $^{132}$Sn are essential for understanding the evolution of the nuclear structure far from the line of stability. Single-particle energies are important characteristics of doubly-magic nuclei and provide stringent tests of nuclear models. The self-conjugate doubly-magic $^{100}$Sn nucleus and its neighbors were targets of many experiments using a variety of probes such as in-beam γ-ray spectroscopy, β-decay, β-delayed proton emission, p-decay and α-decay. Despite this effort nothing is known about excited states in the immediate neighbors of $^{100}$Sn such as $^{101}$Sn, $^{99}$In, and $^{100}$In. Consequently, the relative positions of single-particle states at $^{100}$Sn are unknown. The $^{101}$Sn nucleus β-decays with a half-life of 1.9(3) s [1]. About 15% of $^{101}$Sn β decays are followed by proton emission. This offers a unique opportunity to use the Recoil-Decay Tagging method [2] to search for in-beam γ-ray transitions in $^{101}$Sn.

CP1012, *Frontiers in Nuclear Structure, Astrophysics, and Reactions: FINUSTAR 2*
edited by P. Demetriou, R. Julin, and S. V. Harissopulos
© 2008 American Institute of Physics 0-7354-0532-5/08/$23.00

## THE $^{101}$SN EXPERIMENT

A search for γ-ray transitions in $^{101}$Sn was carried out at the ATLAS facility of the Argonne National Laboratory. $^{101}$Sn nuclei were produced using the $^{46}$Ti($^{58}$Ni,3n) reaction. The beam energy of 192 MeV was close to the Coulomb barrier offering small recoil emission angles and relatively low background due to other reaction channels. Reaction products recoiling from the target (recoils) were separated from the unreacted beam and dispersed according to their mass-to-charge state ratio in the Fragment Mass Analyzer (FMA). The recoil's position at the FMA focal plane was measured in a Parallel Grid Avalanche Counter (PGAC) followed by an ionization chamber (IC) used for energy loss measurement. Recoils with mass number A=101 and charge states $23^+$ and $24^+$ were accepted by mass slits. After passing through the PGAC-IC combination, the recoiling nuclei were implanted into a 140 μm-thick Double-sided Silicon Strip Detector (DSSD). The front and the back sides of the DSSD were divided into 80, 400 μm-wide horizontal and vertical strips forming 6400 pixels. The decays of the implanted nuclei were also observed in the DSSD and in Si detectors surrounding the DSSD: a 300 μm-thick, large area Si detector (LASD) positioned behind the DSSD and a box consisting of four 300 μm-thick Single-sided Si Strip Detectors (SSSD) placed in front of the DSSD. Using spatial and temporal relations in the DSSD, implants were correlated with decays. Characteristic properties of the decays were used to associate prompt γ rays with specific nuclei. More experimental details can be found in Ref. [2].

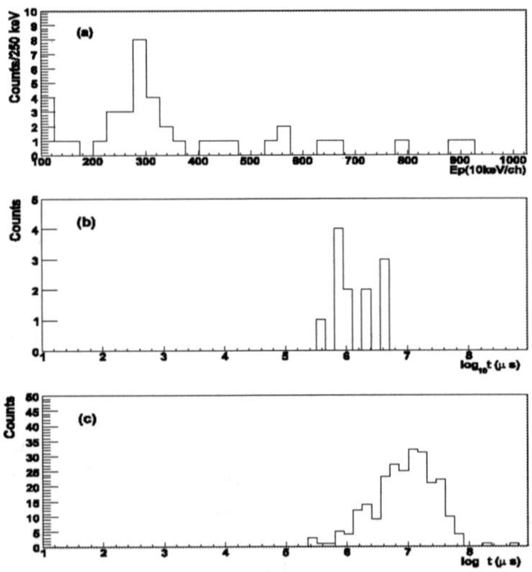


**FIGURE 1.** (a) Energy spectrum for decay events which took place within 5s after implantation. (b) Time spectrum for decay events associated with the 172 keV γ-ray line. (c) Time spectrum corresponding to long lived β-decay events.

The energy spectrum of decay events in the DSSD within 5s of implantation and coincident with β particles in Si detectors surrounding the DSSD (see Fig. 1(a)) is similar to the $^{101}Sn$ proton spectrum in Ref. [1]. Fig. 2(a) shows the γ-ray spectrum correlated with decay events between 1.5 and 4.5 MeV, with decay times shorter than 5s, which were coincident with betas or were scattered between neighboring DSSD strips. A γ-ray line at 172 keV, which can be easily seen in Fig. 2(a), is not present in the random spectrum corresponding to long-lived β emitters shown in Fig. 2(b). The decay time distribution corresponding to the 172 keV events indicates a half-life of 1.3(5)s (see Fig. 1(b)), in agreement with previous measurement [1], which is about 8 times shorter than the time characterizing random coincidences.

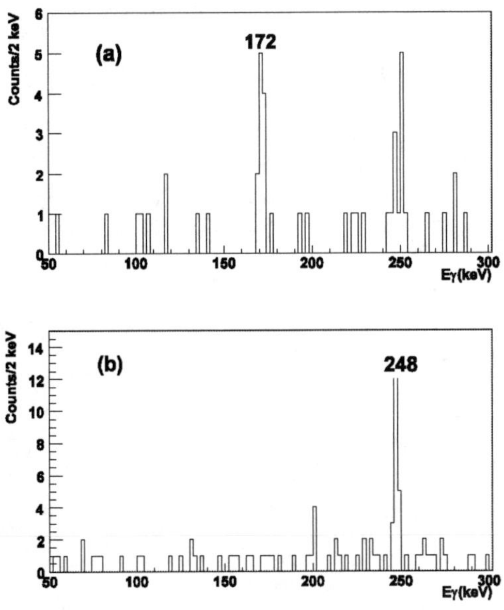


**FIGURE 2.** Gamma rays correlated with decay events (a) in the energy range 1.5-4.5 MeV, coincident with betas or scattered between DSSD stips, which took place within 5s after implantation and (b) below 0.5 MeV corresponding to β decays. The 248 keV line is the strongest line in $^{101}Ag$ which is the strongest reactions channel.

Consequently, we assign the 172 keV line to $^{101}Sn$ and interpret it, in agreement with the systematics of proton-rich N=51 isotones, as the transition between the two single-neutron $g_{7/2}$ and $d_{5/2}$ orbitals. No other transitions were found in $^{101}Sn$. It was attributed to the fact that they are expected to have high energy comparable with the energy required to excite the core, i.e. about 3 MeV.

## DISCUSSION

The relative single-particle energies can be used to test various forms of the mean-field nuclear potential. In Ref. [4], single-particle energies for the doubly-magic nuclei

$^{56}Ni$, $^{100}Sn$, $^{132}Sn$, and $^{208}Pb$ were calculated using the Woods-Saxon potential with the so-called "universal'" set of parameters, the folded Yukawa potential, and the self-consistent Hartree-Fock potential obtained with the Skyrme III phenomenological nucleon-nucleon interaction, resulting in $g_{7/2}$-$d_{5/2}$ energy differences in $^{101}Sn$ of 1.45 MeV, 1.09 MeV, and 0.47 MeV, respectively. The self-consistent potential reproduces the measured $g_{7/2}$-$d_{5/2}$ splitting best. Surprisingly, when a Skyrme interaction was refitted to more recent data across the nuclidic chart, including several single-particle energies, a larger difference of ~1 MeV was obtained [5]. This illustrates limitations of current mean-field potentials in predicting single-particle energies. Recently, the energy separation between the $g_{7/2}$ and $h_{11/2}$ neutron orbitals in the N=51 isotones was reproduced by considering the tensor nature of the nucleon-nucleon interaction [6], and an excitation energy of ~300 keV was predicted for the $\nu g_{7/2}$ state in $^{101}Sn$.

With the measured energy splitting between the $g_{7/2}$ and $d_{5/2}$ neutron orbitals established, a shell-model calculation of energies of multi-particle configurations in light Sn isotopes was performed. The full N=50-82 shell was available for valence neutrons and the following relative single-particle energies were used in the calculation: $E(d_{5/2})=0.0$, $E(g_{7/2})=0.172$, $E(s_{1/2})=1.55$, $E(d_{3/2})=1.65$, and $E(h_{11/2})=3.55$ MeV. The effective neutron-neutron interaction, which is the most important ingredient in the calculations after single-particle energies, was taken from Ref. [7], where it was derived from the Bonn C meson-exchange nucleon-nucleon potential in the one-boson-exchange approximation.

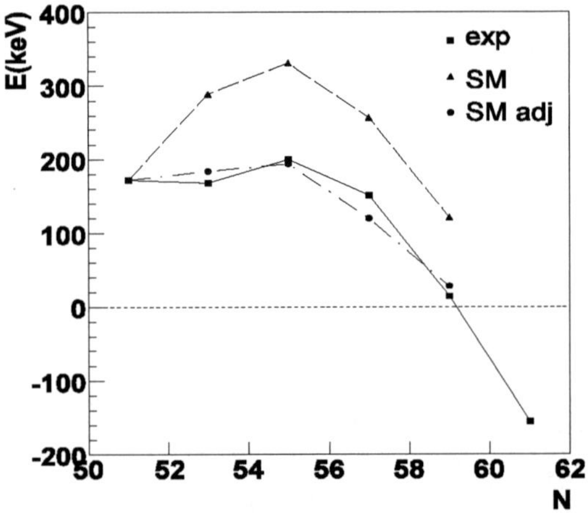


**FIGURE 3.** Shell-model calculations of the energy splitting between the lowest $7/2^+$ and $5/2^+$ states in light odd-A Sn isotopes.

In Fig. 3 the experimental and calculated energies for the two lowest states in the odd Sn isotopes from $^{103}Sn$ to $^{109}Sn$ are compared. The shell model predicts a pair of low-lying $7/2^+$ and $5/2^+$ states, with a calculated energy difference that follows the experimental trend closely, but which is about 200 keV too large. Almost perfect agreement with the data can be achieved if the $(g_{7/2})^2_{0+}$ matrix element is reduced from -1.4 MeV to -1.1 MeV. This adjustment preserves the agreement between experiment and theory for the $2^+$, $4^+$, and $6^+$ energies in the even Sn isotopes.

## OUTLOOK

The experiment presented in this contribution represents the very first measurement of relative single-particle energies at $^{100}$Sn. Tagging on β-delayed protons can be extended to other exotic nuclei near $^{100}$Sn. Observation of higher-lying states in $^{101}$Sn involving core excitations is within reach. Proton-neutron interactions can be studied in the β-delayed proton emitter $^{100}$In, which contains one proton hole and one neutron outside $^{100}$Sn. Also, several high-spin isomers in nuclei along the N=Z line, such as $^{94}$Ag, are known or are predicted to exhibit a significant β-delayed proton branch. In order to facilitate tagging with these often long lived activities a DSSD with even higher granularity is required.

## ACKNOWLEDGMENTS

This work was supported by the U.S. Department of Energy, Office of Nuclear Physics under contract No. DE-AC02-06CH11357.

## REFERENCES

1. O. Kavatsyuk et al., Eur. Phys. J. **A31**, 319 (2007).
2. E.S. Paul et al., Phys. Rev. C **51**, 78 (1995).
3. D. Seweryniak et al., Phys. Rev. Lett. **99**, 022504 (2007).
4. G.A. Leander et al., Phys. Rev. C **30**, 416 (1984).
5. B.A. Brown, Phys. Rev. C **58**, 220 (1998).
6. T. Otsuka et al., Phys. Rev. Lett. **95**, 232502 (2005).
7. M. Hjorth-Jensen, T.T.S. Kuo, and E. Osnes, Phys. Rep. **261**, 125 (1995).

# Quadrupole collectivity of neutron-rich nuclei around $^{132}$Sn


Th. Kröll[*], T. Behrens[*], R. Krücken[*], V. Bildstein[*], T. Faestermann[*], R. Gernhäuser[*], M. Mahgoub[*], P. Maierbeck[*], M. Münch[*], W. Weinzierl[*], F. Ames[†], D. Habs[†], O. Kester[†], R. Lutter[†], T. Morgan[†], M. Pasini[†], K. Rudolph[†], P. Thirolf[†], J. Diriken[**], M. Huyse[**], O. Ivanov[**], P. Mayet[**], N. Patronis[**], I. Stefanescu[**], J. Van de Walle[**], P. Van Duppen[**], O. Niedermaier[‡], H. Scheit[‡], D. Schwalm[‡], J. Eberth[§], F. Finke[§], D. Martin[§], P. Reiter[§], A. Scherillo[§], M. Seidlitz[§], N. Warr[§], D. Weisshaar[§], J. Iwanicki[¶], P. Butler[||,††], J. Cederkäll[||], E. Clément[||], P. Delahaye[||], L. M. Fraile[||], G. Georgiev[||], U. Köster[||], T. Sieber[||], D. Voulot[||], F. Wenander[||], S. Franchoo[‡‡], A. Hurst[††], A. Ekström[§§], P. E. Kent[¶¶], K.-H. Speidel[***], J. Leske[***,†††], S. Schielke[***], A. Jungclaus[‡‡‡], V. Modamio[‡‡‡], J. Walker[‡‡‡], L. Coquard[†††], M. Pantea[†††], N. Pietralla[†††], T. Davinson[§§§], S. Nardelli[¶¶¶] and the REX-ISOLDE and MINIBALL collaborations

[*]*Physik-Department E12, Technische Universität München, Garching, Germany*
[†]*Fakultät für Physik, Ludwig-Maximilians-Universität München, Garching, Germany*
[**]*Instituut voor Kern- en Stralingsfysica, Katholieke Universiteit Leuven, Belgium*
[‡]*Max-Planck-Institut für Kernphysik, Heidelberg, Germany*
[§]*Institut für Kernphysik, Universität zu Köln, Germany*
[¶]*Środowiskowe Laboratorium Ciężkich Jonów, Uniwersytet Warszawski, Poland*
[||]*CERN, Genève, Switzerland*
[††]*Oliver Lodge Laboratory, University of Liverpool, United Kingdom*
[‡‡]*Institut de Physique Nucléaire, Orsay, France*
[§§]*Fysiska Institutionen, Lunds Universitet, Sweden*
[¶¶]*Department of Physics, University of York, United Kingdom*
[***]*Helmholtz-Institut für Strahlen- und Kernphysik, Universität Bonn, Germany*
[†††]*Institut für Kernphysik, Technische Universität Darmstadt, Germany*
[‡‡‡]*Departamento de Física Teórica, Universidad Autónoma de Madrid, Spain*
[§§§]*Department of Physics and Astronomy, University of Edinburgh, United Kingdom*
[¶¶¶]*INFN and Dipartimento di Fisica, Università di Camerino, Italy*



**Abstract.** We report on the "safe" Coulomb excitation of neutron-rich Cd, Xe, and Ba isotopes in the vicinity of the doubly-magic nucleus $^{132}$Sn. The radioactive nuclei have been produced by ISOLDE at CERN and postaccelerated by the REX-ISOLDE facility. The $\gamma$-decay of excited states has been detected by the MINIBALL array. The presented preliminary results for the $B(E2)$ values are consistent with expectations from phenomenological systematics and will be compared with theoretical calculations.



**Keywords:** Coulomb excitation, gamma spectroscopy, neutron-rich nuclei
**PACS:** 25.70.De, 21.10.Re, 27.60.+j



CP1012, ***Frontiers in Nuclear Structure, Astrophysics, and Reactions:*** *FINUSTAR 2*
edited by P. Demetriou, R. Julin, and S. V. Harissopulos


© 2008 American Institute of Physics 0-7354-0532-5/08/$23.00

## MOTIVATION

The explanation of the magic numbers for nuclei in the valley of stability was one of the milestones in the understanding of nuclear structure. Important interest nowadays is the question if they persist or will be altered going away from the valley of stability because the residual interactions responsible for their existence may change. We report here on Coulomb excitation studies of nuclei around $^{132}$Sn, the heaviest doubly-magic nucleus off stability which is accessible for experiments so far.

In our naive understanding of collective states large $B(E2)$ are accompanied with low excitation energies. The product between the $B(E2; 0^+_{gs} \rightarrow 2^+_1)$ value and the excitation energy $E(2^+_1)$ in even-even nuclei has a smooth behaviour near to the valley of stability and can be described by a simple phenomenological formula [1]. A significantly improved description could be obtained by multiplying Raman's version of Grodzins' formula [2] with an isospin dependent modification [3].

An indication for a deviation from this simple picture came some years ago from $B(E2)$ values for neutron-rich Sn and Te nuclei in the vicinity of $^{132}$Sn [4] which were lower than expectations from systematics. As theoretical explanation a reduced neutron pairing above $N = 82$ has been proposed. QRPA calculations were able to reproduce the small $B(E2)$ value for $^{136}$Te [5]. Following this argumentation collectivity in these very neutron-rich nuclei is built up mainly by neutrons resulting in low $B(E2)$ values. Other theoretical calculations using the Monte Carlo Shell Model [6] or the shell model with a realistic interaction based on the CD Bonn NN-potential [7] reproduced the low $B(E2)$ value only partially. Meanwhile, a new Coulomb excitation measurement indicates 30% larger $B(E2)$ values [8]. Furthermore, a lifetime measurement for the first $2^+$ state in $^{136}$Te using the ultrafast timing technique even results in a 50% larger $B(E2)$ value [9].

Aiming to extend such investigations, we studied neutron-rich Cd, Xe, and Ba isotopes at the REX-ISOLDE facility at CERN.

## EXPERIMENTAL METHOD

The radioactive nuclei have been produced at the ISOLDE facility at CERN by the 1.4 GeV proton beam from the PS Booster impinging on a $UC_x$ target. For some isotopes neutron induced fission has been applied using a neutron converter target to reduce the proton-rich isobaric contaminants. For each beam the target and the ion source have been optimised for elemental selectivity. The extracted ions were mass separated, cooled and bunched in the Penning trap REX-TRAP, and sent to an EBIS for charge breeding. Eventually, the energy was boosted to around 2.85 MeV/u by the REX-LINAC [10]. The REX-ISOLDE facility will be upgraded to HIE-ISOLDE in the next years [11].

We employed $\gamma$-spectroscopy following "safe" Coulomb excitation of the radioactive beams. The $\gamma$-rays deexciting the populated states were detected by the highly efficient MINIBALL spectrometer, consisting of 8 triple clusters of six-fold segmented HPGe detectors equipped with fully digital electronics [12], in coincidence with the scattered particles. Since most ISOL beams, in our case all but the Xe beams, are not pure beams, several methods to determine the beam composition have been applied.

The $B(E2)$ values of radioactive beams are determined relative to those of the target

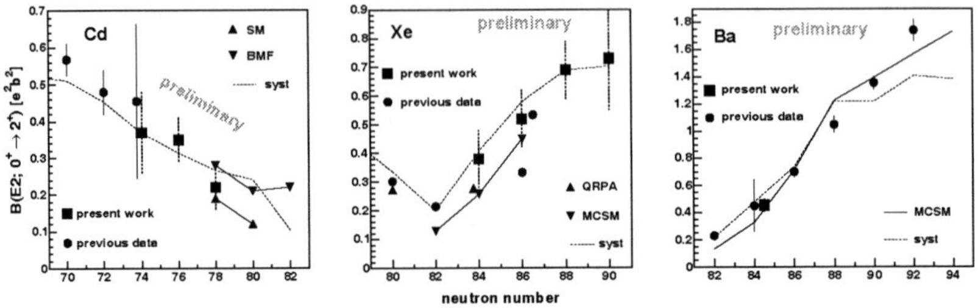


**FIGURE 1.** $B(E2)$ values for Cd (left), Xe (middle) and Ba (right) isotopes in comparison with previous data, phenomenological systematics, and theoretical models. For details and references see text.

whose electromagnetic properties are well known. For the beam nuclei the $B(E2)$ values have been varied until the cross sections, calculated with an extended version of the CLX code [13], agree with the experimental values.

More details of Coulomb excitation experiments at REX-ISOLDE with MINIBALL can be found also in further contributions to this conference [14].

## PRELIMINARY RESULTS FOR $B(E2)$ VALUES

The analysis is still in progress and we will present here only preliminary results [15].

For the Cd isotopic chain, the last known $B(E2)$ value is for $^{122}$Cd, however only within an error of 50% [16] (note that a wrong value is given in Raman's compilation [2]). We determined the $B(E2)$ values for $^{122,124,126}$Cd. As it can be seen in Fig. 1 (left), the values agree well with the simple systematical formula.

Recent interest in this region comes from nuclear astrophysics, as the nuclei below $^{132}$Sn are so-called waiting point nuclei of the r-process. From the study of the $\beta$-decay of $^{130}$Cd it was deduced that its binding energy is lower than expected from mass models [17]. This has been interpreted as signature of a quenching of the $N = 82$ shell gap. With this assumption the $A \approx 130$ peak in the r-process abundances can be described better. On the other hand, in a recent study of the decay of the $8^-$ isomer a good agreement with the $(\nu g_{9/2})^{-2}$ description and, hence, no evidence for a shell quenching was found [18]. A beyond mean field (BMF) approach with projection on particle number and angular momentum predicts for the neutron-rich Cd isotopes rather a prolate deformation than a spherical shape. In Fig. 1 (left), the theoretical predictions are included. The shell model (SM) using a realistic interaction derived from the CD Bonn NN-potential without assuming a shell quenching predicts a steep decrease of the $B(E2)$ values approaching $N = 82$ [19]. Clearly larger $B(E2)$ values are predicted by the BMF approach [18].

For the Xe isotopes previously no $B(E2)$ had been measured above $N = 82$, except the two contradicting values for $^{140}$Xe [2, 20]. We determined $B(E2)$ values for $^{138,140,142,144}$Xe. Our preliminary values, as shown in Fig. 1 (middle), clearly follow the systematical formula. For $^{140}$Xe, our result supports the larger value from literature [20]. The experimental values are compared with predictions from QRPA calculations [5] as well as from the Monte Carlo Shell Model (MCSM) [6]. Both models give lower $B(E2)$ values for $^{138}$Xe, however still within the preliminary error.

Aiming to extend this studies to more deformed nuclei, we started with the $^{140,142}$Ba isotopes in a first beam time in July 2007. Within the $N = 84$ isotonic chain, for $^{140}$Ba a $B(E2)$ value has been measured previously but with a large error which we will improve. Attempts to measure also the very exotic $^{148,150}$Ba failed because of technical problems. Our preliminary $B(E2)$ value for $^{140}$Ba, previous experimental data [2, 9], the systematical formula, and predictions from the MCSM [6] are shown in Fig. 1 (right).

## g-factor measurements

A more direct probe for an increased neutron content in the wave functions are g-factors. The relevant neutrons occupy the $f_{7/2}$ orbital and their contribution has a negative sign resulting in a small, maybe even negative, g-factor different from the collective $Z/A$ value [21]. As Te beams are not yet available with sufficient intensities, we studied the g-factors of the first $2^+$ states in the $^{138}$Xe and $^{140}$Ba $N = 84$ isotones.

For the $^{138}$Xe, we performed an experiment applying the transient field method to be sensitive to the sign of the g-factor [21]. The data are still under analysis [22]. The Ba will be studied applying recoil-in-vacuum method [23]. The data set for $^{142}$Ba, for which the g-factor of the $2_1^+$ state is known, will serve as a calibration. This analysis will be applied also to the Xe data sets.

## Transfer reactions

With Coulomb excitation mainly the collective properties of nuclei are studied. Nevertheless the obtained results are very well described by modern theoretical calculations, there is an increasing interest to investigate also the single particle structure of nuclei by nucleon transfer reactions. At REX-ISOLDE, one- or two-neutron transfer reactions in inverse kinematics are planned using deuterium or tritium targets.

The emitted protons are detected by a new array of position sensitive Si detectors which covers a large angular range to determine the angular distributions accurately. The set-up fits inside MINIBALL to detect efficiently $\gamma$-rays in coincidence.

As a first experiment in autumn 2007, the shore of the "island of inversion" will be studied with the reaction d($^{30}$Mg,$^{31}$Mg)p [24]. Also studies of neutron-rich Ca isotopes and nuclei in the region around $^{68}$Ni [25] are planned already.

## Summary and outlook

We have presented the preliminary $B(E2)$ values for neutron-rich nuclei in the vicinity of the doubly-magic nucleus $^{132}$Sn. All results are well within the expectations from simple phenomenological systematics. However, predictions based on nuclear models differ significantly and our final results will allow in most cases to distinguish between them [15]. Further information will come from the determination of g-factors.

We will extend our studies to the nucleus $^{128}$Cd to evidence the predicted prolate deformation for neutron-rich Cd isotopes. The systematic investigations will be continued towards more deformed nuclei by the study neutron-rich Ba isotopes where also octupole degrees of freedom are expected to become important.

Such investigations will allow deeper insights into the isospin dependence of the quadrupole collectivity and the underlying physics of their phenomenological systematics in this region of high interest for both nuclear structure and nuclear astrophysics.

## ACKNOWLEDGMENTS

This work is supported by the German BMBF under grant No. 06MT238, by the MLL, by the EU through EURONS (contract No. 506065), and by the DFG cluster of excellence Origin and Structure of the Universe (URL: `www.universe-cluster.de`). REX-ISOLDE and MINIBALL are supported by the German BMBF, the Belgian FWO-Vlaanderen and IAP, the U.K. EPRSC, and the EU, as well as the ISOLDE collaboration.

## REFERENCES

1. L. Grodzins, Phys. Lett. 2, 88 (1962).
2. S. Raman et al., Atomic Data and Nuclear Data Tables 78, 1 (2001).
3. D. Habs, R. Krücken et al., CERN Proposal INTC-P-156 (2002).
4. D. C. Radford et al., Phys. Rev. Lett. 88, 222501 (2002); Proc. of INPC2004 and ENAM'04.
5. J. Terasaki et al., Phys. Rev. C 66, 054313 (2002).
6. N. Shimizu et al., Phys. Rev. Lett. 86, 1171 (2001), Phys. Rev. C 70, 054313 (2004), and J. Phys. Conf. Ser. 49, 178 (2006).
7. A. Covello et al., Proc. of RNB7 (2006).
8. C. Baktash et al., MAFF Workshop (2005).
9. H. Mach et al., contribution to this conference; L. M. Fraile et al., Proc. of INPC2007.
10. O. Kester et al., Nucl. Instr. and Meth. B 204, 20 (2003).
11. K. Riisager et al., contribution to this conference.
12. J. Eberth et al., Prog. Part. Nucl. Phys. 46, 389 (2001).
13. H. Ower, computer program CLX, not published.
14. G. Georgiev et al., J. Van de Walle et al., and A. Ekström et al., contributions to this conference.
15. T. Behrens, PhD Thesis (TU München, 2007); to be published.
16. N. V. Zamfir et al., Phys. Rev. C 51, 98 (1995).
17. I. Dillmann et al., Phys. Rev. Lett. 91, 162503 (2003).
18. A. Jungclaus et al., Annual NUSTAR Meeting, GSI (2007); private communication.
19. A. Scherillo et al., Phys. Rev C 70, 054318 (2004).
20. A, Lindroth et al., Phys. Rev. Lett. 82, 4783 (1999).
21. K.-H. Speidel et al., CERN Proposal INTC-P-166 (2003).
22. J. Leske et al., Annual Spring Meeting of DPG, Giessen (2007).
23. N. Stone et al., Phys. Rev. Lett. 94, 192501 (2005).
24. V. Bildstein et al., ISOLDE Workshop 2007; PhD thesis (TU München).
25. N. Patronis et al., contribution to this conference.

# Probing the single particle structure around $^{54}Ca$ with one-neutron knock-out

P. Maierbeck*, R. Gernhäuser*, R. Krücken*, T. Kröll*, H. Alvarez-Pol†, F. Aksouh**, T. Aumann**, K. Behr**, E. A. Benjamim†, J. Benlliure†, V. Bildstein*, M. Böhmer*, K. Boretzky**, M. J. G. Borge‡, A. Brünle**, A. Bürger§,¶, M. Caamano†, E. Casarejos†, A. Chatillon**, L. V. Chulkov**, D. Cortina Gil†, J. Enders‖, K. Eppinger*, T. Faestermann*, J. Friese*, L. Fabbietti*, M. Gascon†, H. Geissel**, J. Gerl**, M. Gorska**, P.G. Hansen[1]††, B. Jonson‡‡, R. Kanungo§§,**, O. Kiselev**,¶¶, I. Kojouharov**, A. Klimkiewicz**, T. Kurtukian†, N. Kurz**, K. Larsson**,‡‡, T. Le Bleis**, K. Mahata**, L. Maier*, T. Nilsson‡‡,‖, C. Nociforo**, G. Nyman‡‡, C. Pascual-Izarra‡, A. Perea‡, D. Perez†, A. Prochazka**,***, C. Rodriguez-Tajes†, D. Rossi¶¶, H. Schaffner**, G. Schrieder‖, S. Schwertel*, H. Simon**, B. Sitar***, M. Stanoiu**, K. Sümmerer**, O. Tengblad‡, H. Weick** and S. Winkler*

**E12, Physik Department, TU München, Germany*
*†USC, Santiago de Compostela, Spain*
***GSI, Darmstadt, Germany*
*‡CSIC, Madrid, Spain*
*§ISKP, Bonn, Germany*
*¶CEA, Saclay, France*
*‖Institut für Kernphysik, TU Darmstadt, Germany*
*††MSU, East Lansing, USA*
*‡‡Chalmers University of Technology, Göteborg, Sweden*
*§§TRIUMF, Vancouver, Canada*
*¶¶Johannes Gutenberg Universität, Mainz, Germany*
****Comenius University, Bratislava, Slovakia*

**Abstract.** The nuclei $^{47}Ca$ and $^{55}Ti$ were populated in one-neutron knock-out reactions at relativistic energies. Momentum distributions of the residual nuclei as well as $\gamma$-ray spectra were measured at the GSI fragment separator (FRS). Preliminary results of the ongoing analysis including cross sections and spin/parity assignments are presented.

**Keywords:** shell structure, neutron knock-out on $^{56}Ti$ and $^{48}Ca$ at relativistic beam energies
**PACS:** 21.10.Jx,23.20.Lv,25.60.Gc,27.40+z

## INTRODUCTION

While the spherical shell closures are well established for stable nuclei up to $^{208}Pb$, it has become apparent in recent years that the shell structure changes as one goes away from

[1] deceased

CP1012, *Frontiers in Nuclear Structure, Astrophysics, and Reactions: FINUSTAR 2*
edited by P. Demetriou, R. Julin, and S. V. Harissopulos
© 2008 American Institute of Physics 0-7354-0532-5/08/$23.00

the valley of stability, in particular in very neutron rich systems. One cause leading to local modifications of the shell structure is related to the action of the residual interaction between orbitals around the Fermi energy. As a result new (sub-)shell closures can emerge (e.g. for neutron-rich oxygen isotopes [1]) or even the ordering of the shells can change as in the so called 'island of inversion' for neutron rich nuclei in the Magnesium (Z=12) region [2]. For Calcium isotopes (Z=20), a new shell closure for neutron number N=34 has been predicted [3].

To study the structure of neutron-rich nuclei knock-out reactions are a powerful tool. In knock-out experiments the momentum distribution of the residual nuclei reflects the momentum distribution of the knocked-out nucleon in the projectile nucleus and thus depends on its orbital angular momentum[1]. By measuring this momentum distribution and quantifying the cross sections for different final states in the knock-out reaction, spectroscopic factors for the single-particle occupancies of the incoming nucleus can be determined. Due to the high energies thick targets can be used and experiments can be performed down to rates of a few particles per second. Therefore, such experiments are an ideal tool to probe the single-particle structure of nuclei at the limits accessible at modern in-flight facilities. The use of relativistic beams in these studies is advantageous due to the fact that the theoretical assumptions made for the calculations of the momentum distributions hold very well.

## EXPERIMENTAL SETUP

For the experiment, the four-stage fragment separator (FRS) at GSI (Darmstadt, Germany) was used as a two-stage spectrometer. Each part contains two 30° dipole stages and was equipped with scintillators and photomultipliers for time-of-flight (TOF) and multiple sampling ionisation chambers (MUSIC) for energy-loss measurements. This allowed an event by event identification of the incoming nuclei at the intermediate focus (S2) and of the nuclei detected at the final focal plane (S4). In the focal plane of the S2 focus, the target (1720 mg/cm$^2$ $^9$Be) for the knock-out reactions was placed. Behind the target, at an average angle of 40° with respect to the beam axis, eight triple cluster detectors of the MINIBALL gamma-ray spectrometer [4] were mounted. Each high purity germanium detector crystal is electrically dividd into six segments. They were used to tag excited states populated in the knock-out reaction by measuring the deexcitation via gamma-ray emission of these states. The absolute efficiency in the laboratory frame of the MINIBALL setup was determined to be 2.8% at an energy of 344.3 keV and to 1.6% at an energy of 1332 keV by $^{152}$Eu and $^{60}$Co source measurements. Due to the large velocity of the residual nuclei ($\beta \approx 0.7$), Doppler correction was mandatory for a quantitative analysis. Using the segmentation of the MINIBALL clusters a resolution of 44 keV (FWHM) at a laboratory $\gamma$-energy of $\approx$ 580 keV was achieved.

Six time projection chambers (TPC), four at S2 and two at S4, provided the position and the angle of the flight path of the incident beam as well as the reaction residues through the experimental setup. This high accuracy position measurement at the two focal planes together with the ion-optical matrices of the FRS allowed a precise measurement of the longitudinal momentum distribution induced in the knock-out reaction.

With a $^{48}$Ca primary beam of 450 AMeV and the reaction $^{48}$Ca$\rightarrow$$^{47}$Ca the properties of the setup and the methodes of analysis have been adjusted with good event statistics. For this measurement, the stable $^{48}$Ca beam was directly injected into the FRS setup.

The radioactive $^{56}$Ti beam for the $^{56}$Ti→$^{55}$Ti knock-out reaction was produced by the fragmentation of 500 A MeV $^{86}$Kr in a production target (1625 mg/cm$^2$ $^{9}$Be) placed in front of the fragment separator. The detector setup described above was used to identify the nuclei of interest on an event by event basis.

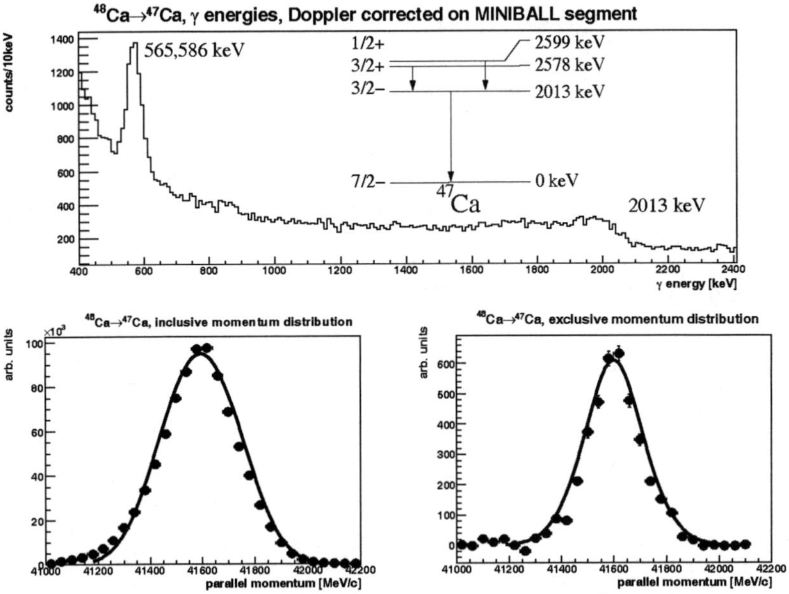


**FIGURE 1.** *Upper panel:* $\gamma$-spectrum of the $^{48}$Ca→$^{47}$Ca reaction after Doppler correction on MINIBALL segment. *Lower panel:* Inclusive (left) and exclusive momentum distribution of the $^{48}$Ca→$^{47}$Ca reaction. (see text for details)

## $^{48}$CA→$^{47}$CA

The one-neutron knock-out reaction $^{48}$Ca→$^{47}$Ca was used to determine the properties of the setup and the to test the reliability analysis methods with good statistics and by comparing the results (cross sections, spectroscopic factors) with with those from previous experiments. In the following, preliminary results will be presented. The data sample was taken in 6 hours of beam time. Approximately $90 \cdot 10^6$ incoming $^{48}$Ca nuclei induced $8.7 \cdot 10^5$ single-neutron knock-out reactions in the target placed at the intermediate focal plane of the FRS.

In the upper panel of figure 1 the $\gamma$-spectrum obtained with the MINIBALL setup is shown after Doppler correction using the segmentation of the hit crystal to enlarge the spatial granularity. In the knock-out reaction neutrons can be removed from the occupied $s_{1/2}$, $d_{3/2}$ and $f_{7/2}$ orbitals leading to the population of the $3/2^+$ and $1/2^+$ states at about 2.6 MeV and the $7/2^-$ ground state in $^{47}$Ca. The transitions of the $3/2^+$ and $1/2^+$ states to the $3/2^-$ intermediate state are clearly separated from the background, but could not be resolved. The deexcitation of the 2013 keV state to the ground state is also clearly separates from the background.

The lower left panel of figure 1 shows the inclusive momentum distribution of the residual $^{47}$Ca nuclei after the knock-out reaction. Superimposed is a theoretical calculation of the reaction,scaled in height to fit the data. In this calculation, the reaction channels to the $f_{7/2}$ ground state and to the $d_{3/2}$ and $s_{1/2}$ excited states were considered. All spectroscopic factors were set to 1, assuming a perfectly closed N=28 shell. The theoretical curve fits already quite well to the data and spectroscopic factor will be extracted in the forthcoming analysis.

An exclusive momentum distribution is shown in the lower right panel of figure 1. To obtain this, only events with a measured $\gamma$-ray in the 565/586 keV region were considered and a suitable background was subtracted. The superimposed calculation is the sum of the $d_{3/2}$ and $s_{1/2}$ reaction channels and scaled to fit the data. As in the inclusive case, the spectroscopic factors were assumed to be equal 1 and the calculated momentum distribution reproduces the experimentally obtained one.

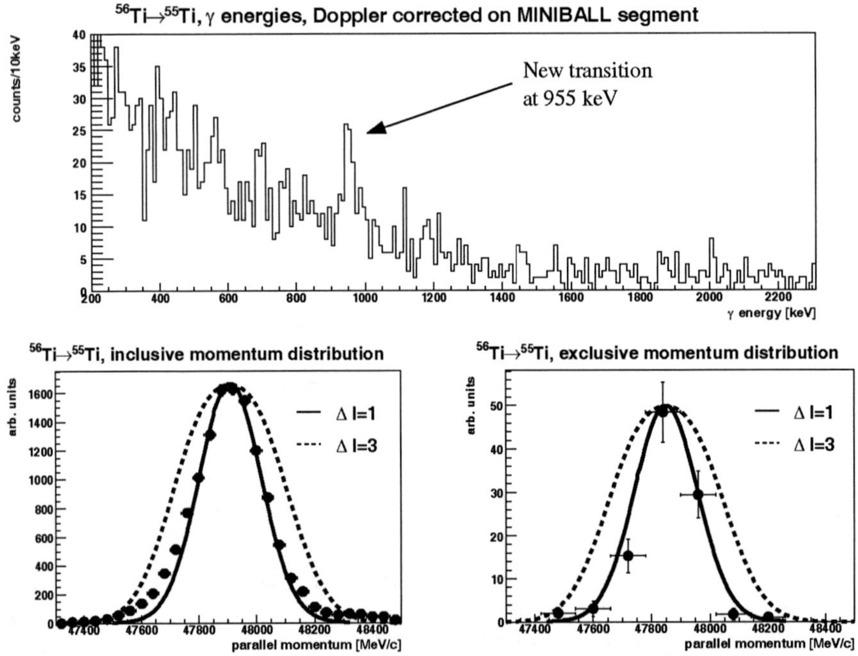


**FIGURE 2.** *Upper panel:* $\gamma$-spectrum of the $^{56}$Ti→$^{55}$Ti reaction after Doppler correction on MINIBALL segment. *Lower panel:* Inclusive (left) and exclusive momentum distribution of the $^{56}$Ti→$^{55}$Ti reaction. (see text for details)

## $^{56}$TI→$^{55}$TI

As described above, $^{56}$Ti nuclei were produced by the fragmentation of a 500 AMeV $^{86}$Kr beam in front of the FRS. An average rate at the intermediate FRS focus of $\approx 2$ $^{56}$Ti nuclei per second yielded an integral of $1.6 \cdot 10^6$ ions in 8.5 days of beam time. At the final focus of the FRS $1.3 \cdot 10^4$ residual nuclei of the $^{56}$Ti→$^{55}$Ti knock-out reaction were identified. The analysis was performed with the scheme established for the $^{48}$Ca→$^{47}$Ca knock-out reaction.

In figure 2 preliminary results of the experiment are shown. The upper panel contains the $^{55}$Ti $\gamma$-spectrum after Doppler correction according to the segment position of the hit MINIBALL crystal. A new transition was found at 955 keV. The lower two panels contain the inclusive and exclusive (cut on the $\gamma$ transition with background subraction) momentum distributions. The superimposed lines are calculations for a angular momentum transfer in the knock-out reaction of pure Δl=1 (solid lines) and Δl=3 (dashed lines), respectively. Clearly, the ground state and the excited state both have dominant l=1 components and, naturally, the $p_{1/2}$ orbital will be closer below the fermi surface than $p_{3/2}$. Thus the ground state of $^{55}$Ti is clearly dominated by the $p_{1/2}$ configuration and the predicted spin assignment of $J^{\pi} = 1/2^{-}$ [3], is for the first time experimentally confirmed. The preliminary spin assignment of Ref. [5], which was based on comparison between experimental and theoretical energies and branching rations, is also confirmed.

## SUMMARY

We have presented results of the one-neutron knock-out reactions on $^{48}$Ca and $^{56}$Ti. Preliminary inclusive and exclusive momentum distributions and $\gamma$-spectra of the residual nuclei were extracted from the data. The good agreement between theoretical and experimental momentum distributions shows that knock-out reactions at relativistic energies are a reliable tool for the investigation of single particle structure in medium-mass nuclei. In the $\gamma$-spectrum of the knock-out product $^{55}$Ti, we identified a so far unobserved transition at 955 keV. By comparing the preliminary momentum distributions with theoretical predictions and experimental results, already a clear spin/parity assignment of $1/2^{-}$ for the ground state and $3/2^{-}$ for the state deexcited by the 955 keV transition could be made. The level sequence is consistent with theoretical predictions [3].

## ACKNOWLEDGMENTS

This work was supported by the BMBF under contract 06MT238, by the DFG cluster of excellence *Origin and Structure of the Universe* (`http://www.universe-cluster.de`) and by the European Commission within the Sixth Framework Programme through I3-EURONS (contract no. RII3-CT-2004-506065).

## REFERENCES

1. D. Cortina-Gil et al., *J. Phys. G* **32**, 1629 (2005).
2. E. K. Warburton et al., *Phys. Rev. C* **41**, 1147 (1990).
3. M. Honma et al., *Phys. Rev. C* **65**, 061301, (2002).
4. J. Eberth et al., *Prog. Part. Nucl. Phys.* **46**, 289 (2001).
5. Zhu et al., *Phys. Lett. B* **650**, 135 (2007).

# SPIRAL2 at GANIL: Status and Perspectives

S.Gales

*GANIL (DSM-CEA/IN2P3-CNRS)*
*Blvd. Henri Becquerel, F-14076 Caen cedex, France*

**Abstract.** To pursue the investigation of a new territory of nuclei with extreme N/Z called "terra incognita" several projects, all aiming at the increase by several orders of magnitude of the RIB intensities are now under discussions worldwide. In Europe, two major new projects have been approved recently FAIR@GSI using the so-called "in-flight" method and SPIRAL2@GANIL, based on the ISOL method. Both projects were selected in the European Strategic Roadmap For research Infrastructures (ESFRI). The main goal of SPIRAL2 is clearly to extend our knowledge of the limit of existence and the structure of nuclei deeply in the medium and heavy mass region (A=60 to 140) which is to day an almost unexplored continent. SPIRAL 2 is based on a high power, CW, superconducting driver LINAC, delivering 5 mA of deuteron beams at 40MeV (200KW) directed on a C converter+ Uranium target and producing therefore more $10^{13}$ fissions/s .The expected radioactive beams intensities for exotic species in the mass range from A=60 to A=140, of the order of $10^{6}$ to $10^{10}$pps will surpass by two order of magnitude any existing facilities in the world. These unstable atoms will be available at energies between few KeV/n to 15 MeV/n. The same driver will accelerate high intensity (100μA to 1 mA), heavier ions up to Ar at 14 MeV/n producing also proton rich exotic nuclei. In applied areas SPIRAL2 is considered as a powerful variable energy neutron source, a must to study the impact of nuclear fission and fusion on materials. The intensities of these unstable species are excellent opportunities for new tracers and diagnostics either for solid state, material or for radiobiological science and medicine. The "Go" decision has been taken in May 2005. The investments and personnel costs amount to 190 M€, for the construction period 2006-2012. Construction of the SPIRAL2 facility is shared by ten French laboratories and a network of international partners. Under the 7FP program of European Union called "Preparatory phase for the construction of new facilities ", the SPIRAL2 project has been granted a budget of about 4M€ to build up an international consortium around this new venture. The status of the construction of SPIRAL2 accelerator and technical R&D programs for physics instrumentation (detectors, spectrometers) in collaboration with EU and International partners will be presented.

**Keywords:** physics with secondary radioactive beams ISOL methods, nuclear structure and reactions far from stability, detectors and instrumentation.
**PACS:** 21, 23, 25, 25. 60.-t, 25.70.-z, 26, 29, 29.20, 29.38.-c

## INTRODUCTION

In the recent years, Rare Isotope Beams (RIB), i.e. beams of synthesized radioactive isotopes, have been recognized by the international scientific community as one of the most promising avenue for the development of fundamental nuclear physics and astrophysics, as well as in applications of nuclear science. The importance of the RIB's has been underlined by NuPECC (Nuclear Physics European Collaboration Committee - an expert committee of the European Science Foundation)

CP1012, *Frontiers in Nuclear Structure, Astrophysics, and Reactions: FINUSTAR 2*
edited by P. Demetriou, R. Julin, and S. V. Harissopulos
© 2008 American Institute of Physics 0-7354-0532-5/08/$23.00

which recommend the construction of two new-generation complementary RIB facilities in Europe. One based on the in-flight fragmentation (IFF) as proposed for the FAIR facility at GSI (Darmstadt, Germany) [1] and the other on isotope-separation on-line (ISOL) method. In ISOL method, a thick target is bombarded with a primary light or heavy-ion beam or with a secondary neutron beam in case of fission targets. The produced radioactive nuclei are post-accelerated to energies from keV to tens of MeV. The ISOL method provides high-intensity beams of high optical quality comparable to stable beams. The ultimate-generation RIB ISOL facility will be realized when the EURISOL facility [2] will be built. Because of the time-line for EURISOL, NuPECC recommends the construction of intermediate-generation facilities that will benefit the EURISOL project in terms of R&D and that will give the community opportunities to perform research and applications with RIB. Among the proposed intermediate facilities SPIRAL2 at GANIL (Caen, France) [3], meets in terms of physics potential, site and size of the investments, the criteria of European dimension.

The French Ministry of Research took a decision on the construction of SPIRAL 2 in the end of May 2005. Its construction cost, estimated to 190 M€ (including personnel and contingency), will be shared by French founding agencies CNRS/IN2P3 and CEA/DSM, the Region of Basse Normandie and the European partners. Funding from EU 7th framework and from others partnership countries are expected to contribute for about 20% to this budget. The operation of the facility will cost 8.5 M€ per year.

In October 2006, ESFRI (European Strategy Forum on Research Infrastructures) published a list of selected 34 Research Infrastructure projects corresponding to major needs of the European scientific community (all fields of science included) for the next decade. The FAIR and SPIRAL 2 are among the selected projects and were recommended as Research Infrastructures necessary to maintain Europe's position at the cutting-edge of world research.

As a result of the ESFRI selection process , under the 7FP program of European Union called "Preparatory phase for the construction of new facilities ", the SPIRAL2 project has been granted a budget of about 4M€ to build up an international consortium around this new venture.

The first beams are expected by the end of 2011, beginning of 2012. As an added bonus, the facility will also produce over $10^{15}$ neutrons per second, making it the world's most powerful source of fast neutrons for several years. SPIRAL 2 will reinforce European leadership in the field of exotic nuclei and will serve a community of about 1000 scientists.

## DESCRIPTION OF THE FACILITY

The quest for Rare Isotope Beams (RIB), which are orders of magnitude more intense than those currently available, is the main motivation behind the SPIRAL 2 project.The new SPIRAL2 facility is a major improvement of the SPIRAL1 facility. More than 20 laboratories from 10 countries from 2001 to 2005 have developed the conceptual and the technical design of SPIRAL2. More than 200 scientists and engineers have contributed to the scientific case and the design of the facility [3-8].

The SPIRAL 2 facility (Fig. 1) is based on a high power, superconducting driver LINAC, which will deliver a high intensity, 40 MeV deuteron beam as well as a variety of heavy-ion beams with mass over charge ratio equals to 3 and energy up to 14.5 MeV/n. Using a carbon converter, the 5 mA deuteron beam and a uranium carbide target, a fast-neutron induced fission is expected to reach a rate of up to $10^{14}$ fissions/s. The RIB intensities in the mass range from A=60 to A=140 will surpass by one or two order of magnitude any existing facilities in the world. A direct irradiation of the $UC_x$ target with beams of deuterons, $^{3,4}He$, $^{6,7}Li$, or $^{12}C$ can be used if higher excitation energy leads to higher production rate for a nucleus of interest.

The neutron-rich fission RIB could be complemented by beams of nuclei near the proton drip line provided by fusion-evaporation or transfer reactions. For example, a production of up to $8x10^4$ atoms of $^{80}Zr$ per second using a 200 $\mu$A $^{24}Mg^{8+}$ beam on a $^{58}Ni$ target should be possible. Similarly, the heavy- and light-ion beams from LINAG on different production targets can be used to produce high-intensity light RIB with ISOL technique.

The extracted RIB will be subsequently accelerated to energies of up to 20 MeV/n. (Typically 6-9 MeV/n. for fission fragments) by the existing CIME cyclotron. Thus using different production mechanisms and techniques SPIRAL 2 would allow performing experiments in a wide range of neutron- and proton-rich nuclei far from the line of stability (Fig. 2).

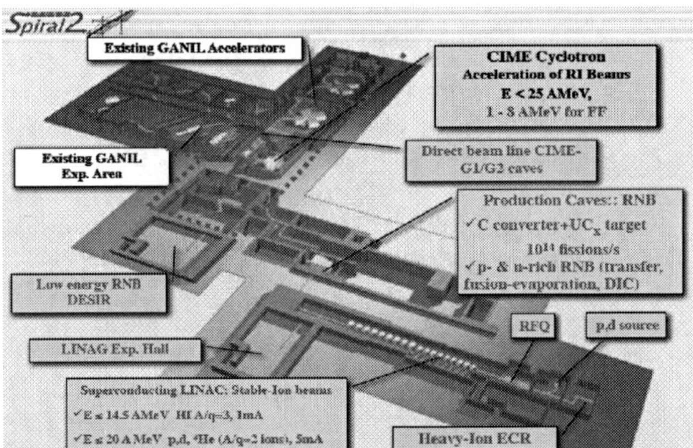


**FIGURE 1.** Schematic layout of the SPIRAL 2 facility at GANIL.

Recent papers describing the technical and instrumentation developments related to the SPIRAL2 facility can be found in Refs. [9-12].

SPIRAL2 has also a remarkable potential for neutron-based research both for fundamental physics and various applications. In particular, in the neutron energy range from a few MeV to about 35 MeV this research would have a leading position for the next 10 years if compared to other neutron facilities in operation or under construction worldwide.

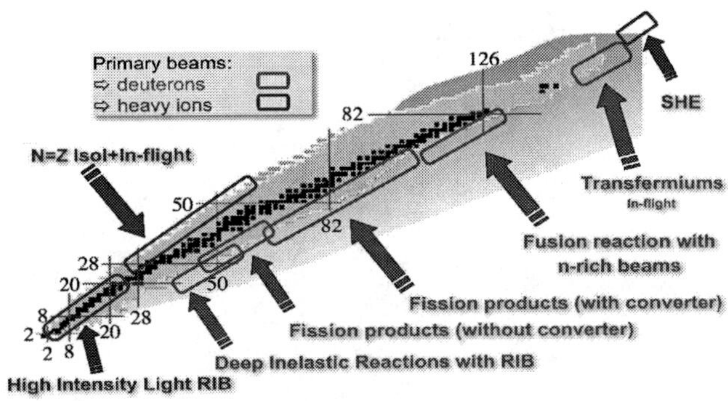


**FIGURE 2.**Regions of the chart of nuclei, far from stability accessible at SPIRAL2.

# PERFORMANCES OF THE SPIRAL2 FACILITY

## Intense stable beams from LINAG

Very intense stable heavy ions beams will be available at the exit of the LINAG accelerator. These heavy ions (from He to Ca in a first step, then up to U), will reach intensities from 1pμA up to 1pmA using new PHOENIX ion sources in combination with A/Q=1/3 RFQ injector and later A/Q=1/6 for heavier ions (Kr to U) as illustrated in Fig. 3.

The construction of the PHOENIX A/q=3 ion source has started at Grenoble (LPSC Grenoble-GANIL collaboration) [13-15]. In addition a recent MoU has been signed between Argonne National Laboratory, Accelerator Division and GANIL-SPIRAL2 project. Under the agreed terms ANL has started the design of a new RFQ injector with A/Q= 6, devoted to the production of intense stable beams with mass higher than A= 40.

Reaching such beams intensities will open a wide range of physics experiments not accessible to day. Super Heavy Elements synthesis and spectroscopy, spectroscopy at or beyond the proton drip line, multinuclear transfer and deep inelastic scattering, study of mechanisms and reaction product distributions, production and study of isomers, ground state properties of rare nuclei.

A new device, S3 (for Super Separator Spectrometer) has been proposed for experiments with the very high intensity stable beams of SPIRAL2 coming from the LINAG accelerator [16-19]. Most of the physics cases listed above can be seen as either the study of rare processes or as the study of a secondary reaction with exotic ions produced in a first step reaction. S3 shall be designed to answer most experimental needs of these physics cases. The crucial point of the device is to cope with the very high intensity beams provided by the LINAG accelerator.

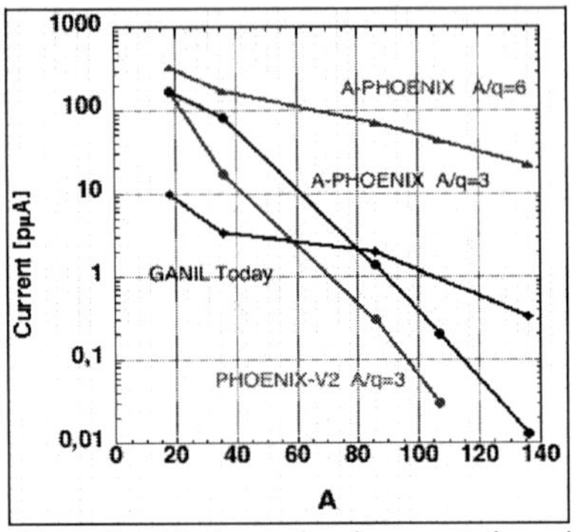


**FIGURE 3.** Expected intensities of stable heavy ion beams at the exit of the LINAG accelerator for various versions of PHOENIX ion sources.

That means:

-the targets should be able to cope with very high power dissipation,

-the reaction products should be separated from the beam ions before they can be detected,

-the full optics of the spectrometer should reduce the background coming from scattered ions and activation products,

Meanwhile, the vast array of different physics that could be possible with S3 implies that it should be compatible with various detection settings .The basic design considered could be the following (see Fig. 4): The primary beam will impinges on a primary target which should be made resistant to the high fluxes (rotating cooled targets, gas or liquid targets...). The separation of the beam can proceed in two steps, the first being a high rejection device and an appropriate beam stopper. After this first step, ions of interest could, if necessary, react on a secondary target. As they deal with moderate fluxes, this secondary target point could have gamma or particle prompt detection. Next to this point, a secondary spectrometer helps with separation if very high purity is needed. It has some mass/momentum resolution and could rotate for reactions products at large angles. The final focal plane can be equipped with the relevant detection system.

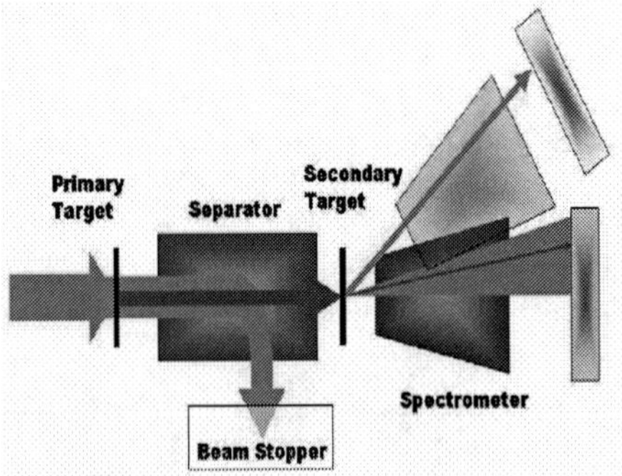


**FIGURE 4.** Basic scheme for S3 design.

## The radioactive ion production system

The SPIRAL2 facility at GANIL will produce radioactive isotopes ranging from the lightest to very heavy elements beyond uranium. Different production mechanisms will be utilized. A lay out of the radioactive production areas is shown in Fig. 5.

The radioactive neutron rich beams will be mainly produced via the fission process [20], with the aim of $5.10^{13}$ fissions/s .The retained solution is the fast neutrons induced fission in a depleted UCx target. The neutrons are generated by the beak-up of 40 MeV deuterons in a thick C target (converter). For the aimed high fission rate the converter should subtend a beam power of 200kW (see Fig. 1). With a length of 80mm and at about 40mm from the entrance of the converter, $5.10^{13}$ fissions/s can be reached in the target volume with an UC depleted target of 11g/cm$^{3.}$ The resulting fission products accumulate in the target, diffuse from the surface from which they evaporate towards an ion source.

Few target-ion sources systems are being developed, often in the framework of international collaborations. Let us mention, an ECR 1+ "MONOBOB" for gaseous elements in test at GANIL[21], a FEBIAD,"IRENA" ion source well suited for short lifetimes at IPN Orsay [13], a Laser Ion source, particularly efficient for its selectivity and for metallic beams. An international collaboration around the Laser Ion Source will be built under the 7$^{th}$ European FP.

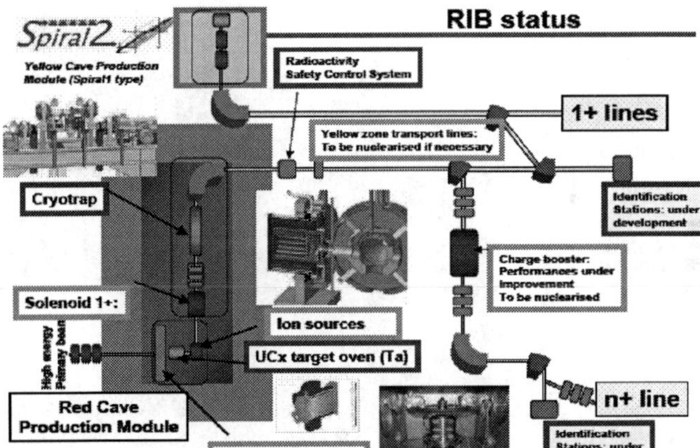


**FIGURE 5**. Lay out of the radioactive production halls for the SPIRAL2 facility.

For neutron rich fission products the whole area will be confined in a red zone (see Fig. 5) with only remote controls and maintenance. At the exit of the target ion source, Cryotraps are being developed in order to catch volatile gaseous atoms which are not ionized. Then, the 1+ ions are separated and charge bred in a ECR ion source in order to be accelerated by the existing CIME cyclotron up to 5-9 MeV/n for masses A=70 to A=140.

The expected intensities (to be validated by measurements) of typical fission fragment beams of Kr and Sn isotopes from SPIRAL2 are compared to existing or planned RIB facilities in Fig. 6.

For SPIRAL2 the RIB intensities in the mass range from A=60 to A=140 will be of the order of $10^6$ to $10^{11}$ part/s. For example, the intensities should reach, $10^{9-10}$ part. /s for $^{132}Sn$ and $10^{11}$ pps for $^{92}Kr$ surpassing by two order of magnitude any existing facilities in the world (see Fig. 6)

Production of rare isotopes are also possible using a second target station (yellow cave production see Fig. 5) similar to SPIRAL1 cave where intense heavy-ion beams from LINAG can induce fusion evaporation reactions leading to rare isotopes for medium-mass, neutron-deficient nuclei. In addition, using transfer and deep-inelastic reactions, SPIRAL2 facility will produce intense secondary beams of light to medium mass nuclei closer to the line of stability.

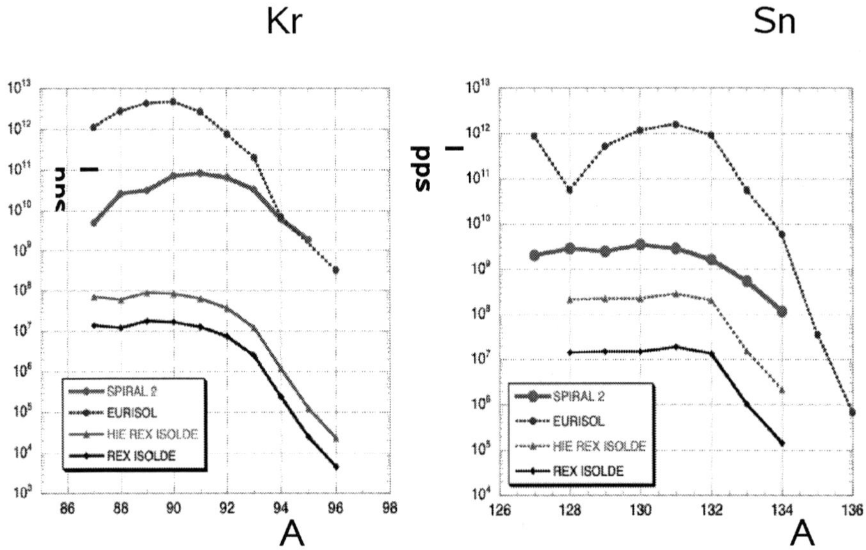


**FIGURE 6.** Expected intensities of Kr and Sn isotopes at SPIRAL2 after acceleration compared to REX-ISOLDE present and future (HE-REX ISOLDE) and to EURISOL predictions.

## SELECTED EXAMPLES OF THE SCIENTIFIC OPPORTUNITIES AT SPIRAL2

As discussed in previous section, the multi-beam driver LINAC will deliver intense heavy ion stable beams and may be dedicated to the synthesis of very heavy and Super-heavy nuclei. A new experimental Hall (AEL) will host the new $S^3$ "Super-Separator-Spectrometer" and a wide range of physics experiments not accessible today will be at reach.

In the same experimental hall a neutron –TOF facility is planned, thanks to the high energy and high intensity neutron flux available at SPIRAL2, the facility will offer a unique opportunity for material irradiations and cross-section measurements both for fission (notably accelerator driven systems (ADS) and Gen-IV fast reactors) and fusion related research, tests of various detection systems and of resistance of electronics components to irradiations, etc.

The DESIR collaboration [22], formed after the SPIRAL2 workshop on low-energy physics at GANIL in July 2005, proposes the construction of an experimental facility to exploit the low-energy "exotic" beams from SPIRAL2. Experiments performed at other ISOL facilities (e.g., ISOLDE or ISAC) show that a high degree of purity is necessary for the ions beams. The most efficient and universal way of achieving isotopic ally pure beams is a high-resolution mass separator.

At SPIRAL2 and its low-energy beam facility DESIR, there is an opportunity of building three kinds of setup to try to cover the whole nuclear chart. A ß-NMR setup will be suitable to study precisely the moments of key cases and, in combination with collinear hyperfine spectroscopy, get unambiguous spin assignments. With a collinear spectroscopy setup, we will have the possibility to study series of isotopes in the intermediate and heavy mass regions. For the heavy elements, a double laser + RF spectroscopy in a Paul trap will be used to study the hyperfine anomaly and higher-order moments up to very high precision.

Shell structure of neutron-rich isotopes near the N=20 and 28 shell are to day the subject of intensive research all around the world. Shell quenching, weakening of the spin-orbit interaction, tensor component in the n-n force are investigated using the rich and powerful tools of nuclear reactions from nucleon transfer [22], in beam gamma spectroscopy [23], Coulomb excitation, fusion –evaporation and deep inelastic processes.

The spectroscopy of Zn, As, Ga, Se isotopes over the magic shell N=50 using collinear laser spectroscopy and reactions will answer this question. Another typical example of an element that has to be studied is of course nickel as close as possible to 78Ni. Is this doubly-magic nucleus still spherical?

A lot of nuclear models predict strong evolution of shapes and/or change of shell structure around neutron number N=60 and mass number A=110. Neutron skin is predicted to occur for the very neutron-rich Sn isotopes. Studying tin isotopes very far from stability will be a very stringent test for these nuclear models.

Nuclei were created during major events that made the evolution of the universe and are the witness of the origin of matter: while the lightest elements were created in the first minutes following the big bang, heavier ones are synthesized in stars. Understanding our world, its origin and its evolution is one goal of nuclear physics. However, these various **nucleosynthesis** processes involve radioactive nuclei. SPIRAL 2 will abundantly produce some of them opening new investigation field.

The future SPIRAL2 facility will contribute prominently to several areas of active research in nuclear astrophysics, such as explosive hydrogen burning, s-process and r-process nucleosynthesis, which are linked to astrophysical observations

For example, radiative capture cross-sections can be determined from spectroscopic factors (or Asymptotic Normalization Coefficients) obtained in transfer reaction or elastic resonant scattering. The (d, p) reaction can be used to simulate the (n,$\gamma$) capture on medium-mass nuclei at SPIRAL2 near the r-process possible paths.

The way the energy of theses s or p states evolve from Z=50 down to Z=40 determines how neutron capture cross sections will evolve along the N=82 shell closure and how the r process synthesis of nuclei will be blocked at this shell closure. The neutron rich isotopes of Sn and Cd nuclei are excellent test cases. The neutron-capture through the soft Pigmy resonance could also play an important role if the energy of the soft pigmy vibration mode is peaked around the $S_n$ value. The way this energy is shifted when moving from Sn to Cd isotopic is certainly an important issue to consider.

SPIRAL2 will be well suited to address these fundamental questions, thanks to the expected large coverage of RIB species, very high intensities and energy range covered by the facility.

The scientific opportunities discussed above are just a few examples among many exciting avenues which can be investigated with SPIRAL2. New exotic shapes and excitation modes, halo-like and molecular structures, single particle properties and new magic numbers, new modes of nuclear decay have been recently observed, while studies of fundamental symmetries, testing and refinement of the Standard Model and of fundamental interactions in very exotic nuclei have all a huge potential of discovery.

## CONSTRUCTION OF THE FACILITY AND INTERNATIONAL COLLABORATIONS

Since September 2005, a project construction group has been formed and the whole project has been organized around three main sections

- The driver accelerator
- The secondary radioactive beams (production and transport)
- The infrastructure (buildings, vacuum and cryogenics, power supplies…)

The project has been divided in each section into work packages.

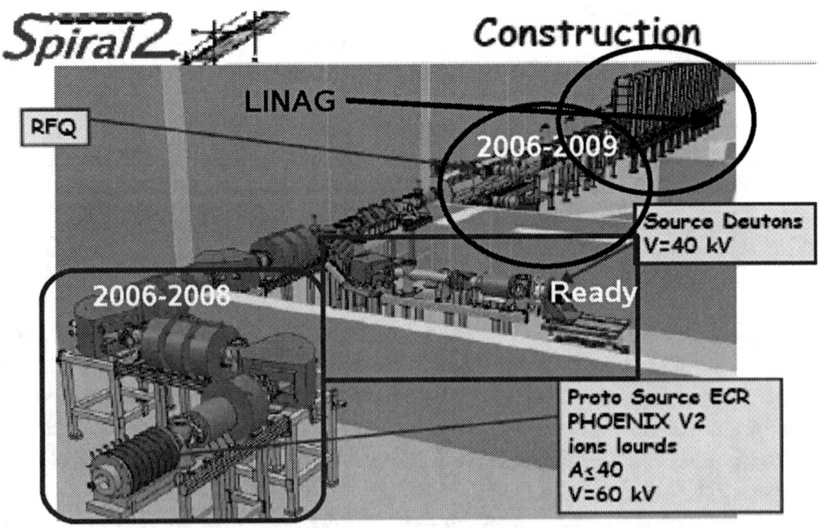


**FIGURE 7.** SPIRAL2 driver accelerator divided into main components.

For the driver accelerator, the work packages are under the responsibility of IPN Orsay and DAPNIA/DSM/Saclay laboratories [25-28]. Figure 7 show the driver accelerator divided into main components and the associated target dates for completion. The R&D on prototypes' for the RFQ, SC cavities and RF couplers are completed and order of all driver components will be completed at the end of 2007.

A SPIRAL2-France consortium had been set to coordinate the efforts of 10 French nuclear physics laboratories from CNRS/IN2P3 and CEA. Moreover SPIRAL2 facility will be built by a large European and International collaboration. Several MoU have been already signed or are in final negotiation stages with JINR Dubna, Israel Romania, Isolde (CERN), Italy, UK ,Germany, Poland, Canada, Japan and US ,both for accelerator components (control system, beam loss monitors, beam lines) and/or construction of components of the RIB production units (neutron converter , beam lines, target-ions sources ,laser ion sources, high-resolution magnetic separator ,....).

At the EU research organization level, SPIRAL2 as well as FAIR have been selected by the ESFRI Committee and a proposal for the construction phase in the framework of the 7$^{th}$ FPRD has been recently granted 4M€ to coordinate the contributions of more than 20 Laboratories from EU, USA, Japan and India.

The process of formation of international collaborations around the SPIRAL 2 research program has started in 2004 and pursued the next years with the organization of 5 specialized workshops held in 2004-2006 (Neutrons for science at SPIRAL 2, 13-14/12/2004,Physics with separated low energy beams at SPIRAL2, 4-5/07/2005,Future prospects for high resolution gamma spectroscopy at GANIL, 4-6/10/2005,Nuclear Astrophysics with SPIRAL2, 17-18/10/2005,SPIRAL2 Reactions, 19-21/10/2005).

In the beginning 2006, under the umbrella of SPIRAL2 Scientific Advisory Committee (SAC) a Call for letters of intent has been launched in March 2006. A very positive and enthusiastic response from the International community has been obtained. 19 Letters of Intents (see Refs. [1, 18]), coming from 600 physicists and 34 countries have been received and evaluated by the SAC in October 2006.

After this evaluation, the formation of about 9 large international collaborations around major instruments is under progress .Next important step is the call for Technical proposals around the spring of 2008.

In terms of schedule for the new facility, the ambition is to start operation of the driver accelerator with intense stable beams before the end of 2011 and to complete the whole project including RIB production and post acceleration of fission fragments in the beginning of 2013.

## CONCLUSION

Nuclear physics has been revolutionized by the recent development of the ability to produce accelerated beams of radioactive nuclei. For the first time it will be possible to study reactions to produce the 6000 to 7000 nuclei we believe exist rather than the 300 stable ones that nature provides. At the larger scale one wants to understand the limits of nuclear stability by producing exotic nuclei with vastly different numbers of neutrons and protons.

SPIRAL2 is major expansion of the SPIRAL facility at GANIL, which will help maintain European leadership in ISOL (Isotope separation on line) development, aiming at two orders of magnitude increase of the secondary beams available for nuclear physics studies.

The technical challenges of the acceleration, targetry and experimental equipments will provide essential knowledge on the road toward EURISOL [28], the most advanced nuclear physics research facility presently imaginable and based on the ISOL principle. It is expected, that the realization of SPIRAL2 will substantially increase the know-how of technical solutions to be applied not only for EURISOL but also in a number of other European/ world projects. The scientific programme, proposes the investigation of the most challenging nuclear and astrophysics questions aiming at the deeper understanding of the nature of matter.

SPIRAL2 will contribute to the physics of nuclear fission and fusion based on the collection of unprecedented detailed basic nuclear data, to the production of rare radioisotopes for medicine, to radiobiology and to material science.

## REFERENCES


1. http://www.gsi.de/fair/reports/btr.html FAIR Baseline Technical report (Sept 2006)
2. http://www.eurisol.org/site01/index.php
3. www.ganil.fr/research/developments/spiral2/
4. "The scientific objectives of the SPIRAL2 project" *rapport WB_SP2_Final-1.pdf* June 2006. http://www.ganil.fr/research/developments/spiral2/index.html and Letter of Intention therein.
5. W. Mittig et al., "LINAG Phase1*", internal report GANIL* (2002).
6. M.-G. Saint-Laurentet al. "SPIRAL phase II, European RTT" R0103 *report GANIL* (2001).
7. M.-H. Moscatello, in "*Seventeenth International Conference on the cyclotrons and their applications*, Tokyo, 2004 October 18-22, p. 518, Particle Accelerator Society of Japan (2005).
8. T. Junquera et al., in "*`9th European Particle Accelerator Conference (EPAC 2004)",* Luzern, July 2004, TUPLT058, p. 1285, (EPS-AG) (2004).
9. M. Lewitowicz, "*Tours Symposium on Nuclear Physics VI"*, Tours (FR) 5th-8th September 2006, Vol. 891, p. 91 AIP Conference Proceedings (2007).
10. M. Jacquemett, E. Petit et al., in ``*XV^th^ International Conference on Electromagnetic Isotopes Separators and Techniques related to their applications EMIS 2007"*, Deauville, 24'h-29th June 2007, to be published.
11. S. Gales, in ``*International Workshop on Nuclear Physics 28th Course radioactive Beams, Nuclear Dynamics and Astrophysics"*, Erice (IT), September 16th-24th 2006, Vol. 59, p. 22, Progress in Particle and Nuclear Physics (2007).
12. M.-G. Saint –Laurent , " Future opportunities with SPIRAL2 at GANIL " to be published
13. M. Cheik Mhamed et al., Rev. Sci. Instr. 77, 03A702 (2006).
14. T. Lamy et al, Rev. Sci. Instr. 77, 03B101 (2006).
15. T. Thullier et al., Rev. Sci. Instr. 77, 03A323 (2006).
16. A.C.C Villari et al., in ``*International Symposium on Exotic Nuclei",* Khanty-Mansiysk (RU) July 17-22 j 2006, AIP Conference Proceedings **912**, p. 436 (2007).
17. A. Drouart et al., in ``*XVth International Conference on Electromagnetic Isotope Separators and Techniques related to their Applications (EMIS2007)"*, France (2007, June 24th-27th), to be published.
18. G. Savard et al., in ``*XV^th^ International Conference on Electromagnetic Isotopes Separators and Techniques related to their applications",* Deauville, 24'h-29th June 2007, to be published.
19. S[3] Collaboration LOI-SPI2-03 ,http://www.ganil.fr/research/developments/spiral2/loi_texts.html
20. M. Fadil et al., in ``*XV^th^ International Conference on Electromagnetic Isotopes Separators and Techniques related to their applications EMIS 2007"*, Deauville, 24'h-29th June 2007, to be published.
21. C. Huet-Equibec et al., Nucl. Instr. Meth. **B 240**, 752 (2005).

22. DESIRCollaborationLOI-SPI2 ,see Ref. 18)
23. L.Gaudefroy et al., Phys. Rev. Lett. **97**, 092501 (2006).
24. B.Bastin et al., Phys. Rev. Lett. **99**, 022503 (2007).
25. R. Ferdinand et al., in ``*10th European Particle Accelerator Conference (EPAC'06)",* Edinburgh, 26-30 June 2006, MOPCH103 (EPS-AG) (2006).
26. P.E.Bernaudin et al., in `` *9th European Particle Accelerator Conference (EPAC 2004)",* Luzern, July 5-9 2004, TUPLT054, p 1276, (EPS-AG) (2004).
27. G. Olry et al., in Proc. of EPAC 2006, Edinburgh, 26-30 June 2006, MOPCH146, European Physical Society Accelerator Group (EPS-AG) (2006).

# The HIE-ISOLDE project

Karsten Riisager

*PH Department, CERN, 1211 Geneva 23, Swizerland*

**Abstract.** The HIE-ISOLDE project is a major staged upgrade of the existing ISOLDE facility. It includes an energy increase of the REX post-accelerator up to 5.5 MeV/u with a future option of going to 10 MeV/u, as well as an upgrade of the REX low energy stage capacity. The beam quality will be improved e.g. with the installation of a RFQ cooler and a new Resonant Laser Ionization System. The driver intensity will be increased, at first from a faster cycling of the PS Booster and later from the new Linac-4.

**Keywords:** Radioactive beam production and acceleration
**PACS:** 28.60.+s, 29.25.Rm, 29.38.Gj

## INTRODUCTION

Radioactive beam production [1] has during the last decades triggered a resurge in the interest in nuclear structure physics. The ISOLDE facility at CERN [2, 3] has through its 40 year history been one of the premier ISOL facilities world-wide, see [4, 5] for details on the ISOL method. The existing facility for accelerated RIB at ISOLDE, REX-ISOLDE [6], exploits innovative techniques using ion traps and new methods for charge breeding of atoms that are unique worldwide. REX (Radioactive beam EXperiment) was originally constructed for specific experiments and was funded by several CERN member states, but is now integrated into the facility and has the potential to transport and accelerate most of the close to 1000 radioisotopes producable at ISOLDE. The radionuclides are created through bombardment of various primary targets with 1.4 GeV protons from the CERN PS Booster. Figure 1 shows the secondary beam intensities measured until 2006, the efficiency of accelerating these beams in REX-ISOLDE is for most elements in the range 5–15 %. Since it became operational in 2001, REX has accelerated 47 different radioactive isotopes [7], which surpasses by far any other European ISOL RIB facility. The HIE-ISOLDE proposal represents a major upgrade of the facility [8] and the physics case for the upgrade is documented in [9], more material can be found through a dedicated website [10].

## THE REX-ISOLDE ENERGY UPGRADE

The REX-ISOLDE postaccelerator [6] has since its start of operation in 2001 delivered in average about 120 eight hour shifts per year for nuclear physics. The concept of a compact linac combined with a cooler, buncher and charge breeder has proven to be a cost efficient and reliable option for the production of radioactive ion beams of energy up to several MeV/u. A clear advantage is the ability to accelerate all ions produced in an ISOLDE target thereby profiting from the many years of target and ion source

CP1012, *Frontiers in Nuclear Structure, Astrophysics, and Reactions: FINUSTAR 2*
edited by P. Demetriou, R. Julin, and S. V. Harissopulos
© 2008 American Institute of Physics 0-7354-0532-5/08/$23.00

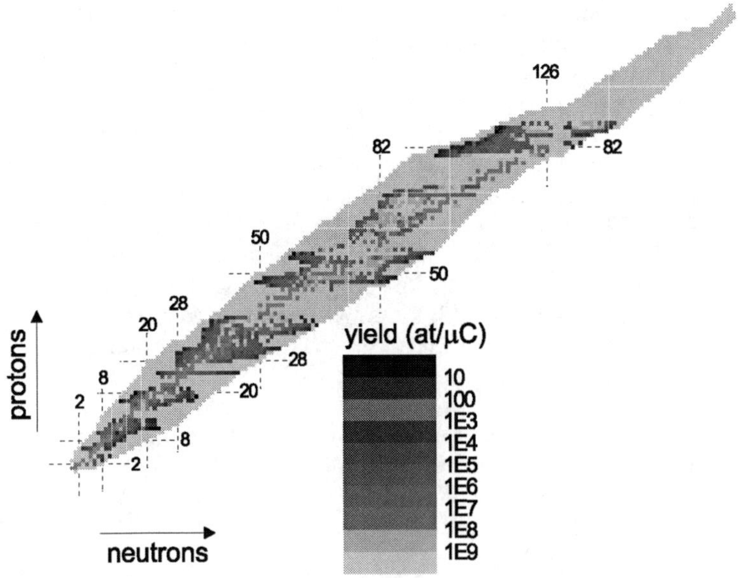


**FIGURE 1.** Yields in atoms per $\mu$C primary proton beam as measured at ISOLDE. The present beam intensity is up to 2 $\mu$A.

development. A good illustration is the recent postacceleration of an isomeric beam [11], but the Miniball gamma-detector array [12] has recently delivered many other Coulomb excitation results from much of the nuclear chart [13, 14, 15]. To continue this in an efficient manner also for the heaviest elements and to allow to explore for more nuclei the transfer experiments presently only performed for the lightest elements [16] requires an energy increase up to at least 5.5 MeV/u. The upgrade proposed within HIE-ISOLDE concerns apart from the top energy also the capacity to handle the full intensity and mass range of available ISOLDE isotopes.

## The REX linac upgrade

The RF group of the AB department made a study in the spring of 2006 of two possible technologies for the REX energy upgrade; a superconducting (SC) option and a continuation of the present room temperature technology. The two options were presented for an international advisory panel which gave a unanimous recommendation for a super conducting linac due to its higher flexibility.

The design is based on quarter wave resonators which can be installed downstream the present linac and eventually replace some of the existing accelerating structure. The energy upgrade is foreseen in three stages; first the maximum energy will be limited to 5.5 MeV/u with energy variability between 2 and 5 MeV/u. Later the final energy will be incremented up to 10 MeV/u, and finally the present low energy section from

1.2 MeV/u will be changed to superconducting cavities. This latter stage will allow to decelerate and to transport the beam to energies lower then 1.2 MeV/u hence increasing the possible use of the machine.

Two gaps quarter wave resonators are chosen as building element of the linac. The reduced numbers of gaps assure a very high flexibility in terms of velocity acceptance and also that a small number of different geometries suffice to cover the whole energy range. The energy range between 1.2 MeV/u and 10 MeV/u corresponds to a reduced velocity between 5.1% and 14.5%; the energy of 5.5 MeV/u corresponds to a reduced velocity of 10.8%. The RF cavities should give 6 MV/m as effective accelerating gradient with a power dissipation less then 7 W. The active lengths are 180 mm for the low beta cavity and 300 mm for the high beta one. This means a total of 1.08 MV and 1.8 MV of voltage gain respectively at the optimum velocity. The final energy of a beam will depend of its A/q, the values quoted above correspond to A/q = 4.5 .

The new extension hall built a few years ago provides an additional space of around $23 \times 17$ m$^2$. The magnetic rigidity for a beam of 10 MeV/u with A/q = 4.5 is about 2 Tm, and in order to optimize the space availability, large bending angles are necessary. At least two beam lines are planned, one of which will include the Miniball set-up. A zero degree spectrometer will be installed later, but is not itself a part of the HIE-ISOLDE project.

## THE REX LOW ENERGY STAGE UPGRADE

The REX low energy stage consists of a Penning trap used for cooling and bunching and an Electron Beam ion source (EBIS) for charge state multiplication. This approach to charge state multiplication was pioneered at REX-ISOLDE and has proven highly successful. The EBIS source operates with UHV vacuum and as a result the extracted ion beam is close to background free with the main source of contaminating ions being the inherent isobaric contamination of the injected radioactive ion beam itself. An upgrade of the REX low energy stage is necessary to enable charge breeding of heavier elements with maintained breeding time (half life limitation) and to handle higher intensities. To do this a higher electron density is required in the EBIS source. The possible upgrades of the present source are limited and the requirements can be met by installing a new EBIS similar to the BNL TestEBIS [17]. This system has higher electron beam current density and total current so both the breeding time (reachable charge state) and the total space charge capacity can be improved with this system.

The alternative option of employing an Electron Cyclotron Resonance charge breeder is studied within EURONS and EURISOL DS. This system is of key importance if physics would require continuous (CW) beams from REX-ISOLDE. Such a development would require the ECR charge breeder to be installed in parallel to the present REX low energy stage and it would also need a new low energy acceleration stage.

## ISOLDE DRIVER BEAM INTENSITY UPGRADE

The approved proton beam based physics programme at PS, SPS and LHC will from 2008 and onwards increase the demand for protons at CERN. A continued operation of ISOLDE at the present proton beam intensity (2 $\mu$A) will require an increase of the number of protons delivered by the PS booster. One solution is to run the PS booster with a shorter cycling than the present 1200 ms, e.g. 900 ms as tested in June 2005 with ISOLDE as a single user. Shared operation would require the replacement of some transfer line power supplies, but more machine studies are needed.

The proposed new injector for the PS Booster, LINAC 4, would result in a doubling of the number of protons per pulse from the PS Booster. This will together with the faster cycling of the PS Booster permit an important intensity upgrade at ISOLDE. In average ISOLDE could receive up to 6 $\mu$A and for limited periods even higher intensities.

To enable ISOLDE to receive the higher proton beam intensities from LINAC 4 a number of modifications of the targets and front-ends has to be done. CERN's Saftey Commission has already made a first study of the safety aspects and the most costly of the required modifications is the construction of gallery above the existing ISOLDE target area in order to provide shielded access to the equipment and minimize the radiation dose received during routine operation and maintenance of the facility. A number of other important changes are also necessary, but benefit to some extent from the on-going EURISOL design study in which CERN plays a key role for target development.

## ISOLDE BEAM QUALITY IMPROVEMENTS

A clear trend in experiments is to require better beam quality: higher intensity, smaller transverse emittances or bunched beams with small longitudinal emittance (energy spread and bunch length). The device best suited to tailor low energy beams transversally and longitudinally is an RFQ cooler which delivers a small transverse emittance beam with the freedom to adapt the time structure and energy spread of the beam to meet specific user requirements. A general purpose RFQ cooler (ISCOOL) [18] with many new features compared to similar systems operating at e.g. the ISOLDE mass experiments and at the cyclotron lab in Jyväskylä has recently been developed at ISOLDE and it will be installed in 2007.

The Resonant Ionization Laser Ion Source (RILIS) at ISOLDE is a quite selective ion source [19] and is the most requested ion source system at the ISOLDE facility. The present RILIS system is based on Copper Vapor Lasers (CVL) pumping dye lasers with non-linear crystals for frequency doubling and tripling. Solid State Lasers (SSL) can replace the CVL allowing most of the currently used ionization schemes still to be used. Regarding operation and maintenance, SSLs are better: they do not require long-time preheating, the power supply control is relatively simple, and the level of electromagnetic noise is much lower compared to CVLs. The development of new ionization schemes and the R&D to improve existing schemes is very time consuming and is today exclusively done at the on-line facility, so only a very limited number of hours are available each year for this work. To circumvent this an off-line LAser Resonance Ionization Spectroscopy laboratory (LARIS) is presently constructed.

## RESOURCES AND PLANNING

The resources required for the HIE-ISOLDE project has been studied by the AB department, the complete project will require 34 MCHF and 128 person-year. The ambition is to complete the first stage of the energy upgrade to 5.5 MeV/u as well as the beam quality improvements by 2010 and to complete the project in 2012 to benefit from the higher intensities available with the start of LINAC 4 operation in that year.

A substantial part of the funding must be obtained from sources external to CERN, at present 3.8 MCHF has already been secured for the first stage. A grant was given in the UK to finance the construction and installation of an RFQ cooler for general use at the high resolution separator at ISOLDE. The off-line testing and installation has been posssible thanks to additional support from other partners. The new RILIS system and the LARIS laboratory have been fully funded by a generous grant from the Wallenberg foundation in Sweden. The lasers for the LARIS laboratory have been inherited from the SILVA project at CEA Saclay.

## ACKNOWLEDGEMENTS

It is a pleasure to acknowledge the important contributions from colleagues at ISOLDE and the ISOLDE collaboration, including Mats Lindroos, Peter Butler and the many contributors to the two HIE-ISOLDE reports.

## REFERENCES

1. H. Geissel, G. Münzenberg, K. Riisager, Ann. Rev. Nucl. Part. Sci. **45**, 163 (1995).
2. E. Kugler, Hyperfine Int. **129**, 23 (2000).
3. http://www.cern.ch/isolde
4. H.L. Ravn, B.A. Allardyce, *Treatise on heavy-ion science*, vol. 8, ed. D.A. Bromley, Plenum, 1989, p. 363.
5. P. Van Duppen, *The Euroschool Lectures on Physics with Exotic Beams*, vol. II, Ed. J. Al-Khalili, E. Roeckl, Lect. Notes Phys. **700**, Springer, Berlin, 2006, p. 37.
6. O. Kester et al., Nucl. Instr. Meth. **204**, 20 (2003).
7. D. Voulot et al., Radioactive beams at REX-Isolde: present status and latest developments, proceedings of EMIS2007, to appear
8. HIE-ISOLDE: the technical options, edited by T. Nilsson and M. Lindroos, CERN report, CERN-2006-13
9. HIE-ISOLDE: the scientific opportunities, edited by P. Butler, M. Huyse, R. Krücken and K. Riisager, accepted for publication as CERN report, 2007
10. http://hie-isolde.web.cern.ch/HIE-ISOLDE/
11. I. Stefanescu et al., Phys. Rev. Lett. **98** , 122701 (2007).
12. J. Eberth et al., Prog. Part. Nucl. Phys. **46** , 389 (2001).
13. O. Niedermaier et al., Phys. Rev. Lett. **94** , 172501 (2005).
14. A.M. Hurst et al., Phys. Rev. Lett. **98**, 072501 (2007).
15. J. Cederkall et al., Phys. rev. Lett. **98**, 172501 (2007).
16. H.B. Jeppesen et al., Phys. Lett. **B642**, 449 (2006).
17. E.N. Beebe et al., Rev. Sci. Instrum. **75**, 1542 (2004).
18. I. Podadera-Aliseda, PhD dissertation, Barcelona, 2006, CERN-THESIS-2006-034
19. V. Fedosseev et al.: Nucl. Instr. Meth. **204**, 353 (2003).

# Neutron capture reactions at DANCE

T.A. Bredeweg and the DANCE Collaboration

*Los Alamos National Laboratory, Los Alamos, NM, 87545, USA*

**Abstract.** The Detector for Advanced Neutron Capture Experiments (DANCE) is a $4\pi$ $BaF_2$ array consisting of 160 active detector elements. The primary purpose of the array is to perform neutron capture cross section measurements on small ($\gtrsim 100$ $\mu$g) and/or radioactive ($\lesssim 100$ mCi) species. The measurements made possible with this array will be useful in answering outstanding questions in the areas of national security, threat reduction, nuclear astrophysics, advanced reactor design and accelerator transmutation of waste. Since the commissioning of DANCE we have performed neutron capture cross section measurements on a wide array of medium to heavy mass nuclides. Measurements to date include neutron capture cross sections on $^{241,243}$Am, neutron capture and neutron-induced fission cross sections and capture-to-fission ratio ($\alpha = \sigma_\gamma/\sigma_f$) for $^{235}$U using a new fission-tagging detector as well as neutron capture cross sections for several astrophysics branch-point nuclei. Results from several of these measurements will be presented along with a discussion of additional physics information that can be extracted from the DANCE data.

**Keywords:** neutron capture, neutron-induced fission, gamma-ray calorimeter
**PACS:** 28.20.Fc, 25.85.Ec, 29.40.Vj, 25.60.Dz

## INTRODUCTION

The accurate measurement of neutron capture cross sections for neutron energies ranging from sub-eV to several hundred keV, particularly on small quantity and/or radioactive samples, is of fundamental importance to numerous fields of scientific study. Applications that would benefit from such measurements include stockpile stewardship, nuclear forensics, s-process nucleosynthesis, the advanced fuel cycle initiative (AFCI), the new, much broader Global Nuclear Energy Partnership (GNEP), generation IV reactor design and nuclear threat reduction initiatives. Measurements on radioactive samples are practically unexplored, leaving large regions of these fields relying on untested theoretical models for cross-section predictions. Past facilities were designed to operate with gram-sized samples making measurements on radioactive samples impossible.

## OVERVIEW OF THE DANCE ARRAY

The Detector for Advanced Neutron Capture Experiments (DANCE) is a 160 element $\gamma$-ray calorimeter with high granularity and nearly $4\pi$ coverage coupled to state-of-the-art data acquisition electronics[1, 2]. The primary purpose of the array is to perform neutron capture measurements on small ($\gtrsim 100$ $\mu$g) and/or radioactive ($\lesssim 100$ mCi) samples. The DANCE array is located on flightpath 14 of the Manuel Lujan Jr. Neutron Scattering Center of the Los Alamos Neutron Science Center (LANSCE). Neutrons are produced through spallation reactions of 800 MeV protons incident on a tungsten neutron production target. The proton pulse width at the production target is approximately 125 ns

CP1012, *Frontiers in Nuclear Structure, Astrophysics, and Reactions: FINUSTAR 2*
edited by P. Demetriou, R. Julin, and S. V. Harissopulos
© 2008 American Institute of Physics 0-7354-0532-5/08/$23.00

full-width at half maximum with a repetition rate of 20 Hz. Flight-path 14 views a partially coupled light water moderator that produces a neutron spectrum with a Maxwellian peaked near thermal energies which drops off as $1/E_{\mathrm{n}}$ above approximately 1 eV[3, 4]. Neutron energies are determined by time-of-flight relative to the proton pulse arrival at the production target. Calibration of neutron energy from the measured time-of-flight is performed by analyzing neutron capture resonances in a thin gold target.

## Data Acquisition

The $BaF_2$ signals are digitized for software pre-processing using Acqiris model DC265 CompactPCI/PXI compliant fast transient digitizers[1]. The DC265 has a maximum sampling rate of 500 mega-samples per second (MS/s) and a base memory depth of 128 kilo-samples per input channel. Each DC265 digitizer card consists of four semi-independent signal input channels and a common external trigger input. The digitizer cards are mounted in Acqiris model CC108 6U CompactPCI crates. Each crate can accommodate six DC265 digitizer cards plus one single board computer (SBC) running Redhat Enterprise Linux 3.0[2] for digitizer control and readout. The digitizers are accessed from software through a kernel level device driver and a C/C++ interface library, both of which are provided by Acqiris.

Two digitizer channels are used per detector, which can be used to look at high and low gains or short and long times. Thus each digitizer crate can support twelve $BaF_2$ detectors. It therefore requires fourteen Acqiris crates, each with its own SBC, to accommodate the 160 $BaF_2$ detectors. A fifteenth crate with one Acqiris model DC270 digitizer card and an SBC is used for three neutron beam monitors for flux normalization.

## GEANT4 Model of the DANCE Array

In order to better understand the efficiency and response functions for the DANCE array it was decided to construct a model of the array using the GEANT4 simulation toolkit[5, 6]. The model includes the following materials: 1) aluminum beam pipe, 2) 6 cm thick $^{6}$LiH neutron moderator shell, 3) each $BaF_2$ crystal with 1 mm of polyvinyl chloride wrapping material, 4) 1 mm thick quartz photomultiplier window butted against the back face of each crystal, 5) aluminum holders for each $BaF_2$/PMT detector element, and 6) external aluminum support structures ( 5 cm thick)[7]. A partial image of the GEANT4 model is provided in Panel a) of Fig. 1.

The GEANT4 model was validated by simulating decay cascades for the three $\gamma$-ray sources ($^{22}$Na, $^{60}$Co and $^{88}$Y) that we regularly use for energy calibrations, and comparing the results to the experimentally measured spectra. The results for $^{88}$Y are shown in Panel b) of Fig. 1. The two pairs of curves are for cuts on DANCE cluster

---

[1] Acqiris USA, http://www.acqiris.com/

[2] Red Hat, Inc., http://www.redhat.com/

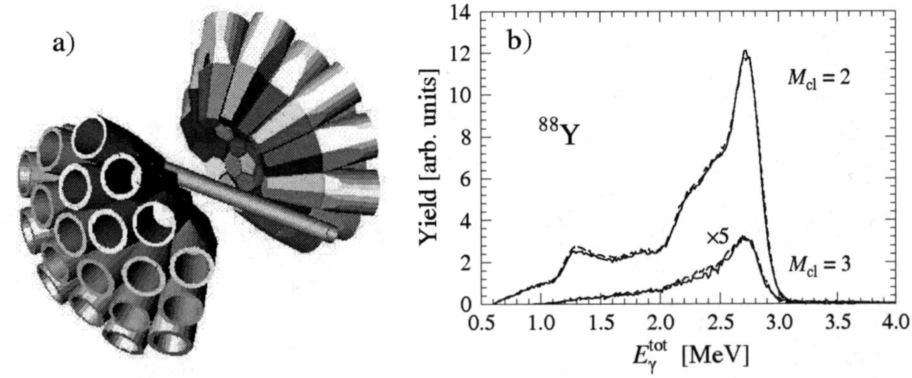


**FIGURE 1.** Panel a) Image of the GEANT4 model of the DANCE array showing forty of the $BaF_2$ detector elements and the neutron beam pipe. The $^6LiH$ shell, remaining detector elements and aluminum support structures have been removed for clarity. Panel b) Experimental (solid line) and simulated (dashed line) total $\gamma$-ray energy spectra for an $^{88}Y$ source. Spectra for cluster multiplicities $M_{cl} = 2$ and 3 are shown.

multiplicity[3] ($M_{cl}$) of 2 and 3. In each pair the experimental data are indicated by the thick line and the results of the simulation are indicated by the thin line. The agreement is excellent for all multiplicities and all three sources and, in most cases, the simulated spectra are indistinguishable from the experimental data. This model has since been used to determine detection efficiencies for production targets by using $\gamma$-ray cascades generated by a Monte Carlo simulation using the DICEBOX code[8].

## Auxiliary Detectors

Measurement of the neutron capture cross section for many of the actinides is complicated by the presence of low energy neutron-induced fission. Efforts to separate prompt $\gamma$-rays arising from fission from those following capture, based on $\gamma$-ray multiplicity and summed $\gamma$-ray energy, have resulted in only partial success, and that only by using an existing evaluation of the fission cross section for the isotope of interest. The addition of a fission-tagging detector around the target can be used to characterize the $\gamma$-ray spectrum and multiplicity arising from fission. The characterization of the fission component can then be used to perform a "background" subtraction, corrected for the efficiency of the fission-tagging detector, to enable good separation of the neutron capture events.

The PPAC fission-tagging detector is a dual parallel-plate avalanche counter which is operated in transmission mode. A photograph of the PPAC fission-tagging detector and removable target/cathode foil along with an exploded schematic is provided in

---

[3] A cluster is a group of adjacent DANCE crystals registering simultaneous hits within a small coincidence window, usually ~10–20 ns. Combined data analysis and GEANT4 simulations indicate that the DANCE cluster multiplicity closely tracks the actual $\gamma$-ray multiplicity of the de-excitation cascade[7].

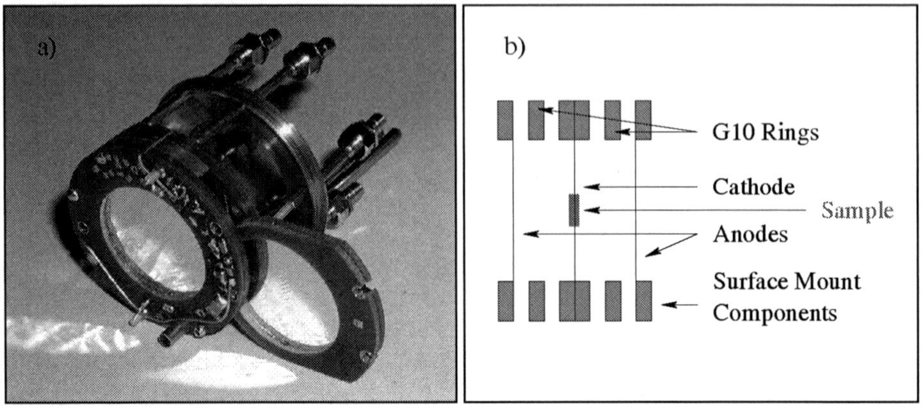


**FIGURE 2.** Panel a) Picture of a PPAC fission-tagging detector and the removable target/cathode foil partially inserted. The surface mount components for the first amplification stage can be seen on the front ring. The neutron beam passes through a 2.5 cm opening in the center of the G10 frame. Panel b) Exploded schematic of the PPAC detector showing the arrangement of the two anode foils and the cathode foil with the sample.

Fig. 2. The actinide sample is electro-deposited directly on the center cathode foil, which is mounted on a G10 fiberglass frame. Fission fragments are detected in two 3.2 mm low pressure gas regions on either side of the target/cathode foil. Isobutane is used for the fill gas with an operating pressure of 6-10 Torr. Signals are collected independently from the two anode foils. Each signal is amplified through two onboard MAR-6 monolithic microwave IC amplifiers[4] or one onboard MAR-6 amplifier and an external A81-1 amplifier[5] located approximately one meter from the detector. The later configuration was developed to eliminate feedback oscillations between the cascaded MAR-6 amplifiers when run at high gain. Semi-rigid coax cables are used to carry the detector bias, test inputs and output signals through the beam pipe to a break-out box located just downstream of the DANCE array. Similarly, detector gas supply and return are carried through 0.25 cm I.D. stainless steel tubing running between the detector and break-out box. An external gas handling system, with LabVIEW[6] based monitoring and control capabilities, controls gas flow and pressure. The active foils (anodes and cathode) located in the neutron flight path are made of thin metalized Mylar (400-550 $\mu$g/cm2 total thickness) to limit the background from capture on these materials and scattered neutrons that are then captured in the $BaF_2$ crystals. The removable target/cathode foil allows for reuse of a single detector for several targets and associated background blanks. This design also allows easy recovery of rare or otherwise valuable sample material for use in other experiments. The relatively low cost of this design also allows for dedicated detectors to be used to limit handling requirements for high activity samples such as

[4] Mini-circuits RF/IF Microwave Components, http://www.minicircuits.com/
[5] M/A-COM Inc. RF/Microwave Products, http://www.macom.com/
[6] National Instruments Corp., http://www.ni.com/

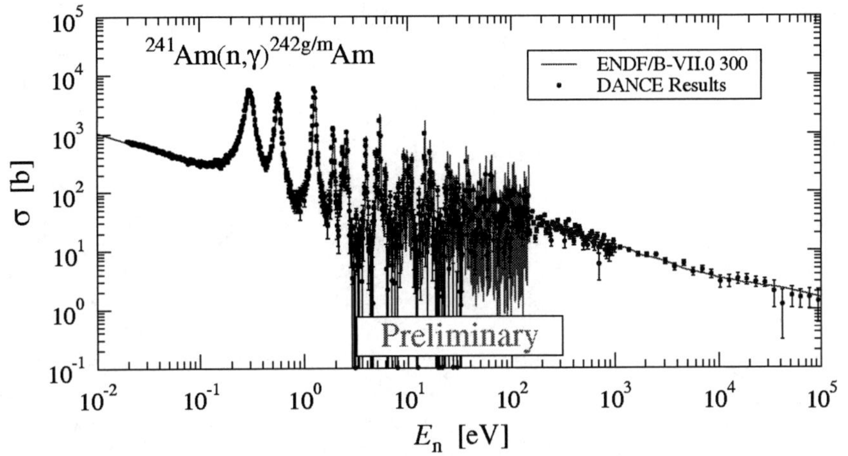


**FIGURE 3.** Preliminary results for neutron capture cross section measurement for $^{241}$Am obtained with the DANCE array (points). The current ENDF/B-VII cross section is included for comparison (line).

$^{238}$Pu and $^{242m}$Am[9]. The decision to use fission-tagging for a particular actinide will depend on the strength of the fission channel relative to capture and the desired level of accuracy for the measurement. For several of the threshold fissioners, such as $^{241}$Am and $^{243}$Am[10], the fission cross section is two orders of magnitude or more below the the capture cross section in the neutron energy region available for flight-path 14 (thermal to $\approx$ 500 keV). In this case fission-tagging might not be necessary for a 5% measurement, but would be required for a 1% measurement. The efficiency of the fission-tagging detector can be empirically determined from the data by comparing 2D histograms of cluster multiplicity versus summed $\gamma$-ray energy in the DANCE array for tagged and untagged events.

## NEUTRON CAPTURE CROSS SECTIONS

Uncertainties present in the current evaluations for neutron-induced reactions on many nuclides have prompted a call for new, high-fidelity measurements of the pertinent cross sections[11]. The neutron capture cross section for $^{241}$Am is of great importance to a number of applied fields including advanced reactor design and accelerator transmutation of waste. Unfortunately the values provided by the three main evaluations, ENDF/B-VII, JEFF-3.1 and JENDL-3.3, differ by as much as 20% in some energy regions.

The DANCE measurement was made with 700 $\mu$g of isotopically enriched $^{241}$Am electro-deposited onto a thin titanium backing. Preliminary results for the $^{241}$Am(n,$\gamma$) measurement are compared with the current ENDF/B-VII evaluated cross section in Fig. 3. A more detailed comparison shows that the new DANCE results tend to fall between the extremes in the three evaluations mentioned above. We are currently working to put the DANCE data on an absolute scale before final publication and release to the nuclear data evaluators.

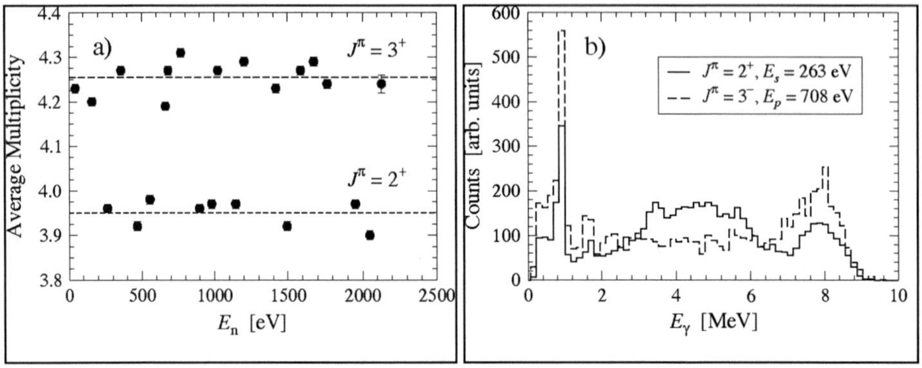


**FIGURE 4.** Panel a) Average DANCE cluster multiplicities for $s$-wave neutron capture resonances from the $^{95}$Mo(n,$\gamma$)$^{96}$Mo reaction in the neutron energy range from 40 to 2100 eV. Values along the lower line correspond to spin $J = 2$ in the compound nucleus, and values along the upper line to spin $J = 3$. Panel b) Measured $\gamma$-ray energy spectra for an $s$-wave neutron capture resonance at $E_n = 263$ eV ($J^\pi = 2^+$) and a $p$-wave resonance at $E_n = 708$ eV ($J^\pi = 3^-$).

## STATISTICAL PROPERTIES OF NUCLEI

Recently there have been efforts to extract additional information from the DANCE data, beyond the original cross section measurements. For certain initial target nuclei it is possible to deduce the spin states of isolated neutron capture resonances by examining the multiplicity distribution of the $\gamma$-ray cascade of the resulting compound nucleus. The technique was first proposed in the 60's by Coceva, *et al.*[12] and later refined into multiplicity spectrometry by Muradyan *et al.*[13]. To date the multiplicity technique has been applied to DANCE data for $^{94,95}$Mo(n,$\gamma$) [14] and $^{147}$Sm(n,$\gamma$)[15].

For the case of neutron capture on $^{95}$Mo ($J^\pi = \frac{5}{2}^+$) the average cluster multiplicity was extracted for each $s$-wave resonance from 40 to 2100 eV. The resulting average multiplicities ($\langle M_{cl} \rangle \approx \langle M_\gamma \rangle$) fall into two distinct groups corresponding to the two final state spins ($J^\pi = \frac{5}{2}^+ \pm \frac{1}{2} = 2^+$ and $3^+$) of the compound nucleus $^{96}$Mo. The results are shown in Panel a) of Fig. 4. This technique allowed for definite spin assignments for 75% of the $s$-wave neutron capture resonances and tentative spin assignments for the remaining 25%[14].

In this analysis it was crucial to separate $s$-wave resonances from $p$-wave resonances. This was achieved by examining the shapes of the $\gamma$-ray energy spectra for cluster multiplicity $M_{cl} = 2$ for each resonance. An example for $s$ and $p$-wave resonances is provided in Panel b) of Fig. 4. The measured $\gamma$-ray energy spectra correspond to two-step cascades as described in[14].

## SUMMARY

The DANCE array is a powerful new tool to measure neutron capture cross sections on a wide range of nuclides of interest to applied fields such as generation-4 reactor design, advanced fuel cycles, astrophysics, science based stockpile stewardship and threat reduction. In addition, new ideas are being tested to use DANCE to probe nuclear structure and answer other basic physics questions.

## ACKNOWLEDGMENTS

Work was performed under the auspices of the U.S. Department of Energy by the University of California (under contracts W-7405-ENG-36 [LANL] and W-7405-ENG-48 [LLNL]) and Los Alamos National Security, LLC (under contract DE-AC52-06NA25396). This work has benefited from the use of the Los Alamos Neutron Science Center.

## REFERENCES

1. M. Heil, et al., *Nucl. Instr. and Meth. A* **459**, 229–246 (2001).
2. J. Wouters, et al., "Acquisition-analysis system for the DANCE (Detector for Advanced Neutron Capture Experiments) $BaF_2$ $\gamma$-ray calorimeter," in *Proceedings of the 14th Conference On Real Time (RT2005), Stockholm, Sweden, June 4-10, 2005*, Published in IEEE Trans. Nucl. Sci. 53, 2006, pp. 880–885.
3. G. Muhrer, et al., *Nucl. Instr. and Meth. A* **527**, 531–542 (2004).
4. G. Muhrer, et al., *Nucl. Instr. and Meth. A* **547**, 555–568 (2005).
5. S. Agostinelli, et al., *Nucl. Instr. and Meth. A* **506**, 250–303 (2003).
6. J. Allison, et al., *IEEE Trans. Nucl. Sci.* **53**, 270–278 (2006).
7. M. Jandel, et al., "GEANT4 simulation of the DANCE array," in *Proceedings of the 19th International Conference on the Application of Accelerators in Research and Industry (CAARI 2006), Fort Worth, Texas, August 20-25, 2006*, published in Nucl. Instr. and Meth. B **261**, 2007, pp. 1117–1121.
8. F. Bečvář, *Nucl. Instr. and Meth. A* **417**, 434–449 (1998).
9. W. Parker, et al., "Review of Livermore-Led Neutron Capture Studies Using DANCE," in *Proceedings of the International Conference on Nuclear Data for Science and Technology (ND2007), Nice, France, April 22-27, 2007*, to be published by Springer-Verlag, 2008.
10. D. Vieira, et al., "Neutron Capture and (n,2n) Measurements on $^{241}$Am," in *Proceedings of the International Conference on Nuclear Data for Science and Technology (ND2007), Nice, France, April 22-27, 2007*, to be published by Springer-Verlag, 2008.
11. G. Aliberti, et al., *Ann. Nucl. Energy* **33**, 700–733 (2006).
12. C. Coceva, et al., *Nucl. Phys. A* **117**, 586 (1968).
13. G. Muradyan, et al., *Nucl. Sci. and Eng.* **90**, 60 (1985).
14. S. Sheets, Ph.D. thesis, North Carolina State University (2007).
15. P. Koehler, et al., *Phys. Rev. C* **76**, 025804 (2007).

# Proton-induced fission on actinide nuclei at energies between 20 and 70 MeV

S. Isaev*, R. Prieels*, Th. Keutgen*, J. Van Mol*, Y. El Masri* and P. Demetriou†

* *FNRS and UCL, Université catholique de Louvain, B-1348 Louvain-la-Neuve, Belgium*
† *Institute of Nuclear Physics, NCSR Demokritos, 153.10 Athens, Greece*

**Abstract.** $^{232}$Th, $^{237}$Np, $^{238}$U, $^{239}$Pu, and $^{241}$Am targets were bombarded with proton beam of 26.5 and 62.9 MeV energies. In this contribution some preliminary results are presented and discussed. Comparison with the predictions of the nuclear-reaction model code TALYS is also shown.

**Keywords:** Fission, Actinides, Neutrons, Multiplicity, Masses
**PACS:** 24.75.+i, 28.20.-v, 25.85.Ge, 28.41.-i, 89.30.Gg

## INTRODUCTION

Considerable effort has been devoted over the last decades to feasibility studies of accelerator driven systems (ADS) for transmutation of long-lived radioactive wastes. In view of the increasing concern over the energy problem, novel nuclear techniques combining reactor and accelerator technology are currently of utmost interest for the nuclear industry. At the same time, the new technologies require a more extensive and reliable database of nuclear fission properties. High-energy neutron- and proton-induced fission data on sub-actinide and actinide nuclei are needed for new generation power plants (fast reactors) and ADS applications. On the theoretical front, fission data are also required for a better understanding of the nuclear processes involved and for the improvements of theoretical models and prediction codes.

## EXPERIMENT

A proton beam of 26.5 and 62.9 MeV ($E_p$), delivered by the Louvain-la-Neuve cyclotron CYCLONE, impinges on actinide targets located in the center of a reaction chamber which is surrounded by 69 large liquid- scintillator cells positioned in a $4\pi$ spherical arrangement of 4m diameter, constituting the so-called multidetector DEMON [1]. This setup allows the determination of the neutron energy and angular distributions and neutron multiplicity spectra associated with the fission process. Figure 1 displays the detector positions inside the reaction chamber. Two microchannel plate-Si diode detector assemblies GJ1 and GJ2, measure, in single mode, the time of flight (TOF) and the energy, of the fission fragments (FF) emitted by the target. Simultaneously, two large active surface (20X20 cm$^2$), X and Y position sensitive Multi-Wire Proportional gas Counters (MWPC), are located at mean $\Theta$ angles of $45^o$ and $135^o$ on both sides of the beam axis, to detect, in coincidence, the FFs emitted back-to-back in the center-of-mass

CP1012, *Frontiers in Nuclear Structure, Astrophysics, and Reactions: FINUSTAR 2*
edited by P. Demetriou, R. Julin, and S. V. Harissopulos
© 2008 American Institute of Physics 0-7354-0532-5/08/$23.00

frame (c.m.) (for more details see [2]). Their high angular (0.4$^o$) and time (500 ps) resolutions lead to a good determination of both FF velocities and angular distributions and thus mass distributions.

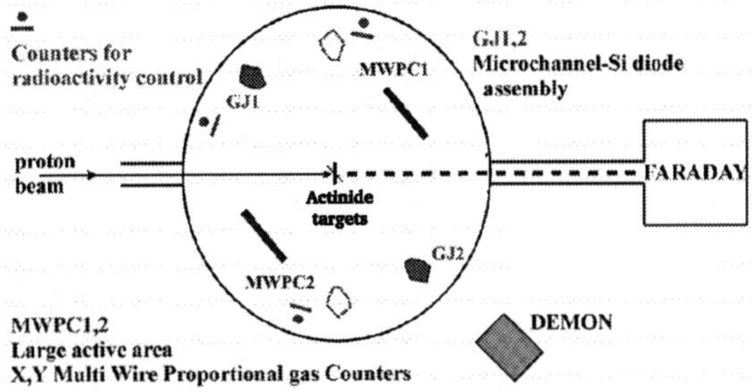


**FIGURE 1.** The experimental set-up, top view.

Some typical measured raw spectra are shown in Fig. 2. Binary fission is a back-to-back process in the rest frame. When boosted by a proton projectile, the 180$^o$ angle between both the produced FFs is reduced. Due to the small proton mass compared to the actinide nuclear masses, the recoil velocity of the initial composite nucleus (CN) is very small and the FFs are emitted quasi back-to-back in the laboratory system (lab) (see the second panel of Fig. 2). In this panel, one clearly observes that the transferred linear momentum (LMT) in these reactions is neither zero or full. In fact, the experimental fission folding angle is apparently in-between these two extreme cases clearly suggesting the occurence of incomplete fusion and more specifically, of preequilibrium (PE) particle emission.

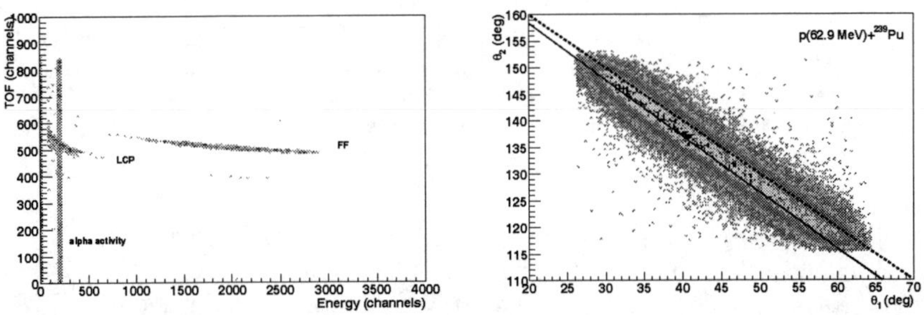


**FIGURE 2.** First panel: Energy-TOF bidimensional spectrum derived from one microchannel plate-Si diode detector in the experiment p(62.9 MeV) + $^{239}$Pu. The start time is initiated by the cyclotron radio-frequency signal. Second panel: Correlation spectrum between the FF $\Theta_1$ and $\Theta_2$ emission angles. The dotted line (or full line) corresponds to the hypothetic case of Zero (or Full) LMT of the proton at 62.9 MeV.

## ANALYSIS AND DISCUSSION

As shown in Fig. 3, the experimental total fission cross sections ($\sigma_f$) agree relatively well with the predictions of TALYS[3] at both bombarding energies. The small differences between the $\sigma_f$ measured at $E_p$ = 26.5 and 62.9 MeV (see the right panel of Fig. 3) suggest that $\sigma_f$ has already reached its saturation point below the lowest $E_p$.

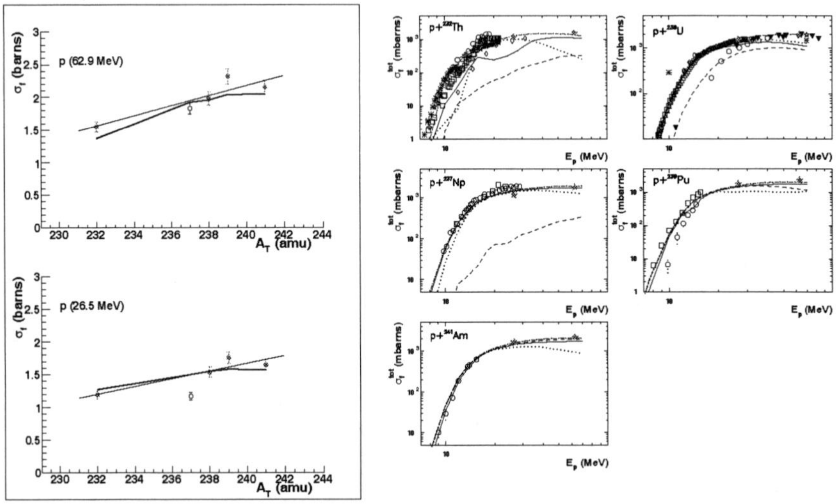


**FIGURE 3.** Left panel : Experimental ($\sigma_f$) (open circles), the error bars and the corresponding TALYS predictions (broken thick curve) as a function of the target mass numbers ($A_T$) at both $E_p$. The dotted line represents the best $\chi^2$-fit to the experimental points assuming a function of type $[a.A_T^{2/3}+1.]$. Right panel: Total fission cross sections for proton-induced fission on $^{232}$Th, $^{238}$U, $^{237}$Np, $^{239}$Pu and $^{241}$Am available in EXFOR data base (open circle) [4] compared with the newly measured $\sigma_f$ (red stars) at $E_p$ = 26.5 and 62.9 MeV and TALYS calculations using the Maslov [5] (solid line), RLDM [6] (dashed line), RFRM [7] (dot-dashed line), HFB [8, 3] (dotted line) fission barrier parameters.

Mass distributions of primary FFs were reconstructed for all actinide targets and are plotted in Fig. 4 together with the mass yields obtained from TALYS. Mixed (asymmetric and symmetric) distributions are observed at both $E_p$ with a clear increase of the symmetric contribution at 62.9 MeV. The latter effect seems to be more pronounced in the experimental mass distributions, however, they are not yet fully symmetric which suggests that shell effects are not completely washed out even at $E_p$ = 62.9 MeV. The cases of $^{238}$U, $^{239}$Pu are puzzling since the theoretical mass yields are more symmetric at 26.5 MeV which is clearly at odds with the experimental results. Note that we only compare the shapes of the experimental and theoretical FF mass distributions and not their absolute magnitudes, as they have not been normalized to the same observable and also because the experimental ones are affected by many experimental artifacts such as energy losses in materials, energy and angular stragglings, particle emissions, detector resolutions, and iterative approximations. Neutron yields and energy spectra were measured in the $4\pi$ geometry in coincidence with the MWPCs. The results of typical moving-source fits to the neutron kinetic energy and angular distribution spectra for $^{241}$Am target at different spatial angles and at both $E_p$ assuming PE, CN and FF1 and FF2 contributing sources are shown in Fig. 5. The neutron contributions from the FFs dominate at angles close to those of the detected FFs i.e. 45$^o$ and 135$^o$. Experimental values of different mean neutron multiplicities, i.e. $\nu^{PE}$, $\nu^{CN}$, $\nu^{post}_{sym}$ ($\nu^{FF1}_{sym}$ + $\nu^{FF2}_{sym}$), $\nu^{post}_{asym}$ and $\nu^{total}$ at the two $E_p$ are shown in Fig. 6. $\nu^{total}$ is compared with the TALYS predictions obtained with Sierk's fission barriers [7]. One observes that the PE contribution remains relatively small, while the CN contribution is quasi-independent of $A_T$ and significantly increases (by a factor $\approx$ 3) with increasing $E_p$. The post-scission multiplicities weakly increase with $A_T$ and are slightly higher at $E_p$ = 26.5 than at 62.9 MeV but they remain quasi-identical whenever symmetric or asymmetric FF mass partitions are selected. The total neutron multiplicities (around 4.7 at 26 MeV) and (about 6.8 at 62.9 MeV) are in agreement, within the error bars, with TALYS calculations at 62.9 MeV, but are partly underestimated at 26.9 MeV. Also shown in Fig. 6 is a comparison of the experimental and theoretical TKE mean values as a function of $A_T$ at the two $E_p$. The agreement is excellent with both TALYS and the systematics of Viola *et al.* [9].

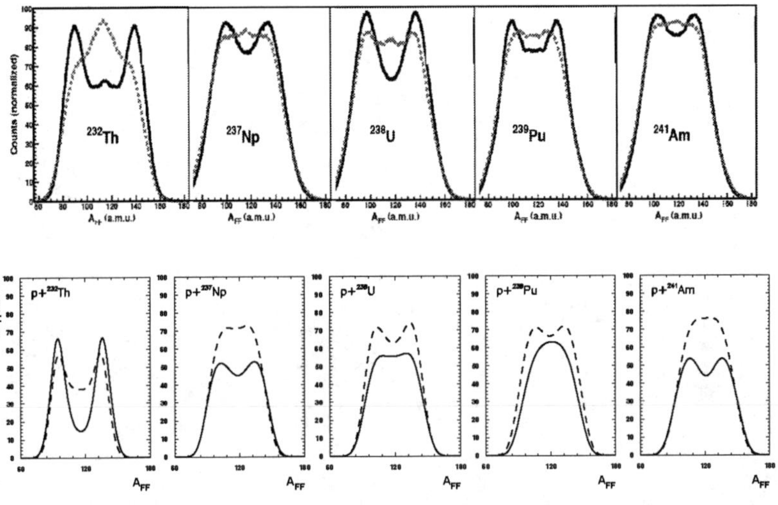


**FIGURE 4.** Upper panel: Mass distribution of FFs normalized to the same integral at 26.5 MeV (full line) and 62.9 MeV (dotted line) proton beams. Distributions corresponding to 26.5 MeV exhibit the strongest asymmetric behaviour. Lower panel: Mass yields of FFs obtained with TALYS at 26.5 MeV (full line) and 62.9 MeV (dotted line) incident energy.

The analysis is still in progress, so all the results presented herein are preliminary. From the theoretical point of view, more work is also needed to improve the description of the data using realistic input parameters and to understand the observed discrepancies.

## REFERENCES

1. Th. Keutgen et al., *Phys. Rev. C* **70**, 014611 (2004).
2. S. Isaev et al., *Phys. Rev.* (2007), to be published.

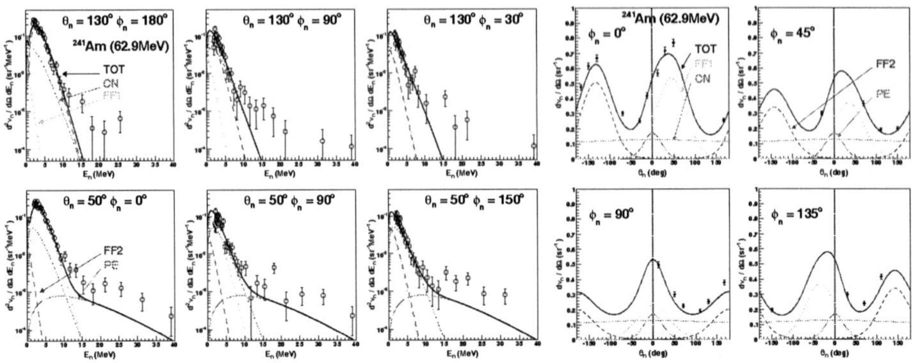


**FIGURE 5.** Left: Some neutron multiplicity (energy) spectra at 62.9 MeV for the $^{241}$Am target at different $\Theta_n$ and $\phi_n$ angles adjusted by four moving neutron-emitting sources. Right: Typical neutron angular distribution fits for the same target at $E_p$ = 62.9 MeV for different $\phi_n$ angles (in- and out-of the reaction plane i.e. $\phi_n = 0^o$ and $\phi_n = 180^o$ respectively).

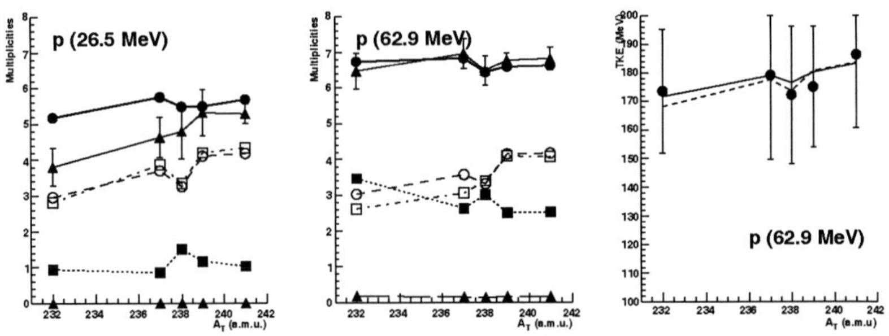


**FIGURE 6.** Left: Neutron multiplicity versus the mass number of the five actinide targets at the two $E_p$. The four broken lines represent the partial contributions of the moving emitting sources. For clarity the experimental errors are only shown for the total neutron multiplicities. The upper curve without error bars is the TALYS prediction. Right: Experimental FF TKEs and corresponding RMS values are compared with mean values obtained from TALYS code (full line) and Viola *et al.* [9] systematics (dotted line) at 62.9 MeV beam energy.

3. A.J. Koning, S Hilaire, M.C. Duijvestijn, *AIP Conference Proceedings 769* p. 1154 (2004).
4. EXFOR database (2005), URL {`www.nds-iaea.org`}.
5. Handbook for calculations of nuclear reaction dataL Reference Input Parameter Library (2001), URL `www.nds-iaea.org/RIPL-2/`.
6. S. Cohen and F. Plasil and W. J. Swiatecki, *Ann. of Phys.* **82**, 557 (1974).
7. A. J. Sierk, *Phys. Rev. C* **33**, 2039 (1986).
8. S. Goriely, M. Samyn, and J. M. Pearson, *Phys. Rev. C.* **75**, 064312 (2007).
9. V. E. Viola, K. Kwiatkowski, and M. Walker, *Phys. Rev. C* **31**, 1550 (1985).

# Towards Universal Predictions with the TALYS Code

S. Hilaire [1], S. Goriely [2], A.J. Koning [3] and E. Bauge [1]

[1] *CEA/DAM Ile-de-France, Service de Physique Nucléaire, BP 12, F-91680 Bruyères-le-Châtel, FRANCE.*
[2] *Institut d'Astronomie et d'Astrophysique, Université libre de Bruxelles, Campus de la plaine CP 226,B-1050, Brussels, BELGIUM.*
[3] *Nuclear Research and Consultancy Group NRG, P.O. Box25, 1755 ZG Petten, THE NETHERLANDS.*

**Abstract.** The increasing need for cross sections far from the valley of stability, for astrophysical as well as advanced technological applications poses a challenge for nuclear reaction models. So far, nuclear reaction cross sections calculations have relied on more or less phenomenological approaches, depending on several parameters that are adjusted, to available experimental data, or deduced from systematical relations. While such calculations are expected to be reliable for nuclei not too far from the experimentally known regions, it is clearly preferable to use more fundamental approaches, based on sound physical bases, when dealing with very exotic nuclei. Thanks to the high computer power available today, all the ingredients required to model a nuclear reaction can now be microscopically (or semi-microscopically) determined starting from a nucleon-nucleon effective interaction as sole input, and are available to the nuclear physics community. We have recently implemented all these microscopic ingredients in the TALYS nuclear reaction code, and we are now able to perform fully microscopic cross section calculations. We will discuss both the quality of these ingredients and the impact of using them instead of the usually adopted phenomenological parameters.

**Keywords:** Nuclear reaction models and methods, Nuclear Forces.
**PACS:** 24.10.-i, 21.30-x

## INTRODUCTION

The TALYS code results from the collaboration between NRG Petten, the Netherlands and CEA Bruyères-le-Châtel, France initiated in 1998, with the objective to provide a complete and accurate simulation of nuclear reactions in the 1 keV-200 MeV energy range, through an optimal combination of reliable nuclear models, flexibility and user-friendliness. The development of TALYS has followed what we can call the TALYS mantra: ***«first completeness then quality»***. This should certainly not suggest that we use toy models to arrive at some quick and dirty results. Indeed, several reaction mechanisms coded in TALYS are based on theoretical models whose implementation is only possible with the current-day computer power. It rather means that, in our quest for completeness, we try to divide our effort equally among all nuclear reaction types. The precise description of all possible reaction channels in a single calculational scheme is such an enormous task that we have chosen, to put it bluntly, not to devote several years to the theoretical research and absolutely perfect

CP1012, *Frontiers in Nuclear Structure, Astrophysics, and Reactions: FINUSTAR 2*
edited by P. Demetriou, R. Julin, and S. V. Harissopulos
© 2008 American Institute of Physics 0-7354-0532-5/08/$23.00

implementation of one particular reaction channel which accounts for only a few millibarns of the total reaction cross section. Instead, we aim to enhance the quality of TALYS equally over the whole reaction range and always search for the largest shortcoming that remains after the last improvement.

TALYS-1.0 has been released at the conference on Nuclear Data for Science and Technology which was held in NICE in 2007 [1]. The previous official release was the beta version TALYS-0.64 at the previous Nuclear Data conference held in Santa Fe in 2004. During these three years we could summarize the modifications/improvements that have occurred by changing the initial mantra into: ***«first completeness, now quality»***. In other words, TALYS-0.64, had somehow reached the searched completeness, and, since then, we have begun to improve the quality of the models contained in the code.

## WHAT "NOW QUALITY" MEANS !

One may believe that TALYS-0.64 was not good enough to reproduce experimental cross section data. In practice, the achieved completeness provides one with basic default results that can be obtained specifying only the projectile, the target, and the incident projectile energy. This is due to an extensive database or/and to the use of systematics giving the unspecified parameters of the various models chained to produce the final cross section. These default models are indeed generally based on analytical expressions depending on parameters which can be adjusted to fit the experimental data. If, in practice, such fitting procedures enable one to fit experimental cross sections quite well, even for the complicated case of a fissile nucleus (see fig.1 in ref.[1]), the analytical expressions rely on too simple approximations to be able to represent the complexity of a nuclear reaction from a ***universal*** point of view. Our quest for **quality** means in fact that we are now beginning to replace such analytic models by microscopic or semi-microscopic ones to reach roughly the same level of accuracy as with the phenomenological approaches when experimental data are available together with a greater confidence in the quality of the predictions when no data are available.

A good illustration is shown for instance when looking at the nuclear level densities. Indeed, all usually employed analytical expressions assume at least (i) that the spin distribution is given by the derivative of a Gaussian distribution and (ii) that there is an equipartition of positive and negative parity levels. These two approximations are clearly wrong for low energies and it has been shown [2-4] that departures from these two approximations could have a direct impact on the cross section predictions. Therefore, we have added in TALYS-1.0, the combinatorial level densities determined as in ref.[3] which are tabulated as function of the excitation energy, spins and parities. We are indeed convinced that the predictive power of such microscopic level densities is higher than that of the usually employed analytical expression especially when dealing with nuclei far from the valley of stability where the parameters of the latter are adjusted. Such microscopic level densities are still being improved [5] to suppress some approximation made in ref.[3], and will of course replace at term those of ref.[3] which are already included in the database which TALYS relies on. Since the microscopic models are not perfect either, one could multiply microscopic level densities by some adjustment function [3] if one needs to improve the fit to experimental data.

Another ingredient which governs the prediction of nuclear cross sections is the optical model potential (OMP). Again, the problem is similar to the level density problem. When experimental data are available, a phenomenological OMP reproduces much better the observables than a microscopic or semi-microscopic OMP. Following this idea, a global OMP has been developed [6] and selected as the default one in TALYS-0.64. An improvement of TALYS-1.0 is the inclusion of the global deformed OMP of ref.[7] which is used as the default for actinides. With this scheme, coupled-channels calculations for various types of deformation (symmetric-rotational, harmonic-vibrational, vibration-rotational, and asymmetric-rotational) can be automatically performed. However, this does not guaranty that extrapolations of these phenomenological OMPs far from the valley of stability is reliable. An alternative to this weakness is to use the semi-microscopic OMP of ref.[8] which has proven its predictive power. This semi-microscopic OMP is based on the knowledge of nuclear matter densities which are calculated within the Hartree-Fock-Bogolyubov (HFB) approach and tabulated in the TALYS structure database. As for the level densities, we believe that OMPs derived from sound physical bases, which in turn can compete with highly parameterized phenomenological models to reproduce experimental data globally, are certainly more reliable when going far from the regions where measurements have been performed.

Last but not least, is the gamma ray strength function. Again, several models have been included in TALYS-0.64 depending on adjustable parameters. A new option in TALYS-1.0 is the possibility to replace these phenomenological expressions by the microscopic HFB plus QRPA predictions from ref.[9]. As the level densities, or the radial matter densities, these microscopic gamma strengths are tabulated and stored in the database. As shown in ref.[10], such microscopic gamma ray strength may provide better results when compared to experimental data than the default phenomenological models.

All the previously mentioned "ingredients" must be known to predict nuclear reactions induced by any light particle on a non fissile nucleus. A good way of testing the reliability of the microscopic models consists in comparing what is obtained when using these models instead of the phenomenological ones. It is important to remember that two options are possible when using the phenomenological approaches. Either, one adopts global phenomenological models (i.e. without any adjustments of the model parameters to fit the experimental data) or one uses local phenomenological models which means that the parameters of such models have been fine tuned to fit the experimental cross section. Of course, the latter comparison is somehow biased since we compare the results of a fit with a purely microscopic prediction, but it is also a good way, if the microscopic approach performs well, to show the reliability of the microscopic inputs. Such a comparison is performed in fig.(1), where we compare, for several channels, the cross sections obtained after fine tuning the various models' parameters (full line) with the cross sections obtained using the microscopic inputs (dash-dot lines) as well as the default - i.e. without any adjustment of any kind - ones (dashed grey lines). It is important to mention that the semi-microscopic OMP [8] is used in the two first cases, while the global OMP of ref.[6] is adopted in the last case (dashed grey lines). As can be observed, the microscopic inputs provide quite good cross sections, almost as good as the fine tuned ones, and works even better than the default parameters. There are certainly cases where the microscopic inputs will provide worse results than the default parameters, but we believe that, in general, the quality of the microscopic ingredients is reasonably good, and, certainly better than the phenomenological extrapolations when going far from the experimentally accessible regions.

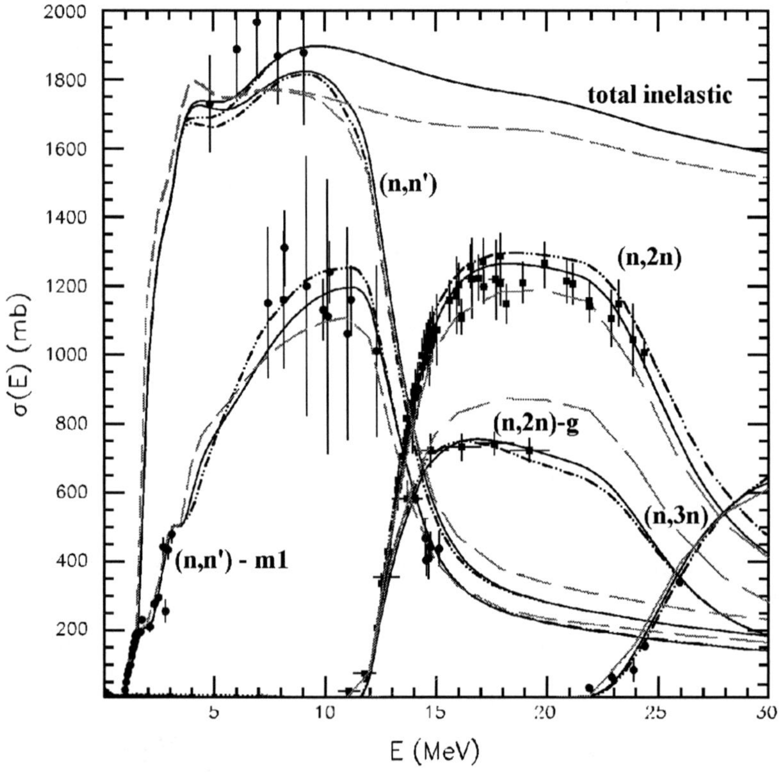


**FIGURE 1.** Excitation functions for an incident neutron on a 89-Yttrium target using several models implemented in the TALYS code (see text for more details).

## WHAT REMAINS TO BE DONE !

As discussed and illustrated in the previous section, the current status of the TALYS code is such that reliable predictions can be obtained using purely microscopic ingredients for non fissile nuclei. For fissile nuclei, as mentioned previously, it is possible (see fig.1 in ref.[1]) to fit correctly fission cross sections, but a purely microscopic calculation is not yet feasible. To reach such an ambitious goal, we need several ingredients among which microscopic fission barriers, and a way to exploit them in actual calculations, which means solving the problem of calculating fission penetrability out of a fission path without assuming a Hill-Wheeler shape as is usually done. Concerning the first point, we already have access to the microscopic fission barriers of ref.[11], and we believe that, as a starting point, the treatment of ref.[11] could be applied together with the microscopic level densities calculated on top of each barriers.

Another important issue is the semi-microscopic OMP [8]. Right now, even if this model can be used in TALYS, it is only automated for spherical nuclei. A future improvement which we plan is to include the treatment for deformed nuclei. We already have the microscopic nuclear densities tabulated in the TALYS database, so that only the deformed coupled channel description needs to be implemented.

## CONCLUSION

The TALYS code has reached the completeness initially expected and we are now improving the quality of the database on which the code relies. Thanks to the high computer power available nowadays, systematic microscopic calculation have been performed [3,5,9,12-14] not only for fundamental research but also for practical applications. We had anticipated the eventual arrival of new microscopic models when we started to develop TALYS, and we are now incorporating them in order to see how good (or maybe bad) they do when compared to experimental cross sections. The final goal is to be able to make a real link between the fundamental effective nucleon-nucleon interaction and the cross section calculation.

## REFERENCES


1. A.J. Koning, S. Hilaire and M.C. Duijvestijn, to be published in *Proceedings of the International Conference on Nuclear Data for Science and Technology*, EDP-NICE 2007.
2. S. Goko et al., Phys. Rev. Lett. **96**, 192501 (2006).
3. S. Hilaire and S. Goriely, Nucl. Phys. **A779**, 63 (2006).
4. S. Hilaire, S. Goriely and A.J. Koning, to be published in *Proceedings of the International Conference on Nuclear Data for Science and Technology*, EDP-NICE 2007.
5. S. Hilaire and S. Goriely, these proceedings.
6. A.J. Koning and J.P. Delaroche, .Nucl. Phys. **A713**, 231 (2003).
7. E. Sh. Soukhovitskii, S. Chiba, J.-Y. Lee, O. Iwamoto and T. Fukahori, *J. Phys. G.* **30**, 905 (2004).
8. E. Bauge, J.P. Delaroche and M. Girod, Phys. Rev. **C63**, 024607 (2001).
9. S. Goriely, E. Khan and M. Samyn, Nucl. Phys. **A739**, 331 (2004).
10. H. Utsunomiya et al., these proceedings.
11. M.Sin et al. to be published in *Proceedings of the International Conference on Nuclear Data for Science and Technology*, EDP-NICE 2007.
12. S. Goriely, J.M. Pearson, these proceedings.
13. S. Hilaire and M. Girod, Eur. Phys. J. **A33**, 237 (2007).
14. S. Hilaire and M. Girod, these proceedings.

# Advancing Along the N = Z Line: A 30 Year Odyssey

C. J. (Kim) Lister

*, Argonne National Laboratory, Argonne, IL 60439, USA*

**Abstract.** N = Z nuclei have always had a special role in nuclear physics due to the symmetry of their wavefunctions, and their anomalously high binding energy. Beyond $^{56}$Ni they gain new significance, as they lie close to the proton dripline, and approaching $^{100}$Sn, they define the dripline. They are interesting for their structure, for rp-process nucleosynthesis, and as laboratories for weak interaction physics. In this paper I review some currently interesting topics and open questions.

**Keywords:** N = Z nuclei, nuclear shapes, pairing correlations, and isomers.
**PACS:** 21.10-k, 27.50+e

## INTRODUCTION

Morinaga's group at Munich [1] really started the push towards the N=Z line in middle mass nuclei during the 1970's. Their innovative use of heavy ion reactions opened this terra incognita and indeed started the exploration away from the valley of stability in proton-rich nuclei. However, like many explorers, they underestimated the difficulty of their quest. It has taken more than 30 years to reach the middle mass N=Z line for spectroscopy and allow the investigation of the neighboring mirror systems. It has required the development of techniques which are a million times more sensitive than Morinaga's group originally employed. My thesis advisor, John Sharpey Schafer, saw the potential of this research and goaded me in this direction, but I expected it to be a post-doc project not a research career. However, now, using modern stable beam accelerators, large gamma detector arrays like Gammasphere [2], and spectrometers like the Fragment Mass Analyzer [3] we can reach sensitivity levels to perform spectroscopy on nuclei produced at the10's nb level [4-6]. The next generation of gamma tracking arrays like AGATA and GRETA will allow us to progress to nuclei below the 1nb level. This sensitivity permits detailed spectroscopic studies all along the N = Z line, perhaps to $^{100}$Sn itself [6].

The huge improvements in stable-beam techniques over the years are increasingly being challenged by the use of radioactive beams, both produced in high velocity fragmentation reactions and from reacceleration following production by ISOL techniques. It is clear that we are fast approaching the time when the radioactive beams will be in the ascendency as their intensity, quality and general availability increase. It is also the case that the use of these beams is in its infancy and there are many avenues for technical improvements of the experiments. However, for today,

CP1012, *Frontiers in Nuclear Structure, Astrophysics, and Reactions: FINUSTAR 2*
edited by P. Demetriou, R. Julin, and S. V. Harissopulos
© 2008 American Institute of Physics 0-7354-0532-5/08/$23.00

the competition between stable and radioactive probes is in the balance, and very important information is emerging from both sources.

There are many interesting physics topics one might discuss, but I will concentrate on three areas of rapid recent development: the change of nuclear shapes along the N=Z line, the possible influence of isomers on the rp-process nucleosynthesis reaction rates, and the study of T = 1 and T = 0 neutron-proton pairing correlations.

## NUCLEAR SHAPES ALONG N = Z

Above $^{56}$Ni one enters the fpg shell model space which spans nucleon numbers 28-50. This space is big enough for considerable collectivity to develop, and indeed, nuclei with some of the largest groundstate deformations known are found in this area, namely with prolate deformation of $\beta_2 > 0.4$ [7-10]. The space is a blessing and a curse; a blessing as unusual new collective effects have been found, and a curse as the shell model space is too large to be diagonalizable in the foreseeable future. It is on the cutting edge of tests for current big basis shell models, both conventional [11, 12], and truncated by various methods [13,14]. At present we often have to resort to more approximate approaches to interpret the data, though rapid progress is now being made. These include microscopic-macroscopic [15-17] calculations, relativistic mean field approaches [18], and Hartree-Fock-based models [19-21]. The fact that neutrons and protons are filling the same single particle orbits as one proceeds along the N = Z line has many consequences. One effect is very sharp shape polarization, a driving topic of interest when the region was first entered experimentally [22,23], and one that continues to attract much attention to this day.

The shape polarization is driven by occupancy of the $g_{9/2}$ intruder orbital [13]. However, more than just the $g_{9/2}$ state is involved, and inclusion of other N = 4 shell orbitals are needed to properly describe the collectivity [13]. If the N = 4 orbits are significantly occupied then large prolate deformation develops and Nilsson-like approaches are very successful [15]. If the orbit is not significantly occupied, below N = Z = 33, then nuclei are collective, but soft in shape [24,25]. On the borderline, in selenium isotopes (Z = 34), the conditions may be appropriate for collective oblate rotation, one of the very few places where this occurs for a nuclear groundstate [26,27]. Experimentally, it appears that in $^{68}$Se an oblate groundstate is more bound than prolate configurations by ~1 MeV [27], allowing a region of rather pure oblate rotation to occur. This interpretation has been questioned [28], and definitive measurements of electromagnetic properties are needed to resolve the issue. These are currently in progress at GANIL and Legnaro, and planned for ANL. The resolution of this issue reflects the experimental progress in studying these nuclei, moving from the initial discovery of low-lying states, to the detailed spectroscopy which is now possible. The combination of measuring state lifetimes by traditional Doppler Shift methods and absolute yields in radioactive beam Coulex has proven very effective for investigating quadrupole collectivity in detail.

Above selenium, in the krypton isotopes, even more violent shape coexistence occurs. This shape competition provides a fine and discriminating issue to test our understanding of nuclear binding. In $^{72}$Kr the first excited state is a $J^{\pi} = 0^{+}$ shape isomer [29]. The shapes, and the degree of mixing between them is a cutting edge topic. High quality electromagnetic investigations [30-34] help clarify the issues. Theoretically, Hartree-Fock approaches [20,21] have been essential for understanding the competition. Still, there is a long way to go: the role of tri-axial excitation seems critical [16,30], but as yet not really explored.

As the $g_{9/2}$ orbit fills, the prolate deformation-driving effects decrease rapidly, and the nuclei become less deformed. It seems the decrease in groundstate deformation is not monotonic, and after the Z = 40 gap in the prolate-deformed Nilsson sequence is passed, the nuclei rapidly become less deformed. The upper periphery of deformation appears to be $^{82,83}$Nb [35,36], beyond which the nuclei return to soft shapes, with collectivity decreasing as $^{100}$Sn is approached. The loss of collectivity, and the return to shell-model like behavior has yet to be charted. Of particular current interest (mainly due to technical feasibility) is the study of high spin isomers which are spin-traps arising from multi-hole configurations in $^{100}$Sn [37].

## ISOMERS AND WAITING POINTS IN N = Z NUCLEI

Rp-process nucleosynthesis was originally thought to terminate close to $^{56}$Ni. However, it has become increasingly clear that in explosive scenarios the process continues across the fpg-shell to somewhere just above $^{100}$Sn where α-recycling causes termination. The scenario most often considered for this process is on the surface of a neutron star, accreting material from a companion normal star. As pressures and temperatures rise and the abundance of surface hydrogen and helium increases, a thermonuclear runaway reaction process occurs causing X-ray bursts and rapid nucleosynthesis through (p,γ) and (a,γ) reactions. When the proton dripline is reached, this rapid capture must stop and await β-decay to stabilize the seed nuclei before capture can continue. The structure and masses of the N = Z nuclei become critical in defining the reaction rates and timescales. For example, $^{68}$Se cannot capture protons as $^{69}$Br is unbound with a half-life of < 30 ns [38]. $^{68}$Se is relatively well bound, and has a β-decay lifetime of 34 s. Thus, if the 1 and 2-proton capture rates are low, this becomes a waiting point for the entire synthesis of heavier elements by the rp-process. Rather exact (< 100 keV) mass measurements are needed to ascertain the reaction Q-values and rates. Direct, precise measurements are difficult on short-lived isotopes and it is only recently that suitable techniques have been developed. For this type of investigation, "on-line" Penning traps have revolutionized mass determinations of nuclei far from stability. Penning traps at Argonne [39,40], MSU [41] and Isolde [42] have all made major recent contributions in this region. However, questions remain about the unbound odd-A nuclei across the dripline, like $^{69}$Br and $^{73}$Rb. Observing the short lived proton radioactivity (or radioactivities) from these nuclei would go far in resolving this puzzle.

The issue of isomers is important in this domain too. An isomer in an even-even waiting point nucleus can offer a "jumping-off" point for (p,γ) reactions and a method of by-passing the waiting points [43]. However, to significantly influence the overall process, the isomers have to be long-lived *in the hostile neutron star surface environment*, and be significantly populated in the (p,γ) production reactions coming from below. Searches have been made for such isomers in $^{68}$Se [27,43] without success, and currently it appears unlikely that they play a crucial role in rp-synthesis.

However, the low-lying $^{72}$Kr shape isomer [29] shows that such scenarios should not be ruled out, and new long-lived isomers may yet emerge.

## T = 1 AND T = 0 PAIRING CORRELATIONS

Perhaps the biggest progress in N=Z nuclear spectroscopy has been in developing a better understanding of neutron-proton pairing correlations. Heavy nuclei always have substantial neutron excesses, and are dominated by T=1 neutron-neutron and proton-proton pairing that influence almost every facet of nuclear behavior. The fpg shell model space is sufficiently large for collective pairing fields to develop, and along N = Z both T=1 and T=0 neutron-proton pair fields have been discussed.

Initially, emphasis was focused on masses and mass differences. Analysis [44,45] indicated that the systematic trends were consistent with T = 1 neutron-proton pairing on an equal footing with other T = 1 pair fields, but left little room for T = 0 pairs. This led to the notion that perhaps T=0 pairs were more robust against rotation, so they may become important at high spin. If so, delays in "backbending" would be expected. After considerable effort [31,32,46] it has become clear that although backbending delays do occur, the nuclei are shape-unstable, and so the delays cannot be clearly attributed to any anomalous pairing effects. Recent work [47] on the most stable rotor, $^{76}$Sr, indicate that T = 0 pairs are not in evidence, even when the other T = 1 pair fields have broken down.

The odd-odd N = Z systems have proven to be the "smoking guns" for neutron-proton correlations, as originally suggested by Warner and his co-workers [48] and supported by extensive new data [49,50]. Now, from $^{66}$As (N = Z = 33) [51] to $^{86}$Tc (N = Z = 43) [52], spanning across the middle of the fpg shell, clear evidence for T = 1 pairing has emerged. All these nuclei show unusually low level densities for odd-odd nuclei and the emergence of a "pairing-gap" [53]. Indeed, were the T = 0 field equal to the T = 1 fields in strength, then Chasman [54] has predicted a degenerate (J = 0, T=1) and (J = 1 T = 0) doublet. This is clearly not the case experimentally. Further, the fact that all the T = 0 states seem to lie near 1 MeV, or above, indicates that there is little or no collective T = 0 pair field.

Again, much work remains to be done. We need to find the lowest ($J^{\pi} = 1^{+}$, T = 0) states in these odd-odd nuclei. This can probably be best achieved through spectroscopy of the β-decay of their $T_z$ = -1 even-even parents. And, at the end of the day, pair fields are best studied not with gamma spectroscopy, but by measuring level occupancies through direct reactions. This requires radioactive beams, but is an exciting prospect for the future.

## CONCLUSIONS

Spectroscopy of N = Z nuclei has come a long way in 30 years. This really was terra incognita in the 1970's. Now it has been explored, and its main features charted. We are close to a detailed understanding of these nuclei, though I have pointed out where issues remain and more effort is needed. "Non-standard" physics has emerged. The violent shape changes, and shape co-existence, are not found along the valley of stability, nor the effects of neutron-proton pairing correlations. Perhaps, in the long term, this exploration has been most important as it has taught us how to go about studying the new terrain in the nuclear landscape far from stability, and how to develop new techniques which allow us to range into exotic regions. It has provided tools for identifying deviations from what we expect based on extrapolations. It has certainly positioned us for the challenge of tomorrow: reaching far out into the new terra incognita of neutron rich nuclei. There, the discovery potential is very high and the prospects of new phenomena caused by highly decoupled neutron and proton matter is very good. However, for this next exploration, a new generation of radioactive beam accelerators is essential.

## ACKNOWLEDGEMENTS

This review has covered research from around the world and represents a true international endeavor. I have concentrated on what I know best, and have been most involved in; the experimental research with Gammasphere at Argonne. I apologize to the researchers in this field who I might not have had space to mention. I would like to thank David Jenkins, Paddy Regan, and Georg Bollen for sharing their results prior to publication. I would also like acknowledge the dedicated team who maintain Gammasphere as the leading gamma ray spectrometer in the world; Mike Carpenter, John Greene, Torben Lauritsen, Robert Janssens, John Rohrer, Darek Seweryniak and Shaofei Zhu. This research was supported by the U.S. Department of Energy, Office of Nuclear Physics, under contract DE-AC02-06CH11357.

## REFERENCES

1. H. Morinaga et al., Z. Phys. A281, 409 (1977).
2. I. Y. Lee, Nucl. Phys. A520, 641c (1990).
3. C. N. Davids et al., Nucl. Instrum. Methods B70, 358 (1992).
4. P. Reiter et al., Phys. Rev. Lett. 95, 032501 (2005).
5. D. Seweryniak et al., Phys. Rev. Lett. 99, 082502 (2007).
6. D. Seweryniak et al., Phys. Rev. Lett. 99, 022504 (2007).
7. C. J. Lister et al., Phys. Rev. C 42, R1191 (1990).
8. B. Rubio et al., Phys. Rev. Lett. 92, 232501 (2004).
9. Ph. Dessagne et al., Eur. Phys. J. A 20, 405 (2004).
10. D. A. Eastham et al., Phys. Rev. C 36, 1583 (1987).
11. B. A. Brown, T. Duguet, T. Otsuka, D. Abe, and T. Suzuki, Phys. Rev. C 74, 061303 (2006).
12. T. Otsuka, T. Suzuki, R. Fujimoto, H. Grawe, and Y.Akaishi, Phys. Rev. Lett. 95, 232502 (2005).
13. M. Hasegawa et al., to be published in Phys. Lett. (2007).
14. Y. Sun, Eur. Phys. J. A 20, 133 (2004).
15. W. Nazarewicz et al., Nucl. Phys. A435 397 (1985).
16. P. Moller et al., Los Alamos preprint LA-UR-07-6469 and to be published in At. Data and NDT (2007).
17. A. V. Afanasjev and S. Frauendorf, Phys. Rev. C 71, 064318 (2005).
18. A. V. Afanasjev, Int. J. Mod. Phys. E16, 275 (2007).
19. A. Petrovici, Int. J. Mod. Phys. E15, 1477 (2006).
20. A. Petrovici, K. W.Schmid, and A. Faessler, Nucl. Phys. A708, 190 (2002).
21. M. Bender, P. Bonche, and P.-H. Heenen, Phys. Rev. C 74, 024312 (2006).
22. R. B. Piercey, J. H. Hamilton et. al., Phys. Rev. Lett. 47, 1514 (1981).

23. C. J. Lister, B. J. Varley, H. G. Price, J. W. Olness, Phys. Rev. Lett. 49, 308 (1982).
24. P. J. Ennis et al., Nucl. Phys. A535, 392 (1991); Erratum Nucl. Phys. A560, 1079 (1993).
25. K. Starosta et al., Phys. Rev. Lett. 99, 042503 (2007).
26. R. Bengtsson. Nuclear Structure of the Zirconium Region, Bad Honnef, Germany, Ed. Ebeth, Meyer and Sistemich, Publ. Springer, Berlin (1988) p. 17.
27. S. M. Fischer et al., Phys. Rev. Lett. 84, 4064 (2000).
28. A. M. Hurst et al., Phys. Rev. Lett. 98, 072501 (2007).
29. E. Bouchez et al., Phys. Rev. Lett. 90, 082502 (2003).
30. E. Clement et al., Phys. Rev. C 75, 054313 (2007).
31. S. M. Fischer, C. J. Lister, and D. P. Balamuth, Phys. Rev. C 67, 064318 (2003).
32. C. Andreoiu et al., Phys. Rev. C 75, 041301 (2007); Erratum Phys. Rev. C 75, 049901 (2007).
33. A. Goergen et al., Eur. Phys. J. A26, 153 (2005).
34. A. Gade et al., Phys. Rev. Lett. 95, 022502 (2005); Erratum Phys. Rev. Lett. 96, 189901 (2006).
35. L. S. Caceres, P. Reagan et. al., Acta Phys. Pol. B38, 1271 (2007).
36. S. M. Fischer et al., Phys. Rev. C 75, 064310 (2007).
37. I. Mukha et al., Nature (London) 439, 298 (2006).
38. Z. Janas et al., Phys. Rev. Lett. 82, 295 (1999).
39. J. A. Clark et al., Phys. Rev. Lett. 92, 192501 (2004).
40. J. A. Clark et al., Phys. Rev. C 75, 032801 (2007).
41. P. Schury, G. Bollen et al., Phys. Rev. C 75, 055801 (2007).
42. A. Kankainen, J. Aysto et al., Eur. Phys. J. A 29, 271 (2006).
43. A. Wohr et al., Nucl. Phys. A742, 349 (2004).
44. P. Vogel, Nucl. Phys. A662, 148 (2000).
45. A. O. Macchiavelli et al., Phys. Rev. C 61, 041303 (2000).
46. G. DeAngelis et al., Phys. Lett. B415, 217 (1997).
47. P. J. Davies et al., Phys. Rev. C 75, 011302 (2007).
48. D. Rudolph, D. D. Warner et al., Phys. Rev. Lett. 76, 376 (1996).
49. S. M. Fischer et al., Phys. Rev. C 74, 054304 (2006).
50. A. N. Steer et al., Nucl. Instrum. Methods A565, 630 (2006).
51. G. DeAngelis, Private Communication (2004) and R.Grzywacz, Nucl. Phys. A682, 41c (2001).
52. A. B. Garnsworthy, P. H. Reagan et al., Nucl. Phys. A787, 491c (2007).
53. D. G. Jenkins et al., Phys. Rev. C 65 064307 (2002).
54. R. R. Chasman, Phys. Lett. B553, 204 (2003); Phys. Lett. B577, 47 (2003).

# High-Spin States of $^{84,85}$Br: Mapping the Proton Sub-Shells towards $^{78}$Ni

A. Astier[a], M.-G. Porquet[a], Ts. Venkova[b,a], I. Deloncle[a], F. Azaiez[c], A. Buta[c], D. Curien[c], O. Dorvaux[c], G. Duchêne[c], B.J.P. Gall[c], F. Khalfallah[c], I. Piqueras[c], M. Rousseau[c], M. Meyer[d], N. Redon[d], O. Stézowski[d], R. Lucas[e], and A. Bogachev[f]

*[a]CSNSM IN2P3-CNRS and Université Paris-Sud, 91405 Orsay, France*
*[b]INRNE, BAS, 1784 Sofia, Bulgaria*
*[c]IPHC, IN2P3-CNRS and Université Louis Pasteur, 67037 Strasbourg Cedex 2, France*
*[d]IPNL, IN2P3-CNRS and Université Claude Bernard, 69622 Villeurbanne Cedex, France*
*[e]CEA/Saclay, DSM/DAPNIA/SPhN, 91191 Gif-sur-Yvette Cedex, France*
*[f]JINR, Joliot-Curie 6, 142080, Dubna, Moscow region, Russia*

**Abstract.** The $^{84,85}$Br nuclei have been produced as fission fragments in the fusion reaction $^{18}$O+$^{208}$Pb at 85 MeV bombarding energy and studied with the Euroball IV array. The high-spin states of the odd-odd $^{84}$Br nucleus have been identified for the first time. From angular correlation analysis, spin values have been assigned to most of the $^{85}$Br excited states up to 4 MeV. All observed states in $^{84,85}$Br can be described by various proton excitations involving at least the two sub-shells ($\pi f_{5/2}$ and $\pi p_{3/2}$) located just above the Z=28 shell closure.

**Keywords:** Fusion-fission, high-spin states, γ-γ coincidences, level scheme, shell model.
**PACS:** 21.60.Cs, 23.20.Lv, 25.85.Ge, 27.50.+e

## INTRODUCTION

The characterization of the N=50 shell gap, particularly its evolution with a very-large neutron excess, represents currently an intense effort of the nuclear physics community (see for instance [1-7]). In this intermediate-mass region the shell-model calculations suffer from a fragmentary determination of some basic inputs, i.e. the single particle energies and the two-body matrix elements. For instance the prediction of the evolution of the N=50 gap at very large neutron excess needs the knowledge of the residual interaction energies for various πν configurations, issued from the *fp* protons (28<Z<38) and the *dg* neutrons (N~50), which are mainly not known at the present time. Measurements of the level structure of some moderately neutron-rich nuclei provide data which can be compared with the results of state-of-the-art shell-model calculations, leading to the determination of these crucial parameters.

Whereas the lack of suitable stable beam-target combinations prevents the use of fusion-evaporation reactions in order to populate high-spin states of these neutron-rich nuclei, fusion-fission reactions –combined to a powerful prompt γ-spectroscopy– prove to be fruitful. We have thus undertaken the high-spin state study of many

CP1012, *Frontiers in Nuclear Structure, Astrophysics, and Reactions: FINUSTAR 2*
edited by P. Demetriou, R. Julin, and S. V. Harissopulos
© 2008 American Institute of Physics 0-7354-0532-5/08/$23.00

isotopes ($_{37}$Rb, $_{36}$Kr, $_{35}$Br, and $_{34}$Se) located around the N=50 shell closure. One can for instance refer to our previous publications concerning $^{84}Se_{50}$ [7] and $^{87}Kr_{51}$ [8]. Recently, the $^{84,85}$Br isotopes have been thoroughly examined, and some of the results are reported in this paper (for full details see reference [9]).

## EXPERIMENTAL PROCEDURES

$^{84,85}$Br isotopes have been produced as fission fragments in the fusion reaction $^{18}O+^{208}Pb$ at 85 MeV bombarding energy (provided by the Vivitron accelerator of IReS). $\gamma$-rays were detected by the Euroball IV array, composed of 15 "Cluster" Ge detectors at backward angles, 26 "Clover" Ge detectors around 90°, and 30 "Tapered" detectors at forward angles. A 100 mg/cm$^2$ target of $^{208}$Pb was used to stop the recoiling nuclei. The data were recorded in an event-by-event mode with the requirement that a minimum of three unsuppressed Ge detectors fired in prompt coincidence. A set of $4x10^9$ three and higher-fold events was available for a subsequent analysis. More than 100 nuclei being produced at high spin in such an experiment, single-gated spectra are useless in most of the cases. The selection of one particular nucleus needs at least two energy conditions. The identification of transitions depopulating high-spin levels which are completely unknown is based on the fact that prompt $\gamma$-rays emitted by complementary fragments are detected in coincidence. That was the case for the $^{84}$Br nucleus, unknown at high spins.

The timing information from the Ge detectors have been also used to look for isomeric states. During the experiment, for the purpose of time calibration, part of the data has been registered using a fast trigger signal delayed by 20 ns. This has allowed us to precisely align the timing responses of every channel to a common one displaying 0.239 ns/channel. Several spectra representing the difference in the times associated to two transitions have then been built, either in prompt coincidence or located on both sides of a well-known isomeric state. The method has proven to be very successful, all measured half-lives being in perfect agreement with the known ones. We could for instance reproduce half-life values such as $T_{1/2}$=21.7(5) ns in $^{95}$Sr (1238 keV level) and $T_{1/2}$=103(3) ns in $^{97}$Zr (556 keV level). The range covered in our analysis is 10-300 ns, 10 ns being given by the intrinsic detector resolution.

## RESULTS AND DISCUSSION

The $^{85}Br_{50}$ level scheme has been recently published [10]. In our work we have extended our knowledge of the $^{85}$Br states up to 5 MeV excitation energy and spin I~21/2 [9]. Moreover, from angular correlation analysis, spin and parity values have been assigned to most of the observed states up to 4 MeV. In particular, we have established that the 1859 keV level is unambiguously the $I^\pi=9/2^+$state which is expected from the promotion of the odd proton to the $\pi g_{9/2}$ sub-shell. And thanks to our time measurement, no delayed component was observed for this level, leading to the upper limit of 10 ns for its half-life.

The configurations of the first excited states of the light N=50 isotones are mainly due to the four proton sub-shells located between Z=28 and Z=50, namely $\pi f_{5/2}$, $\pi p_{3/2}$,

$\pi p_{1/2}$ and $\pi g_{9/2}$. The two first sub-shell are obviously involved in the $3/2^-$ ground state and $5/2^-$ first excited state of $^{85}$Br, its odd proton being located in one of the two sub-shells, $\pi p_{3/2}$ or $\pi f_{5/2}$, close in energy. Negative-parity states up to spin $I^\pi=11/2^-$ can be explained by the re-orientation of the proton angular momenta within the *fp* sub-shells ($[\pi f_{5/2}\text{-}\pi p_{3/2}]^{+7}$ configuration). Then the promotion of one pair into the $\pi p_{1/2}$ sub-shell leads to a maximum spin $I^\pi=13/2^-$ ($[\pi f_{5/2}\text{-}\pi p_{3/2}]^{+5}$ $[\pi p_{1/2}]^{+2}$ configuration).

The $9/2^+$ level at 1859 keV has the $[\pi f_{5/2}\text{-}\pi p_{3/2}]^{+6}$ $[\pi g_{9/2}]^{+1}$ configuration. Above this level, all the observed states can be explained by the re-orientation of the proton angular momenta within the *fp* sub-shells (including also $\pi p_{1/2}$): they result of the coupling of one proton in the $\pi g_{9/2}$ sub-shell and one broken pair within the $\pi f_{5/2}\text{-}\pi p_{3/2}$ sub-shells (giving $I_{max}=17/2^+$) or two broken pairs ($I_{max}=21/2^+$).

Concerning the odd-odd $^{84}Br_{49}$ nucleus, very few exited states were known prior to this work [11]: a $6^-$ state ($T_{1/2}$=6.0 min) lies at 320(100) keV excitation energy above the $2^-$ ground state ($T_{1/2}$=31.8 min). In order to identify the completely unknown transitions depopulating yrast states of $^{84}$Br, we have first looked into spectra gated only by the first transitions of its main complementary fragment ($^{137}$Cs). We have, in a second time, analyzed spectra in double coincidence with one transition in $^{137}$Cs and one new transition, or with two transitions of $^{84}$Br. The coincidence relationships have been carefully analyzed in order to build the level scheme shown in fig. 1.

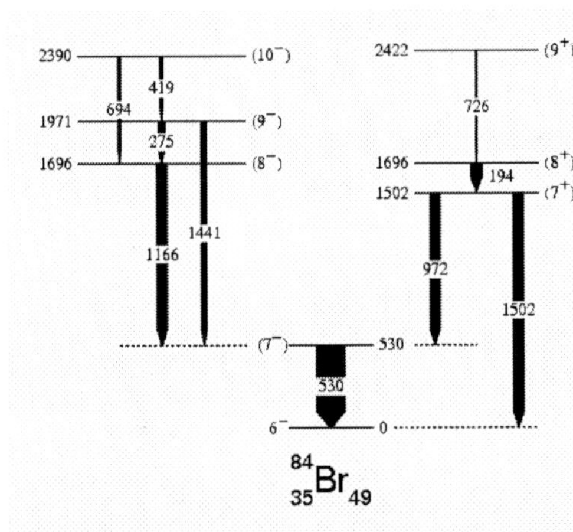


**FIGURE 1:** Level scheme of $^{84}$Br obtained in this work. The structures are built upon the $6^-$ isomer ($T_{1/2}$=6.0 min) lying at 320(100) keV excitation energy above the $2^-$ ground state ($T_{1/2}$=31.8 min).

In the odd-odd $^{84}$Br nucleus (Z=35, N=49), four configurations are expected below 2 MeV, giving four multiplets of states: $\pi p_{3/2}\otimes\nu g_{9/2}$ ($3^-<I^\pi<6^-$), $\pi f_{5/2}\otimes\nu g_{9/2}$ ($2^-<I^\pi<7^-$), $\pi p_{1/2}\otimes\nu g_{9/2}$ ($I^\pi=4^-$ et $5^-$), and $\pi g_{9/2}\otimes\nu g_{9/2}$ ($0^+<I^\pi<9^+$). When the residual proton-neutron interaction is taken into account, the relative order of the actual states depends on the nature (particle or hole) of the odd nucleons (see for instance [12]). More precisely, when the two odd nucleons have the same character (particle-particle or hole-hole), the levels from the multiplet with extreme spins are the most lowered in energy by the residual interaction. On the contrary, for particle-hole configurations, the states with medium spin values are energetically favoured. Whereas the odd neutron of $^{84}$Br is obviously $[\nu g_{9/2}]^{-1}$, the nature of the odd proton depends on its sub-shell. To determine it, we can start with the simpler case of the neighbouring $^{86}$Rb isotone (Z=37). It is easy to see that in this nucleus $\pi p_{3/2}$ and $\pi f_{5/2}$ are of hole type, whereas $\pi g_{9/2}$ is of particle type. The experimental results on the level spins in $^{86}$Rb are in very good agreement with such simple arguments. Then, assuming a successive filling of the proton orbits between Z=35 and Z=37, we could expect that only the $\pi p_{3/2}$ excitation would become of particle type in $^{84}$Br. But in this case the $6^-$ state should not be isomeric since the $5^-$ state should be favoured. This is obviously at variance with the experimental result, the spin value of the long-lived isomeric state being $I^{\pi}=6^-$. In order to go further, one can remark that depending on the proton-pair distribution among the two-sub-shells (which are very close in energy), $\pi f_{5/2}$ and $\pi p_{3/2}$, the nature of the odd proton occupying the latter can easily flip from particle, $[\pi f_{5/2}]^6 \otimes [\pi p_{3/2}]^{+1}$, to hole, $[\pi f_{5/2}]^4 \otimes [\pi p_{3/2}]^{-1}$. The residual-interaction energies in the $\pi p_{3/2} \otimes \nu g_{9/2}$ configuration have been calculated as a function of the $\pi p_{3/2}$ occupation, using the same method as in [13]. They are displayed in fig. 2. It can be noticed that the lowest state of the multiplet is the $6^-$ level as soon as the $\pi p_{3/2}$ sub-shell is half-full. Hence the long-lived $6^-$ isomeric state in $^{84}$Br can be interpreted as an experimental evidence of the proton pair scattering among the *fp* sub-shells.

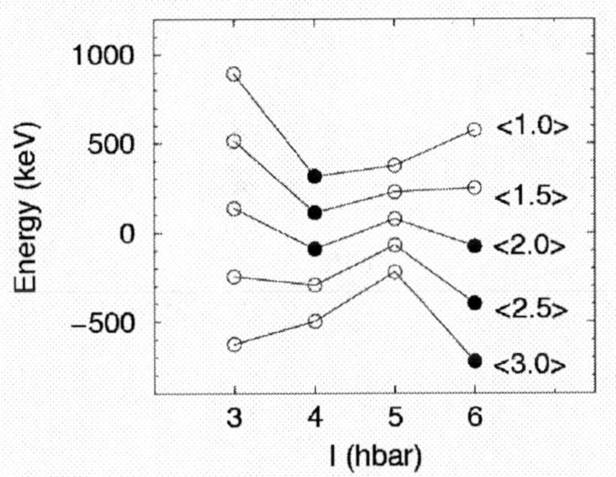


**FIGURE 2:** Energies of residual-interaction in the $\pi p_{3/2} \otimes \nu g_{9/2}$ configuration as a function of the angular momentum I, calculated for five values of the $\pi p_{3/2}$ occupation. The full symbol indicates the lowest state of the multiplet for each case.

By analogy to the $^{86}$Rb neighbouring isotone, one can assume that the state located at 530 keV above the $6^-$ isomer in $^{84}$Br is certainly the $7^-$ state coming from the $[\pi f_{5/2}]^{-1} \otimes [\nu g_{9/2}]^{-1}$ configuration. For the higher excitations, the $\pi g_{9/2}$ excitation and the breaking of one *fp* proton pair are in the same energy range. Therefore the $[\pi g_{9/2}]^{+1} \otimes [\nu g_{9/2}]^{-1}$ configuration can explain the states at 1696, 1971, and 2390 keV above the $6^-$ isomer, whereas the states at 1696, 1971, and 2390 keV above the $6^-$ isomer are most probably coming from the re-orientation of the proton angular momenta within the *fp* sub-shells, the maximum spin value being $10^-$.

## SUMMARY

The $^{84,85}$Br nuclei have been produced as fission fragments in the fusion reaction $^{18}$O+$^{208}$Pb at 85 MeV bombarding energy and studied with the Euroball IV array. The high-spin states of $^{84}$Br have been identified for the first time, using the fact that prompt $\gamma$-rays emitted by complementary fragments are detected in coincidence. The already-known high-spin level scheme of $^{85}$Br has been improved thanks to timing measurements and analysis of angular correlations. Spin and parity values have been assigned to most of the observed states. The excited states observed in $^{85}$Br have been mostly interpreted in terms of proton excitations within the $\pi f_{5/2}$-$\pi p_{3/2}$ sub-shells. Similarly, the excited states of the odd-odd $^{84}$Br nucleus have been discussed in terms of both neutron and proton excitations, involving the sub-shells located close to the Fermi levels, $\nu g_{9/2}$ and $\pi f_{5/2}$, $\pi p_{3/2}$, $\pi g_{9/2}$.



## ACKNOWLEDGMENTS

The Euroball project was a collaboration between France, the United Kingdom, Germany, Italy, Denmark and Sweden. The Euroball IV experiment has been supported by the EU under contract HPRI-CT-1999-00078 (EUROVIV). This work was partly supported by the collaboration agreement Bulgarian Academy of Sciences – CNRS under contract n° 16946. We thank the crew of the Vivitron. We are very indebted to M.-A. Saettle for preparing the Pb targets, P. Bednarczyk, J. Devin, J.-M. Gallone, P. Médina, and D. Vintache for their help during the experiment.


## REFERENCES


1. C. Mazzochi *et al.*, Phys. Lett. B **622**, 45 (2005).
2. J.M. Daugas *et al.*, Phys. Lett. B **476**, 213 (2000).
3. O. Perru, Thèse de l'Université Paris XI, Orsay (December 2004); O. Perru *et al.*, Eur. Phys. J. A **28**, 307 (2006); D. Verney *et al.*, Braz. J. Phys. **34**, 979 (2004).
4. J.S. Thomas *et al.*, Phys. Rev. C **71**, 021302 (2005).
5. E. Padille-Rodal *et al.*, Phys. Rev. Lett. **94**, 122501 (2005).
6. Y.H. Zhang *et al.*, Phys. Rev. C **70**, 024301 (2004).
7. A. Prévost *et al.*, Eur. Phys. J. A **22**, 391 (2004).
8. M.-G. Porquet *et al.*, Eur. Phys. J. A **28**, 153 (2006).
9. A. Astier *et al.*, Eur. Phys. J. A **30**, 541 (2006).
10. N. Fotiades *et al.*, Phys. Rev. C **71**, 064312 (2005).
11. ENSDF database, http://www.nndc.bnl.gov/ensdf/.
12. K. Heyde, *The Nuclear Shell Model*, second edition (Springer-Verlag, 1994).
13. M.-G. Porquet *et al.*, Eur. Phys. J. A **20**, 245 (2004).

# Nuclear Structure far from stability at the N=50 Shell Closure


E. Sahin*,†, G. de Angelis†, A. Gadea†, G. Duchene**, T. Faul**, D.R. Napoli†, E. Farnea‡, J.J. Valiente-Dobón†, R. Orlandi†, D. Mengoni‡, F. Della Vedova†, R.P. Singh§, C. Ur‡, F.Recchia†, M.N. Erduran*, M. Bostan*, S. Erturk¶, S. Aydin‖, K. T. Wiedemann††, A. M. Stefanini†, S. Lenzi‡, S. Lunardi‡, N. Marginean†, L. Corradi†, E. Fioretto†, G. Montagnoli‡, F. Scarlassara‡, P. Mason‡, S. Szilner‡‡, D. Ackermann §§, G. Pollarolo¶¶, J. R. B. Oliveira††, T. Byrski**, D. Curien**, B.Gall**, F. Haas**, O. Dorwaux**, J. Robin**, M. Nespolo‡, F. Azaiez***, G. De France†††, A. Algora‡‡‡, X. Liang§§§, F. Nowacki**, Y.H. Zhang¶¶¶ and X.H Zhou¶¶¶

*Department of Physics,Istanbul University, 34134 Istanbul, Turkey
†INFN, Laboratori Nazionali di Legnaro, Italy
**IPHC, IN2P3/CNRS et Univesitè Louis Pasteur, BP 28,F-67037 Strasburg CEDEX 2, France
‡Dipartimento di Fisica, Universitá di Padova and INFN, Sezione di Padova, Padova,Italy
§Inter-University Accelerator Center, New Delhi-67, India
¶Department of Physics, Nigde University ,Nigde, Turkey
‖Department of Physics, Cukurova University, Adana, Turkey
††University of São Paulo, Sãu Paulo, Brazil
‡‡Ruder Boskovic Institute, HR-1002 Zagreb,Crotia
§§Kernphysik II, Gesellschaft für Schwerionenforschung mbH - GSI, Planckstr.1, D-64291 Darmstadt, Germany
¶¶Dipartimento di Fisica Teorica, Universitá di Torino and INFN, Sezione di Torino, Italy
***IPNO, IN2P3/CNRS et Universit Paris-Sud, F-91406 Orsay, France
†††GANIL Caen Ž013 France
‡‡‡ Institute of Nuclear Research, H-4001 Debrecen, Pf. 51, Hungary
§§§ Electronic Engineering and Physics, University of Paisley, Paisley PA1 2BE, United Kingdom
¶¶¶Institute of Modern Physics, Chinese Academy of Sciences, Lanzhou, P. R. China



**Abstract.** Medium and high spin states of the nuclei close to the N=50 shell closure are investigated using deep inelastic and multi-nucleon transfer reactions. New excited states of the nuclei $^{81}$Ga, $^{82}$Ge and $^{83}$As at N=50 shell closure have been identified by means of the $^{82}$Se+$^{238}$U reaction. Gamma rays have been acquired using the PRISMA-CLARA setup of the Laboratori Nazionali di Legnaro, Italy (LNL).



**Keywords:** Neutron-rich, shell model, deep inelastic, multi-nucleon transfer, single-particle state
**PACS:** 21.10.Pc,21.60.Cs,25.70.Hi,29.40.Cs


## INTRODUCTION

Neutron-rich nuclei close to shell gaps are particularly interesting since, when compared with the shell-model prediction, they allow to search for anomalies into the shell structure [5, 6, 7]. These nuclei are difficult to reach, particularly in the high-spin states,


CP1012, *Frontiers in Nuclear Structure, Astrophysics, and Reactions: FINUSTAR 2*
edited by P. Demetriou, R. Julin, and S. V. Harissopulos


© 2008 American Institute of Physics 0-7354-0532-5/08/$23.00

since they can not be produced via fusion-evaporation reactions. The study of the N=50 isotones towards the doubly closed $^{78}$Ni nucleus, which is the main aim of this work, will possibly enable to understand the size and eventually the evolution of the N=50 energy gap far from stability. Neutron-rich nuclei close to the N=50 shell closure have been produced using the deep inelastic and multi-nucleon transfer reaction $^{82}$Se + $^{238}$U. We will report here on the results obtained for the $^{83}_{33}As_{50}$, $^{82}_{32}Ge_{50}$, and $^{81}_{31}Ga_{50}$.

## EXPERIMENTAL DETAILS

In order to identify characteristic gamma rays to the different nuclei, three experiments were performed at LNL in April 2004, February 2006, and February 2007. Gamma rays have been acquired using the CLARA [4] $\gamma$-detector array in coincidence with the reaction products selected into the PRISMA [1, 2, 3] magnetic spectrometer. The reaction $^{82}$Se + $^{238}$U was used to populate the nuclei around N=50. The $^{82}$Se ions were accelerated at an energy of 505 MeV for the first two experiments and 515 MeV for the last experiment by the combination of the Tandem-XTU and the superconducting LINAC ALPI accelerators. The target was $^{238}UO_2$, isotopically enriched, and its thickness was 400 $\mu$m/cm$^2$ for the first two experiments while a 1000 $\mu$m/cm$^2$ thick metallic $^{238}$U target was used for the last experiment. In order to detect projectile-like nuclei, produced via multi nucleon transfer reactions, PRISMA was positioned at an angle of 64° ($\theta_{PRISMA}$). The $\gamma$-rays following the de-excitation of the reaction products were detected by the CLARA $\gamma$ array consisting of 23 Clover detectors, placed in the hemisphere opposite to the entrance of the spectrometer. Due to the limited acceptance of the spectrometer and also to the limited gamma ray efficiency of the CLARA array, the statistics for $\gamma$-$\gamma$ coincidence data was too low to provide coincidence information for the most exotic reaction channels. Where $\gamma$-$\gamma$ coincidence data from the CLARA array were not sufficient, new information was obtained from data collected in a complementary "thick-target" experiment using the GASP detector array.

## RESULTS AND DISCUSSIONS

The analysis procedure of the PRISMA-CLARA data aims to perform trajectory reconstruction, identification of the atomic number Z, of mass number A, and absolute value of the velocity using position, time (and TOF), and energy information provided by PRISMA spectrometer. The Doppler correction for $\gamma$-rays, in coincidence with the projectile-like ions, is performed on event-by-event basis, taking the information of the recoil velocity vector determined by the reconstruction of the ion trajectories. In Figure 1a mass distributions for the different elements with atomic numbers ranging from Z=34 to Z=28 are shown. The average mass resolution obtained is of about $\Delta A/A$=1/140 allowing a good identification of all isotopes seen in the mass spectra. After a detailed analysis of the data from PRISMA-CLARA, we could newly identify characteristic gamma rays of several neutron-rich nuclei in the N=50 region. As an example we are presenting here the excited states of the $^{83}_{33}As_{50}$, $^{82}_{32}Ge_{50}$, and $^{81}_{31}Ga_{50}$ nuclei selected by their A and Z values. The Doppler-corrected $\gamma$ spectra of those nuclei are shown in Figure 1b. $\gamma$-$\gamma$ coincidence data from the thin target experiments performed with PRISMA-CLARA as well as data from the thick target experiment with GASP have been used to construct the level schemes of $^{83}$As and $^{82}$Ge nuclei, while the level scheme of the $^{81}$Ga nucleus was simply based on the systematics. For the multipolarity assignments of the observed $\gamma$ transitions, where possible, we have determined the asymmetry ratio defined as:

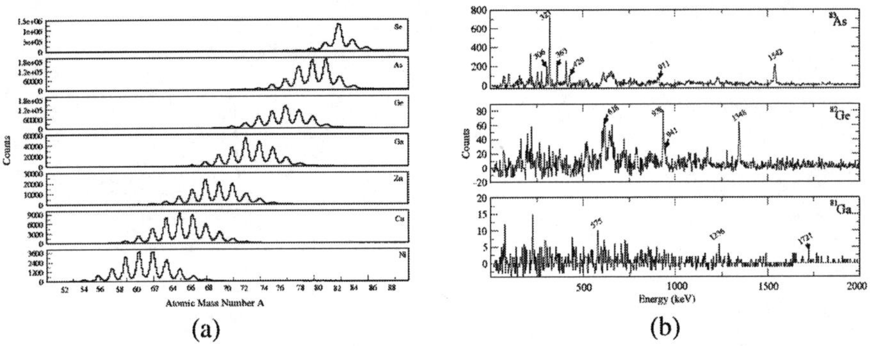


**FIGURE 1.** (a) Total mass yield spectra from the PRISMA spectrometer obtained for the different elements produced in the $^{82}$Se + $^{238}$U thin target reactions and (b) Doppler-corrected $\gamma$ spectra of $^{83}_{33}$As$_{50}$, $^{82}_{32}$Ge$_{50}$, and $^{81}_{31}$Ga$_{50}$, measured with the CLARA Ge array

$$R_{Asymm} = \frac{I_\gamma(\theta_2)}{I_\gamma(\theta_1)} \qquad (1)$$

where $I_\gamma(\theta_1)$ and $I_\gamma(\theta_2)$ are the intensities of the same $\gamma$ transition detected at $\theta_1$ and $\theta_2$ angles, respectively. Here, $\theta$, corresponds to the angle between the Clover detector and the entrance direction of PRISMA. In order to determine typical asymmetry ratio ($R_{Asymm}$) the values obtained from transitions of known multipolarities have been used as reference. The $\theta_1$ value was fixed at 98.8° while $\theta_2$ had different values ranging from 174.1° to 98.8°. Using the extracted asymmetry ratios from the experimental data, we obtained the plot reported in the Figure2. Figure 3a shows the $R_{Asymm}$ values obtained for the transitions of $^{83}$As. Such values can be grouped according to the expected values from ΔI=2 and ΔI=1 transitions of known multipolarities. Similarly, the measured multipolarities for $^{82}$Ge suggest a ΔI=2 character, when compared to the reference values (see Figure 3b).

## Comparison with the Shell Model Calculations

Figure 4a and Figure 4b show the comparisons of the experimentally observed states with the prediction from shell model calculations performed with the OXBASH shell model code [8]. Due to the strong attraction (a result of the tensor interaction) between the particles filling the $\pi f_{5/2}$ and the $\nu g_{9/2}$ orbits and repulsion between the particles filling the $\pi p_{3/2}$ and the $\nu g_{9/2}$, one expects an inversion of the relative position of the $\pi f_{5/2}$-$\pi p_{3/2}$ effective single-particle states with respect to the usual order where the $\pi p_{3/2}$ state is the lowest in excitation energy. Such inversion is confirmed

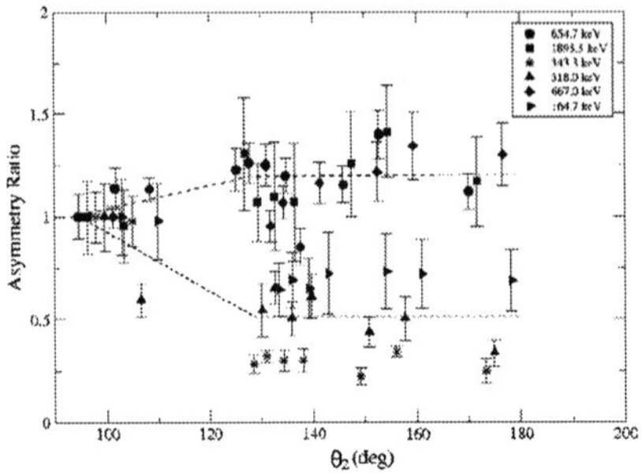


**FIGURE 2.** Dependence of the $R_{Asymm}$ ratios on $\theta_2$ for the reference $\gamma$ ray transitions of various Se isotopes.

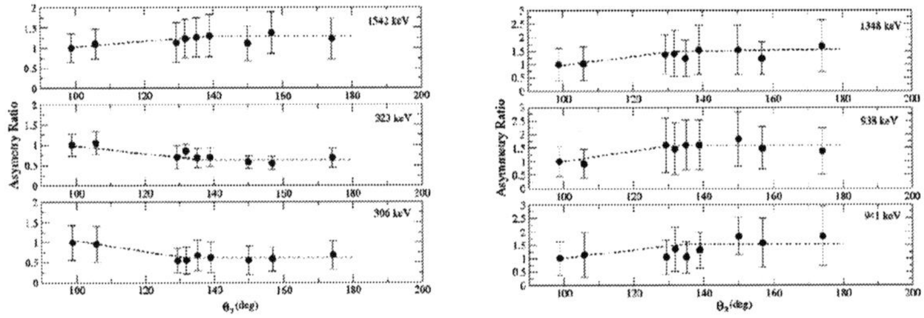


(a) (b)

**FIGURE 3.** (a) and (b) Dependence of the $R_{Asymm}$ ratios on $\theta_2$ for $\gamma$ transitions of $^{83}$As and $^{82}$Ge, respectively.

by the experimental data (Figure.4b). It is interesting to mention here the good agreement obtained between the low lying excited states and the predicted values of the shell model calculation using the effective interaction reported in Ref. [9]. At higher angular momentum (I>5,6$\hbar$) more disagreement is observed. In Figure 5 the experimentally observed excited states of $^{82}$Ge are compared with the prediction of shell model calculation performed varying the N=50 shell gap (E($d_{5/2}$)-E($g_{9/2}$)). Low spin states are mainly unaffected since they can be build using the proton valence particles. For higher spin (I=5,6$\hbar$) the breaking of the N=50 core becomes more important showing a reduction of the excitation energy as the neutron energy gap is reduced. One can notice that a good agreement for the newly assigned excited states $5^+$ and $6^+$ is obtained for an energy gap of 3.8 MeV. This indicates a persistence of the energy gap at Z=32 when compared to the gap value of about 4 MeV expected for $^{78}$Ni.

## ACKNOWLEDGMENTS

We thank the technical staff of the TANDEM-ALPI accelerator complex of the LNL and of the GSI target laboratory. Many thanks to all PRISMA, CLARA, and GASP collaborators for their supportive works.

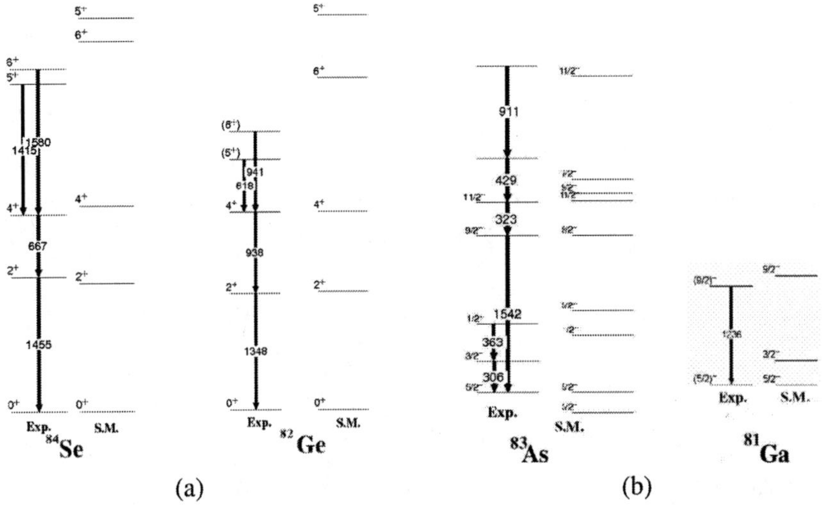


**FIGURE 4.** (a) and (b) The experimental level schemes and the shell model calculations for $^{84}$Se-$^{82}$Ge and $^{83}$As-$^{81}$Ga, respectively.

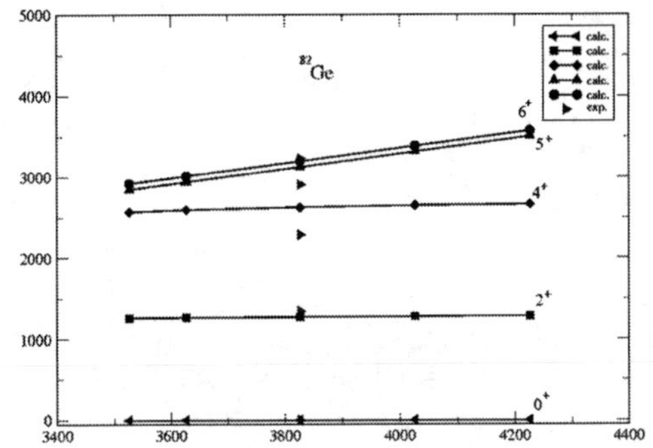


**FIGURE 5.** The excited states of $^{82}$Ge nucleus compared with shell model calculation as a function of N=50 shell gap (E($d_{5/2}$)-E($g_{9/2}$)). One can notice that a good agreement is obtained for an energy gap of 3.8 MeV

# REFERENCES


1. A. M. Stefanini et al., Nucl. Phys. A**701**, 217c (2002).
2. G. Montagnoli et al., Nucl. Instr. Methods Phys. Res. A**547**, 455 (2005).
3. S. Beghini et al., Nucl. Instr. Methods Phys. Res. A**551**, 364 (2005).
4. A. Gadea et al., Eur. Phys. J. A**55**, 193 (2004).
5. I. Hamomoto et al., Nucl. Phys. A**683**, 255 (2001).
6. T. Otsuka et al., Phys.Rev.Lett.**95**, 232502 (2005).
7. T. Otsuka et al., Prog.Theor.Phys. Supp.**146**, 6 (2002).
8. B.A. Brown et al., MSU-NSCL Repor No. 524, 1985.
9. A.F. Lisetskiy et al., Phys. Rev. **C70**, 044314 (2004).

# Recent results on the $^{12}C + ^{12}C$ reactions

T. Spillane[*,†], F. Raiola[**], S. Zeng[‡], C. Rolfs[†], D. Schürmann[†], F. Strieder[†], H.-W. Becker[§], C. Bordeanu[¶], L. Gialanella[||], M. Romano[||] and J. Schweitzer[*]

[*]*University of Connecticut, Storrs, USA*
[†]*Institute für Experimentalphysik III, Ruhr-Universität Bochum, Germany*
[**]*Department of Physics and Joint Institute for Nuclear Astrophysics, University of Notre Dame, Notre Dame, USA*
[‡]*Chinese Institute of Atomic Energy, Beijing, China*
[§]*Fakultät für Physik und Astronomie, Ruhr Universität Bochum, Germany*
[¶]*Horia Hulubei National Institute for Physics and Nuclear Engineering (IFIN-HH), Bukarest, Romania*
[||]*Dipartimento di Scienze Fisiche, Universita Federico II, Napoli and INFN, Napoli, Italy*

**Abstract.** The set of fusion reactions $^{12}C + ^{12}C$ play a critical role in several astrophysical processes. In a recent experiment, the fusion reactions $^{12}C + ^{12}C$ have been studied at $E_{CM}$ = 2.10 to 4.75 MeV by $\gamma$-ray spectroscopy using a C target of ultra-low hydrogen contamination. The deduced astrophysical $\tilde{S}(E)$ factor exhibits previously unknown resonances at E $\leq$ 3.0 MeV, in particular a strong and narrow resonance at E = 2.14 MeV, the high-energy tail of the Gamow peak. Current extrapolations of the reaction rate, which assume that the astrophysical $\tilde{S}(E)$ factor varies smoothly with energy within the Gamow window, are thus subject to significant uncertainty.

**Keywords:** $^{12}C(^{12}C,\alpha)^{20}Ne$, $^{12}C(^{12}C,p)^{23}Na$, $E_{c.m.}$ = 2.10 - 4.75 MeV, astrophysical S-factor
**PACS:** 24.30.-v, 26.20.+f, 27.30.+t

The fusion reactions $^{12}C(^{12}C,\alpha)^{20}Ne$ (Q = 4.62 MeV) and $^{12}C(^{12}C,p)^{23}Na$ (Q = 2.24 MeV) are referred to as carbon burning in stellar environments. The rates of these reactions are critical not only to the nucleosynthesis and subsequent abundances of $^{20}Ne$ and $^{23}Na$, but also to the basic evolution of a star, e.g., whether a star proceeds to the heavy-ion burning branches following hydrogen and helium burning and whether a white dwarf evolves into a type Ia supernova. Thus, the cross section of these reactions must be known with high accuracy down to the Gamow energy $E_G = 1.5 \pm 0.3$ MeV which corresponds to a temperature of $5 \times 10^8$ K [1].

Extensive experimental work on the $^{12}C + ^{12}C$ reactions has provided data over a wide energy range, down to a minimum energy of 2.25 MeV in the center-of-mass using charged-particle or $\gamma$-ray spectroscopy [2, 3, 4, 5, 6, 7, 8, 9]. Inspection of the data reveals several notable inconsistencies, however. Throughout the energy range investigated all experiments show qualitatively similar resonance structure, but disagree significantly (up to 100 keV) as regards the energies at which these structures are found [8]. Additionally, and possibly of greater significance to the extrapolation of the cross section into the Gamow window, there is considerable uncertainty in the general trend of the cross section at the lowest energies, below 3.0 MeV. It is expected that the gross energy dependence of the cross section will be dominated by barrier penetration effects at energies below the Coulomb barrier (6.7 MeV for $^{12}C + ^{12}C$). Thus the modified

CP1012, *Frontiers in Nuclear Structure, Astrophysics, and Reactions: FINUSTAR 2*
edited by P. Demetriou, R. Julin, and S. V. Harissopulos
© 2008 American Institute of Physics 0-7354-0532-5/08/$23.00

astrophysical factor $\tilde{S}$(E),

$$\sigma(E) = \frac{\tilde{S}(E)}{E} exp(-87.21E^{-1/2} - 0.46E) \tag{1}$$

with E in units of MeV should exhibit minimal energy dependence. The second term in the exponential represents the second order term in the barrier penetration approximation within the WKB framework. This term is non-negligible for heavy-ion charged-particle induced reactions, in contrast to lighter ion reactions where it is generally omitted. Hence, in order to expose the strictly nuclear nature of the reaction, the distinction between $\tilde{S}$(E) and the more common S(E) is necessary [2].

## THE $^{12}$C + $^{12}$C REACTIONS

Of the possible $^{12}$C + $^{12}$C reactions (figure 1), the $^{12}C(^{12}C,\alpha)^{20}Ne$ and $^{12}C(^{12}C,p)^{23}Na$ channels dominate the total reaction yield within the Gamow window due to a combination of positive Q-values and relatively small Coulomb barriers. The $\gamma$-ray yields of these reactions represent sizable fractions of the total fusion yield, which includes feeding of the ground states $p_0$ and $\alpha_0$. Specifically, the 440 and 1634 keV lines represent on average 55 $\pm$ 5% and 48 $\pm$ 5% of the $p$ and $\alpha$ yields respectively at E $>$ 2.8 MeV [7]. Furthermore, it has been found that the two $\gamma$-rays exhibit nearly isotropic angular distibutions [5]. We adopt these values in our analysis at all energies. While we restricted our focus to the $p$ and $\alpha$ reaction channels, we were able to observe the 451 keV $\gamma$-ray from $^{12}C(^{12}C,n\gamma)^{23}Mg$ at high energies. This line is sufficiently resolved from the 440 keV line of the $p$ channel. Finally, the 2076 keV state of $^{23}$Na can contribute to the 1634 keV line via transition to the 440 keV state of $^{23}$Na ($E_\gamma$ = 1636 keV). This contribution was monitored by the concurrent 2076 $\rightarrow$ 0 keV transition of $^{23}$Na and found to be negligible (less than 4%).

## TARGET CONTAMINATION

The fusion cross section of $^{12}$C + $^{12}$C at 2.25 MeV is of nano-barn order of magnitude, thus the primary experimental challenge is to measure a sufficient number of reactions for a meaningful statistical analysis. This difficulty is compounded by the presence of $^{1}$H and $^{2}$H contamination in the C targets which produce intense background in both the charged-particle and $\gamma$-ray spectra. This is particularly problematic in $\gamma$-ray spectroscopy where the transition from the first excited states of $^{20}$Ne and $^{23}$Na ($E_\gamma$ = 1634 and 440 keV respectively) are completely swamped by the Compton plateaus of the $E_\gamma$ = 2.36 line from $^{1}H(^{12}C,\gamma)^{13}N$ and $E_\gamma$ = 3.09 line from $^{2}H(^{12}C,p\gamma)^{13}C$ as shown in figure 2. Thus, experimental progress to lower energies requires targets of ultralow hydrogen contamination. Such measurements with $\gamma$-ray spectroscopy are reported here.

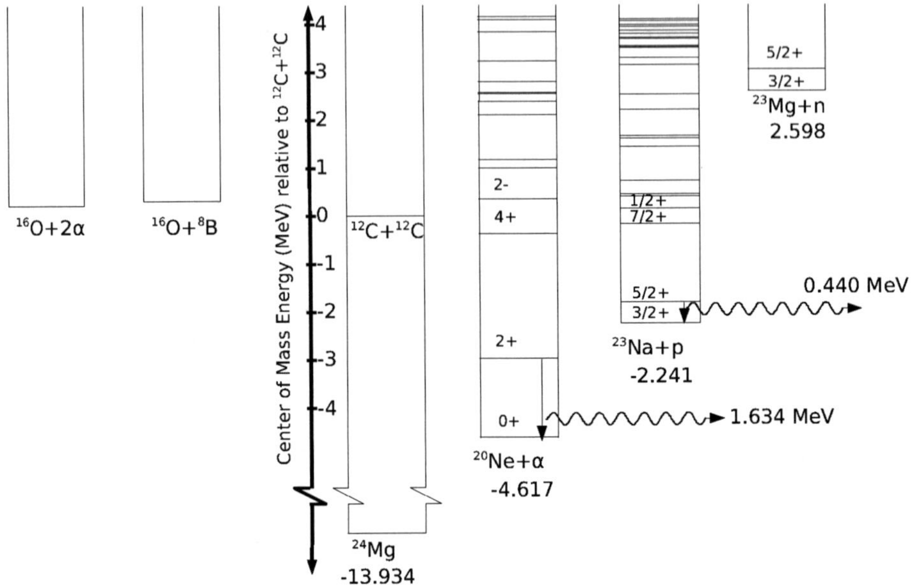


**FIGURE 1.** Energy level diagram of the $^{12}C + {}^{12}C$ exit channels.

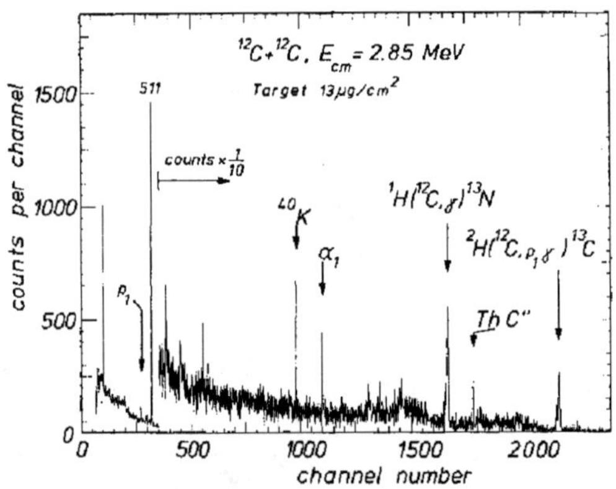


**FIGURE 2.** $^{1}H$ and $^{2}H$ contamination in the $\gamma$-ray spectrum. The lines $p_1$ and $\alpha_1$, from the transitions from the first excited states of $^{20}Ne$ and $^{23}Na$ are nearly overwhelmed by the $^{1}H(^{12}C,\gamma)^{13}N$ and $^{2}H(^{12}C,p\gamma)^{13}C$ background (Kettner *et.al.* 1977).

## EXPERIMENTAL DETAILS

The beam delivery and experimental process are described elsewhere [10]. Briefly, the 4 MV Dynamitron tandem provided the $^{12}C$ beam at up to 40 particle $\mu A$ at low incident energies. The beam is characterized by high purity (better than one part in $10^{11}$ for O and other light ions [11]), high energy resolution (2 keV [12]), and an energy calibration known to better than 3 keV. The beam was tightly collimated onto the target and secondary electrons were suppressed, such that beam current could be integrated using the electrically insulated target mount as a Faraday cup, with an estimated error of 3%. A Ge $\gamma$ detector was placed at 0° to the beam axis in close geometry (detector front face to target distance = 2 cm). A shield of 15 cm Pb was placed around the detector and target chamber and a cosmic ray veto was performed using a plastic scintillator (thickness = 4 cm) placed above the setup. The shielding and cosmic ray veto combined to reduce $\gamma$-ray background by a factor of 800 near $E_\gamma$ = 1600 keV.

The target was a graphite foil (1.0 mm thick, 20 x 20 $mm^2$ in area) of natural isotopic composition obtained from Goodfellow with a quoted purity of 99.8%. In the absence of external target cooling, the intense $^{12}C$ beam heated the graphite target to an estimated temperature of 700 °C. By observing the aforementioned $\gamma$-rays produced by interaction of the beam with contaminants, we found that the level of contamination in the target was reduced to a negligible level within about 20 minutes of exposure to the beam. Furthermore, no re-contamination of the target was observed while the target remained under vacuum, thus allowing for subsequent indirect water cooling of the target.

## RESULTS

The reaction yield of the infinitely thick C target, $Y^\infty$(E), was obtained in several runs from E = 2.10 to 4.75 MeV, with energy steps of $\Delta$ = 12.5 to 25 keV. The cross section was extracted from the thick target yield by a point to point differentiation; i.e., the yield difference between two adjacent points $Y^\infty(E)$ and $Y^\infty(E-\Delta)$ was calculated. The result is

$$Y(E) = \phi\Delta\varepsilon(E)^{-1}\sigma(E_{eff}) \quad (2)$$

where $E_{eff}$ is the effective energy over the step $\Delta$ [1] and $\varepsilon(E)$ is the stopping power [13] with a quoted error of 4.6%. The use of thick targets avoids the problems of carbon depositions and target sputtering which are believed to be responsible for much of the disagreement among previous experiments [8]. At the lowest energies room background becomes significant relative to the 440 and 1634 keV lines, thus a background spectrum was subtracted (beam off, 5 days running time). The resulting cross sections $\sigma(E_{eff})$, i.e. the weighted average of several runs, are presented in figure 3 in the form of $\tilde{S}$(E) as defined in equation 1.

The data display resonance structure throughout the range of the study, with a strong and previously unknown resonance at 2138 $\pm$ 6 keV. Assuming a Breit-Wigner shape the thick target yield at resonance exhibits a step profile characterized by the function

$$\omega\gamma_R \frac{\lambda^2}{2\varepsilon}\frac{M+m}{M} \arctan\left[\frac{(E-E_R)}{(0.5\Gamma_R)}\right] \quad (3)$$

where $\Gamma_R$ is the width of the resonance, $\omega\gamma_R$ is the strength, M and m are the masses of the target and projectile nuclei respectively, and $\lambda$ is the de Broglie wavelength. Fits to

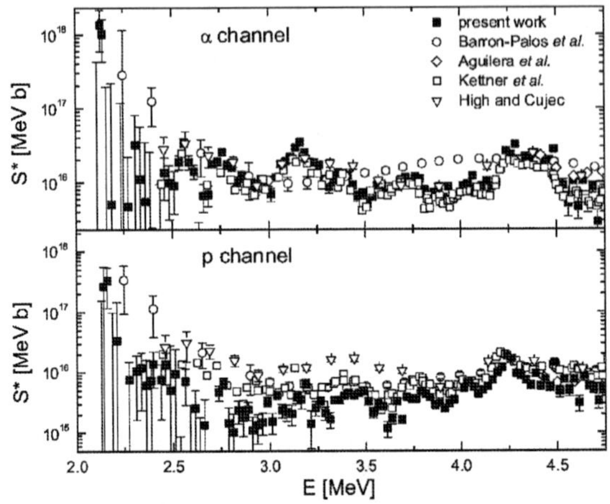


**FIGURE 3.** $\tilde{S}$(E) factor of the fusion process $^{12}C + {}^{12}C$ for the $\alpha$ and proton channels compared to previous data obtained by $\gamma$-ray spectroscopy.

the experimental thick target yields result in the values $\omega\gamma_R = 0.11 \pm 0.03$ and $0.02 \pm 0.03$ for the $\alpha$ and $p$ channels and $\Gamma_R < 12$ keV for both channels.

Typically, extrapolations of the reaction rate of the C + C reactions through the Gamow window have assumed a constant $\tilde{S}$(E), normally using the average value from available experiments (e.g. [14]). The weighted mean values of $\tilde{S}$(E) from the present work are $1.0 \times 10^{16}$ MeV b and $0.4 \times 10^{16}$ MeV b for the $\alpha$ and $p$ channels respectively. The sum of these values is a factor of two smaller than the corresponding value $\tilde{S}(E) = 3.0 \times 10^{16}$ MeV b assumed in [14]. However, the presence of stark resonance structure within the Gamow window demonstrates the uncertainty implicit in this method of extrapolation. The resonance at 2.138 MeV introduces a peak in the reaction rate near $T_9 = 0.8$, within the temperature region of stellar carbon burning (figure 4). Thus extrapolation via average $\tilde{S}$(E)-factor risks underestimating the reaction rate precisely in the critical temperature range.

## ACKNOWLEDGMENTS

The authors thank the Dynamitron Tandem Laboratory (RUBION) for technical and other support.

## REFERENCES

1. C. E. Rolfs, and W. S. Rodney, *Cauldrons in the Cosmos*, The University of Chicago Press, 1988.
2. J. R. Patterson, H. Winkler, and C. S. Zaidins, *Astrophys. J.* **157**, 367 (1969).
3. H. Spinka, and H. Winkler, *Nucl. Phys. A* **233**, 456–494 (1974).
4. M. G. Mazarakis, and W. E. Stephens, *Phys. Rev. C* **7**, 1280 (1973).

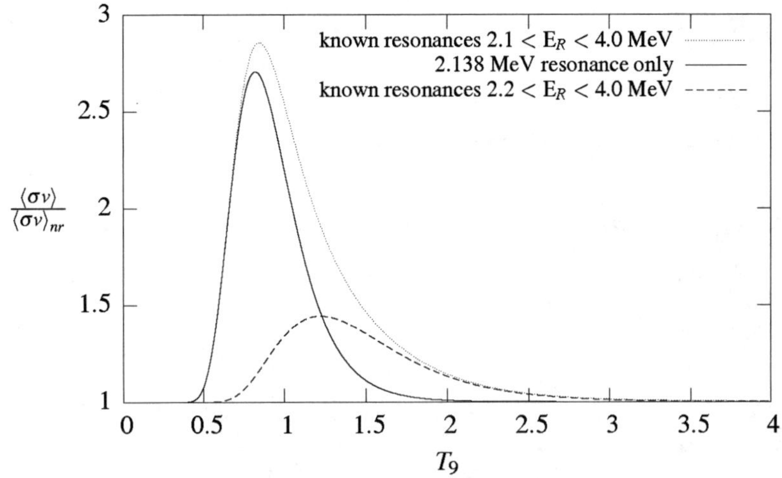


**FIGURE 4.** The effect of the 2.138 MeV resonances on the reaction rate $\langle \sigma v \rangle$, relative to the non-resonant reaction rate $\langle \sigma v \rangle_{nr}$, deduced from averaged $\tilde{S}(E)$ factor at non-resonant energies. The collective effects of nearby resonances also shown.


5. K. U. Kettner, H. Lorenz-Wirzba, C. Rolfs, and H. Winkler, *Phys. Rev. Lett.* **38**, 337–340 (1977).
6. M. D. High, and B. Cujec, *Nucl. Phys. A* **282**, 181 (1977).
7. H. W. Becker, K. U. Kettner, C. Rolfs, and Z. Trautvetter, *Phys. Rev. A* **303**, 305 (1981).
8. E. F. Aguilera, P. Rosales, E. Martinez-Quiroz, G. Murillo, M. Fernandez, H. Berdejo, D. Lizcano, A. Gomez-Camacho, R. Policroniades, A. Varela, E. Moreno, E. Chavez, M. E. Ortiz, A. Huerta, T. Belyaeva, and M. Wiescher, *Physical Review C (Nuclear Physics)* **73**, 064601 (2006).
9. L. Barrón-Palos, E. F. Aguilera, J. Aspiazu, A. Huerta, E. Martínez-Quiroz, R. Monroy, E. Moreno, G. Murillo, M. E. Ortiz, R. Policroniades, A. Varela, and E. Chávez, *Nuclear Physics A* **779**, 318–332 (2006).
10. T. Spillane, F. Raiola, C. Rolfs, D. Schurmann, F. Strieder, S. Zeng, H.-W. Becker, C. Bordeanu, L. Gialanella, M. Romano, and J. Schweitzer, *Phys. Rev. Lett.* **98**, 122501 (2007).
11. D. Rogalla, M. Aliotta, C. A. Barnes, L. Campajola, A. D'Onofrio, E. Fritz, L. Gialanella, U. Greife, G. Imbriani, A. Ordine, J. Ossmann, V. Roca, C. Rolfs, M. Romano, C. Sabbarese, D. Schürmann, F. Schümann, F. Strieder, S. Theis, F. Terrasi, and H. P. Trautvetter, *Eur. Phys. J. A* **6**, 471–477 (1999).
12. S. Wüstenbecker, H. Ebbing, W. H. Schulte, H. Baumeister, H. W. Becker, B. Cleff, C. Rolfs, H. P. Trautvetter, G. E. Mitchell, J. S. Schweitzer, and C. A. Peterson, *Nucl. Instrum. Meth. Phys. Res., Sect. A* **279**, 448–466 (1989).
13. J. F. Ziegler, *computer code SRIM version 2006, http://www.srim.org/* (2006).
14. G. R. Caughlan, and W. A. Fowler, *Atomic Data and Nuclear Data Tables* **40**, 283 (1988).

# $^{12}C(\alpha,\gamma)^{16}O$ Thermonuclear Reaction Rate : Present Status And Perspectives

A. Lefebvre-Schuhl

*Centre de Spectrométrie Nucléaire et de Spectrométrie de Masse (CSNSM), UMR8609, CNRS/IN2P3; Université Paris-Sud, F-91405 Orsay-Campus, France*

**Abstract.** I review quite briefly for reasons of length, the reaction $^{12}C(\alpha,\gamma)^{16}O$ that follows core hydrogen burning and governs, with the triple-α process, most of the universal abundances of $^{12}C$ and $^{16}O$, including new and previously unpublished work. The present status of the $^{12}C(\alpha,\gamma)^{16}O$ cross section knowledge and its consequences on the reaction rate uncertainties are presented with some perspective to reach the desired precision at stellar temperatures.

**Keywords:** Hydrostatic stellar nucleosynthesis, $^{4}$He-, $^{12}$C-induced reactions.
**PACS:** 25.55.–e, 24.30.–v, 26.20.-f, 27.20.+n

## INTRODUCTION

The radiative capture $^{12}C(\alpha,\gamma)^{16}O$ reaction is one of the most important thermonuclear reactions in nonexplosive astrophysical sites and follows the production of $^{12}C$ by the $\alpha(2\alpha,\gamma)^{12}C$ triple-α process in stellar cores which have exhausted their hydrogen. In helium burning, the ratio of both reaction rates determines directly the ratio of carbon and oxygen abundances at the end of helium burning. It has a strong impact on the abundances of the heavier elements, produced in later hydrostatic burning stages and on the final destiny of a massive star as a neutron star or a black hole.

In spite of its importance, the cross section of this reaction at stellar temperature is still uncertain [1-5]. For helium burning temperatures (T~$2\times10^{8}$ K), the Gamow peak is situated at 300 keV and any measurement of the cross section at such energy is excluded, at least with today's experimental techniques because of its extremely low value of about 0.01 fb. The theoretical extrapolation, to the Gamow peak energy range, of experimental values obtained at higher energies is particularly difficult because of the superposition of the various contributions still too uncertain or even just estimated: the radiative capture in the thermonuclear energy range is composed of direct capture (DC) to various states of $^{16}O$, each one consisting in a few multipolar contributions. For example, *E*1 and *E*2 radiative capture contributes about equally in the dominating direct capture process to the ground state. At present time, extrapolations are performed via R- or K-matrix formalism (even microscopic cluster calculations need to be normalized via such formalisms and each contribution has to be extrapolated separately.

CP1012, *Frontiers in Nuclear Structure, Astrophysics, and Reactions: FINUSTAR 2*
edited by P. Demetriou, R. Julin, and S. V. Harissopulos
© 2008 American Institute of Physics 0-7354-0532-5/08/$23.00

Since the previous reviewing work of Buchmann and Barnes [5], new direct and indirect investigations of this reaction have been published. Results and consequences are discussed in the following.

## STATUS OF $^{12}C(\alpha,\gamma)^{16}O$ REACTION

Indirect methods such as $\alpha$-transfer reaction ([6] and references quoted therein), or Coulomb dissociation ([7] and references quoted therein), provide alternative ways for extracting the total astrophysical S(E) factor of the $^{12}C(\alpha,\gamma)^{16}O$ reaction. Among these indirect methods and as regards to direct ones, only the $\beta$-delayed $\alpha$-transfer $^{16}N$ decay (see for example the discussion in [5] and references quoted therein) was constraining the astrophysical S(E) factor at stellar core temperature ($S(E_0)$). Direct methods used to study the $^{12}C(\alpha,\gamma)^{16}O$ reaction cross section in regular or inverse kinematics constrained separately various contributions (to excited states and E1 and E2 to the ground state of $^{16}O$) or the total S-factor. To separate the E1 and E2 contributions, the most precise way consists in measuring $\gamma$-ray angular distributions. It was used down to $E_{cm}$= 0.89 MeV ($E_{cm}$ is the center-of-mass energy of the reaction), while total S-factor measurements were performed using a recoil separator down to $E_{cm}$= 1.9 MeV.

For the first time, Schürmann et al. [8] measured recently the total cross-section of $^{12}C(\alpha,\gamma)^{16}O$ reaction at the Bochum 4 MV Dynamitron Tandem accelerator by a direct and ungated detection of the $^{16}O$ recoils, in inverse kinematics and in the $E_{cm}$ energy range 1.9- 4.9 MeV. $^{12}C$ ion beams with particle currents of up to 10 $\mu$A were guided into a windowless $^4He$ gas target. The recoil nuclei were detected in the European Recoil separator for Nuclear Astrophysics (ERNA, [9] and references quoted therein). A particular attention was paid to reach the 100% transmission of the recoils in the selected charge state (equilibrium charge state distribution was obtained with an Ar post-target stripping system). This is an essential condition for measuring reliably the absolute capture cross-section using a recoil mass separator. Results were compared with the sum of the different amplitudes obtained in the best R-matrix fit to the experimental data available at that time, performed by Kunz et al [4]. In this analysis the interference effects of the cascade transitions were neglected. Even if a good agreement was obtained in absolute S(E) values between previous work and the present data on top of the E = 2.4, 3.2, and 4.3 MeV resonances, large differences can be observed at energies around E = 3.0 and 4.0 MeV [8]. A detailed study of the influence of the data is in progress with a new R-matrix code [10] including fits to the different amplitudes and to the total cross-section [11]. In conclusion, the ERNA approach provided new data of the $^{12}C(\alpha,\gamma)^{16}O$ S factor at energies above 1.9 MeV which represent new informations for an improved $S(E_0)$ determination.

Assunçao et al. [12] and Fey et al. [13] performed new direct measurements in regular kinematics using intense $\alpha$-particle beams of up to 600 $\mu$A from the Stuttgart DYNAMITRON, in two complementary experiments using high efficient $\gamma$-ray detection setups. For the first time, a $4\pi$-germanium-detector setup has been used to measure the angular distribution of the $\gamma$ rays at all angles simultaneously: nine EUROGAM high-purity Ge (HPGe) detectors in close geometry [12], were actively shielded individually with bismuth-germanate crystals (BGO). Of critical importance were the high purity, homogeneity and stability of the high beam power $^{12}C$ targets. Special efforts were made to improve their quality via isotopic implantation, and their depth-profile analysis via scanning by Rutherford backscattering. These studies were limited to the ground-state transitions. Angular distributions were measured with two different γ-ray setups, in the $E_{cm}$ energy range 0.89-2.78 MeV and in the laboratory angular ranges 30°–130° [12] or 15°–140° [13]. From these distributions, astrophysical E1 and E2 S-factor functions vs energy were calculated. Especially true for the E2 capture cross section, reduced error bars on these new data help to reduce significantly the uncertainty in the studied energy range. R-matrix calculations were performed using the code ERMA of Kunz [3,4]. They were based, in particular, on these E1 and E2 capture data, the scattering phases of [14,15], and the data of the β-delayed α-decay of $^{16}N$ [16]. For E1 a three level fit and for E2 a five level fit were performed [4,13].

DC to the ground state remains the contribution that dominates by about an order of magnitude the $^{12}C(\alpha,\gamma)^{16}O$ cross section at central Helium-burning stellar temperature. As regards the difficulty of such measurements, it is easy to understand the lack of precise results for the DC to excited states of $^{16}O$. Up to 2006, only a few data were available (see [5] and references quoted therein), and gave partial estimations or limits of their contributions to the total S factor. At the DRAGON recoil separator at ISAC [17], Matei et al [18] performed measurements of the cross section of the DC to the $J^{\pi}=0^{+}$, 6.05 MeV level of $^{16}O$ for the $E_{cm}$ energy range 2.22-5.42 MeV using $^{12}C$ beams of typically 30–50 pnA intensities. Around the windowless gas target, the setup included a BGO-detector array for γ-ray detection. In the procedure to get the cross section values, the acceptance of the recoil separator had to be normalized using GEANT simulations. The transition strength between resonances has been identified in R-matrix fits as resulting both from E2 contributions as well as E1 radiative capture but no angular distribution were available to extrapolate separately these contributions, inducing less constraints on the R-matrix analysis. Results of the extrapolation of the total cross section to low energies showed that this transition is likely the most important contribution of the DC to excited state for $^{12}C(\alpha,\gamma)^{16}O$: in these R-matrix results, it represents about 15% of the total.

A comparison of the most recently published R-matrix calculations results is given

**TABLE 1.** Results in keVb obtained for the various contributions to the astrophysical S factor at stellar energies ($S(E_0)$) from: a) the NACRE compilation [2], b) Hammer et al. [19] and Kunz et al. [4], and c) Buchmann and Barnes [5]. Contributions of the DC to the excited states of $^{16}O$ are referred by their energy (6.05, 6.92 and 7.12 MeV). Error bars are quoted between parentheses.

| $S(E_0)$ | a) | b) | c) |
|---|---|---|---|
| E1 to the ground state | 79 (21) | 77 (17) | 80 (20) |
| E2 to the ground state | 120 (60) | 81 (22) | 53 (+13,-18) |
| To excited states: 6.92 & 7.12 | | 4 (4) | |
| To excited states: 6.92 | | | Up to a few 10 |
| To excited states: 6.05 | | | 25 (+16, -15) |
| Total | | 162 (39) | |

in the Table 1. In that comparison, one has to take care of the differences between the cited R-matrix calculations, such as the treatment of direct components. These results for the contribution of the E2 direct capture to the ground state were obtained considering the best $\chi^2$ minimizations but resulted in different signs of interferences for b and c calculations.

Attention should be paid on the non-yet-understood partial disagreement between the phases of the interference pattern between the E1 and E2 derived from the ground-state angular-distribution and the elastic diffusion measurements analysis [12,14,15]. Results of columns b and c were obtained taking into account the elastic diffusion results considered as more precise by the authors. In the past, some analyses were using the angular distribution results.

## CONCLUSIONS AND PERSPECTIVES

In conclusion, this comparison underline that the precision on the knowledge of the $^{12}C(\alpha,\gamma)^{16}O$ reaction rate has been strongly increased by the new experiments since the Nacre compilation [2] and the last reviewing work [5], and that new R-matrix calculation should be performed taking into account the ground- and excited-state contributions together with the total S-factor data. Present knowledge of interference signs for the contribution of the E2 direct capture to the ground state and evaluation of the contributions to excited states induce still now large uncertainties on the total S factor at stellar energies. Resulting precision on the $^{12}C(\alpha,\gamma)^{16}O$ cross section at 300 keV remains far from satisfying if a precision of 10% or better is required by theoretical astrophysicists [20].

To improve such results, certainly better data with a separation of the E1 and E2 component, i.e., angular-distribution measurements, are most desirable as well as an extension of the energy range of the data. Angular distributions of the DC to excited states and additional angular distributions of the DC to the ground state would better constrain the interference pattern in the R-matrix analysis and the extrapolations at stellar energies using the various multipolar excitation functions separately, for the ground and the excited states. That first requirement seemed unreachable with today's available techniques. The recent development of new scintillation detectors [21] opens new perspectives. Some research and development (R&D) studies have to be performed to evaluate their ability to discriminate the various $\gamma$-ray transitions through the $^{16}O$ levels at 6.05, 6.92 and 7.12 MeV. Using a $4\pi$ $\gamma$-detector array with such high efficiency in coincidence technique will allow to decrease drastically the $\gamma$-ray background events and to avoid effects described in H. Makii et al [22].

Up to now, attempts were performed using $\gamma$-ray coincidences with the $^{16}O$ recoil nucleus [ERNA, DRAGON]. The use of extended gas target limited the $\gamma$-ray angular distribution studies. R&D studies on dense jet-gas target for example, will demonstrate the possibility to study $\gamma$-ray angular distributions. As these experiments were limited to $E_{cm}$ above 1.9 MeV, additional work has to be done on recoil separators to reach energies lower than 1 MeV keeping in mind the importance of the recoil-nucleus transmission knowledge. Other attempts were also recently performed using a few NaI(Tl) $\gamma$-ray detectors [23] or the Karlsruhe $4\pi$-BaF2-detector array [24] in coincidence with a pulsed $^4He$ beam demonstrating the feasibility of such measurements below 2 MeV. Finally new R&D on accelerator have to be performed to recover the ability to get the intense stable beams either of $^{12}C$ or $^4He$ needed to reach lower energies.

Precision improvements will come into reach, either in regular- or in inverse-kinematics direct studies, by such consequential experimental implementations. These implementations will present serious challenges to the detector technology and to the ingenuity of the experimentalists.

## ACKNOWLEDGMENTS

My special thanks go to the EUROGAM experiment collaboration with whom I got experienced on this field and to the ERNA collaboration for detailed and helpful discussions. I am obliged to P. Aguer for extensive advice and to C. Angulo and P. Descouvemont for helpful discussions. My work was partly supported by the IN2P3-CNRS and the France-UK (IN2P3/EPSRC) Loan Pool for the EUROGAM detectors.

## REFERENCES

1. T. A. Weaver and S. E. Woosley, Phys. Rep. **227**, 65 (1993).
2. C. Angulo et al., Nucl. Phys. **A 656**, 3 (1999) NACRE-compilation.
3. R. Kunz et al. Phys. Rev. Lett **86**, 3244 (2001).
4. R. Kunz et al., Astrophys. J. **567**, 643 (2002) and R.W. Kunz, Ph.D. thesis, Universität Stuttgart, 2002; URL:http://elib.uni-stuttgart.de/opus/volltexte/2003/1266/.
5. L. Buchmann and C. A. Barnes, Nucl. Phys. **A 777**, 254 (2006).
6. A. Belhout et al. Nucl. Phys. **A 793**, 178 (2007).
7. F. Fleurot et al., Phys. Lett. **B 615**, 167 (2005).
8. D. Schürmann et al., Eur. Phys. J. **A 26**, 301 (2005).
9. L. Gialanella et al., Nucl. Inst. and Meth. **A 522**, 432 (2004).
10. R. Kunz, private communication.
11. D. Schürmann Ph.D. thesis, Ruhr-Universität Bochum, Germany 2007, URL: http://www.brs.ub.ruhr-uni-bochum.de/netahtml/HSS/Diss/SchuermannDaniel/.
12. M. Assunção et al., Phys. Rev. **C 73,** 055801 (2006).
13. M. Fey et al. Nucl. Phys. **A 718**, 131c (2003) and M. Fey, Ph.D. thesis, Universität Stuttgart, Germany 2004, URL: http://elib.uni-stuttgart.de/opus/volltexte/2004/1683/.
14. M. D'Agostino Bruno et al., Nuovo Cimento Soc. Ital. Fis. **27**, 1 (1975).
15. R. Plaga et al., Nucl. Phys. **A 465**, 291 (1987).
16. R. E. Azuma *et al.,* Phys. Rev. **C 50**, 1194 (1994).
17. D.A. Hutcheon et al., Nucl. Inst. and Meth. **A 498**, 190(2003).
18. C. Matei et al., Phys. Rev. Lett. **97,** 242503 (2006).
19. J. W. Hammer et al., Nucl. Phys. **A 752**, 514c (2005); Nucl. Phys. **A 758**, 363c (2005).
20. S. E. Woosley et al. Nucl. Phys. **A 718,** 3c (2003).
21. Paris (Photon Array for studies with Radioactive Ion and Stable beams) collaboration, private communication.
22. H. Makii et al., Phys. Rev. **C 76**, 022801(R) (2007).
23. H. Makii et al., Nucl. Phys. **A 758**, 371c (2005).
24. R. Plag et al., Nucl. Phys. **A 758**, 415c (2005).

# AGB fluorine nucleosynthesis studied by means of Trojan-horse method: the case of $^{15}N(p,\alpha)^{12}C$

R.G. Pizzone*, C. Spitaleri*, M. La Cognata*, S. Cherubini*, V. Crucillà*, M. Gulino*, L. Lamia*, S.M.R. Puglia*, G.G. Rapisarda*, S. Romano*, S. Sergi*, A. Tumino* and R. Tribble†, L. Trache†, C. Fu†, V. Goldberg†, A. Mukhamedzhanov†, G. Tabacaru†

*Dipartimento di Metodologie Fisiche e Chimiche per l'Ingegneria - Università di Catania and Laboratori Nazionali del Sud - INFN, Catania, Italy*
† *Cyclotron Institute, Texas A&M University, College Station, TX 77843, USA*

**Abstract.** The low-energy bare-nucleus cross section for $^{15}N(p,\alpha)^{12}C$ is extracted by means of the Trojan-horse Method applied to the $^{2}H(^{15}N,\alpha)^{12}C)n$ reaction at $E_{beam}$=60 MeV. The astrophysical $S$-factor is deduced and compared to the direct data in the same energy region. A fair agreement with direct data down to 80 keV is found if energy resolution effects are taken into account.

**Keywords:** Stars: abundances – Stars: Asymptotic Giant Branch – Stars: Fluorine Nucleosynthesis – Nuclear Reactions: Astrophysical $S$-Factor
**PACS:** 24.50.+g,26.20.Np

## INTRODUCTION

An open question in nuclear astrophysics is the $^{19}F$ abundance in the Milky Way and its production sites. This isotope is synthesized in type II supernovae explosions, in Wolf-Rayet stars and during the AGB phase of stellar evolution [14]. In particular spectroscopic observations show that in low-mass AGB stars fluorine abundance is enhanced with respect to the solar one. Large efforts have been devoted in understanding $^{19}F$ nucleosynthesis since fluorine abundance observed in giants can constrain AGB star models [9]. In particular a detailed study of nuclear reaction rates involved in $^{19}F$ production and destruction can play an important role. A key isotope for $^{19}F$ production in AGB stars is $^{15}N$ from which $^{19}F$ is synthesized via $^{15}N(\alpha,\gamma)^{19}F$ in the intershell region during a thermal pulse. The $^{15}N(p,\alpha)^{12}C$ reaction is then of primary importance since it removes both proton and $^{15}N$ nuclei from $^{19}F$ production chain [9].

The astrophysical $S$-factor of the $^{15}N(p,\alpha)^{12}C$ reaction has been measured down to 73 keV [16, 13] and a Breit-Wigner extrapolation has been performed to derive its trend down to zero energy [13]. This extrapolation has also been employed to calculate the reaction rate [1] to be introduced in astrophysical stellar evolution codes. Of course this simple extrapolation does not take into account the presence of subthreshold or low-energy resonances, neither the enhancement of the $S$-factor due to the presence of atomic electrons. Indeed the electron screening effect introduces at least a 10% increase of the $S$-factor at 80 keV for the $^{15}N+p$ system [2], which can modify the low energy behaviour of the $S$-factor.

CP1012, ***Frontiers in Nuclear Structure, Astrophysics, and Reactions:*** *FINUSTAR 2*
edited by P. Demetriou, R. Julin, and S. V. Harissopulos
© 2008 American Institute of Physics 0-7354-0532-5/08/$23.00

In order to derive the low-energy bare-nucleus $S$-factor, which is the parameter of interest for astrophysics, with no need of extrapolation, a number of indirect techniques has been developed, e.g. the Coulomb dissociation [4] and the ANC (asymptotic normalization coefficient) method [10], and the Trojan-horse method (THM) [3, 17, 5]. In particular, the THM allows to extract a charged-particle two-body cross section at astrophysical energies free of Coulomb suppression and electron screening enhancement. In the next section the method is briefly discussed while its application to the experimental study of the $^{15}$N$(p,\alpha)^{12}$C reaction is described in subsequent sections.

## TROJAN HORSE METHOD

The THM theory is extensively discussed in many papers [17, 5]. Recently the validity of the method has been discussed by [12]. Here only the essential details are briefly reviewed.

A reaction $A+a \rightarrow C+c+s$, with nucleus $a$ having a strong $x \oplus s$ cluster structure, can proceed via a number of reaction mechanisms. In THM application the reaction mechanism is selected where the Trojan horse nucleus $a$ breaks up into a nucleus $x$, acting as the participant to the subreaction $A+x \rightarrow C+c$, and a nucleus $s$ which can be regarded as a spectator. This direct reaction mechanism gives the dominating contribution to the cross section in a restricted region of the three-body phase space when the momentum transfer to the spectator $s$ is small (QF conditions). The QF reaction is well described in the Impulse Approximation. In particular in the Plane-Wave Impulse Approximation (PWIA) the cross section of the three-body reaction can be easily factorized as follows [6]:

$$\frac{d3\sigma}{dE_c d\Omega_c d\Omega_C} \propto KF\, |\Phi(\vec{p}_s)|^2 \cdot \left(\frac{d\sigma}{d\Omega_{c.m.}}\right)^{\text{off}} \tag{1}$$

where KF is a kinematical factor, $\Phi(\vec{p}_s)$ is the Fourier transform of the radial wave function $\chi(\vec{r})$ for the $x-s$ intercluster relative motion, $[(d\sigma/d\Omega)_{c.m.}]^{\text{off}}$ is the off-energy-shell differential cross section for the two-body $A(x,c)C$ reaction at the center of mass energy $E_{c.m.}$. Since $|\Phi(\vec{p}_s)|^2$ is known from nuclear clustering studies the product $KF \cdot |\Phi(\vec{p}_s)|^2$ is evaluated via a Montecarlo calculation and therefore $[(d\sigma/d\Omega)_{cm}]^{\text{off}}$ is deduced from a measurement of $d3\sigma/dE_c d\Omega_c d\Omega_C$.

In the present work the cross section of the $^{15}$N$(p,\alpha)^{12}$C reaction is derived by selecting the QF contribution to the $^{2}$H$(^{15}$N$,\alpha^{12}$C$)n$ three-body process. Because of the approximations adopted, the absolute value of the two-body cross section cannot be extracted. However, the normalization factor is determined by scaling THM data to the direct ones which are available at energies above the Coulomb barrier. In particular, an improved analysis of the same partial data set as the one reported in [8] is discussed, in which the effect of finite energy resolution is taken into account.

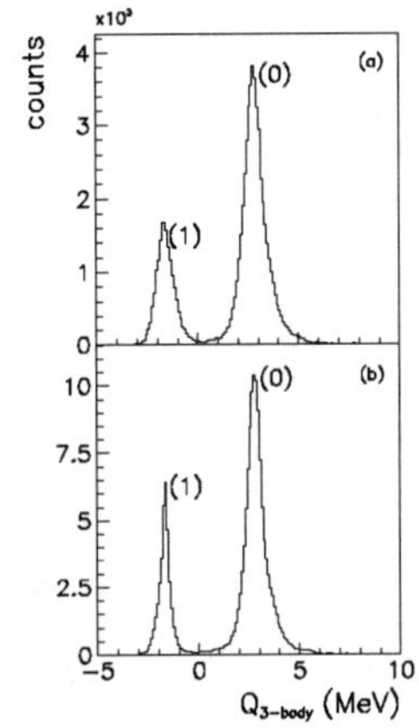


**FIGURE 1.** $Q-$value spectrum for the three-body process for two different detector pairs.

## DATA ANALYSIS AND RESULTS

The experiment was performed at Texas A&M University Cyclotron Institute. Further details are reported in [7].As it can be seen from THM prescriptions before proceeding to the extraction of the cross section and of the astrophysical $S$-factor of the $^{15}$N($p$,$\alpha$)$^{12}$C reaction two steps are necessary: (i) by investigation of the kinematics of the three-body process the $^{2}$H($^{15}$N,$\alpha_0$$^{12}$C)$n$ channel has to be disentangled from other reactions taking place in the same target; (ii) from the study of the three-body reaction dynamics and in particular from the analysis of the correlation between the $\alpha_0 - ^{12}$C coincidence yield and the neutron momentum the presence of the QF mechanism is deduced. Its contribution is then separated from the one of different reaction mechanisms feeding the same exit channel.

In order to single out the contribution of the $^{2}$H($^{15}$N,$\alpha_0$$^{12}$C)$n$ reaction, the one which provides the $S$-factor relevant for astrophysics (since the $\alpha_1$ channel shows a much lower cross section than the $\alpha_0$ channel [15]), first a number of conditions is imposed to the data set. In detail, coincidences are selected by introducing a window on the Time-to-Amplitude Converter (TAC) parameter and the carbon locus is identified in telescope A. An additional cut is introduced in the $E_{^{12}C}$ vs. $E_\alpha$ 2D spectrum to single out the kinematic locus of the three-body reaction. Finally the $Q-$value spectrum of the three-body process is deduced through energy conservation (see figure 1). Two peaks show up, corresponding to the $^{2}$H($^{15}$N,$\alpha_0$$^{12}$C)$n$ channel (the one at higher energy, $Q_{3-body} = 2.74$ MeV) and to the $^{2}$H($^{15}$N,$\alpha_1$$^{12}$C)$n$ channel ($Q_{3-body} = -1.7$ MeV). The absence of additional peaks and the good agreement between the theoretical and experimental $Q-$values make us confident that additional reaction channels have been disentangled from the $^{2}$H($^{15}$N,$\alpha_0$$^{12}$C)$n$ coincidence yield and that an accurate calibration procedure has been performed. Only the events belonging to the high energy peak are retained for further analysis.

An extensive study of the processes feeding the exit channel is a necessary step

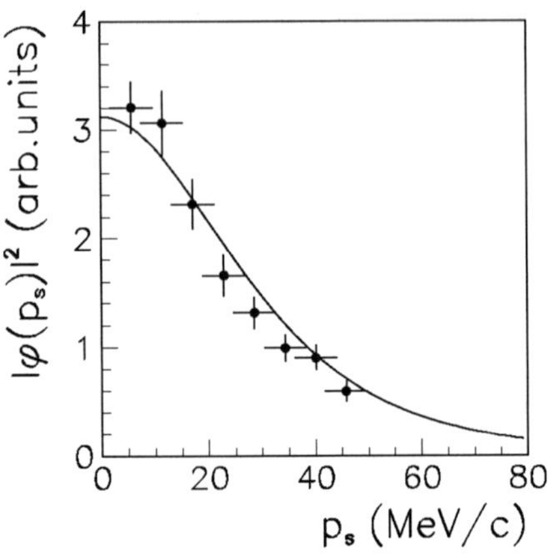


**FIGURE 2.** The momentum distribution deduced from the present data set compared with the behaviour of the Hulthen function.

in order to disentangle the QF component from other reaction mechanisms. In fact the analysis of the experimental results is often complicated by the presence of other reaction mechanisms yielding the same particles in the final state, e.g. sequential decay (SD) [7]. SD contributions due to the $^{13}C^* \rightarrow {}^{12}C + n$ and $^{5}He^* \rightarrow {}^{4}He + n$ decays can be singled out by studying the $E_{\alpha-n}$ vs. $E_{\alpha-{}^{12}C}$ and $E_{{}^{12}C-n}$ vs. $E_{\alpha-{}^{12}C}$ 2D plots and identifying the kinematic loci appearing in the spectra due to the excited levels of such intermediate nuclei [8, 7]. It is found that in the chosen phase space region (within a neutron momentum window $|\vec{p}_s| < 40$ MeV/c) SD processes give negligible contributions to the coincidence yield in the astrophysically relevant energy region [8]. On the other hand the excited states of the $^{12}C+\alpha$ system can be formed through a QF reaction mechanism or through a sequential one. The procedure to discriminate the two is more complicated and relies on the study of angular correlation spectra and of the $E_{c.m.}$ spectra as a function of the neutron momentum [8, 7]. Indeed, if direct break-up occurs, the experimental momentum distribution should resemble the one neutron has inside deuteron, given by the Fourier transform of the intercluster wave function in the PWIA approach. The momentum distribution deduced from the present data set is displayed in figure 2. The theoretical one, the square of the Hulthén wave function in momentum space, is superimposed to the data (black line in figure 2). The normalization constant, which is the only adjustable parameter, has been fitted to the experimental distribution. The nice agreement between the two demonstrates that in the $|\vec{p}_s| < 40$ MeV/c momentum window the QF contribution is dominant thus neutron can be regarded as a spectator to the $^{15}N(p,\alpha)^{12}C$ subreaction.

Before extracting the $S(E)$−factor for astrophysical application, it is necessary to perform appropriate validity tests on the indirect two-body cross section. Thus angular distributions and excitation function are deduced and compared to the ones in literature [16, 13]. Once the S(E)-factor is extracted following THM prescriptions [7] normalization to direct data is deduced by equating the areas subtended by the 314 keV peak, due to the 12.44 MeV $1^-$ $^{16}O$ excited state, in both data sets. In order to perform a coherent

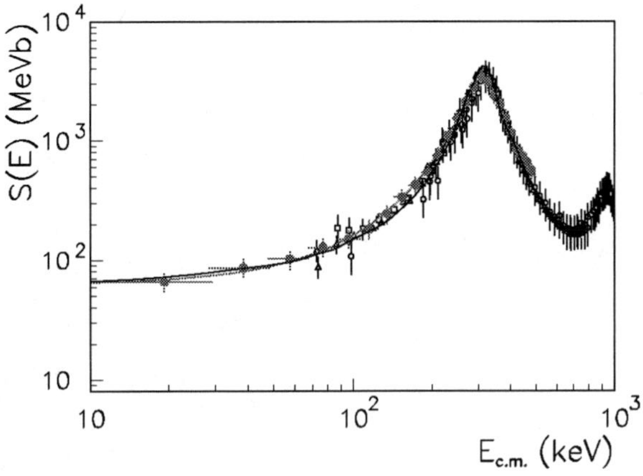


**FIGURE 3.** Astrophysical S(E)-factor extracted via THM (full dots) compared with direct data (black symbols) from [13].

comparison, energy resolution of direct data was reduced to 40 keV too. The results for the astrophysical S(E) factor are shown in figure 3. THM and direct data are displayed as full dots and a full line respectively. A good agreement is apparent in the energy region where direct data are available.

## REFERENCES


1. C. Angulo et al., Nucl. Phys. **A 656**, 3 (1999).
2. H.J. Assenbaum et al., Z. Phys. **A 327**, 461 (1987).
3. G. Baur, Phys. Lett. **B 178**, 135 (1986).
4. G. Baur and H. Rebel, J. Phys. **G 20**, 1 (1994).
5. S. Cherubini et al., Astroph. J. **457**, 855 (1996).
6. M. Jain et al., Nucl. Phys. **A 153**, 49 (1970).
7. M. La Cognata et al., Phys. Rev. C, in press.
8. M. La Cognata et al., Eur. Phys. J. **A 27**, 249 (2006).
9. M. Lugaro et al., Astroph. J. **615**, 934 (2004).
10. A. Mukhamedzhanov et al., Phys. Rev. **C 56**, 1302 (1997).
11. A. Mukhamedzhanov and R.E. Tribble, Phys. Rev. **C 59**, 3418 (1999).
12. A. Mukhamedzhanov et al., 2006, nucl-th/0602001
13. A. Redder et al., Z. Phys. **A 305**, 325 (1982).
14. A. Renda et al., Mnras 354, 575 (2004).
15. C. Rolfs and W.S. Rodney, Nucl. Phys. **A 235**, 450 (1974).
16. A. Schardt et al., Phys. Rev. **86**, 527 (1952).
17. C. Spitaleri, Prob. of Fund. Mod. Phys. II, World Scientific, Singapore, 21 (1990).

# Two experimental approaches in nuclear astrophysics using neutrons

Y. Parpottas*, C. Iliadis†,** and H. Tsertos*

*Department of Physics, University of Cyprus, Nicosia 1678, Cyprus.
†Department of Physics and Astronomy, University of North Carolina, Chapel Hill, North Carolina 27599-3255, USA.
**Triangle Universities Nuclear Laboratory, Durham, North Carolina 27708-0308, USA.

**Abstract.** High resolution measurements using the ($^3$He,$n$) reaction are used to determine excitation energies of astrophysically important states. Further, spins can be assigned to these states by comparing the measured differential cross sections to Hauser-Feshbach cross sections. The astrophysical reaction rates are then calculated using the experimental values. Reference [1] is an example of such measurements.

Absolute values of the astrophysical $S$ factors and cross sections are calculated from measurements using the $(d,n)$ reaction at low incident energies. Reference [2] is an example of such measurements.

**Keywords:** ($^3$He,$n$), $^{25}$Al$(p,\gamma)^{26}$Si, measured $E_n$, Hauser-Feshbach cross sections, measured differential cross sections, calculated astrophysical reaction rates, $(d,n)$, yields, angular distribution, deduced astrophysical $S$ factor
**PACS:** 25.45Hi, 25.40Lw, 25.55Hp, 27.30+t

## Astrophysical reaction rates using ($^3$He, $n$) reactions

The astrophysical reaction rates of many proton-capture reactions has not yet been measured directly, due to difficulties of producing radioactive beams. At the present time, these reaction rates must thus be determined from indirect spectroscopic information. High resolution measurements, involving determination of resonance energies, partial widths and spin-parity, reduce these reaction rates uncertainties.

Excitation energies can be accurately determined using ($^3$He,$n$) reactions due to the excellent neutron energy resolution achieved in such measurements. This is due to the thin targets, long flight paths and low bombarding energies used in these experiments. Neutron detection experiments could result in better resolution than charged-particle detection experiments since neutrons could be detected at any distance without losing energy using time-of-flight techniques. These experiments are capable of seeing all states regardless of their structure including unnatural parity states. This is because at low incident energies ($^3$He,$n$) reactions favor the compound nuclear reaction mechanism, particularly when the outgoing neutron energy is low. Spins can be also assigned using ($^3$He,$n$) reactions by comparing the measured differential cross sections to Hauser-Feshbach calculations.

Experiments [1, 3] using the ($^3$He,$n$) reaction were conducted at the Ohio University 4.5 MV Van de Graaff Tandem Accelerator. The $^3$He$^{++}$ pulsed beam was delivered to 10 keV thin targets supported by stopping Ta disks. Neutrons were detected in a three-


© 2008 American Institute of Physics 0-7354-0532-5/08/$23.00

detector array of 12.7-cm-diameter and 5.1-cm-thickness liquid scintillators placed at 10 m from the target in the 30-meter-long well-shielded tunnel. The bombarding energies were chosen so that the time of flight for the excitation energies of interest were close to the maximum practical value, optimizing in this way the energy resolution for these neutron groups. The neutron energy resolution was approximately model, using physically reasonable inputs, with the following equation:

$$\Delta E = 2E_n \left[ \left(\frac{\Delta x}{x}\right)^2 + \left(\frac{\Delta t}{t}\right)^2 + \left(\frac{\Delta E_b}{E_n}\right)^2 \right]^{1/2} \tag{1}$$

where $\Delta x = 0.0254$ m (detector thickness), $x = 10$ m (flight path), $\Delta t = 1.5$ ns (time resolution), and $\Delta E_b = 0.005$ MeV (beam-energy resolution, including the effect of energy loss in the target), while the time-of-flight $t$ and neutron energy $E_n$ vary over the spectrum.

The pulse height, pulse shape discrimination (PSD) and time information were processed and saved for further analysis. Cuts were applied on the pulse shape and height optimized for best n-$\gamma$ discrimination. The time spectra were calibrated using the method described in Ref. [4] providing an accuracy of approximately 1 part in $10^4$. The excitation energies were deduced from the neutron peak centroids. A 4-keV systematic uncertainty was assigned to high excitation energies (statististical uncertainties were 1 keV or less).

The differential cross section spectra were extracted from the neutron ones according to the following equation:

$$\frac{d\sigma}{d\Omega} = \frac{Y}{I \eta \, \Delta x \, \varepsilon \, \Delta\Omega} \tag{2}$$

where $Y$ is the number of detected neutrons, $\varepsilon$ is the absolute efficiency of a neutron detector, $\Delta\Omega$ is the solid angle fraction subtended by a neutron detector, $I$ is the number of particles that strikes the target with thickness $\Delta x$, and $n$ is the nuclei per unit volume. The quantity $I$ is determined by the accumulated charge during the measurement time while $\eta \, \Delta x$ is given by

$$\eta \, \Delta x = \frac{P N_A \rho \, \Delta x}{M} \tag{3}$$

where $P$ is the isotopic abundance of target atoms in the prepared foil, $N_A$ is the Avogadro's number, $M$ is the target molar mass and $\rho \, \Delta x$ is the target areal mass density.

The experimental differential cross sections were compared with the Hauser-Feshbach (HF) cross sections. This theory predicts the angular distribution of the differential cross section of the emitted particles from the various exit channels of a compound nucleus. It takes into account the formation of the compound nucleus and the various residual nuclei of exit channels in states of different $J^{\pi}$ and the level density in the region where compound nuclear levels overlap. The HF code [5] is based on the theory and formulas in Ref. [6]. HF is sensitive to $J$ but not to $\pi$ changes; therefore the measurements can lead to a value for $J$ but rely on other results for inferring $\pi$.

The HF differential cross sections of every state were converted into the lab system, reduced, and compared with the experimental ones. The HF results should be reduced because the HF theory assumes that the cross sections in the non-elastic channels come entirely from the decay of the compound nucleus, so that only the elastic channels contribute in both direct (shape elastic) and compound processes [7]. In practice, however, there are frequently direct contributions to the non-elastic channels as well. This affects the compound nucleus cross section even in channels with no direct component since the flux going into the direct non-elastic reactions does not enter the compound nucleus. Therefore, in the analysis of these interactions a reduction factor $R$ should be defined. This reduction factor denotes the percentage of the reaction cross section which is compound. Lack of knowledge of all levels in every exit channel or low density parameter values could also lead to $R < 1$, while the opposite to $R > 1$.

Differential cross sections for states above few MeV were compared with the corresponding HF calculations after the reduction. Low-lying states should be excluded since they have significant direct contribution. Spins can be assigned to higher excited levels verifying possible existing assignments or choosing a specific $J^\pi$ in the case that more than one $J^\pi$'s are reported by previous studies (e.g. Fig. 6, Ref. [1]).

Accurate determination of the resonances (energies, spin-parity) that contribute to the total reaction rates reduce further the uncertainties. The three $^{25}$Al(p,$\gamma$)$^{26}$Si resonant reaction rates at 152 keV ($1^+$), 394 keV ($3^+$), 428 keV ($0^+$) important to the total rates were calculated from spectroscopic information of the $^{24}$Mg($^3$He,$n$)$^{26}$Si reaction [1]. The non-resonant and tail contribution rates were taken directly from Refs. [8, 9] for temperatures $T_9 = 0.03 - 1.5$. These rates were, previously, calculated studying the $^{29}$Si($^3$He,$^6$He)$^{26}$Si reaction [10]. In the latter study, only two of the above resonances were experimentally found at 160 keV and 427 keV; the third resonance was observed in $^{28}$Si($p$, $t$)$^{26}$Si reaction measurements [11], at 398 keV. Spins and parities were assigned to the two resonances found in Ref. [10], [160 keV ($1^+$), 427 keV ($3^+$)] from mirror nuclei considerations while spin-parity was assigned to the resonance in Ref. [11], 398 keV ($0^+$), through distorted-wave Born approximation (DWBA) analysis for $\theta$ = 20°-85°. Recently, the ($p$,$t$) study [11] was extended to smaller angles [12] and it was found that the 396 keV resonance is consistent with a multi-step process $^{28}$Si($p$, $d$)$^{27}$Si($\frac{1}{2}^+$)($d$, $t$)$^{26}$Si($3^+$) populating a $3^+$ state. Then the three resonant rates at 155 keV ($1^+$), 394 keV ($3^+$), 428 keV ($0^+$) important to the total rates were recalculated [12]. The latest study lend credibility to the spin-assignment procedure followed using the ($^3$He,$n$) reaction and, consequently, to the calculation of the astrophysical reaction rates.

## Astrophysical $S$ factor for ($d$, $n$) reactions at low incident energies

The low energy behavior of the astrophysical $S$ factor for many reactions is important in nuclear astrophysics. This is necessary to perform a reliable extrapolation to the energies of interest (in the so-called Gamow window) and eventually to calculate the astrophysical reaction rates.

The slope and absolute value of the $S$ factor and the absolute cross section for two ($d$,$n$) reactions at low incident energies were measured using the Triangle Universities Nuclear Laboratory's (TUNL) 80 keV atomic beam polarized ion source (ABPIS). This source produced unpolarized deuteron beams at 45, 60 and 80 keV for the $^7$Li($d$,$n$)$^8$Be reaction [13] and vector polarized deuteron beams at 120, 140, 160 keV (since the target chamber could be biased up to -80 keV) for the $^{11}B(d,n)^{12}C$ reaction [2]. In the latter case, unpolarized yields were obtained by combining the 'spin-1' and the 'spin-2' yields [2]. The targets were thick enough to stop the beam. Neutrons were detected in six 12.7-cm-diameter and three 11.4-cm-diameter BC501A liquid scintillators of 5.1 cm thickness. The 12.7-cm-diameter detectors were placed approximately at 40 cm from the target face with their centers at forward angles (0°-90°) and the three 11.4-cm-diameter detectors were placed approximately at 60 cm from the target face with their centers at backward angles (112°-150°). Pulse shape discrimination techniques were utilized to enable separation of neutrons from gammas.

Response functions for each type of detector and for each detected neutron group energy are necessary to fit the pulse height spectra and extract the yields. These functions were previously measured in TUNL's Shielded Source Area (SSA) using monoenergetic neutron beams produced via the $^2H(d,n)^3He$ reaction. A response function to fit the background neutron group from the $^2H(d,n)^3He$ reaction was also measured. This background peak was identified by studying the rise in its yield in conjunction with the integrated beam current, corresponding to the buildup of deuterium in the target.

The neutron spectra with PSD cuts for each angle were fitted with a combination of the measured response functions, as shown in Fig. 1 of Ref. [2], to extract the angle-integrated yields for each neutron group. These yields were normalized ($C$) by the detector efficiency, the correction factor (explained in Ref. [2]), the fraction of the solid angle subtended by the detector and the total number of incident deuterons. The angular distribution for each neutron group was fitted to an expansion of Legendre polynomials to obtain the angle-integrated yields at each energy.

The measured energy-dependent yield because of the stopping target can be written as

$$Y(E_d) = C \int_{E_d}^{0} \frac{\sigma(E)}{STP(E)}\, dE, \tag{4}$$

where $\sigma(E)$ is the energy-dependent cross section and $STP(E)$ is the stopping power of the target for deuterons. The constant $C$ is the normalization factor described above.In Eq. 4, the cross section can be written in terms of the astrophysical $S(E)$ factor as

$$\sigma(E_{c.m.}) = \frac{S(E_{c.m.})}{E_{c.m.}}\, e^{-2\pi\eta}, \tag{5}$$

where $\eta$ is the Sommerfeld parameter and $2\pi\eta$=$31.29 Z_1 Z_2 (\mu/E_{c.m.})^{1/2}$. $Z_1$ and $Z_2$ are the projectile and target charges, respectively, $\mu$ is the reduced mass in amu, and $E_{c.m.}$ is the center-of-mass energy in keV. At these low energies, the $S$ factor was assumed to be a linear function of energy:

$$S(E_{c.m.}) = S_0 + S_1 E_{c.m.}. \tag{6}$$

The measured yields at each incident energy were fitted simultaneously to determine the best fit parameters ($S_0$ and $S_1$) by evaluating the integrals in Eq. 4. The calculated or fitted yields and the constant $C$ in Eq. 4 were used to determine the absolute $S$ factor and, subsequently, the absolute cross section from Eq. 5.

## ACKNOWLEDGMENTS

The ($^3$He,$n$) reaction measurements were supported by the U.S. D.O.E Grant No. DE-FG02-88ER40387. The $(d,n)$ reaction measurements were partially supported by the U.S. D.O.E under Grants Nos. DE-FG02-97ER41046, DE-FG02-97ER41033, and DE-FG02-97ER41042. We gratefully acknowledge the co-authors of Refs. [1, 2] for the help in the completion of the measurements.

## REFERENCES

1. Y. Parpottas, S. M. Grimes, S. Al-Quraishi, C. R. Brune, T. N. Massey, J. E. Oldendick, A. Salas, and R. T. Wheeler, *Phys. Rev. C* **70**, 065805 (2004).
2. Y. Parpottas, M. W. Ahmed, M. A. Blackston, R. H. France III, B. A. Perdue, R. M. Prior, A. Sabourov, M. C. Spraker, and H. R. Weller, *Phys. Rev. C* **74**, 015804 (2006).
3. Y. Parpottas, S. M. Grimes, S. Al-Quraishi, C. R. Brune, T. N. Massey, J. E. O'Donnell, J. E. Oldendick, A. Salas, and R. T. Wheeler, *Phys. Rev. C* **72**, 025802 (2005).
4. C. A. Baker, C. J. Batty and L. E. Williams, *Nucl. Instr. and Meth.* **59**, 125 (1968).
5. S. M. Grimes, *Code and Input Description for HF*, Ohio University Report (unpublished, INPP-04-01).
6. S. M. Grimes, J. D. Anderson, J. W. McClure, B.A. Pohl, and C. Wong, *Phys. Rev. C* **10**, 2373 (1974).
7. P. E. Hodgson, *Nuclear Reactions and Nuclear Structure* Clarendon Press, Oxford, 1971.
8. C. Iliadis, L. Buchmann, P.M. Endt, H. Herndl, and M. Wiescher, *Phys. Rev. C* **53**, 475 (1996).
9. C. Iliadis, J. M. D'Auria, S. Starrfield, W. J. Thompson, and M. Wiescher, *Astrophys. J., Suppl. Ser.* **134**,151 (2001).
10. J. A. Caggiano, W. Bradfield-Smith, R. Lewis, P. D. Parker, D. Visser, J. P. Greene, K. E. Rehm, D. Bardayan, and A. E. Champagne, *Phys. Rev. C* **65**, 055801 (2002).
11. D. W. Bardayan, J. C. Blackmon, A. E. Champagne, A. K. Dummer, T. Davinson, U. Greife, D. Hill, C. Iliadis, B. A. Johnson, R.L. Kozub, C. S. Lee, M. S. Smith, and P. J. Woods, *Phys. Rev. C* **65**, 032801(R) (2002).
12. D. W. Bardayan, J. A. Howard, J. C. Blackmon, C. R. Brune, K. Y. Chae, W. R. Hix, M. S. Johnson, K. L. Jones, R. L. Kozub, J. F. Liang, E. J. Lingerfelt, R. J. Livesay, S. D. Pain, J. P. Scott, M. S. Smith, J. S. Thomas, and D. W. Visser, *Phys. Rev. C* **74**, 045804 (2006).
13. A. Sabourov, M. W. Ahmed, M. A. Blackston, A. S. Crowell, C. R. Howell, B. A. Perdue, K. Sabourov, A. Tonchev, and H. R. Weller, *Phys. Rev. C* **73**, 015801 (2006).

# Recent developments in Coulomb breakup calculations

P. Capel

*Physique Quantique, C.P. 165/82 and Physique Nucléaire Théorique et Physique Mathématique, C.P. 229, Université Libre de Bruxelles (ULB), B 1050 Brussels, Belgium*

**Abstract.** The theory of reactions applied to Coulomb breakup of loosely-bound projectiles is reviewed. Both the Continuum Discretized Coupled Channel (CDCC) and time-dependent models are described. Recent results about sensitivity of breakup calculations to the projectile wave function are reviewed. Analyses of the extraction of radiative-capture cross section from Coulomb breakup measurements are presented. Current developments in breakup theory are also mentioned.

**Keywords:** Nuclear reaction theory, Coulomb dissociation, halo nuclei, radiative capture
**PACS:** 24.10.-i, 25.60.Gc, 25.70.De, 25.40.Lw

## INTRODUCTION

The development of radioactive ion beams in the early 80s, enabled physicists to study nuclear structure far from stability. This led, among other things, to the discovery of halo nuclei [1]. It also permitted the measurement of reactions involving unstable nuclei.

Halo nuclei present a very unusual structure. They can be seen as a core containing most of the nucleons to which one or two nucleons are loosely bound. Due to their small separation energy, these nucleons have a high probably of presence at large distance from the core, far beyond the range of the nuclear interaction. This peculiar structure is not well reproduced by ab initio nuclear-structure models. It is thus a stringent test for these models. Halo nuclei are therefore the subject of many studies [2, 3, 4].

Coulomb breakup is one the reactions used to probe the halo structure [5]. It consists of the dissociation of the core-halo structure through interaction with a heavy target. The main advantage of this reaction lies in the fact that the dominant Coulomb interaction reduces the uncertainties associated with the nuclear interaction.

Coulomb breakup is also used as an indirect technique to determine the cross sections of radiative capture reactions at stellar energies [6, 7]. This can be done if one assumes that the breakup is due to the exchange of one virtual photon of the same multipolarity as in the radiative process.

In both applications, an accurate model of the reaction coupled to a realistic description of the projectile is needed to extract valuable information from experiments.

After a brief review of the theoretical framework used to model Coulomb breakup, some of the recent results obtained about the sensitivity of these calculations to the description of the projectile are reviewed. Then the astrophysical application of breakup is discussed. Finally the current developments in theoretical models is mentioned.

CP1012, *Frontiers in Nuclear Structure, Astrophysics, and Reactions: FINUSTAR 2*
edited by P. Demetriou, R. Julin, and S. V. Harissopulos
© 2008 American Institute of Physics 0-7354-0532-5/08/$23.00

## THEORETICAL FRAMEWORK

### Projectile description

We consider the description of the breakup of a projectile $P$ impinging on a target $T$. The projectile is assumed to have a two-body structure: a pointlike and structureless fragment $f$ (mass $m_f$ and charge $Z_f e$) loosely bound to a structureless core $c$ (mass $m_c$ and charge $Z_c e$). The target is described by a particle of mass $m_T$ and charge $Z_T e$. The internal structure of the two-body projectile is described by the Hamiltonian $H_0$

$$H_0 = -\frac{\hbar^2}{2\mu}\Delta + V_{cf}(\vec{r}), \tag{1}$$

where $\vec{r}$ is the relative coordinate of the fragment to the core, and $\mu = m_c m_f/(m_c+m_f)$ is the reduced mass of the $c$-$f$ system. The potential $V_{cf}$ simulates the interaction between the core and the fragment. It is composed of a Coulomb term plus a nuclear term, which comprises a central part and a spin-orbit coupling term

$$V_{cf}(\vec{r}) = V_C(r,R_0) + V_0(r) + \vec{L}\cdot\vec{I}V_{LI}(r), \tag{2}$$

where $\vec{L}$ is the orbital momentum of the $c$-$f$ relative motion, and $\vec{I}$ is the spin of the fragment. The spin of the core is usually neglected. The parameters of the potential are adjusted to reproduce the bound states of the system and some of its resonances.

In the $lj$ partial wave, the eigenstates of $H_0$ are

$$H_0\phi_{ljm}(E,\vec{r}) = E\phi_{ljm}(E,\vec{r}), \tag{3}$$

where $j$ is the total angular momentum resulting from the coupling of the orbital momentum $l$ and the fragment spin $I$, $m$ is its projection. The negative-energy states correspond either to physical bound states or to Pauli-forbidden states simulating the presence of fragment-like particles in the core. They are normed to unity. The positive energy states describe the continuum of the two-body projectile, i.e. the scattering of the fragment by the core. Their radial part $r^{-1}u_{lj}$ is normalized as

$$u_{lj}(k,r) \underset{r\to\infty}{\longrightarrow} [\cos\delta_{lj}(k)F_l(k,r) + \sin\delta_{lj}(k)G_l(k,r)], \tag{4}$$

where $k = \sqrt{2\mu E/\hbar^2}$ is the momentum of the relative motion of the core and the fragment, $F_l$ and $G_l$ are respectively the regular and irregular Coulomb functions [8], and $\delta_{lj}$ is the nuclear phase-shift.

### CDCC

The reaction of a loosely bound two-body projectile on a target can be approximated to a three-body scattering problem. Typically, target excitation is neglected and the three-body wavefunction is expanded in terms of the intrinsic motion of the core and the fragment $c+f$ within the projectile $u_{lj}(k,r)$, and the motion of the projectile relative to the target $f_{LjJ}(k,R)$. Here $\vec{R}$ is the vector connecting the center of mass of the projectile with the target, $L$ is the corresponding orbital angular momentum, and $J$ is the total angular momentum of the system.

For modeling breakup it is important to have a good description of the projectile continuum. One of the most successful method for elastic breakup is Continuum Discretized Coupled Channel method (CDCC) [9] first developed for deuteron breakup but today widely applied to exotic nuclei (e.g. [10, 11, 12, 13, 14]). In the standard CDCC, the projectile continuum is included through a discretization into energy bins. The bin wavefunction are usually approximated by the integral over the momentum interval, weighted by the scattering phase shift $\delta_{lj}$ in the following way :

$$\widetilde{u}_{lj,i}(r) = \sqrt{\frac{2}{\pi N}} \int_{k_i}^{k_{i+1}} e^{-i\delta_{lj}(k)} u_{lj}(k,r) dk \,. \tag{5}$$

The single-energy scattering wave function $u_{lj}(k,r)$ was introduced in the previous section. $N$ is a normalization factor.

The three-body Schrödinger equations can then be reduced to a coupled channel equation in $R$,

$$\left[-\frac{\hbar^2}{2\mu}\left(\frac{d^2}{dR^2} - \frac{L(L+1)}{R^2}\right) + V^J_{\alpha:\alpha}(R) + E_i - \mathscr{E}\right] f_{\alpha J}(R) = \sum_{\alpha' \neq \alpha} i^{L'-L} V^J_{\alpha:\alpha'}(R) f_{\alpha' J}(R), \tag{6}$$

where the subscript $\alpha$ represents all relevant quantum numbers and $E_i$ is the average energy of continuum bin $[k_i, k_{i+1}]$ (or $E < 0$ for the bound states), and $\mathscr{E}$ is the total energy of the system. The coupling potential $V^J_{\alpha:\alpha'}(R)$ consists of the core-target and fragment-target interactions averaged over the projectile states [15]. One can solve Eq. (6) exactly: the solution $\Psi^{CDCC}$ includes all possible rearrangements within the projectile continuum. One can also solve Eq. (6) iteratively: the one-step iteration of Eq. (6) is equivalent to the standard DWBA, where the optical potential between the projectile and the target is replaced by the folding potential $V^J_{0:\alpha'}(R)$ [16].

## Time-dependent model

The time-dependent description of the reaction relies on the semiclassical approximation [17] in which the projectile-target relative motion is treated classically, while the internal motion of the projectile is described quantum mechanically. The target is assumed to follow a classical trajectory in the projectile rest frame. The projectile is therefore seen as evolving in a time-dependent potential which simulates its interaction with the target. The wave function $\Psi$ describing its internal structure is solution of the following time-dependent Schrödinger equation

$$i\hbar\frac{\partial}{\partial t}\Psi(\vec{r},t) = [H_0 + V(\vec{r},t)]\Psi(\vec{r},t), \tag{7}$$

where $H_0$ [see Eq. (1)] describes the internal structure of the projectile, and $V$ is the time-dependent potential that simulates the $P$-$T$ interaction. The latter reads

$$V(\vec{r},t) = V_{cT}[r_{cT}(t)] + V_{fT}[r_{fT}(t)] - \frac{(Z_c + Z_f)Z_T e^2}{R(t)}, \tag{8}$$

where the time-dependent coordinate $\vec{R}$ describes the classical trajectory followed by the target in the projectile rest frame. It is a hyperbola, which can be approximated fairly well by straight lines at high energies. The vectors $\vec{r}_{cT}$ and $\vec{r}_{fT}$ correspond respectively to the core-target and fragment-target coordinates. In Eq. (8), $V_{cT}$ and $V_{fT}$ are optical potentials which model the interaction between the target and the projectile constituents.

Eq. (7) is solved considering the projectile initially in its ground state $\phi_{l_0 j_0 m_0}$ of energy $E_0$. Different techniques have been proposed to solve this equation numerically [18, 19, 20, 21, 22, 23]. The resolution is performed for different trajectories. For each trajectory, characterized by impact parameter $b$, we deduce the breakup probability $P_{\rm bu}$ by projecting the output wave function $\Psi(\vec{r},t \rightarrow +\infty)$ onto the positive eigenstates of $H_0$, which describe the projectile after dissociation. The breakup cross section $\sigma_{\rm bu}$ is then obtained by summing this probability over all impact parameters. At high energies, the solution of Eq. (7) can be approximated using the first-order perturbation theory [17].

## SENSITIVITY TO THE PROJECTILE DESCRIPTION

Since the early days of radioactive-beam experiments, breakup reactions have been an important source of information on the structure of nuclei near the drip-line [2, 3, 4, 7]. It is well known that breakup is sensitive to the binding energy of the fragment to the core [24]. However, is it possible to extract spectroscopic information from Coulomb dissociation?

Usually, a spectroscopic factor (SF) of the ground-state single-particle wave function (3) is obtained from the comparison between the theoretical prediction and the data [5, 25, 26, 27]. Nevertheless, uncertainty remains as for the part of the projectile wave function which is actually probed by breakup reactions. It has indeed been suggested that these reactions being peripheral, only the asymptotic normalization coefficient (ANC) could be obtained from these measurements [28, 29].

Recently different theoretical works have been done to evaluate the sensitivity of breakup calculations to the bound-state wave function [30, 31, 32, 33]. The interesting result obtained by these different groups, using various reaction models, is that breakup is not sensitive to the form factor of the wave function within the range of the nuclear interaction. Breakup calculations have been performed using initial wave functions that exhibit identical asymptotics, but are very different in the interior of the projectile. No difference has been observed. One concludes that breakup calculations are sensitive only to the asymptotic part of the wave function, i.e. the ANC. This means in particular that Coulomb dissociation is not sensitive to the whole normalization of the single-particle wave function, i.e. the SF.

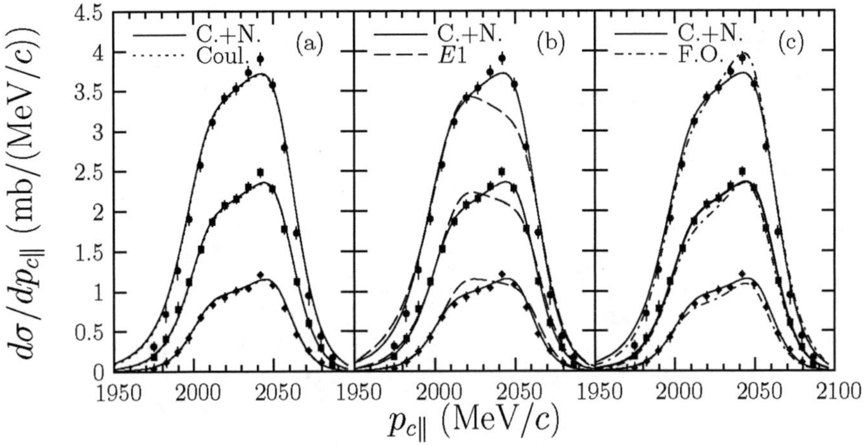


**FIGURE 1.** Parallel-momentum distribution of $^7$Be after breakup of 44*A*MeV $^8$B on lead. Dynamical calculation with full Coulomb plus nuclear *P-T* interaction (full lines) is compared to calculations (a) with Coulomb interaction (dotted lines), (b) with $E1$ interaction (dashed lines), (c) at first-order approximation with Coulomb *P-T* interaction (dash-dotted lines). Experimental data are shown as well [37].

## APPLICATION TO NUCLEAR ASTROPHYSICS

The difficulty of the direct measurement of radiative-capture cross section at stellar energies has raised interest in indirect methods where the time-reversed reaction is simulated by virtual photons in the Coulomb field of a heavy nucleus [6, 7]. The radiative-capture cross section can be extracted if one assumes the breakup to be due to a single virtual photon of the same multipolarity as the radiative capture.

Though appealing, the breakup method faces a number of difficulties. First, the nuclear interaction between the projectile and the target may interfere with the Coulomb interaction. Second, higher-order effects, i.e. transitions from the initial bound state to the continuum through several steps, are not negligible [34]. Finally, the multipolarity of the transition may differ between the breakup and the radiative-capture [34]. To illustrate these effects, I present recent theoretical analyses [14, 35] of $^8$B Coulomb breakup measurements [36, 37, 38] performed with the aim of extracting the astrophysical $S_{17}$ factor of the $^7$Be(p,$\gamma$)$^8$B radiative capture. This reaction plays a significant role in the study of high-energy neutrinos emitted by the sun [39].

Fig. 1 summarizes the results obtained in Ref. [35]. It depicts the breakup cross section of $^8$B impinging on lead at 44*A*MeV as a function of the $^7$Be longitudinal momentum. In Fig. 1(a), the full lines correspond to calculations performed using both nuclear and Coulomb projectile-target interactions. The agreement with the data [37] is excellent, indicating that a mere two-body description of $^8$B is sufficient for breakup calculations.

To investigate the role played by the nuclear interaction in the breakup mechanism, calculations are also performed with purely Coulomb *P-T* interactions [dotted lines in Fig. 1(a)]. The corresponding cross sections are identical to the previous ones, indicating that nuclear interaction is negligible when data are restricted to forward angles.

The influence of the $E2$ term of the Coulomb interaction can be evaluated by comparing the full calculation with a calculation that contains only the $E1$ strength [see Fig. 1(b)]. Both results differ significantly. The former (full lines) is in excellent agreement with the data, while the latter (dashed lines) exhibits a reverse asymmetry. This confirms that $E2$ transitions are required to reproduce $^8$B breakup data [34, 37].

In Fig. 1(c), higher-order effects are analysed. The dynamical calculation is compared to a first-order calculation (dash-dotted lines). The latter exhibits too strong an asymmetry. As noted by previous analyses [34], higher-order effects reduce the asymmetry.

These results confirm that nuclear interactions can be suppressed by selecting the data at forward angle. Unfortunately, $E2$ transitions and higher-order effects play a significant role in the Coulomb breakup of $^8$B, whereas the $^7$Be(p,$\gamma$)$^8$B radiative capture is purely $E1$ at stellar energy, and occurs in one step. It seems therefore difficult to infer the accuracy of this indirect method of extracting radiative-capture cross section.

## FUTURE DEVELOPMENTS

Recently efforts have been made to improve the simple two-body description of the projectile. Summers *et al.* have extended the CDCC model to include a multi-configuration description of the projectile [40, 41]. This description corresponds to the particle-rotor model, in which the internal structure of the core is simulated by a rotational Hamiltonian [42, 43]. This allows to include several core states in the projectile description. Not only is that description improved, but processes of excitation or de-excitation of the core during the reaction are thus coherently included. This eXtended CDCC (XCDCC) has been applied to the breakup of $^{11}$Be on $^9$B [40, 41]. The good agreement with the experiment suggests it as a promising technique for Coulomb breakup calculations.

Another improvement of breakup model is the extension of CDCC towards three-body projectile, i.e. two-neutron halo nuclei [44, 45, 46]. The results obtained with these new reaction models are, until now, limited to elastic scattering. Nevertheless, they show that both the three-body description of the projectile and the breakup channel have a significant influence on calculations. Better agreement with the data is obtained when they are taken into account in the reaction model.

Recently, new reaction models have been developed. Margueron *et al.* have proposed a correction to the eikonal approximation to extend it to Coulomb breakup [47]. The eikonal approximation [48] is indeed known to diverge for Coulomb dominated reactions. This first-order correction solves this divergence. This new model has been successfully applied for the Coulomb breakup of $^{11}$Be [49, 32] and $^8$B [50].

Another extension of the eikonal model has been proposed by Baye *et al.* [51]. This new Dynamical Eikonal Approximation (DEA) unifies the time-dependent approach and the eikonal model. It includes interferences between different trajectories, missing in the semi-classical approximation. It also includes the internal dynamics of the projectile, which is neglected in the usual eikonal model. This extends the time-dependent approach to differential observables, and the eikonal model to heavy targets. The excellent agreement obtained with experiment for the breakups of $^{11}$Be [52] and $^8$B [35] confirms the validity of this new model.

## CONCLUSION

Coulomb breakup of exotic nuclei is used as a probe of their structure. An accurate reaction model coupled to a realistic description of the projectile is needed to extract valuable information from experiments.

Several reaction models exist. The CDCC model [9, 10, 14] and time-dependent [18, 19, 23] have been described in this presentation. Both of them indicate that breakup is sensitive only to the tail of the projectile wave function, i.e. its ANC [30, 31, 32, 33]. Extraction of SF from breakup measurements is therefore questionable.

Coulomb breakup is also used as an alternative to direct measurements of radiative-capture cross sections at stellar energies [6, 7]. Unfortunately, it seems that higher-order effects and interferences between Coulomb multipoles are significant. These issues hinder the extraction of radiative-capture cross sections from breakup data.

The developments of new reaction models will probably help us to better understand the physics of Coulomb breakup, and to extend the range of theoretical predictions.

## ACKNOWLEDGMENTS

This text presents research results of the Belgian program P6/23 on interuniversity attraction poles initiated by the Belgian-state Federal Services for Scientific, Technical and Cultural Affairs (FSTC). I also acknowledge the support of the Fund for Scientific Research (F. R. S.-FNRS), Belgium.

## REFERENCES

1. I. Tanihata, H. Hamagaki, O. Hashimoto, S. Nagamiya, Y. Shida, N. Yoshikawa, O. Yamakawa, K. Sugimoto, T. Kobayashi, D. E. Greiner, N. Takahashi, and Y. Nojiri, *Phys. Lett. B*, **160**, 380 (1985).
2. P. G. Hansen, A. S. Jensen, and B. Jonson, *Annu. Rev. Nucl. Part. Sci*, **45**, 591 (1995).
3. I. Tanihata, *J. Phys. G*, **22**, 157 (1996).
4. G. Baur, K. Hencken, and D. Trautmann, *Prog. Part. Nucl. Phys.*, **51**, 487 (2003).
5. T. Nakamura, S. Shimoura, T. Kobayashi, T. Teranishi, K. Abe, N. Aoi, Y. Doki, M. Fujimaki, N. Inabe, N. Iwasa, K. Katori, T. Kubo, H. Okuno, T. Suzuki, I. Tanihata, Y. Watanabe, A. Yoshida, and M. Ishihara, *Phys. Lett. B*, **331**, 296 (1994).
6. G. Baur, C. A. Bertulani, and H. Rebel, *Nucl. Phys. A*, **458**, 188 (1986).
7. G. Baur, and H. Rebel, *Annu. Rev. Nucl. Part. Sci*, **46**, 321 (1996).
8. M. Abramowitz, and I. A. Stegun, *Handbook of Mathematical Functions*, Dover, New-York, 1970.
9. M. Kamimura, M. Yahiro, Y. Iseri, H. Kameyama, Y. Sakuragi, and M. Kawai, *Prog. Theor. Phys. Suppl.*, **89**, 1 (1986).
10. J. A. Tostevin, F. M. Nunes, and I. J. Thompson, *Phys. Rev. C*, **63**, 024617 (2001).
11. J. A. Tostevin, D. Bazin, B. A. Brown, T. Glasmacher, P. G. Hansen, V. Maddalena, A. Navin, and B. M. Sherrill, *Phys. Rev. C*, **66**, 024607 (2002).
12. M. Takashina, S. Takagi, Y. Sakuragi, and Y. Iseri, *Phys. Rev. C*, **67**, 037601 (2003).
13. N. C. Summers, and F. M. Nunes, *Phys. Rev. C*, **70**, 011602(R) (2004).
14. K. Ogata, S. Hashimoto, Y. Iseri, M. Kamimura, and M. Yahiro, *Phys. Rev. C*, **73**, 024605 (2006).
15. F. M. Nunes, A. M. Mukhamedzhanov, C. C. Rosa, and B. Irgaziev, *Nucl. Phys. A*, **736**, 255 (2004).
16. F. M. Nunes, and I. J. Thompson, *Phys. Rev. C*, **57**, R2818 (1998).
17. K. Alder, and A. Winther, *Electromagnetic Excitation*, North-Holland, Amsterdam, 1975.
18. T. Kido, K. Yabana, and Y. Suzuki, *Phys. Rev. C*, **50**, R1276 (1994).

19. H. Esbensen, G. F. Bertsch, and C. A. Bertulani, *Nucl. Phys. A*, **581**, 107 (1995).
20. S. Typel, and H. H. Wolter, *Z. Naturforsch. Teil A*, **54**, 63 (1999).
21. V. S. Melezhik, and D. Baye, *Phys. Rev. C*, **59**, 3232 (1999).
22. M. Fallot, J. A. Scarpaci, D. Lacroix, P. Chomaz, and J. Margueron, *Nucl. Phys. A*, **700**, 70 (2002).
23. P. Capel, D. Baye, and V. S. Melezhik, *Phys. Rev. C*, **68**, 014612 (2003).
24. S. Typel, and R. Shyam, *Phys. Rev. C*, **64**, 024605 (2001).
25. R. Palit, P. Adrich, T. Aumann, K. Boretzky, B. V. Carlson, D. Cortina, U. D. Pramanik, T. W. Elze, H. Emling, H. Geissel, M. Hellström, K. L. Jones, J. V. Kratz, R. Kulessa, Y. Leifels, A. Leistenschneider, G. Münzenberg, C. Nociforo, P. Reiter, H. Simon, K. Sümmerer, and W. Walus, *Phys. Rev. C*, **68**, 034318 (2003).
26. U. D. Pramanik, T. Aumann, K. Boretzky, B. V. Carlson, D. Cortina, T. W. Elze, H. Emling, H. Geissel, A. GrÃijnschloss, M. Hellström, S. Ilievski, J. V. Kratz, R. Kulessa, Y. Leifels, A. Leistenschneider, E. Lubkiewicz, G. Münzenberg, P. Reiter, H. Simon, K. Sümmerer, E. Wajda, and W. Walus, *Phys. Lett. B*, **551**, 63 (2003).
27. N. Fukuda, T. Nakamura, N. Aoi, N. Imai, M. Ishihara, T. Kobayashi, H. Iwasaki, T. Kubo, A. Mengoni, M. Notani, H. Otsu, H. Sakurai, S. Shimoura, T. Teranishi, Y. X. Watanabe, and K. Yoneda, *Phys. Rev. C*, **70**, 054606 (2004).
28. L. Trache, F. Carstoiu, C. A. Gagliardi, and R. E. Tribble, *Phys. Rev. Lett.*, **87**, 271102 (2001).
29. L. Trache, F. Carstoiu, C. A. Gagliardi, and R. E. Tribble, *Phys. Rev. C*, **69**, 032802(R) (2004).
30. S. Typel, and G. Baur, *Nucl. Phys. A*, **759**, 247 (2005).
31. P. Capel, and F. M. Nunes, *Phys. Rev. C*, **73**, 014615 (2006).
32. A. Garcìa-Camacho, A. Bonaccorso, and D. M. Brink, *Nucl. Phys. A*, **776**, 118 (2006).
33. P. Capel, and F. M. Nunes, *Phys. Rev. C*, **75**, 054609 (2007).
34. H. Esbensen, and G. F. Bertsch, *Nucl. Phys. A*, **600**, 37 (1996).
35. G. Goldstein, P. Capel, and D. Baye, *Phys. Rev. C*, **76**, 024608 (2007).
36. T. Kikuchi, T. Motobayashi, N. Iwasa, Y. Ando, M. Kurokawa, S. Moriya, H. Murakami, T. Nishio, J. Ruan (Gen), S. Shirato, S. Shimoura, T. Uchibori, Y. Yanagisawa, T. Kubo, H. Sakurai, T. Teranishi, Y. Watanabe, M. Ishihara, M. Hirai, T. Nakamura, S. Kubono, M. Gai, R. France III, K. Hahn, T. Delbar, P. Lipnik, and C. Michotte, *Phys. Lett. B*, **391**, 261 (1997).
37. B. Davids, D. W. Anthony, S. M. Austin, D. Bazin, B. Blank, J. Caggiano, M. Chartier, H. Esbensen, P. Hui, C. Powell, H. Scheit, B. Sherrill, M. Steiner, P. Thirolf, J. Yurkon, and A. Zeller, *Phys. Rev. Lett.*, **81**, 2209 (1998).
38. B. Davids, D. W. Anthony, T. Aumann, S. M. Austin, T. Baumann, D. Bazin, R. R. C. Clement, N. C. Davids, H. Esbensen, P. A. Lofty, T. Nakamura, B. M. Sherrill, and J. Yurkon, *Phys. Rev. Lett.*, **86**, 2750 (2001).
39. J. N. Bahcall, *Neutrino Astrophysics*, Cambridge University Press, Cambridge, 1989.
40. N. C. Summers, F. M. Nunes, and I. J. Thompson, *Phys. Rev. C*, **73**, 031603 (2006).
41. N. C. Summers, F. M. Nunes, and I. J. Thompson, *Phys. Rev. C*, **74**, 014606 (2006).
42. F. M. Nunes, I. J. Thompson, and R. C. Johnson, *Nucl. Phys. A*, **596**, 171 (1996).
43. F. M. Nunes, J. A. Christley, I. J. Thompson, R. C. Johnson, and V. D. Efros, *Nucl. Phys. A*, **609**, 43 (1996).
44. T. Matsumoto, E. Hiyama, K. Ogata, Y. Iseri, M. Kamimura, S. Chiba, and M. Yahiro, *Phys. Rev. C*, **70**, 061601 (2004).
45. T. Matsumoto, T. Egami, K. Ogata, Y. Iseri, M. Kamimura, and M. Yahiro, *Phys. Rev. C*, **73**, 051602 (2006).
46. M. Rodríguez-Gallardo, contribution to the DREB conference, RIKEN (2007).
47. J. Margueron, A. Bonaccorso, and D. M. Brink, *Nucl. Phys. A*, **703**, 105 (2002).
48. R. J. Glauber, in *High Energy Collision Theory*, edited by W. E. Brittin and L. G. Dunham, "Lecture in Theoretical Physics" Vol. 1 (Interscience, New York, 1959), p. 315.
49. B. Abu-Ibrahim, and Y. Suzuki, *Prog. Theor. Phys.*, **112**, 1013 (2004).
50. A. García-Camacho, G. Blanchon, A. Bonaccorso, and D. M. Brink, *Phys. Rev. C*, **76**, 014607 (2007).
51. D. Baye, P. Capel, and G. Goldstein, *Phys. Rev. Lett.*, **95**, 082502 (2005).
52. G. Goldstein, D. Baye, and P. Capel, *Phys. Rev. C*, **73**, 024602 (2006).

# Enhanced photoneutron cross sections for zirconium isotopes near threshold: evidence for extra M1 strength?


H. Utsunomiya*, S. Goriely†, A.J. Koning**, S. Hilaire‡, S. Goko§, A. Makinaga*, T. Kaihori*, T. Kondo*, T. Yamagata*, H. Akimune*, H. Toyokawa¶, T. Matsumoto¶, H. Harano¶ and Y.-W. Lui‖

*Department of Physics, Konan University, Okamoto 8-9-1, Higashinada, Kobe 658-8501, Japan
†Institut d'Astronomie et d'Astrophysique, Université Libre de Bruxelle, Campus de la Plane, CP-226, 1050 Brussels, Belgium
**Nuclear Research and Consultancy Group, P.O. Box 25, NL-1755 ZG Petten, The Netherlands
‡Département de Physique Théorique et Appliquée, Service de Physique Nucléaire, B.P. 12 - F-91680 Bruyères-le-Châtel, France
§Japan Atomic Energy Agency, Tokai-mura, Naka, Ibaraki 319-1195, Japan
¶National Institute of Advanced Industrial Science and Technology, Tsukuba 305-8568, Japan
‖Cyclotron Institute, Texas A & M University, College Station, Texas 77843, USA



**Abstract.** Photoneutron cross sections were measured for $^{91}$Zr, $^{92}$Zr and $^{94}$Zr near the neutron separation energy with quasi-monochromatic $\gamma$ rays. Some systematic extra strength is found around the neutron threshold and identified, through a coherent analysis of $^{92}$Zr($\gamma$,n)$^{91}$Zr and $^{91}$Zr(n,$\gamma$)$^{92}$Zr cross sections, as being most probably a strong M1 resonance. The microscopic HFB+QRPA model for the E1 strength with the additional contribution of a strong M1 resonance at 9 MeV is found to explain all the reaction cross section data available. With such an M1 resonance strength tuned on the $^{92}$Zr($\gamma$,n)$^{91}$Zr reaction, the photoneutron cross section of $^{91}$Zr and $^{94}$Zr are also found to be in good agreement with our new experimental data in the vicinity of the threshold and of the M1 peak energy. The present extra component in the photoneutron channel is qualitatively consistent with the M1 giant resonance reported in the inelastic proton scattering.



**Keywords:** ($\gamma$,n) cross sections for $^{91,92,94}$Zr, E1 and M1 $\gamma$ strength function
**PACS:** 25.20.-x, 21.10.Pc


## INTRODUCTION

Photonuclear reactions are an excellent probe of the electromagnetic response of nuclei. Recently, the E1 response near the neutron separation energy has drawn much attention in nuclear astrophysics in conjunction with the p-process nucleosynthesis [1, 2]. Measurements of photoneutron cross sections immediately above the neutron separation energy [3] supplemented with the quasi-particle random phase approximation (QRPA) calculation [4, 5] have enabled us to investigate the E1 $\gamma$ strength function of direct relevance to the p-process. Despite the experimental efforts in the ($\gamma$,n) and ($\gamma$,$\gamma$') channels [2], little is known, however, regarding the magnetic dipole strength function.

Magnetic-dipole strength was observed for $^{90}$Zr in the (e,e') [6], the (p,p') [7, 8, 9, 10], and the ($\gamma$,$\gamma$') [11] experiments at excitation energies $E_x$ = 8 - 10 MeV. While the strength is weak and fragmented in (e,e'), it is strong like the M1 giant resonance in


CP1012, *Frontiers in Nuclear Structure, Astrophysics, and Reactions: FINUSTAR 2*
edited by P. Demetriou, R. Julin, and S. V. Harissopulos


© 2008 American Institute of Physics 0-7354-0532-5/08/$23.00

**TABLE 1.** Target properties of the zirconium samples

| target nucleus | chemical form | quantity [mg] | isotopic composition [%] | | | | |
|---|---|---|---|---|---|---|---|
| | | | 90 | 91 | 92 | 94 | 96 |
| $^{91}Zr$ | $ZrO_2$ | 1335.3 | 5.83 | 90.4 | 3.2 | 0.5 | 0.05 |
| $^{92}Zr$ | $ZrO_2$ | 1186.5 | 3.7 | 2.8 | 91.4 | 1.9 | 0.2 |
| $^{94}Zr$ | $ZrO_2$ | 1315.6 | 3.7 | 1.1 | 2.1 | 92.6 | 0.5 |

(p,p'). The M1 excitation in $^{90}$Zr found in (p,p') is rather consistent with the excitation of the Gamow-Teller resonance in the analogue (p,n) reaction [12, 7, 9]. The strong M1 strength is also observed for $^{92}$Zr, $^{94}$Zr, and $^{96}$Zr in the very same energy region in the inelastic proton scattering [9]. It is interesting to note that although it lies below the neutron separation energy ($S_n$ = 11.97 MeV) in $^{90}$Zr, the M1 strength lies above $S_n$ at 8.635 MeV for $^{92}$Zr, 8.220 MeV for $^{94}$Zr, and 7.854 MeV for $^{96}$Zr. Since our knowledge of M1 states in the photoneutron channel is very limited [13], it is interesting to investigate how M1 states emerge in the excitation function of ($\gamma$,n) reactions on these zirconium isotopes both experimentally and theoretically. Such investigations may help us to improve our understanding of the electromagnetic response of nuclei in connection with the p-process nucleosynthesis.

## EXPERIMENT

We measured the photoneutron cross sections for $^{91}$Zr, $^{92}$Zr, and $^{94}$Zr near neutron threshold with quasi-monochromatic $\gamma$-ray beams produced from laser inverse-Compton scattering (LCS) in the electron storage ring TERAS at the National Institute of Advanced Industrial Science and Technology. Target properties of the zirconium samples are listed in Table I. Each target sample in the chemical form $ZrO_2$ was encapsulated in an aluminum container of 8mm inner diameter. A Nd:$YVO_4$ Q-switch laser was operated at 20 kHz in the second harmonics ($\lambda$=532 nm). The $\gamma$-ray beams had the same macroscopic time structure of 80ms beam-on and 20ms beam-off as that of the laser. The target sample was mounted at the center of a $4\pi$-type neutron detector consisting of 20 $^3$He counters embedded in a polyethylene moderator in a triple-ring configuration. Background neutrons were detected during the 20ms beam-off. The neutron detection efficiency is more than 56% in the neutron energy range below 1 MeV. The so-called ring ratio technique [14, 3] was used to determine the average neutron energies. The LCS $\gamma$-ray beam was measured with a 120% high-purity germanium detector (HPGe). The energy distribution of the LCS beam was determined by a least-squares analysis of the response function of the HPGe detector with the Monte Carlo code EGS4 [15]. The LCS beam was monitored with a large volume (20.3cm in diameter times 30.5cm in length) NaI(Tl) detector. Pile-up spectra were used to determine the number of the incident LCS $\gamma$ rays. Photoneutron cross sections were determined at the average $\gamma$-ray energies with the Taylor expansion method [16]. The systematic uncertainty for the cross section is 4.4% whose breakdown is 3.2% for the neutron detection efficiency and 3% for the number of incident $\gamma$ rays.

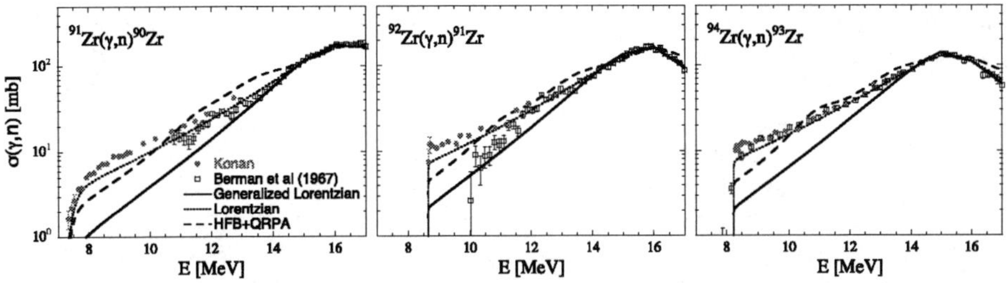


**FIGURE 1.** Comparison of experimental and theoretical photoneutron cross sections. The M1 strength is taken from systematics as described in the text.

## RESULTS AND DISCUSSIONS

Results of the present photoneutron cross section measurement for $^{91}$Zr, $^{92}$Zr, and $^{94}$Zr are shown in Fig. 1. For comparison, the experimental cross sections of [14] are also shown. The present measurement fills previously-unexplored energy regions near neutron threshold from 7.33 to 10.71 MeV for $^{91}$Zr and from 8.66 to 9.97 MeV for $^{92}$Zr, respectively. Note that the present cross sections for $^{92}$Zr are significantly larger than those of [14] below 11 MeV, while for $^{94}$Zr, the data are in good agreement.

Theoretically, the present experimental data have been analyzed on the basis of the TALYS reaction code [17] and different global predictions for the $\gamma$-ray strength function. These concern three different models, namely the Lorentzian model of [18], the Generalized $E$-dependent Lorentzian model of [19] and the Hartree-Fock-Bogolyubov (HFB) plus QRPA model of [5]. Note that in the case of the Lorentzian-type models, the E1 resonance energy, width and strength at maximum are all deduced from the photoabsorption data [14]. In contrast, these quantities are predicted by the global HFB-QRPA model, and in this case, some deviations in the resonance properties could be expected since no renormalization is performed. On top of such an E1 description, the contribution of the M1 strength is usually included on the basis of the global systematics defined in [20, 21], i.e. a Lorentzian function of centroid energy $E_{M1} = 41A^{-1/3}$ MeV, width $\Gamma = 4$ MeV and strength normalized to $f_{M1} = 1.58\ 10^{-9}\ A^{0.47}$ MeV$^{-3}$ at the reference energy of 7 MeV. On the basis of such $\gamma$-ray strength models, it is found that none of them can decently reproduce the present experimental photoneutron cross section close to the neutron separation energy (Fig. 1).

A priori, the underestimate of the $\gamma$-ray strength around the neutron threshold seen in Fig. 1 could be due to a lack of either E1 or M1 strength. In particular a simple increase of the parametric E1 Lorentzian width could solve the discrepancy. There is however some additional data that can help us constraining the low-energy tail of the $\gamma$-ray strength. It concerns the inverse radiative neutron capture available for $^{91}$Zr. The keV-neutron capture cross section is sensitive to the $\gamma$-ray strength below the neutron separation energy. It clearly appears that if the low strength seen in the $^{92}$Zr($\gamma$,n)$^{91}$Zr (Fig. 1, middle panel), is adjusted uniquely on the E1 strength, it is impossible to reproduce the $^{91}$Zr(n,$\gamma$)$^{92}$Zr cross section satisfactorily, the cross section being systematically overestimated. This is particularly the case with the Lorentzian model [18], the width of which could have been adjusted to properly fit the photoneutron cross section. No solution to both channels could therefore be found by increasing simply the E1 strength close to the neutron threshold.

In contrast the increase of the M1 strength could provide an interesting solution or more exactly a confirmation of the presence of the strong M1 resonance already measured. Indeed, a broad giant resonance has been observed by inelastic scattering of 200 MeV protons from the different Zr isotopes [9]. This resonance located systematically around 9 MeV and of width around 2 MeV has been identified as being most probably an M1 giant resonance. For this reason, we have renormalized the above-mentioned M1 resonance considered in the TALYS calculation taking the total strength as a free parameter, but with the centroid energy and width constrained by the measurements, i.e more precisely a peak energy at 9 MeV and a width of 2.5MeV. It was found that with the Lorentzian model [18], any additional M1 strength inevitably leads to an overestimate of the radiative neutron capture cross section. Adopting the Generalized $E$-dependent Lorentzian [19], it is found that the M1 component could fill the gap seen in Fig. 1 (middle panel) around the threshold but not up to the 13 MeV excitation energy (a wider M1 resonance would be needed), and even in that case, (n,$\gamma$) cross section is also overestimated. With the E1 contribution of [19], the only solution found for a coherent description of both channels would be to locate the M1 resonance at higher energy, i.e around 10–11 MeV with a strength about 4 times the one expected from the systematics, i.e a peak cross section of $\sigma_0^{M1} = 10$ mb. Such a high resonance centroid energy is however not in agreement with the inelastic scattering data of [9].

The only way we found to coherently reproduce both the neutron capture and inverse photoneutron cross sections is considering the E1 strength from the HFB+QRPA model [5] and adding the M1 contribution of strength $\sigma_0^{M1} = 7$ mb, i.e some 2.8 times stronger than predicted by the systematics [21]. The corresponding cross sections are compared in Fig. 2 with exprimental data. As a comparison, we also show with the dotted line the cross sections obtained omitting the M1 contribution. In both cases, the M1 is seen to have an important impact on the cross section. In comparison with the Generalized Lorentzian model, the HFB+QRPA model naturally predicts more strength above some 11 MeV photon energy, so that a narrow M1 resonance can bring the missing magnetic-dipole $\gamma$ strength function above the threshold in the photoreaction and at the same time below the threshold in radiative neutron capture (Fig. 2). Note that the neutron capture cross section is also relatively sensitive to the nuclear level density. The model adopted here is the HFB plus combinatorial model [25]. Experimental data exists to constrain the level density in $^{91,92}$Zr, namely the s-wave resonance spacing [21], so that using different level density models was not found to affect the conclusion drawn above regarding the contribution of the M1 resonance.

With such an M1 resonance strength tuned on the $^{92}$Zr($\gamma$,n)$^{91}$Zr reaction, the calculation of the photoneutron cross section were repeated for $^{91}$Zr and $^{94}$Zr. The results shown in Fig. 3 (adopting the E1 strength from the HFB+QRPA calculation of [5]) are seen to be in excellent agreement with the present experimental data, especially around the M1 centroid energy of 9 MeV. At higher energy, some deviations with the data of [14] are obtained, these are due to some additional E1 strength predicted by the global HFB+QRPA model. The coherent description of the photoreactions for these 3 Zr isotopes is an additional confirmation of the presence of this strong M1 resonance.

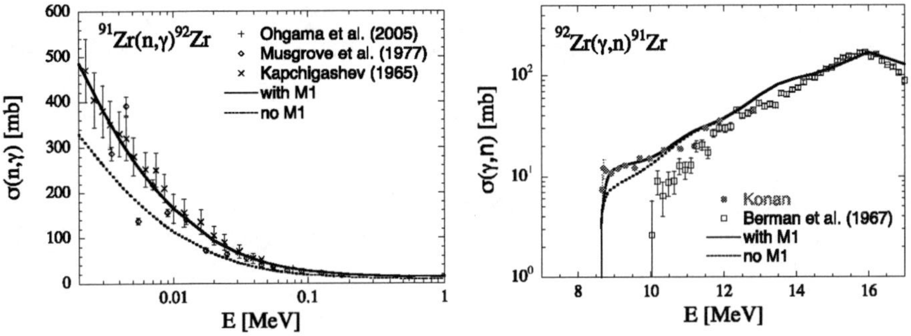


**FIGURE 2.** Left panel: comparison of experimental [22, 23, 24] and theoretical radiative neutron capture cross section $^{91}$Zr(n,$\gamma$)$^{92}$Zr. Right panel: same for the inverse photoneutron cross sections. In both panels, the solid line correspond to the calculation with an M1 strength renormalized as explained in the text, and the dotted line without any contribution of the M1. In all cases, the E1 strength from the HFB+QRPA calculation of [5] is adopted.

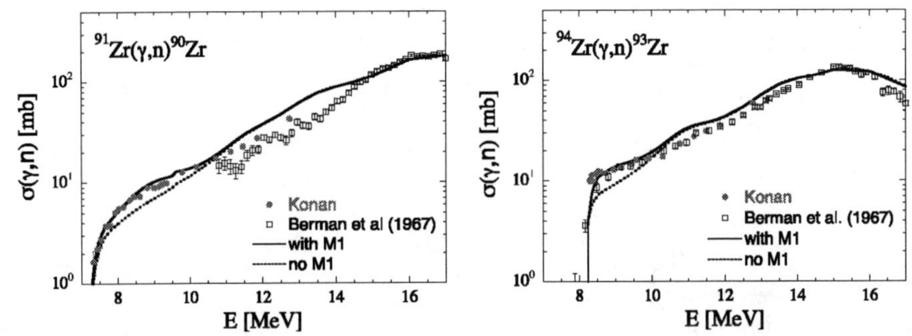


**FIGURE 3.** Left panel: comparison of experimental and theoretical photoneutron capture cross section $^{91}$Zr($\gamma$,n)$^{90}$Zr. Right panel: same for the$^{94}$Zr($\gamma$,n)$^{930}$Zr reaction. In both panels, the solid line correspond to the calculation with an M1 strength renormalized as explained in the text, and the dotted line without any contribution of the M1. In all cases, the E1 strength from the HFB+QRPA calculation of [5] is adopted.

## CONCLUSIONS

The extra $\gamma$ strength was identified for zirconium isotopes in the photoneutron channel. We have identified the extra strength in $^{92}$Zr through a coherent analysis of $^{92}$Zr($\gamma$,n)$^{91}$Zr and $^{91}$Zr(n,$\gamma$)$^{92}$Zr cross sections. Both channels enables us to constrain the $\gamma$-ray strength in a wide energy range around the neutron separation energy. The microscopic HFB+QRPA model for the E1 strength with the additional contribution of a strong M1 resonance at 9 MeV is found to explain all the reaction cross section data available. In contrast, a pure additional E1 strength cannot provide a satisfactory explanation of the data. Even with an additional strong M1 component, the Lorentzian-type models for the E1 strength are not found to give a coherent description of the experimental data. With such an M1 resonance strength tuned on the $^{92}Zr(\gamma,n)^{91}Zr$ reaction, the photoneutron cross section of $^{91}Zr$ and $^{94}Zr$ are also found to be in good agreement with our new experimental data in the vicinity of the threshold and of the M1 peak energy.

Although the present measurements alone do not show conclusively the nature of the extra strength, it is most likely an M1 component in view of the sensitivity of the real-photon probe, which is qualitatively consistent with the M1 giant resonance reported in the inelastic proton scattering. The present extra M1 strength awaits further theoretical studies that could lead to a clarification of its nuclear structure.

## ACKNOWLEDGMENTS

This work is supported by the Japan Private School Promotion Foundation and the Konan-ULB convention. S.G. is FNRS research associate.

## REFERENCES

1. M. Arnould, and S. Goriely, *Phys. Rep.* **384**, 1 (2003).
2. H. Utsunomiya, P. Mohr, A. Zilges, M. Rayet, *Nucl. Phys.* **A777**, 459 (2006).
3. H. Utsunomiya *et al.*, *Phys. Rev. C* **67**, 015807 (2003).
4. S. Goriely and E. Khan, *Nucl. Phys.* **A706**, 217 (2002).
5. S. Goriely, E. Khan, and M. Samyn, *Nucl. Phys.* **A739**, 331 (2004).
6. D. Meuer *et al.*, *Nucl. Phys.* **A349**, 309 (1980).
7. N. Anantaraman *et al.*, *Phys. Rev. Lett.* **46**, 1318 (1981).
8. F.E. Bertrand *et al.*, *Phys. Lett.* **103B**, 326 (1981).
9. G.M. Crawley *al.*, *Phys. Rev. C* **26**, 87 (1982).
10. S.K. Nanda *et al.*, *Phys. Rev. Lett.* **51**, 1526 (1983).
11. R.M. Laszewski, R. Alarcon, and S.D. Hoblit, *Phys. Rev. Lett.* **59**, 431 (1987).
12. D.E. Bainum *et al.*, *Phys. Rev. Lett.* **44**, 1751 (1980).
13. C.D. Bowman *et al.*, *Phys. Rev. Lett.* **25**, 1302 (1970).
14. B.L. Berman *et al.*, *Phys. Rev.* **162**, 1098 (1967).
15. W.R. Nelson, H. Hirayama and W.O. Roger, *The EGS4 Code Systems* SLAC-Report-265, (1985).
16. H. Utsunomiya *et al.*, *Phys. Rev. C* **74**, 025806 (2006).
17. A.J. Koning, S. Hilaire and M.C. Duijvestijn, Proc. Int. Conf. on Nuclear Data for Science and Technology (eds. C. Haight et al.) AIP Conference Vol. 769, p. 1154, 2005.
18. P. Axel, *Phys. Rev.* **126**, 671 (1962).
19. J. Kopecky, and M. Uhl, *Phys. Rev. C* **41**, 1941 (1990).
20. A.G. Bohr and N.R. Mottelson, *Nuclear Structure, Vol. II* (Benjamin London 1975), p.636
21. *Handbook for calculations of nuclear reaction data, RIPL-2* (IAEA-Tecdoc-1506, 2006).
22. K. Ohgama, M. Igashira and T. Ohsaki, J. of Nuclear Science and Technology, **42**, 333 (2005).
23. A.R. Del.Musgrove, J.W. Boldemann, B.J. Allen, J.A. Harvey and R.L. Macklin, Australian Journal of Physics **30**, 391 (1977).
24. S.P. Kapchigashev, Atomnaya Energiya **19**, 294 (1965).
25. S. Hilaire, S. Goriely, Nucl. Phys. **A779**, 63 (2006).

# Cross-Section Measurements of alpha-particle Capture Reactions Relevant to the P Process: The Case of $^{65}Cu(\alpha,\gamma)^{69}Ga$

A. Lagoyannis[a], T. Konstantinopoulos[a], P. Demetriou[a], H.-W. Becker[b], and S. Harissopulos[a]

[a]*Institute of Nuclear Physics, NCSR "Demokritos", 153.10 Aghia Paraskevi, Athens, Greece.*
[b]*DTL – Institut für Experimentalphysik III, Ruhr-Universität Bochum, 40781 Bochum, Germany*

**Abstract.** The cross sections of the $^{65}Cu(\alpha,\gamma)^{69}Ga$ reaction were measured using the $4\pi$ $\gamma$-summing method. Angle-integrated $\gamma$-fluxes were measured with a large volume NaI(Tl) detector. Statistical model calculations were performed using the MOST code. A very good agreement between theory and experiment was found.

**Keywords:** p process, cross sections, $\alpha$-capture reactions, $4\pi$ calorimeter.
**PACS:** 24.60.Dr, 25.40.Lw, 26.50.+x, 27.60.+j, 29.40.Mc

## INTRODUCTION

The origin in the cosmos of the so-called *p* nuclei [1] is one of the most puzzling problems to be solved by any model of heavy-element nucleosynthesis. The class of *p* nuclei consists of 35 proton rich stable nuclei that are heavier than iron and cannot be synthesized by the two neutron-capture processes referred to as *s* and *r* process. So far, all the models of *p*-process nucleosynthesis [1] are able to reproduce most of the *p*-nuclei abundances within a factor of 3, but they fail completely in the case of the light *p* nuclei. Due to the huge number of reactions involved in abundance calculations, the latter have to rely almost completely on the reaction cross-section predictions of the Hauser-Feshbach (HF) theory. It is therefore of key importance, on top of any astrophysical model improvements, to investigate the uncertainties in the nuclear data, and in particular in the nuclear level densities (NLD), nucleon-nucleus optical model potentials (OMP), and $\gamma$-ray strength functions entering the HF calculations.

Comparisons with experimental (p,$\gamma$) and/or (n,$\gamma$) cross sections indicate that the associated reaction rates can be predicted by the HF theory within a factor of two. However, in the case of the very few ($\alpha$,$\gamma$) data, the reaction rates calculated using phenomenological alpha-particle optical potentials can be wrong by a factor of ten or more. These uncertainties might be reduced substantially by putting constraints on the alpha-nucleus optical potentials that are so far poorly known at low energies (E<12 MeV). For this purpose, systematic cross-section measurements of $\alpha$-particle capture reactions at energies from 5 to 12 MeV are necessary. Therefore, we have performed several cross sections measurements of ($\alpha$,$\gamma$) reactions in the Cu-Sn region. This contribution reports on preliminary results obtained for the $^{65}Cu(\alpha,\gamma)^{69}Ga$ reaction.

CP1012, *Frontiers in Nuclear Structure, Astrophysics, and Reactions: FINUSTAR 2*
edited by P. Demetriou, R. Julin, and S. V. Harissopulos
© 2008 American Institute of Physics 0-7354-0532-5/08/$23.00

## EXPERIMENTAL SETUP AND MEASUREMENTS

The cross-section measurements were carried out at the Dynamitron-Tandem-Laboratorium (DTL) of the Ruhr-Universität Bochum, Germany. The beam current on-target was kept lower than 30 nA to minimize possible dead time effects. A self-supporting highly-enriched (99.7%) $^{65}$Cu foil was used as target. Its thickness (162±13 μg/cm$^2$) was determined by means of an RBS analysis carried out at the 5.5 MV T11 Tandem accelerator of NCSR "Demokritos", Athens.

The experimental setup used to determine cross sections in Bochum is described in detail in [2]. It consists of a 4π large volume (12inch x 12 inch) single NaI(Tl) crystal (BICRON) of cylindrical shape with a borehole (Ø35 mm) along its axis. The target was placed in the borehole of the detector, at the center of its axis. The main advantage in using this 4π calorimeter is that its response function leads predominantly to a single peak, called the sum peak, at the sum of the energies of the cascade transitions de-exciting the compound nucleus produced in the entrance channel. As a result, a prominent peak appears in the measured γ spectra at an energy $E_\Sigma = Q + E_{c.m.}$, where Q is the Q value of the reaction and $E_{c.m}$ is the center-of-mass energy. Due to the 4π geometry covered by the summing crystal, the observed γ-ray fluxes are angle integrated. Hence, systematic errors due to γ-angular distribution effects are eliminated. The working principle of our setup, its efficiency calibration and applicability to capture reactions is given in detail in [2]. A typical spectrum measured at E(α)=6 MeV is shown in Fig. 1.

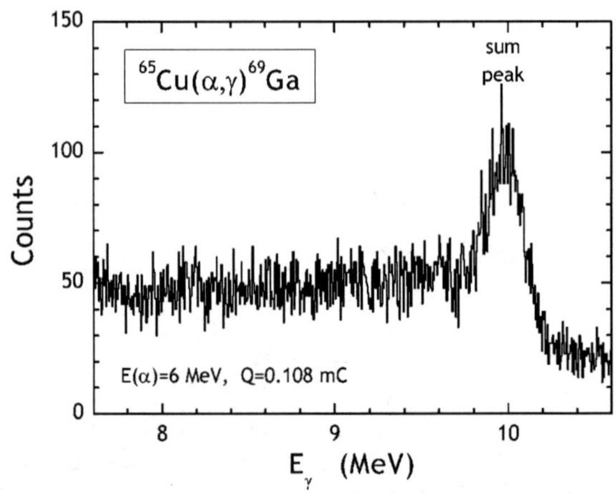


**FIGURE 1.** Sum peak in the γ spectrum taken at E(α)=6 MeV with the 4π NaI(Tl) summing detector.

## DATA ANALYSIS, RESULTS, AND DISCUSSION

The total cross section $\sigma_T$ of $^{65}$Cu(α,γ)$^{69}$Ga was obtained from the intensity $I_\Sigma$ of the sum peak using

$$\sigma_T = \frac{A}{N_A N_{beam}} \cdot \frac{1}{\xi} \cdot \frac{I_\Sigma}{\varepsilon_\Sigma} , \quad (1)$$

where $N_A$ is the Avogadro number, A is the atomic weight (in *amu*) of the target, $N_{beam}$ is the total number of incident beam particles, $\xi$ is the target thickness, and $\varepsilon_\Sigma$ is the efficiency of the detector at the sum-peak energy. The latter quantity depends strongly on the mulitiplicities of the $\gamma$ cascades and is in most cases unknown. In [2], we developed a new method to determine the average multiplicities of the $\gamma$ cascades and the corresponding sum-peak efficiency $\varepsilon_\Sigma$ using the GEANT4 Monte-Carlo code [3]. This method was applied also to derive $\varepsilon_\Sigma$ in the present work. The experimental cross sections of $^{65}Cu(\alpha,\gamma)^{69}Ga$ obtained using Eq. (1) were then corrected for electron-screening effects [4]. These effects were found to result in an increase of $\sigma_T$ by up to 11%.

The electron-screening-corrected cross sections are compared with the predictions of the HF theory in Fig. 2. Calculations were performed by the statistical model code MOST [5]. The code includes all the available experimental information on nuclear masses, deformation, and spectra of low-lying states. The nuclear masses are obtained from the experimental compilation of Audi and Wapstra [6] and the ground state properties (matter density, single-particle level scheme) are predicted from the microscopic Hartree-Fock-BCS model [7]. The E1 transition strength functions are described by the hybrid model of [8], while the M1 transitions are parameterized following [9], with the energies and widths taken from [10]. The nucleon transmission coefficients are obtained from the nucleon-nucleus global phenomenological OMP of [11], while the nuclear level densities (NLD) are taken from the microscopic statistical model of [12].

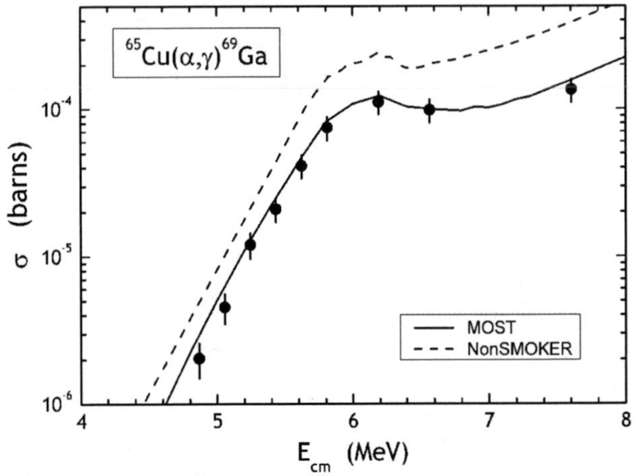


**FIGURE 2.** Comparison of the cross-section data (circles) with the MOST calculations (solid line). The corresponding predictions of the Non-SMOKER code [14] are shown as a dotted curve.

The $(\alpha,\gamma)$ reaction is in general very sensitive to the $\alpha$-nucleus OMP. In our calculations we use the global $\alpha$-OMP III of Demetriou *et al.* [13] based on the double-folding method. This particular optical potential takes into account the dispersive corrections to the real part arising from the volume and surface parts of the imaginary potential. As shown in Fig. 2, the agreement between data and MOST calculations is very good. For completeness, Non-SMOKER predictions are also included in this figure. The latter code uses the phenomenological $\alpha$-OMP of McFadden and Satchler [15].

## REFERENCES


1. M. Arnould, and S. Goriely, Phys. Rep. **384**, 1 (2003).
2. A. Spyrou, H.-W. Becker, A. Lagoyannis, S. Harissopulos, and C. Rolfs, Phys. Rev. C **76**, 01580 (2007)
3. GEANT4: A Simulation Toolkit, S. Agostinelli et al., Nucl. Instrum. Methods A **506**, 250 (2003).
4. C. E. Rolfs, Nuclear Physics News **16**(2), 9 (2006), and *private communication.*
5. S. Goriely, in *Nuclei in the Cosmos V*, edited by N. Prantzos and S. Harissopulos, (Edition Frontières, Paris, 1998), p. 314 (see also http://www-astro.ulb.ac.be).
6. G. Audi, A.H. Wapstra, and C. Thibault, Nucl. Phys. **A 729**, 3 (2003).
7. S. Goriely, F. Tondeur, J.M. Pearson, At. Data. Nucl. Data Tables 77, 311 (2001).
8. S. Goriely, Phys. Lett. **B 436**, 10 (1998).
9. J. Kopecky, and R.E. Chrien, Nucl. Phys. **A 468**, 285 (1987).
10. Reference Input Parameter Library, IAEA-Tecdoc-1034 (1998), (see also http://iaeand.iaea.or.at/ripl).
11. J. P. Jeukenne, A. Lejeune, and C. Mahaux, Phys. Rev. C **16**, 80 (1977).
12. P. Demetriou, and S. Goriely, Nucl. Phys. **A 695**, 95 (2001).
13. P. Demetriou, C. Grama, and S. Goriely, Nucl. Phys. **A 707**, 253 (2002).
14. T. Rauscher and T.-K. Thielemann, At. Data Nucl. Data Tables **75**, 1 (2000).
15. L. McFadden and G. R. Satchler, Nucl. Phys. **84** (1966) 177.

# Microscopic nature of $0^+$ states populated in large abundance in deformed nuclei

N. Lo Iudice*, A. V. Sushkov† and N. Yu. Shirikova†

*Dipartimento di Scienze Fisiche, Universitá di Napoli "Federico II" and Istituto Nazionale di Fisica Nucleare, Monte S Angelo, Via Cintia I-80126 Napoli, Italy
†Bogoliubov Laboratory of Theoretical Physics, Joint Institute for Nuclear Research, 141980 Dubna, Russia

**Abstract.** The low-lying excited $0^+$ states populated in large abundance in several deformed nuclei through two-nucleon transfer reaction experiments are thoroughly investigated within a microscopic multiphonon framework. The approach reproduces satisfactorily the experimental data and provides exhaustive information about the nature of those states.

**Keywords:** Zeroplus; Deformation; Two-nucleon transfer.
**PACS:** 21.10.Jx,21.10.Ky,21.10.Re,21.60.Jz,27.70.+q

## INTRODUCTION

The population of a large number of low-lying excited $0^+$ levels in several nuclei through high resolution $(p,t)$ experiments [1, 2, 3, 4, 5, 6] has stimulated several theoretical investigations aiming at establishing their nature. These states have been studied within the interacting boson model (IBM)[7], especially suitable for systematics, and, microscopically, within a projected shell model (PSM) [8] or by adopting [9, 10] the quasiparticle-phonon model (QPM) [11].

After briefly mentioning the IBM and PSM calculations and results, we will concentrate on the QPM study and show how such an approach allows to gain a deep insight into the nature of all those states.

## THE QPM STUDY

In the QPM, one adopts a Hamiltonian of the form

$$H = H_0 + V_{pair} + V_P + V_M. \tag{1}$$

The one-body piece $H_0$ is composed of a kinetic term plus an axially deformed Woods-Saxon potential, while the two-body potential consists of a monopole pairing $V_{pair}$ and of other two main pieces, $V_P$ and $V_M$, acting respectively in the particle-particle and particle-hole channels. Both pieces are sums of several separable potentials of different multipolarity $\lambda$.

The Hamiltonian is expressed in terms of quasi-particle creation ($\alpha_q^\dagger$) and annihilation ($\alpha_q$) operators by means of the Bogoliubov canonical transformation and, then, in terms

CP1012, *Frontiers in Nuclear Structure, Astrophysics, and Reactions: FINUSTAR 2*
edited by P. Demetriou, R. Julin, and S. V. Harissopulos
© 2008 American Institute of Physics 0-7354-0532-5/08/$23.00

of the RPA phonon operators

$$Q_{i\nu}^{\dagger} = \frac{1}{2}\sum_{q_1q_2}\left(\psi_{q_1q_2}^{i\nu}A_{q_1q_2}^{\dagger} - \phi_{q_1q_2}^{i\nu}A_{q_1q_2}\right) \tag{2}$$

where $A_{q_1q_2}^{\dagger} = \alpha_{q_1}^{\dagger}\alpha_{\bar{q}_2}^{\dagger}$. This procedure yields the interacting phonon Hamiltonian

$$H_{QPM} = \sum_{\nu_i}\omega_{\nu_i}Q_{\nu_i}^{\dagger}Q_{\nu_i} + H_{\nu q}, \tag{3}$$

where $\nu_i = \{i\lambda_i\mu_i\}$. The first term is an unperturbed Boson Hamiltonian diagonal in the basis of the RPA phonon states $\mid \nu_i\rangle = Q_{\nu_i}^{\dagger} \mid 0\rangle$ of energies $\omega_{\nu_i}$. These are coupled by a quasiparticle-phonon piece $H_{\nu q}$.

The interacting phonon Hamiltonian is put in diagonal form through the variational principle with the trial wave function

$$\Psi_{nK^{\pi}=0^+} = \sum_i C_i(n) \mid i, K^{\pi} = 0^+\rangle + \sum_{\nu_1\nu_2} C_{\nu_1\nu_2}(n) \mid \left[\nu_1 \otimes \nu_2\right]^{0^+}\rangle, \tag{4}$$

where $\mid i, K^{\pi} = 0^+\rangle = Q_{i0^+}^{\dagger} \mid 0\rangle$ and $\mid \left[\nu_1 \otimes \nu_2\right]^{0^+}\rangle = \left[Q_{\nu_1}^{\dagger} \otimes Q_{\nu_2}^{\dagger}\right]^{0^+} \mid 0\rangle$ are the one- and two-phonon components, respectively.

This wave function can be used for computing the $(p,t)$ normalized spectroscopic factors

$$\mathscr{S}_n(p,t) = \left(\frac{\Gamma_n(p,t)}{\Gamma_0(p,t)}\right)^2, \tag{5}$$

where

$$\Gamma_n(p,t) = \langle 0_n^+, N-2 \mid \sum_q^{(\tau=n)} a_q a_{\bar{q}} \mid 0_0^+, N\rangle = \sum_i C_i^{(n)}\Gamma_i(p,t) \tag{6}$$

are the QPM $(p,t)$ transfer amplitudes. The $\Gamma_i(p,t)$ include the amplitudes of the *allowed* transitions involving the one-phonon components $\mid i, K^{\pi} = 0^+\rangle$ of the QPM state (4) as well as the *forbidden* transitions from the ground to the two-phonon components $\mid [\nu_1\,\nu_2]^{0^+}\rangle$. The two-phonon contribution, however, came out to be negligible.

## Numerical results

The QPM calculation accounts for all $0^+$ levels in all nuclei. For the sake of illustration, we discuss few typical spectra. Quite interesting is the case of $^{168}$Er, where about 25 $0^+$ and 63 $2^+$ have been populated [3]. The calculation yields an even larger number of levels in both cases. As shown in Fig. 1, the QPM $0^+$ levels below 4 Mev are about

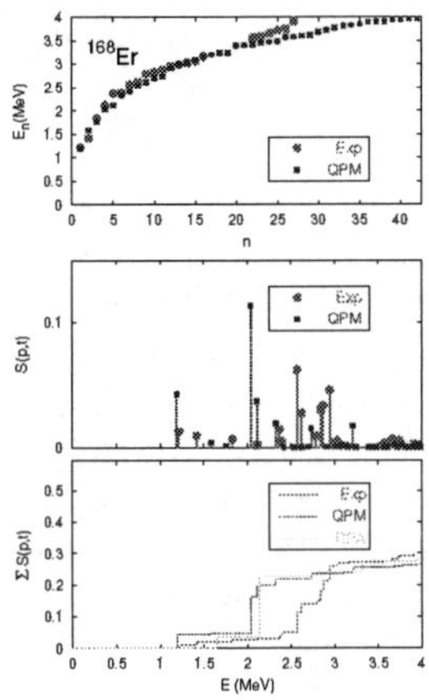


**FIGURE 1.** QPM versus experimental data in $^{168}$Er. These include the energies of the $0^+$ states (top panel), the energy distribution of the $(p,t)$ spectroscopic factors (middle) and their running sum (bottom)

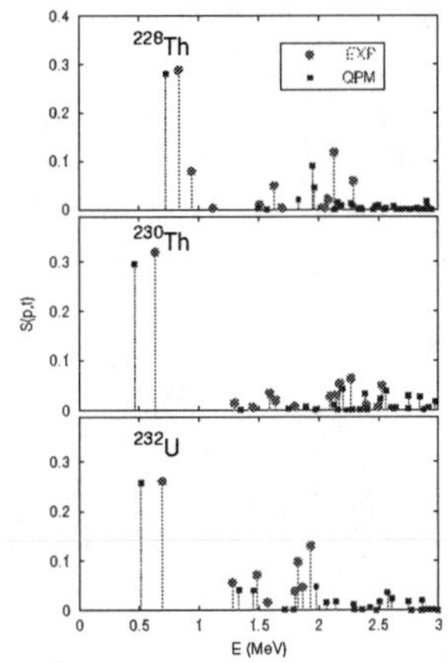


**FIGURE 2.** Experimental versus QPM energy distributions of $(p,t)$ spectroscopic factors in the actinides. The experimental data are taken from [2]

40. All the $0^+$ in excess fall in the range $\sim 3.2 \div 3.5$ MeV and carry a negligible $(p,t)$ strength (second panel).

Almost all the $0^+$ excitations below 3 MeV are described by one-phonon states which, in some cases, are rather fragmented. The weight of the two-phonon components becomes appreciable just above 3 MeV and increases more and more with the energy. The octupole phonons do not play any special role, except for few states above 3 MeV, where they are dominant.

For the $E2$ decay of the $0^+_1$ bandhead to the $2^+$ of the ground band, the QPM yield the strength $B^{(QPM)}(E2;0^+_1 \rightarrow 2^+_0) = 0.11$ W.u., in close agreement with the experiments

[12] $B^{(exp)}(E2;0^+_1 \rightarrow 2^+_0) = 0.08(1)$ W.u. The computed $E2$ decay strengths of the other $0^+$ states are even smaller. Also the normalized monopole transitions are generally weak. The strength collected by the bandhead $0^+_1$ is close to the experimental value [13, 14, 15] $10^3 \rho^2_{exp}(E0;0^+_1 \rightarrow 0^+) = 0.8(8)$. We find therefore an almost complete lack of quadrupole collectivity in all $0^+$ states of $^{168}$Er.

All these states, indeed, are composed almost entirely of pairing correlated $q\bar{q}$ components. The high energy *ph* configurations are practically absent. Pairing, however, acts coherently only into one RPA $0^+$ state. This collects most of the two-nucleon transfer strength $\mathscr{S}(p,t)$. The coupling with the other phonons, however, depletes the $(p,t)$ strength of the collective RPA $0^+$ state, yielding a fragmented spectrum (second panel of Fig. 1). The QPM misses some detailed features of the spectrum. These discrepancies get manifest in the third panel of Fig. 1 showing that both RPA and QPM strengths reach saturation too early with respect to experiments. The total strength, however, is $\sim 0.25$, very close to the measured one.

The QPM reproduces even better the $(p,t)$ strength distribution of the $K^\pi = 2^+$ states in $^{168}$Er [3] as well as the $0^+$ spectra in most of the other nuclei. In the actinides, for instance, the QPM energies and normalized spectroscopic factors $\mathscr{S}(p,t)$ are in good agreement with the data (Fig. 2). Consistently with experiments, the calculation yields correctly one strong peak plus other small ones. This is the typical structure of a (pairing) collective spectrum. Such a property is supported by the analysis of the RPA one-phonon components. Only in one RPA state, pairing acts coherently thereby enhancing the $(p,t)$ strength. We may, therefore, make the reasonable guess that the observed $0^+$ states describe pairing vibrations arising from ground state fluctuations. The damping due to phonon coupling is negligible leaving the $(p,t)$ strength concentrated predominantly into the lowest $0^+$.

Contrary to expectations, the octupole phonons play a rather minor role in the actinides. Although the lowest unperturbed octupole RPA phonons fall at very low-energy, consistently with experiments, the enforcement of the Pauli principle on the octupole two-phonon components, entering the $0^+$ states, has a repulsive effect and spoils the octupole coherence. As a result, these components spread among several QPM $0^+$ states at high energy.

The $E2$ collectivity in the actinides is considerably stronger than in $^{168}$Er. The lowest $0^+$ excited state, in particular, is basically a one-phonon state and collects an appreciable E2 strength, in agreement with the available data. In $^{230}$Th, in fact, the measured E2 strength is $B^{(exp)}(E2;0^+_1 \rightarrow 2^+_0) = 1.10$ W.u [16], close to the computed value $B^{(QPM)}(E2;0^+_1 \rightarrow 2^+_0) = 1.71$ (W.u.).

The monopole transition strength collected by the first $0^+$ is estimated to be $10^3 \rho^2_{exp}(E0;0^+_\beta \rightarrow 0^+_g) = 50(20)$ [15], few times larger than the QPM value and orders of magnitude larger than in $^{168}$Er. Still, neither the $E2$ or the $E0$ transitions are sufficiently strong to qualify the $0^+$ of the actinides explored here as beta vibrational levels, according to the criterion established in Ref. [17].

## CONCLUSION

Our analysis suggests that the appearance of so many low-lying $0^+$ states in deformed nuclei is a manifestation of pairing collectivity. Indeed, almost all low-lying $0^+$ levels are described as one-phonon states. These resulted to be built out of pairing correlated configurations giving rise to a pairing vibrational mode [18], strongly populated through $(p,t)$ reactions.

According to the QPM calculation, the $0^+$ states do not exhibit a pronounced quadrupole collective. This prediction needs to be tested by more reliable measurements of the $E2$ transition strengths.

On the whole, the present QPM calculation provides a fairly satisfactory description of the existing properties of the $0^+$ states in deformed nuclei. In order to assess its complete reliability, however, it is desirable to complete the characterization of these states by systematic measurements of their $E2$ and $E0$ decay strengths.

## ACKNOWLEDGMENTS

This work is partly supported by the Ministero dell' Istruzione, Università e Ricerca (MIUR).

## REFERENCES

1. S. R. Lesher *et al*, *Phys. Rev. C* **66**, 051305(R) (2002).
2. H.-F. Wirth *et al*, *Phys. Rev. C* **69**, 044310 (2004).
3. D. Bucurescu *et al*, *Phys. Rev. C* **73**, 064309 (2006).
4. D. A. Meyer *et al.*, *J. Phys. G: Nucl. Part. Phys.* **31**, S1399 (2005).
5. D. A. Meyer *et al.*, *Phys. Lett.* **B 638**, 44 (2006).
6. D. A. Meyer *et al.*, *Phys. Rev. C* **74**, 044309 (2006).
7. N. V. Zamfir, Jing-ye Zhang, and R. F. Casten, *Phys. Rev. C* **66**, 057303 (2002)
8. Y. Sun, A. Aprahamian, J. Zhang, C. Lee, *Phys. Rev. C* **68**, 061301(R) (2003).
9. N. Lo Iudice, A. V. Sushkov, and N. Yu. Shirikova, *Phys. Rev. C* **70**, 064316 (2004)
10. N. Lo Iudice, A. V. Sushkov, and N. Yu. Shirikova, *Phys. Rev. C***72**, 034303 (2005)
11. V. G. Soloviev, *Theory of Atomic Nuclei: Quasiparticles and Phonons*, Institute of Physics Publishing, Bristol, 1992.
12. T. Härtlein, M. Heinebrodt, D. Schwalm, C. Fahlander, *Eur. Phys. J. A* **2**, 253 (1998)
13. W. F. Davidson, D. D. Warner, R. F. Casten, K. Schreckenbaçh, H. G. Börner, J. Simić, M. Stojanović, M. Bogdanović, S. Koički, W. Gelletly, G. B. Orr, and M. L. Stelts, *J. Phys. G* **7**, 455 (1981).
14. H. Lehmann, J. Jolie, F. Corminboeuf, H. G. Börner, C. Doll, M. Jentschell, R. F. Casten, N. V. Zamfir, *Phys. Rev. C* **57**, 569 (1998)
15. J. L. Wood, E. F. Zganjar, C. De Coester, K. Heyde, *Nucl. Phys. A* **651**,323 (1999).
16. Y. A. Akovali, *Nuclear Data Sheets* **66**, 505 (1992).
17. P. E. Garrett, *J. Phys. G* **27**, R1 (2001).
18. D.R. Bes and R. A. Broglia, *Nucl. Phys.* **80**, 289 (1966).

# Nature of the $K^{\pi} = 4^{+}$ bands in the Os isotopes

P.E. Garrett*, A.A. Phillips*, L. Bettermann†, N. Braun†, D.G. Burke**, G.A. Demand*, T. Faestermann‡, P. Finlay*, K.L. Green*, R. Hertenberger§, K.G. Leach*, R. Krücken‡, M.A. Schumaker*, C.E. Svensson*, H.-F. Wirth‡ and J. Wong*

*Department of Physics, University of Guelph, Guelph, ON N1G2W1 Canada
†Institut für Kernphysik, Universität zu Köln, D-50937 Köln, Germany
**Department of Physics and Astronomy, McMaster University, Hamilton, ON L8S4M1, Canada
‡Physik Department, Technische Universität München, D-85748 Garching, Germany
§Department für Physik, Ludwig-Maximilians-Universität München, D-85748 Garching, Germany

**Abstract.** Levels in $^{186,188}$Os have been investigated using the $(^{3}\mathrm{He},d)$ reaction with 30 MeV $^{3}$He beams. Absolute level-population cross sections have been determined, and angular distributions measured between 5° and 50°. The $4_3^+$ levels are observed to be some of the strongest populated states below 2 MeV excitation energy, and the magnitudes of the $5/2^{+}[402]_{\pi} + 3/2^{+}[402]_{\pi}$ configuration extracted are in line with quasiparticle-phonon model predictions which state that the lowest-lying $K^{\pi} = 4^{+}$ band is predominantly a hexadecapole excitation.

**Keywords:** single-proton transfer, hexadecapole excitations, two-phonon $\gamma$ vibrations
**PACS:** 21.10.Jx,21.10.Re,25.55.Hp,27.70.+q

## INTRODUCTION

The structures of $4_3^+$ states in the Os nuclei have been the subject of debate for the past several decades. Early studies [1, 2] interpreted the $K^{\pi} = 4^{+}$ band as two-phonon $\gamma$ vibrations, based on their nearly harmonic spacing with respect to the $K^{\pi} = 2^{+}$ $\gamma$ vibration. Later, during a study of the $^{190,192}$Os nuclei with the $^{191,193}\mathrm{Ir}(t,\alpha)$ single-proton pickup reactions, the $I,K^{\pi} = 4,4^{+}$ levels were observed [3] to be very strongly populated, and it was suggested that they were hexadecapole vibrations. The hexadecapole hypothesis was tested with $(\alpha,\alpha')$ inelastic scattering measurements [4] where large $E4$ matrix elements were extracted for excitation of the $4_3^+$ levels. The hexadecapole collectivity was confirmed by a $^{192}\mathrm{Os}(p,p')$ measurement [5] that determined $M(E4) = 1080 \pm 270$ $e$fm$^4$. Based on measured $B(E2)$ values in Coulomb excitation, Oshima *et al.* [6] claimed the $4_3^+$ level in $^{192}$Os was the $K^{\pi} = 4^{+}$ two-phonon $\gamma$ vibration, and detailed Coulomb excitation studies of $^{186-192}$Os [7] led Wu and Cline [8] to interpret the $K^{\pi} = 4^{+}$ bands in these nuclei as two-phonon excitations. The large $B(E2; 4_3^+ \rightarrow 2_\gamma^+)$ values of $27 \pm 9$ W.u., $10 \pm 3$ W.u., $7.5 \pm 1.5$ W.u., and $11 \pm 1$ W.u. for $^{186}$Os, $^{188}$Os, $^{190}$Os, and $^{192}$Os are in very good agreement with the expected harmonic ratio of 2 relative to the $B(E2; 2_\gamma^+ \rightarrow 0_{gs}^+)$ values of $10.1 \pm 0.4$ W.u., $5.0 \pm 0.6$ W.u., $5.9 \pm 0.6$ W.u., $5.6 \pm 0.2$ W.u., respectively. Later lifetime measurements [9] confirmed the Coulomb excitation results.

The two-phonon $\gamma$ vibration interpretation, however, is in serious conflict with the single-nucleon transfer results [10]. Uncertainties in the $(t,\alpha)$ reaction mechanism lead

CP1012, *Frontiers in Nuclear Structure, Astrophysics, and Reactions: FINUSTAR 2*
edited by P. Demetriou, R. Julin, and S. V. Harissopulos
© 2008 American Institute of Physics 0-7354-0532-5/08/$23.00

to the suggestion [8] that the magnitude of the hexadecapole component was still an open question, and that the $E2$ matrix elements represented strong evidence that the two-phonon components *dominate* the structure of the $4_3^+$ levels. However, the spectroscopic strengths extracted from the $(t,\alpha)$ reaction data into single-proton states in $^{187,189}$Re and $^{193}$Ir compared well with Nilsson model calculations, and also with results using the $(d,^3\text{He})$ reaction for $^{192}$Os, suggesting that uncertainties in the $(t,\alpha)$ reaction mechanism were well under control [11]. Thus, focusing on the $E2$ matrix elements favours a two-phonon interpretation, whereas focusing on transfer results and $E4$ matrix elements favours a hexadecapole interpretation. In light of this controversy, we have investigated the nature of the $K^\pi = 4^+$ states in $^{186}$Os and $^{188}$Os using the $(^3\text{He},d)$ reaction which has fewer uncertainties in its mechanism than the $(\alpha,t)$ reaction (or correspondingly, the $(t,\alpha)$ reaction).

## EXPERIMENTAL DETAILS AND RESULTS

The experiment was performed at the tandem Van de Graaff accelerator laboratory of the University and Technical University of Munich. Beams of 30 MeV $^3$He ions of up to 1.2 $\mu$A current bombarded targets of $^{185,187}$Re with enrichments of 96.7% and 98.7%, respectively. The $^{185}$Re target had a thickness of 78 $\mu$g/cm$^2$ on a backing of 8 $\mu$g/cm$^2$ $^{\text{nat}}$C, whereas the $^{187}$Re target had a thickness of 84 $\mu$g/cm$^2$ on a backing of 19 $\mu$g/cm$^2$ $^{12}$C, as determined from measurements of the elastic scattering cross sections and distorted wave Born approximation (DWBA) calculations described below. A conservative estimate of 5% was assumed for the uncertainty on the target thicknesses. The number of $^3$He beam particles was determined by integrating the current observed in the Faraday cup. The products of the reaction were momentum analyzed by a Q3D spectrograph with a solid angle of acceptance up to 11.6 msr. The focal plane detector consisted a position-sensitive proportional counter with a readout of a cathode foil, and yielded $\Delta E$ vs. $E$ for particle identification. A total of 9 angles were measured, in 5° steps between 5° and 40°, and at 50°.

Shown in Figure 1 are the low-excitation-energy portions of the deuteron spectra obtained at 40° for the $^{185}$Re target (top panel) and $^{187}$Re target (bottom panel). The energy resolution achieved was 12 keV full width at half maximum (FWHM). Figure 2 displays selected angular distributions obtained with both targets; those for the ground states, the $K^\pi = 2^+$ $\gamma$ band head, and the $K^\pi = 4^+$ band head. The curves are the results of DWBA calculations for a single transferred $l$ value ($l = 2$ for $0^+$ and $4^+$, $l = 0$ for $2^+$) obtained using the code DWUCK4 [12]. The optical parameters used were those of Wirth *et al.* [13] that were found to reproduce the observed angular distributions very well. The magnitudes of the single-particle cross sections are very similar to those calculated using the optical model parameters of Lu and Alford [14] that have been used across the entire deformed rare-earth region (see, for example, the discussion in Burke *et al.* [15]), and are believed to be accurate to within 25%.

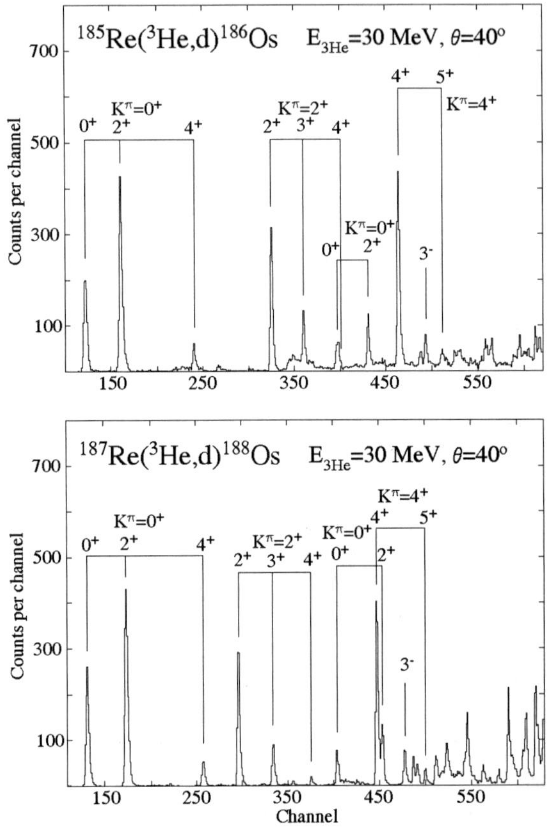


**FIGURE 1.** Low-excitation-energy portions of spectra obtained with the $^{185}$Re($^{3}$He,d)$^{186}$Os (top) and $^{187}$Re($^{3}$He,d)$^{188}$Os reactions (bottom) with 30 MeV $^{3}$He beams.

## DISCUSSION

Starting with a target in a one-quasiparticle state, the transfer of one particle must populate two-quasiparticle components in the final state wave function. Since the target proton is in the $\frac{5}{2}^{+}[402]_{\pi}$ Nilsson state, to first order the final components must have the $\frac{5}{2}^{+}[402]_{\pi} \pm \Lambda_{\pi}^{tr}$ configuration, where $\Lambda_{\pi}^{tr}$ represents the Nilsson orbital that the proton is transferred into. Assuming the target and final state wave functions are pure Nilsson configurations, the theoretical cross section for population of a particular final state in an even-even nucleus is given by

$$\frac{d\sigma}{d\Omega} = Ng^{2} \sum_{j,l} \left[ C_{jl} P_{i} \langle I_{t} K_{t}\, j \Delta K | I_{f} K_{f} \rangle \right]^{2} \left( \frac{d\sigma}{d\Omega} \right)_{DW} \quad (1)$$

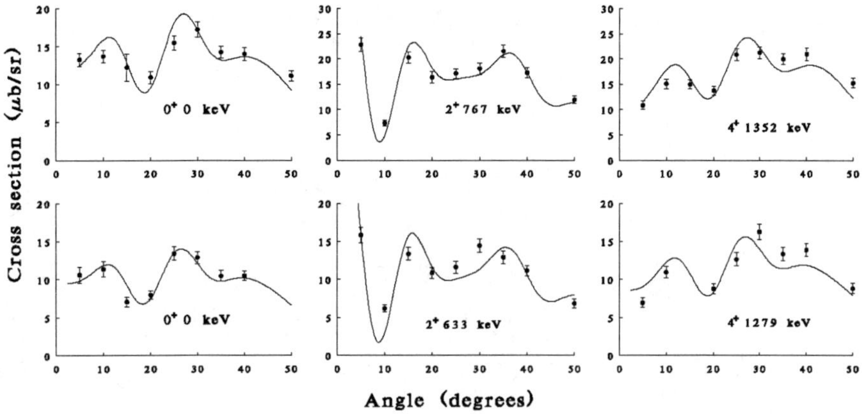


**FIGURE 2.** Sample angular distributions obtained obtained with the $^{185}$Os($^3$He,d)$^{186}$Os (top) and $^{187}$Os($^3$He,d)$^{188}$Os reactions (bottom) with 30 MeV $^3$He beams. The energy and spins of the states are indicated. The curves are DWBA predictions using optical model parameters of Wirth *et al.* [13].

where $I_t$ ($I_f$) and $K_t$ ($K_f$) are the target (final state) spin and projection, $N$ is a normalization factor for the DWBA cross sections, $(d\sigma/d\Omega)_{DW}$, and the summation extends over all possible transferred $j$ and $l$ values leading to the final state. The factor $g^2$ has the value of 1 for usual two-quasiparticle states, but has a value of 2 for $K = 0$ states formed by coupling two nucleons in time-reversed orbitals (e.g., the ground state band). The $C_{jl}$ values are the expansion coefficients of the Nilsson wave function for the transferred nucleon in the spherical basis, and $P_i$ are the pairing factors which, for the present stripping reaction, are $V^2$ for ground state population and $U^2$ otherwise.

Figure 3 displays the experimental angular distributions observed for the $K^\pi = 2^+$ $\gamma$ band head and $K^\pi = 4^+$ band head, with the theoretical prediction of the Nilsson component indicated, calculated using a deformation parameter $\delta = 0.2$. Assuming an emptiness probability $U^2 = 0.85$ for both the $\frac{1}{2}^+[400]$ and $\frac{3}{2}^+[402]$ orbitals, the magnitude of the $\frac{5}{2}^+[402]_\pi - \frac{1}{2}^+[400]_\pi$ component in the $\gamma$ band is 0.35 and 0.29 ($\pm$25%) for $^{186}$Os and $^{188}$Os, respectively. (An estimated 25% systematic uncertainty has been assumed in the magnitude of the DWBA cross sections.) Random phase approximation calculations [16] for the $\gamma$ band in $^{190}$Os give a admixture of 0.22 for the $\frac{5}{2}^+[402]_\pi - \frac{1}{2}^+[400]_\pi$ configuration, in line with the present extracted values. For the $K^\pi = 4^+$ band, the $\frac{5}{2}^+[402]_\pi + \frac{3}{2}^+[402]_\pi$ components are 0.45 and 0.35 ($\pm$25%), respectively. Quasiparticle-phonon model calculations [17] predict that the $4_3^+$ level has a 90% hexadecapole phonon component, resulting in the $\frac{5}{2}^+[402]_\pi + \frac{3}{2}^+[402]_\pi$ configuration comprising 51% and 55% of the level wave function in $^{186}$Os and $^{188}$Os. While the extracted admixtures of these configurations do not rule out a two-phonon contribution in the $4_3^+$ wave functions, it is unlikely that it is the dominant amplitude since to first order it would not be populated in single-nucleon transfer reactions.

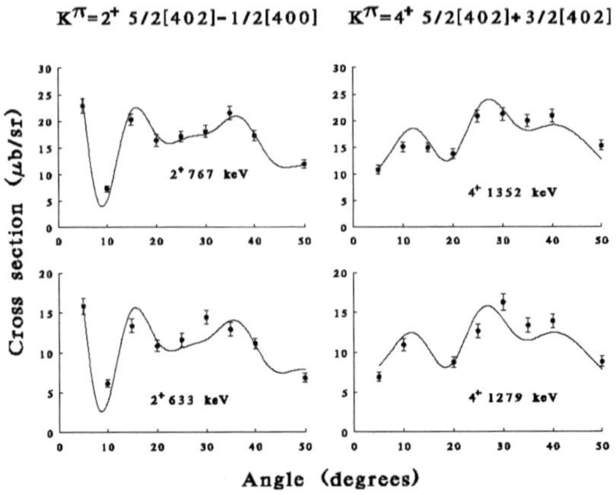


**FIGURE 3.** Predicted angular distributions for the configurations indicated for the $K^{\pi} = 2^{+}$ $\gamma$ band head, and the $K^{\pi} = 4^{+}$ band head for $^{186}$Os (top) and $^{188}$Os (bottom).

## CONCLUSIONS

The single-proton stripping reactions $^{185,187}$Re($^{3}$He,$d$) were used to probe the $4_3^+$ states in $^{186,188}$Os. The $4_3^+$ levels were observed to be some of the strongest peaks populated below 2 MeV, and using a DWBA analysis the extracted magnitudes of the $\frac{5}{2}^{+}[402]_{\pi} + \frac{3}{2}^{+}[402]_{\pi}$ configurations in the wave functions are 45% and 35% for $^{186,188}$Os. These magnitudes are consistent with a hexadecapole interpretation for the $K^{\pi} = 4^{+}$ bands.

This work was supported in part by the Natural Sciences and Engineering Research Council of Canada.

## REFERENCES


1. T. Yamazaki, Nucl. Phys. **44**, 353 (1963); **49**, 1 (1963).
2. H.L. Sharma and N.M. Hintz, Phys. Rev. Lett. **31**, 1517 (1973); Phys. Rev. C **13**, 2288 (1976).
3. R.D. Bagnell, *et al.*, Phys. Lett. **B66**, 129 (1977).
4. D.G. Burke, M.A.M. Shahabuddin, and R.N. Boyd, Phys. Lett. **B78**, 48 (1978).
5. F. Todd Baker, *et al.*, Nucl. Phys. **A501**, 546 (1989).
6. M. Oshima, *et al.*, Nucl. Phys. **A557**, 635c (1993).
7. C.Y. Wu, *et al.*, Nucl. Phys. **A607**, 178 (1996).
8. C.Y. Wu and D. Cline, Phys. Lett. **B382**, 214 (1996).
9. C.Y. Wu, *et al.*, Phys. Rev. C **64**, 014307 (2001).
10. D.G. Burke, Phys. Rev. Lett. **73**, 1899 (1994)
11. D.G. Burke, Phys. Lett. **B406**, 200 (1997).
12. P.D. Kunz, computer code DWUCK4, University of Colorado (unpublished).
13. H.-F. Wirth, *et al.*, Phys. Rev. C **70**, 014610 (2004).
14. M.T. Lu and W.P. Alford, Phys. Rev. C **3**, 1243 (1971).
15. D.G. Burke, J.C. Waddington, and O.P. Jolly, Nucl. Phys. **A688**, 716 (2001).
16. R.D. Bagnell, *et al.*, Phys. Rev. C **20**, 42 (1979).
17. V.O. Nesterenko, *et al.*, Sov. J. Nucl. Phys. **44**, 938 (1987).

# $\beta$-delayed particle studies of the halo nucleus $^{11}$Li and its core $^{9}$Li

M. Madurga*, M.J.G. Borge*, L.M. Fraile*,†, H.O.U. Fynbo**, B. Jonson‡, T. Nilsson‡, G. Nyman‡, Y. Prezado*, K. Riisager**,†, O. Tengblad*, the IS-417 collaboration§ and the TONNERRE collaboration¶

*Instituto de Estructura de la Materia, CSIC, E-28006 Madrid, Spain.
†EP Division, CERN, CH-112 Genève, Switzerland.
**Department of Physics and Astronomy, University of Aarhus, DK-8000, Århus, Denmark.
‡Fundamental Physics, Chalmers University of Technology, S-41296 Göteborg, Sweden.
§Århus-Caen-Göteborg-Madrid-York.
¶LPC-Bucharest-IReS.

**Abstract.** The $\beta$-decay of $^{11}$Li and its core $^{9}$Li has been revisited with the aim to disentangle the very complex decays from unbound states in the daughters. The decay channels for the states fed by $\beta$-decay in $^{9}$Be are established including the 2.43 MeV state for which the analysis of $\alpha\alpha$ and n$\alpha$ spectra seem to favour direct break-up of the level. For the decay of $^{11}$Li the $\beta$-delayed charged particle spectrum can be explained by the contribution of the 10.6 and 18.2 MeV states. The $\alpha\alpha$ correlations indicates that decay through resonant He-isotopes are more likely than phase space distributed 5-body break-up.

**Keywords:** Radioactivity $^{11}$Li and $^{9}$Li ($\beta^-$-decay) from Ta(p,X): Measured $\beta$-delayed $E_\alpha$ and $\alpha$-coincidences; 1 $^{11,9}$Be deduced levels; Partial decay branches
**PACS:** 21.10.-k; 23.40.Hc; 23.40.-s; 27.20.+n

## INTRODUCTION

Nuclear halos have been found in a number of light nuclei, and their structure is understood as resulting from the presence of a bound state close to the continuum, see [1] and references therein. Information about the halo wave function has been gained mainly by reactions where the halo state is destroyed (Coulomb dissociation, diffraction or absorption) or separated from the core (core break-up). Alternatively we can learn about the halo wave function by the well known weak interaction that mediates the beta decay of halo states or feeding of halo states. This method is unique in the case where the halo is an excited state.

A halo can affect the beta decay in different ways; the spatial extension of the halo state might reduce the overlap with daughter states, this effect might be seen either in specific transitions or globally by the beta strength. The halo nucleons might decay more or less independently from the core, that might give rise to specific patterns in the decay going to the continuum states. Beta delayed deuteron emission is typical for nuclei with two-neutron halos as the energy window open for this decay mode depends on the two neutron separation energy, $Q_{\beta d}$(Z,N) = 3.007 - $S_{2n}$ (MeV). This decay mode is energetically allowed in, for example, $^{6}$He, $^{11}$Li, $^{14}$Be and $^{17}$B. It has been observed

CP1012, *Frontiers in Nuclear Structure, Astrophysics, and Reactions: FINUSTAR 2*
edited by P. Demetriou, R. Julin, and S. V. Harissopulos
© 2008 American Institute of Physics 0-7354-0532-5/08/$23.00

and the deuteron spectrum has been measured in $^6$He. In $^{11}$Li the presence of a $\beta$t branch of similar branching ratio and larger energy window has hampered the determination of the deuteron energy spectrum. The beta-delayed deuteron branch, $^{11}$Li($\beta$) $\rightarrow$ $^{11}$Be* $\rightarrow$ d+$^9$Li, was established by looking at the alphas following the delayed decay of $^9$Li with its characteristic half-life [2].

A particular transition could also give information about the composition of the halo state wave function. In the decay of light neutron rich nuclei it has been found [3] that some beta transitions are characterized by large $B_{GT}$ values ranging from 1-6 that can be classified as "superallowed". The term superallowed refers to fast (logft-values $\sim$ 3.5) transitions of essentially single particle strength normally found for Fermi transitions. This pattern has been observed in $^9$Li [4] and less pronounced in $^{11}$Li [5]. These favoured states are observed within a few MeV of the energy of the mother state.

Assuming that the halo state can be factorised into core and halo parts, the beta decay will have a pattern similar to that of the core following eq. 2 given in [6] if $(\vartheta_\beta \, |core>) \, |halo>)$ is, in good approximation, an eigenstate of the daughter nucleus or part of it. The similarity in patterns have been recognized in the decay of the 2n-halo nucleus $^{14}$Be and its core $^{12}$Be [6, 7]. The nucleus $^{12}$Be mainly decays to the ground state of $^{12}$B. The decay of $^{14}$Be goes mainly to the neutron-unbound state at 1.28 MeV above the neutron threshold. This result has been corroborated by experiments done at different facilities MSU, RIKEN and ISOLDE: Theoretically [8], the decay can be explained as due to a two neutron spectator around a decaying $^{12}$Be core.

The decay of $^{11}$Li is very fragmented and superficially similar to the $^9$Li core decay. The aim of this work is to revisit these two decays and measure them using the same experimental set-up. Special emphasis has been given to the analysis of the charged particle coincidences in order to disentangle the intermediate decay channels and to be able to have an accurate deduction of the Gamow-Teller strength at high excitation energies in $^{11}$Li and its core $^9$Li.

## EXPERIMENTAL SET-UP

The experiment took place at ISOLDE, CERN. A Ta target was bombarded by 1 GeV protons from the PS-Booster and connected to a tungsten ionization source. The singly charged beams of $^9$Li and $^{11}$Li were isotopically selected using the GPS isotope separator. A 30 kV extraction voltage was used, in order to transport and finally stop in a thin, 60 $\mu$g/cm$^2$, carbon foil the $^{9,11}$Li$^+$ pure beams .

The set-up consisted of 4 telescopes with DSSSD (16$\times$16 strips) detectors (two of ultra-thin deadlayers [9]) in front of thick silicon pads, mounted on the surfaces of a 10x10x10 cm$^3$ cubic frame, covering 4% of $4\pi$ each. This set-up allows a high angular coverage while having enough spacial resolution. Behind each DSSSD a 1.5 mm thick silicon detectors were placed for $\beta$ detection. The $\beta$-delayed neutrons were detected using the neutron time of flight detector TONNERRE [10]. The array used consisted of 16 plastic scintillator paddles around the experimental chamber covering 20% of $4\pi$.

## ANALYSIS METHOD AND RESULTS

The study of these two decays concentrates on the channel involving charged particles. In the case of $^9$Li they correspond to approximatelly 50% of the decays while in $^{11}$Li they correspond only to 3 % of the decay [11].

The high granularity of the DSSSD detectors allows a precise reconstruction of the momentum of the impinging particles. If we neglect the recoil due to the $\beta$ emission the break-up of states in the daughter fed in the $\beta$-decay occurs nearly at rest. If the excited state breaks into three particles of which two are detected, the conservation of energy and momentum serves to reconstruct energy and direction of the third particle and characterise the daughter state. In both decays several decay channels are open to different intermediate levels. To get the partial decay branches of each level in the daughter, R-matrix calculations have been applied with Monte Carlo simulations to account for the different experimental efficiencies for the two detected particles depending on the intermediate channel.

Inclusion of angular correlation is essential to obtain a good description of the data. The high granularity of the setup has allowed measurements of angular correlations. For non interfering states the direction of the first particle is isotropic and the second is distributed according to an angular correlation function $W(\theta) = \sum_{v=0}^{v_{max}} A_v P_v(cos(\theta))$, where $P_v$ are Legendre polynomials of order $v$ and the coefficients $A_v$ are determined by the angular momenta involved in the process. We studied the angular correlation following the formulation of Biedenharn and Rose [12], details are given in [13]. In this way the spin of several states in $^9$Be has been determined or cross-checked when known.

*The $^9$Li decay*

In the beta decay of $^9$Li all states fed but the g.s. break-up into $n\alpha\alpha$. Earlier $\beta$-decay works indicated that the high lying 11.28 MeV level in $^9$Be was fed in the $\beta$-decay [14]. Later Nyman and coworkers [15] analysed the $\beta\alpha$ spectrum within the R-matrix theory and showed that feeding to the higher 11.8 MeV state in $^9$Be should be included. The feeding to these broad states gave a large $B_{GT}$-value indicating transitions of the "superallowed" type. Our study in full kinematics allowed to fix that $^9$Li mainly decay to the 11.8 MeV state in $^9$Be, whose spin was determined to be 5/2 and confirmed the large $B_{GT}$-value for this $\beta$-transition [4].

Within the same framework we discuss in the following the studies done on the lower energy states of $^9$Be i.e., levels with $I^\pi = 1/2^-$, $3/2^-$ or $5/2^-$ populated in allowed beta transitions from $^9$Li. The main facts are summarised below.

A 1.5(5) % beta-feeding was proposed by Langevin [14] to the 7.94 MeV state that he assumed to be I = 1/2. An analysis of the $\alpha - \alpha$ angular correlation of the region of interest through the $^5$He channel excluded a spin value of 1/2 and favour a spin of 5/2 for this state [16]. This new spin value is in agreement with the unpublished results of $^{10}$B(e,e'p)$^9$Be cited in [11].

The energy, spin and decay branches of the narrow 2.43 MeV state in $^9$Be are well known and the existing experiments agree that only 7% decay to the $^8$Be(g.s.). However, the decay mode of the remaining ca. 93 % is controverted and data from reactions have been analysed both assuming sequential and direct decay, see the recent work of [17] and references therein. In our work from the $\alpha - \alpha$ coincidence, and applying the momentum conservation, the energy and direction of the neutron has been deduced. We have selected the events with reconstructed energy sum $E_{sum} < 0.9$ MeV ($E^*(^9Be) = E_{sum} + 1.57$ MeV) corresponding to the 2.43 MeV level in $^9$Be. The decay of this level was studied using detailed Monte Carlo simulations. Decay through the tail the $^8$Be($2^+$)

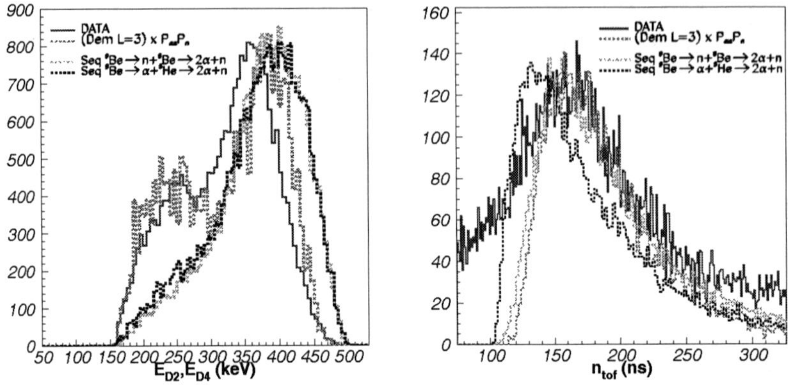


**FIGURE 1.** On the left the alpha-particle energy spectrum from the 2.43 MeV state in $^{9}$Be and on the right the neutron energy spectrum in coincidence with alphas from the 2.43 MeV state in $^{9}$Be. Both spectra have been fitted assuming direct decay or sequential decay via intermediate states in $^{8}$Be($2^{+}$) or $^{5}$He(gs).

state and $^{5}$He ground state were simulated using the R-matrix formalism. Democratic break-up as prescribed by Bochkarev [18] plus Coulomb interaction was considered as well. The results can be seen on the left hand side of figure 1. The neutron spectrum from TONNERRE in coincidence with the alphas from the 2.43 MeV state was fitted using the same decay channels as for the alpha spectrum. The result is shown on the right hand side of figure 1. The simulations favour the democratic break-up described by hyper-spherical harmonics of grand orbital momentum K=3 over the decay through the low energy tail of $^{8}$Be($2^{+}$) as proposed in earlier reaction works, see [17] and references therein.

### *The $^{11}$Li decay*

The $\beta$-delayed charged particle emission of $^{11}$Li has been the subject of several previous studies [2, 5, 14, 19]. The two channels involving $\alpha$ particles (2$\alpha$+3n and $^{6}$He+$\alpha$+n) were previously studied in a coincidence experiment by Langevin et al. [14] using two silicon detectors for coincidence measurements. The 2$\alpha$3n channel was linked to the break-up of the 10.6 and the 18.2 MeV states in $^{11}$Be. The $^{6}$He$\alpha$n channel was established as the $^{6}$He+$\alpha$ break-up of the 9.6 MeV state in $^{10}$Be fed by neutron decay of the 10.6 MeV state in $^{11}$Be. However no precise modeling was done to support this claim. In this work we model the $^{11}$Be*→n+$^{10}$Be→n+$\alpha$+$^{6}$He channel using multiple level single-channel R-Matrix formalism and a Monte-Carlo simulation to take into account the efficiency of the detection system for this decay channel. This allows comparison of the simulations with experimental data at the level of projected energy spectrum as well as at the level of particle-particle correlations.

As the three-body break-up of the 10.6 MeV state in $^{11}$Be involves the emission of two charged particles, $^{6}$He and $\alpha$, we are interested in the events in coincidence. Assuming as in [14] that the decay occurs through the 9.6 MeV state in $^{10}$Be, the energy of the neutron is such that the recoil of the 9.6 MeV state is about 55 keV. In these conditions we expect this channel to be observed predominantly close to back-to-back geometry. The resonance shapes were described using the multiple-level single channel R-Matrix formalism as explained above [15]. Monte-Carlo simulations were done to account for the experimental efficiency of the set up for the different channels. The simulation is in very good agreement with the data, supporting the claim of [14].

The simulation of the 5 body ($2\alpha$+3n) decay of the 10.6 and 18.2 MeV states in $^{11}$Be is not able to reproduce the $\beta\alpha$ from $^{11}$Li with the correct width for the 18.2 MeV state in $^{11}$Be. In addition the shape of the $\alpha\alpha$ coincidence distribution is different from the one obtained from 5-body phase space break-up. The $\alpha\alpha$ scatter plot shows diagonal distributions with 7/4 and 4/7 slopes indicating contribution of decay through the $^{7}$He resonances. New decay channels, involving $^{5}$He and $^{7}$He, might have to be included in order to describe the correlated particle energy spectrum. The contribution of other excited states as suggested in different publications [2, 5, 20] has been considered, but no experimental evidence has been found to include them [21].

## SUMMARY AND OUTLOOK

In this work the delayed charged-particles plus neutron channels following the decay of the two neutron halo $^{11}$Li and its core $^{9}$Li are studied. The states in $^{9}$Be contributing to the $\beta$-decay of $^{9}$Li have been established. Angular correlations have been used for firm spin determination [4, 16]. Both $\alpha\alpha$ and n$\alpha$ spectra of the 2.43 MeV state favour democratic decay.

Our new data for $^{11}$Li show that the suggested pure phase-space break-up [14] cannot explain the structures observed in the coincidence data. Two new decay channels, one of them involving $^{5}$He-$^{6}$He gs resonances and the other involving $^{7}$He(gs), are introduced to explain these discrepancies.

## REFERENCES


1. B. Jonson and K. Riisager, *Philos. Trans. Roy. Soc. London Ser.* **A 356**, 2063 (1998).
2. I. Mukha et al., *Phys. Lett.* **B 367**, 65 (1996).
3. M. J. G. Borge et al., *Z. Phys.* **B 340**, 255 (1991).
4. Y. Prezado et al., *Phys. Lett.* **B 576**, 55 (2003).
5. M. J. G. Borge et al., *Nucl. Phys.* **A 613**, 199 (1997).
6. T. Nilsson, G. Nyman and K. Riisager, *Hyperfine Interactions* **129**, 67 (2000).
7. U. C. Bergmann et al., *Nucl. Phys.* **A 658**, 129 (1999).
8. N. K. Timofeyuk et al., *Nucl. Phys.* **A 600**, 1 (1996).
9. O Tengblad et al., *Nucl. Inst. and Meth.* **A 525**, 458 (2004).
10. A. Buta et al., *Nucl. Inst. and Meth* **A 455**, 412 (2000).
11. D. R. Tilley et al., *Nucl. Phys.* **A 745**, 155 (2004).
12. L. C. Biederharn and M. E: Rose, *Rev. Mod. Phys.* **25**, 729 (1953).
13. M. J. G. Borge et al., *Phys. Scr.* **T 125**, 103 (2006).
14. M. Langevin , et al., *Nucl. Phys.* **A 366**, 449 (1981).
15. G. Nyman et al., *Nucl. Phys.* **A 510**, 189 (1990).
16. Y. Prezado et al., *Phys. Lett.* **B 618**, 43 (2005).
17. P. Papka et al., *Phys. Rev.* **C 75**, 045803 (2007).
18. O.V. Bochkarev et al, *Nucl. Phys.* **A 505**, 215 (1989); *Sov. J. Nucl. Phys.* **52**, 964 (1990).
19. M. Langevin et al., *Phys. Lett.* **B 146**, 176 (1984).
20. H.O.U. Fynbo et al., *Nucl. Phys.* **A 736**, 39 (2004).
21. M. Madurga et al., *in preparation*

# Thermodynamic properties of atomic nuclei with $T < 1$ MeV

M. Guttormsen[*], A.C. Larsen[*], J. Rekstad[*], A. Schiller[†], S. Siem[*], N.U.H. Syed[*] and A. Voinov[**,†]

[*]*Department of Physics, University of Oslo, P.O.Box 1048 Blindern, N-0316 Oslo, Norway*
[†]*Department of Physics and Astronomy, Ohio University, Athens, OH, 45701, USA*
[**]*Frank Laboratory of Neutron Physics, Joint Institute of Nuclear Research, 141980 Dubna, Moscow region, Russia*

**Abstract.** The thermal motion of single quasi-particles represents the largest contribution to nuclear entropy and is essential for the description of thermodynamic properties. The entropy is an approximately extensive (additive) quantity and the concept of thermal quasi-particle entropy is presented. Various applications of thermodynamics in heated nuclei are discussed.

**Keywords:** Level density, entropy, thermodynamics
**PACS:** 21.10.Ma, 24.10.Pa, 65.40.Gr, 64.60.-i

## INTRODUCTION

The motion of nucleon pairs in time-reversed orbitals forms a favored configuration at low excitation energy. These Cooper pairs manifest themselves in different ways: (i) all even-even mass nuclei have zero spin in the ground state, (ii) the number of levels suddenly increases at an excitation energy of $\sim 2\Delta$ ($\Delta$ being the pairing gap parameter) and (iii) odd-even mass defects differ by $\sim \Delta$.

The Cooper pair condensate strongly simplifies the description of nuclear structure close to the ground state. However, the picture becomes much more complicated as Cooper pairs are broken by collective (Coriolis force) or intrinsic (temperature) excitations. At high excitation energy, the almost free motion of quasi-particles on vacant single particle orbitals are responsible for the exponential growth of levels as function of excitation energy.

When the number of levels exceeds 50 - 100 per MeV, spectroscopic methods are useless to resolve all individual levels. As this statistical region emerges, new experimental as well as theoretical approaches have to be applied. A fruitful concept in the statistical regime is the entropy $S(E) = k_B \ln \Omega(E)$, where $\Omega(E)$ is the number of ways the system can be arranged at a certain excitation energy $E$. The Boltzmann's constant $k_B$ is set to unity in the following.

The entropy is the starting point for evaluating thermodynamic properties. In nuclei, the most common ensembles to be considered are the microcanonical (totally isolated system with fixed $E$) and the canonical ensembles (thermal contact with a heat bath of fixed temperature $T$). In this work we will discuss thermodynamic quantities within both ensembles.

CP1012, ***Frontiers in Nuclear Structure, Astrophysics, and Reactions:*** *FINUSTAR 2*
edited by P. Demetriou, R. Julin, and S. V. Harissopulos
© 2008 American Institute of Physics 0-7354-0532-5/08/$23.00

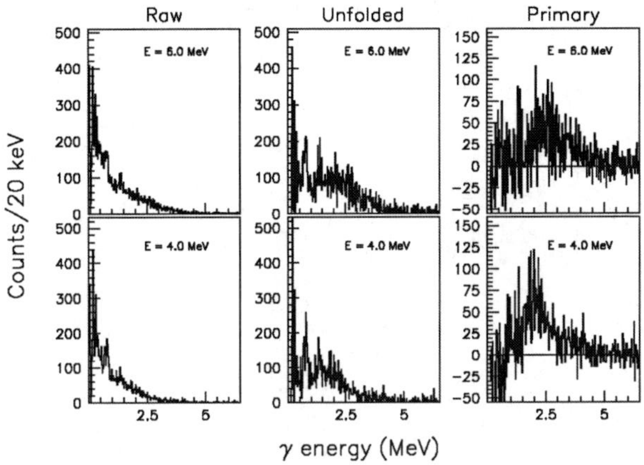


**FIGURE 1.** Gamma-ray spectra at $E = 4$ and 6 MeV of excitation energy for $^{166}$Er. Left panel: raw $\gamma$-ray spectra, central panel: unfolded $\gamma$-ray spectra, and right panel: primary $\gamma$-ray spectra. Data are taken from Ref. [4].

## EXPERIMENTAL METHOD

The Oslo nuclear physics group has developed a method [1] to extract nuclear level density and $\gamma$-ray strength function simultaneously from particle-$\gamma$ coincidences. The experiments are performed with light-particle inelastic scattering or transfer reactions. The charged ejectile is used to tag $\gamma$-ray spectra with excitation energies from the ground state up to the neutron (or proton) binding energy. The particle-$\gamma$ coincidences are measured with the highly efficient CACTUS detector set-up at the Oslo Cyclotron Laboratory (OCL).

The experiments were performed with a 45-MeV $^3$He beam impinging on self-supporting targets. Particle-$\gamma$ coincidences were detected with an array of 28 collimated NaI $\gamma$-ray detectors and eight Si particle telescopes placed at 45° with respect to the beam. The data are sorted into a matrix of $(E, E_\gamma)$ energy pairs, $E$ being the excitation energy. The $\gamma$ spectra from each excitation energy are assumed to follow [1]

$$P(E, E_\gamma) \propto \rho(E - E_\gamma)\mathscr{T}(E_\gamma), \tag{1}$$

where $\rho$ is the level density and $\mathscr{T}$ is the radiative transmission coefficient. In this work we will only focus on the level density $\rho$.

Examples of the recorded $\gamma$ spectra, the so called raw spectra, from two excitation energies are shown in the left panel of Fig. 1. In order to determine the true $\gamma$-energy distribution, the $\gamma$ spectra are corrected for the response of the NaI detectors with the unfolding procedure of Ref. [2], as shown in the central panel. The now corrected $P(E, E_\gamma)$ matrix comprises the $\gamma$-energy distribution of the total $\gamma$ cascade, and make it possible to derive the primary $\gamma$ matrix according to a subtraction technique [3]. Primary $\gamma$ spectra can be seen in the right panel of Fig. 1.

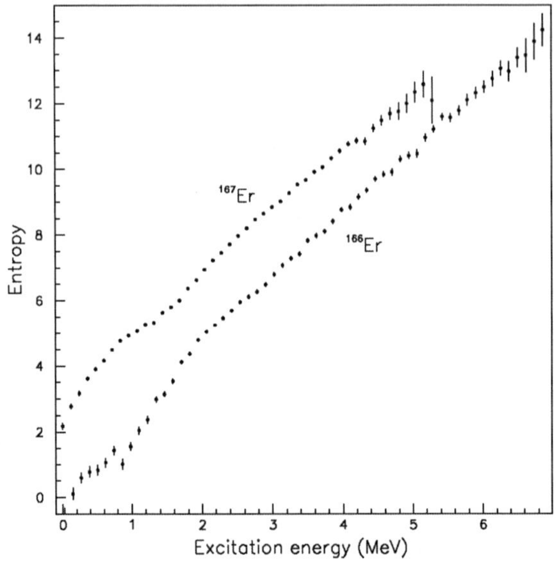


**FIGURE 2.** Entropy as a function of excitation energy in $^{166,167}$Er. Data are taken from Ref. [4].

# THERMODYNAMICS

Figure 2 shows the entropy derived from the level densities of $^{166,167}$Er as function of excitation energy $E$. The level density is transformed into entropy by $S(E) = \ln \rho(E)/\rho_0$, where $\rho_0$ is adjusted to fulfill the condition of the third law of thermodynamics; $S \to 0$ when $T \to 0$, $T$ being the nuclear temperature. The odd $^{167}$Er system reveals an increase of $\Delta S = 1.5 - 2.0$, which is due to the entropy carried by the valence particle outside the heated even-even core [4]. A systematic review [5] of nuclei having known resonance spacing data shows that the odd particle (or hole) for mid-shell nuclei carries entropy of $1.5 - 2.0$. The fact that we find this value to be approximately constant for all $E$ indicates that one could use the concept of single quasiparticle entropy, independent of the number of excited particles in the underlying core.

In the microcanonical ensemble, the temperature and heat capacity can be directly determined from the entropy by

$$T(E) = \left(\frac{\partial S(E)}{\partial E}\right)^{-1} \text{ and } C_V(E) = \left(\frac{\partial T(E)}{\partial E}\right)^{-1}. \tag{2}$$

Figure 3 shows the temperature deduced within the microcanonical ensemble for $^{166,167}$Er. We see that the small bumps in the entropy curves of Fig. 2 are enhanced through the differentiation performed to obtain the temperatures. This is even more pronounced for the heat capacity by the double differentiation of the entropy. The spectacular feature of negative heat capacity is a direct consequence of the decrease in the temperature and has also been observed in other critical systems.

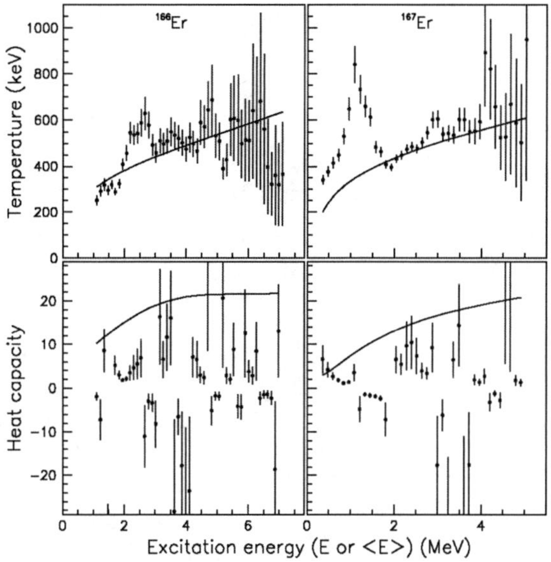


**FIGURE 3.** The temperature (upper panel) and heat capacity (lower panel) of $^{166}$Er (left) and $^{167}$Er (right) from the microcanonical ensemble (data points) and the canonical ensemble (lines). Data are taken from Ref. [4].

An anomalous decrease in the temperature means an unusual increase in the entropy, i.e. a more than normal opening of new domains of the phase space, e.g. opening of new degrees of freedom as more particle pairs are broken. The successive bumps of Figs. 2 and 3 indicate that the transition from a paired to an unpaired phase is a gradual process of breaking up more and more nucleon pairs.

When we interpret the area of negative heat capacity, corresponding to the region where the temperature has a negative slope, as the location of the break up of a nucleon pair, a signal of the break up of the first pair can be seen around $E \sim 1.5$ MeV in $^{167}$Er. This value is close to twice the pairing-gap parameters $2\Delta_n = 1.7$ MeV and $2\Delta_p = 1.5$ MeV, which are the expected cost of breaking a neutron and proton pair, respectively.

Figure 3 also shows the results within the canonical ensemble (solid lines). However, the quantities are severely smoothed out and not useful for detailed investigations.

The tin isotopes have a closed proton shell at $Z = 50$ and are expected to reveal neutron pair breaking as function of excitation energy. In this work we use the technique of Lee and Kosterlitz [6] to investigate the criticality of low temperature transitions. The probability $W$ that a system has energy $E$ at fixed temperature $T$ is given by

$$W(E,T) = \Omega(E)\exp(-E/T) \Big/ \int_0^\infty \Omega(E')\exp\left(-E'/T\right) dE'. \tag{3}$$

If the $-\ln W(E,T)$ function exhibits a characteristic double-minimum structure for $E = E_1$ and $E_2$ at temperature $T_c$, a structural transition is present. It can be easily shown that $\ln W(E,T)$ is closely connected to the linearized free energy $F_c(E,T_c) = E - T_cS(E)$

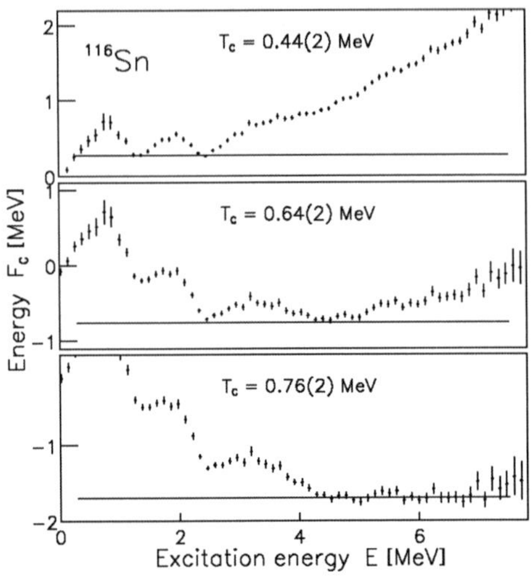


**FIGURE 4.** Preliminary linearized free energy $F_c$ for $^{116}$Sn. The critical temperatures for the breaking of one, two and three or more pairs are found at $T_c = 0.44$, 0.64, and 0.76 MeV, respectively.

giving $F_c(E_1, T_c) = F_c(E_2, T_c)$ for the critical transition.

Figure 4 displays free energies $F_c$ in $^{116}$Sn for certain temperatures $T_c$. In the two upper panels two bumps are seen, which we interpret as the critical temperatures $T_c = 0.44$ and 0.64 MeV for breaking one and two neutron pairs, respectively. When three or more pairs are broken the transition seems to be continuous as function of excitation energy. Thus, we interpret the behavior of $F_C$ above $E = 4$ MeV as a general quenching of the pair correlations at a temperature of $T_c = 0.76$ MeV.

## ACKNOWLEDGMENTS


Financial support from the Norwegian Research Council (NFR) is gratefully acknowledged.


## REFERENCES


1. A. Schiller *et al.*, Nucl. Instr. Methods Phys. Res. A **447**, 498 (2000).
2. M. Guttormsen *et al.*, Nucl. Instr. Methods Phys. Res. **A 374**, 371 (1996).
3. M. Guttormsen *et al.*, Nucl. Instr. Methods Phys. Res. **A 255**, 518 (1987).
4. E. Melby *et al.*, Phys. Rev. **C63**, 044309 (2001).
5. M. Guttormsen *et al.*, Phys. Rev. **C63**, 044301 (2001).
6. J. Lee and J.M. Kosterlitz, Phys. Rev. **B43**, 3265 (1991).

# Coupled Channel Calculations for Reactions with the $^{16}O$+ $^{64}Zn$ System

J. F. P. Huiza[1], E. Crema[1], D. S. Monteiro[1], J. M. B. Shorto[1], R. F. Simões[1], P. R. S. Gomes[2]

[1] *Departamento de Física Nuclear, Instituto de Física da Universidade de São Paulo, SP, Brazil.*
[2] *Instituto de Física, Universidade Federal Fluminense, RJ, Brazil.*

**Abstract.** The experimental quasielastic excitation function has been measured at $\theta_{lab}$=161° for the $^{16}O$+$^{64}Zn$ system and its correspondent quasielastic barrier distributions has been derived. These data were taken at energies near and below the Coulomb barrier. We compared these data with theoretical predictions from coupled channel calculations. Besides, other experimental data available in the literature, for elastic and inelastic scattering and fusion of the same system, were also compared with theoretical calculations. The analysis shows that there is a strong coupling channel effect due to the quadrupole and octupole vibrations of the $^{64}Zn$ target, which affects all reaction channels.

**Keywords:** quasielastic, elastic, inelastic and fusion reaction. Theoretical Coupled Channel Calculations.
**PACS:** 25.70.Bc, 25.70.Jj, 24.10.Eq.

## QUASIELASTIC EXCITACION FUNCTION AND QUASIELASTIC BARRIER DISTRIBUTIONS

Quasielastic barrier distributions [1-3] are obtained by differentiation of the high precision quasielastic excitation function in respect to the energy. Quasielastic processes include elastic and inelastic scattering and transfer channels.

Our experiments were performed at the Pelletron laboratory of the University of São Paulo. The beam energy range covers near and sub-barrier energies. The identification of the quasielastic scattered nuclei was extracted from E-ΔE spectra of a proportional counter + silicon surface barrier detector telescope, placed at $\theta_{LAB}$ = 161°. Monitors were placed at forward angles (±30° and 45°) for normalization purposes. Details of the experiment can be found in ref [3].

The quasielastic excitation function and quasielastic barriers distributions were compared with coupled channel calculations (CCC) using the CCQUEL [4] code. Our theoretical calculations used an optical potential, where its real part is described through the double-folding São Paulo Potential, SPP [5]. As the CCQUEL codes requires potentials with the Woods-Saxon geometry, we performed a Woods-Saxon fit to the parameter-free SPP around the barrier radius (±2 fm), getting the following parameter for the real nuclear potential: $V_0$=127.0 MeV, $r_0$ = 1.06 fm and $a_0$=0.69 fm. The results of the fit are shown in Fig. 1. A very short-range imaginary potential was

CP1012, *Frontiers in Nuclear Structure, Astrophysics, and Reactions: FINUSTAR 2*
edited by P. Demetriou, R. Julin, and S. V. Harissopulos
© 2008 American Institute of Physics 0-7354-0532-5/08/$23.00

used, in order to avoid artificial absorption at the nuclear surface: $V_i$=50.0 MeV, $r_i$ = 0.6 fm and $a_i$=0.4 fm .

Figures 2 and 3 show the derived quasielastic excitation function and barrier distribution, respectively.

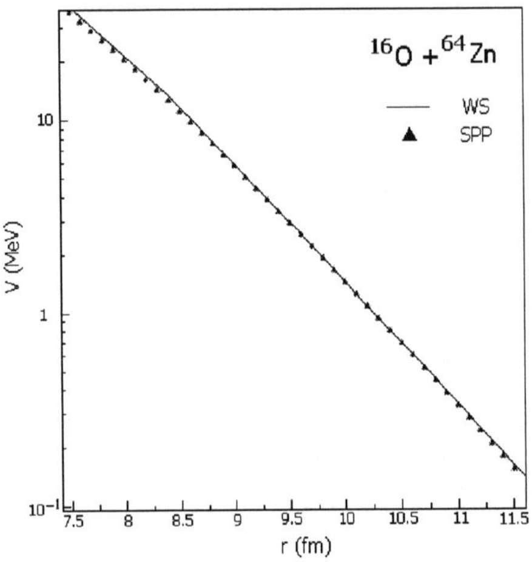


**FIGURE 1.** Fit of Woods-Saxon potential to the parameter-free São Paulo Potential.

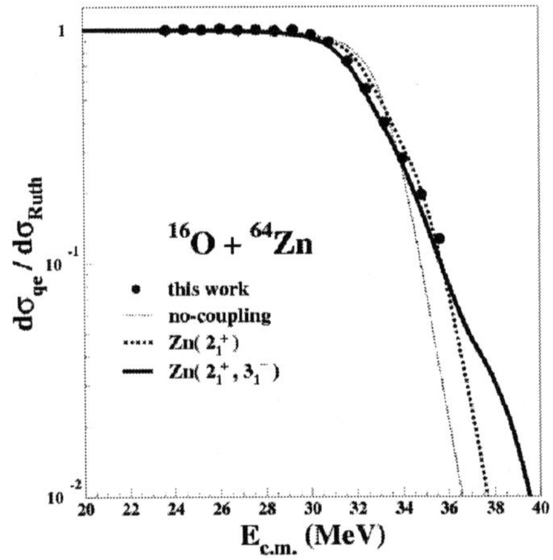


**FIGURE 2.** Experimental quasielastic excitation function, and theoretical predictions obtained by coupling inelastic states of the target.

We performed calculations with several target excited states and the projectile $2_1^+$ and $3_1^-$ states. The most important states in the coupling scheme are the collective quadrupole state $2_1^+$ ($E^*$ = 0.992 MeV, $\beta_2$= 0.311, $r_0$= 1.06 fm) and the collective vibrational state $3_1^-$ ( $E^*$ = 2.998 MeV, $\beta_2$=0.333, $r_0$=1.06 fm). The effects of these two states can be seen in the Fig. 2, where the quasielastic excitation function is shown. The coupling of the $2_1^+$ state enhances the quasielastic cross section significantly at the highest energies. When the $3_1^-$ octupole state is also included in the coupling scheme, an almost exact fit of the experimental data is obtained.

Although the effect of the couplings at energies above the barrier can be easily observed from the excitation function of Fig. 2, this is not true for the energy region close to the barrier. At this energy region, the coupling effects can be better observed from the quasielastic barrier distribution, as shown in Fig. 3. When the $2_1^+$ quadrupole state is coupled, the barrier distribution decreases approximately 25% in comparison with the no-coupling calculation, and approaches the behavior of experimental data. When the additional $3_1^-$ octupole state of the target is included in the coupling scheme, the barrier position is shifted by 1 MeV. With this lowering of the barrier; the calculations reproduce the quasielastic experimental barrier distributions. The additional inclusion of $2_1^+$ and $3_1^-$ excited states of the projectile in the coupling scheme produces only a shift on the quasielastic barrier distribution, as shown in Fig. 4. This is an well-known renormalization effect [6, 7].

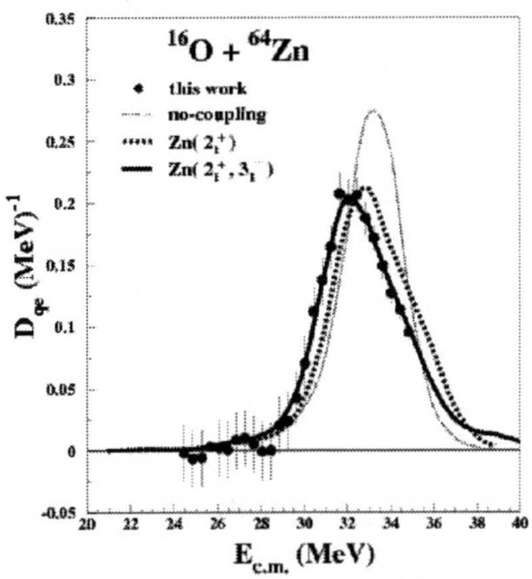


**FIGURE 3.** Experimental quasielastic barrier distribution, and theoretical predictions obtained by coupling inelastic states of the target.

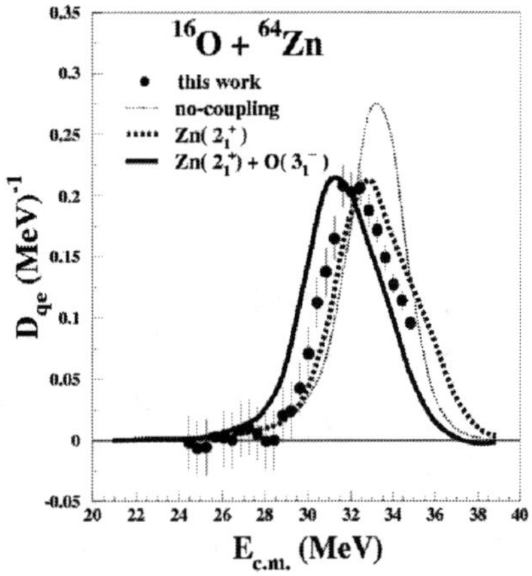


**FIGURE 4.** Experimental quasielastic barrier distribution, and theoretical predictions obtained by coupling inelastic states of the target and projectile.

# ELASTIC ANGULAR DISTRIBUTIONS

Elastic scattering angular distributions from ref [8] were compared to coupled channel calculations using the CCQUEL code. The same values for the real and imaginary potentials used in the previous the quasielastic analysis were assumed. Fig. 5 shows this comparison at 34.0 MeV. Results from the no-coupling calculation are far from the data. When the $2_1^+$ state is coupled, calculations get closer to the data, especially at backward angles. The inclusion of the $3_1^-$ octupole state of the target in the calculations leads to a good agreement with the data. As for the quasielastic scattering, the coupling of these two one-phonon states shifts the elastic scattering angular distribution, corresponding to the lowering of the barrier height, when compared with the no-coupling predictions.

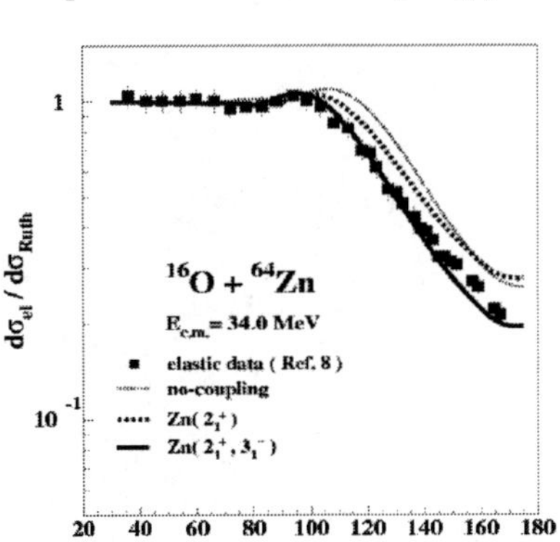


**FIGURE 5.** Elastic scattering angular distribution and results of CCC.

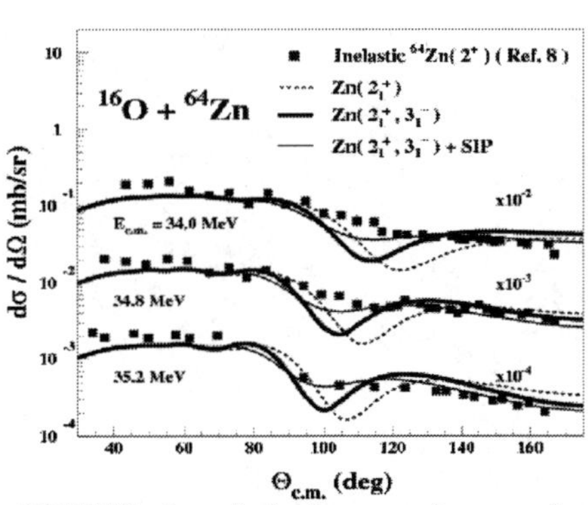


**FIGURE 6.** Inelastic scattering angular distribution of $2_1^+$ state of the target and results of CCC.

## INELASTIC ANGULAR DISTRIBUTIONS AND INELASTIC EXCITATION FUNCTION

Inelastic scattering data for the same system, from ref [8, 9], were compared with coupled channel calculations. Fig. 6 shows three inelastic angular distributions for the $2_1^+$ excited state of the target. In this figure one can observe the theoretical predictions when only the $2_1^+$ quadrupole vibrational state is considered in the coupling scheme, then when two collective states, the $2_1^+$ quadrupole and the $3_1^-$ octupole states, are considered. Calculations which take into account these two vibrational states predict the behavior of the experimental data of the angular distributions at various energies. In Fig. 7 we show data of the excitation function of the $2_1^+$ collective state of target [9]. One can observe a valley at about 31.5 MeV, corresponding to the amplitude of the interference minimum [9]. In this figure it is shown results of calculations when quadrupole and octupole vibrational states of target are coupled. A good description of the data is obtained.

We could not include other inelastic channels in the coupled channel calculations, due to the limitations of the CCQUEL code. Therefore, in order to simulate the inclusion of other inelastic channels, we extended the previous imaginary potential from the volume region to the surface region. The values of the parameters of this long range imaginary potential that give a good description of the inelastic angular distributions and inelastic excitation functions are $V_i$=105 MeV, $r_i$=0.96 fm and $a_i$=0.67 fm. The effects of these hidden coupled channels are shown in Figs. 6 and 7 by the thin solid line denoted by SIP (surface imaginary potential). The predictions of these coupled channel calculations are in good agreement with the data.

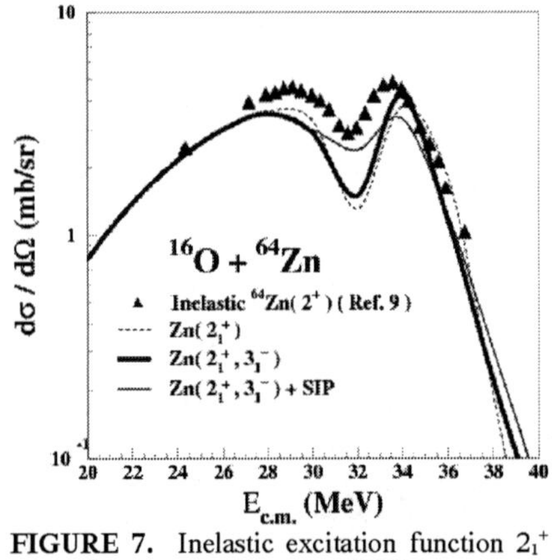


**FIGURE 7.** Inelastic excitation function $2_1^+$ state of the target and results of CCC.

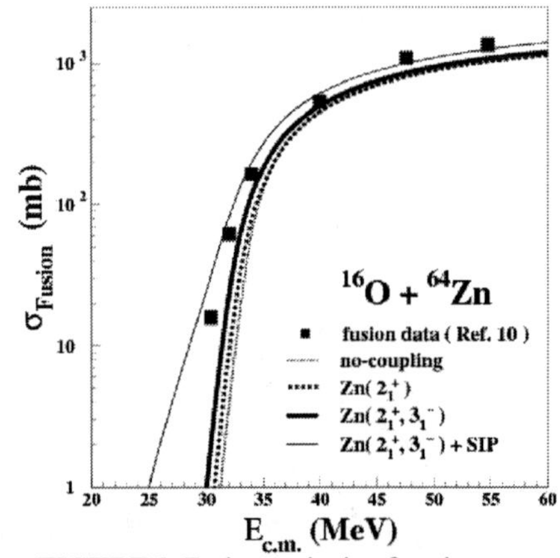


**FIGURE 8.** Fusion excitation function.

## FUSION

Finally, we compared the available data for fusion cross sections for this system [10] with coupled channel calculations, similar to those performed for quasielastic, elastic and inelastic scattering. Results are shown in Fig. 8. The calculations underestimate the data, especially at sub-barrier energies. Only when we include additional channels, by the artificially introduced surface imaginary potential (SIP), we may be able to describe reasonably well the behavior of the data.

## CONCLUSIONS

In this work we have presented results from the measurement of a precise quasielastic excitation function, which was used to derive the quasielastic barrier distribution for the $^{16}O$ +$^{64}Zn$ system, at energies near the Coulomb barrier. Our data and other data available in the literature for this system were compared with predictions from coupled channel calculations which used the parameter-free double folding São Paulo Potential as the real bare nuclear potential between the projectile and the target. A comprehensive coupled channel analysis for several reactions processes (quasielastic, elastic, inelastic and fusion) shows that the most important inelastic channels in the coupling scheme are the collective quadrupole and octupole vibrational states of the target. This means that $^{64}Zn$ has a strong deformation during the interaction projectile-target. Finally, the quasielastic barrier distribution has proved to be a powerful tool to describe internal detail analysis of reaction mechanisms at energies close to the Coulomb barrier.

## ACKNOWLEDGMENTS

This work was partially supported by the Fundação de Amparo à Pesquisa do Estado de São Paulo (FAPESP), the Conselho Nacional de Desenvolvimento Cientifico e Tecnológico (CNPQ) and the Fundação de Amparo à Pesquisa do Estado do Rio de Janeiro (FAPERJ).

## REFERENCES

1. R.F. Simões, D.S. Monteiro, J.M.B. Shorto, E. Crema, Phys Lett. **B 527**, 187 (2002).
2. H. Timmers et al., Nucl. Phys. **A 584,** 190 (1995) ; Nucl. Phys. **A 633,** 421 (1998).
3. J. F. P. Huiza, E. Crema, D.S. Monteiro, J.M.B. Shorto, R.F. Simões, N. Added and P.R.S. Gomes, Phys. Rev. **C 75,** 064601 (2007).
4. K. Hagino and N. Rowley (unpublished).
5. L. C. Chamon et al., Phys. Rev. **C 66,** 014610/1-13 (2002).
6. C. R. Morton et al., Phys. Rev. **C 60,** 044608/1-11 (1996).
7. J. R. Leigh et al., Phys. Rev. **C 52,** 3152 (1995).
8. C. Tenreiro et al., Phys. Rev. **C 53,** 2870 (1996).
9. C S. Salem-Vasconcelos et al.., Phys. Rev. **C 50,** 927 (1994).
10. C. Tenreiro, PhD thesis, IFUSP, SP, Brasil, 1987.

# First Evidence of di-proton decay from $^{18}$Ne excited states

C. Sfienti[*,†], M. De Napoli[†], G. Raciti[†], E. Rapisarda[†], G. Cardella[†], F. Amorini[†] and F. Giacoppo[†]

[*]*Gesellschaft für Schwerionenforchung, D-64291 Darmstadt, Germany*
[†]*Dipartimento di Fisica, Universitá degli studi di Catania and INFN-Sezione di Catania, 64, Via S. Sofia, I-95123 Catania, Italy*

**Abstract.** Two-proton radioactivity studies have been performed on excited states of $^{18}$Ne produced by $^{20}$Ne fragmentation at the FRIBS facility of the Laboratori Nazionali del Sud and excited via Coulomb excitation on a $^{209}$Pb target. Decay of $^{18}$Ne levels has been studied by complete kinematical reconstruction. Present results shows unambiguously that the 6.15 MeV $^{18}$Ne state two-proton decay proceeds through a true $^{2}$He resonance (diproton).

**Keywords:** Reaction induced by unstable nuclei, Coulomb excitation, Decay by proton emission, Properties of nuclei-nuclear energy levels, Charged-particle spectroscopy
**PACS:** 25.60.-t, 23.50.+z, 24.30.-v, 25.40.Ny

## INTRODUCTION

One of the most exciting new phenomena at the proton drip-line is the possible occurrence of two-proton decay, which has been predicted about 40 years ago [1]. This decay mode is in competition with the sequential 2p decay, where the two protons are emitted sequentially via an intermediate state. The two-proton decay may, in turn, proceed via an uncorrelated (*democratic*) emission of the two protons, or a correlated (*diproton*) emission, where a $^{2}$He cluster is emitted.
No firm evidence exists today for the observation of $^{2}$He emission. In the case of the ground state and first excited state of $^{6}$Be [2], as well as for the ground state of $^{12}$O [3], sequential two-proton emission through the tail of an intermediate state was observed. Much better cases for a correlated emission to take place are expected in the medium mass region (A=40-60), where the Coulomb barrier of nuclei is strong enough to ensure longer half-lives. Indeed the first observation of two-proton decay came from $^{45}$Fe at ground state [4] and from a metastable state in $^{94}$Ag [5]. Simultaneous emission of two protons has also been observed from the decay of the 6.15 MeV, $I^{\pi}=1^{-}$ level in $^{18}$Ne, produced in the fusion reaction $^{17}$F+p [6]. However, the energy and angular distribution data do not distinguish between *diproton* and *democratic* emission.
Therefore, in order to disentangle between these two decay mechanisms an experiment at the Laboratori Nazionali del Sud (LNS) was performed using a $^{18}$Ne beam at about 30 AMeV, produced with the FRIBs facility [7, 8]. The $^{18}$Ne levels were populated by Coulomb excitation on a $^{209}$Pb target and their decay has been studied by complete kinematical reconstruction.

CP1012, *Frontiers in Nuclear Structure, Astrophysics, and Reactions: FINUSTAR 2*
edited by P. Demetriou, R. Julin, and S. V. Harissopulos
© 2008 American Institute of Physics 0-7354-0532-5/08/$23.00

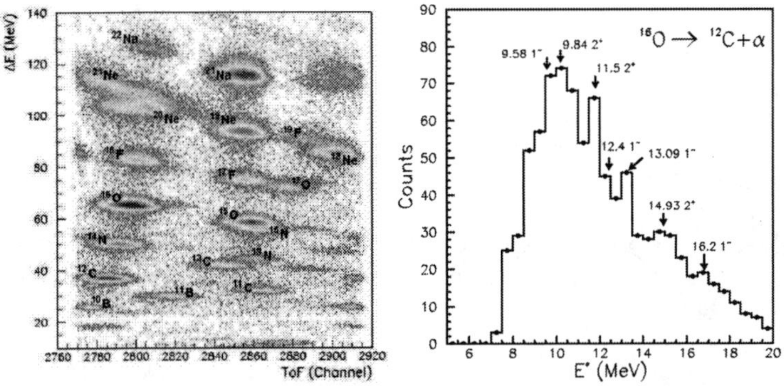


**FIGURE 1.** Left: Energy-loss in the Si-Strip *tagging* detector versus ToF. Right: Experimental excitation energy spectrum of $^{16}O$ from the $^{12}C+\alpha$ events

# EXPERIMENTAL SETUP AND KINEMATIC RECONSTRUCTION PROCEDURE

The secondary beam has been produced by the fragmentation of a primary stable $^{20}Ne$ beam at 45 AMeV delivered by the Superconducting Cyclotron (SC) on a $^{9}Be$, 500 $\mu$m thick, production target. The secondary ions have been separated in-flight in the fragment separator of the LNS (Ref. [7, 8]) operated at a B$\rho$ setting optimized for $^{18}Ne$. A primary current of 300 enA produced a total RIBs rate of $10^5$ ions/sec at the exit of the fragment separator. The secondary beam has been transported with a 60% of transmission up to the scattering chamber [7, 8]. The $^{18}Ne$ rate was about 9% of the total RIBs mixture rate, i.e. about $5.4\times10^3$ pps.
The measurement of energy, charge, mass and position of each incoming radioactive ion before impinging on the secondary target was derived, event-by-event, from the energy loss measured by a 16×16 X-Y Si-Strip detector 5×5 cm$^2$ of area and 300$\mu$m thick [7] and the time-of-flight measured with respect to the radio frequency (RF) signal provided by the SC (Fig .1). The $^{18}Ne$ incident energy on the $^{209}Pb$ target (200$\mu$m thick) was in the range of 32-35 AMeV as evaluated from the measured energy loss in the Si-Strip.

The detection system, setup at about 80 cm from the interaction target, consisted on two Si-CsI hodoscopes: the first composed by 81 two-fold telescopes ( 300 $\mu$m Silicon detectors 1×1 cm$^2$ of active area followed by a 10 cm long CsI(Tl)) covered the angular range $\theta_{lab} = \pm5°$. The second one consisted on 89 three-fold telescopes (50 $\mu$m + 300 $\mu$m Silicon detectors 3×3 cm$^2$ active areas followed by a 6 cm long CsI(Tl)) covering the angular range between ±5° and ±21.5°. This setup allowed the measurement, event-by-event, of the energies and angles of all outgoing decay particles.
The projectiles have been selected by gating on the ΔE-ToF plot provided by the Si-Strip *tagging* detector (see Fig. 1 - left panel) and then a complete reconstruction of the decay kinematics in the center-of-mass system (CM) was performed. From the velocities and angles of the decay products measured in the hodoscopes, the CM velocity, i.e. the velocity of the decaying nucleus, was determined. From energy conservation, the CM energy, including Q-value and energy loss in the target, must be almost equal to the pro-

jectile energy, reconstructed from the ΔE measured in the Si-Strip. Through an iterative procedure the interaction point in the target could then be determined [9]. Therefore it was possible to reconstruct, event-by-event, the invariant mass in the CM system and thus the excitation energy spectrum. Background produced by reactions on the Si-Strip tagging detector has been measured in empty target runs and subtracted from the data.
The whole procedure was checked for the well-known $^{16}$O stable nucleus. Figure 1 (right panel) shows the excitation energy spectrum as determined from the fully measured $^{12}$C+$\alpha$ events produced by the $^{16}$O* decay. The well-known $1^-$ and $2^+$ excited levels of $^{16}$O [10] are recognized despite the poor resolution. The experimental resolution is about 500 keV, mainly dominated by the error in the determination of the depth of the interaction point in the thick Pb target. The spectrum presents only the $1^-$ and $2^+$ levels as expected since only these levels can be populated via Coulomb excitation.
The measured efficiency corrected cross section of the 13.05 MeV state is about 0.87±0.22mb in good agreement with the recent predictions on Coulomb excitation cross section of Ref. [11]

## $^{18}$NE TWO-PROTON DECAY

The same procedure was then applied to the fully measured $^{17}$F+p and $^{16}$O+2p events produced by the selected $^{18}$Ne secondary beam. The excitation energy spectrum of the $^{17}$F+p events is shown in figure 2 (top-left panel). The peaks corresponding to the decay of the 5.09 MeV($2^+$), 5.15 MeV($2^+$) and 6.15 MeV($1^-$) excited states [10] are recognized although partially superimposed.

The excitation energy spectrum corresponding to the $^{16}$O+2p events is plot on the top of the previous in the same figure 2. The presence of the 6.15 MeV peak confirms the evidence for simultaneous two-proton decay of this $^{18}$Ne state reported in Ref. [6]. For this level, the sequential 2p decay is indeed energetically forbidden. Whether the presence of two protons in the final state of the known 7.06 MeV($1^-$,$2^+$), 7.91 MeV($1^-$,$2^+$) and 8.5 MeV ($1^-$,$2^+$) levels indicate a two-proton decay mechanism, deserve further investigations since, for these levels, the sequential 2p decay channel is available. Moreover, together with the known states, the Coulomb excitation allowed to populate high-lying levels around 10.7 MeV, 11.2 MeV, 12.5 MeV, 13.7MeV and 14.5 MeV still unknown and promising possible future candidates for two-proton radioactivity. Recent study of the decay of excited states of $^{17}$Ne, performed at GANIL, seem indeed to indicate that one or several high-lying states could decay by a correlated 2p emission, despite the fact that the sequential branch is widely open [12].
In order to understand whether the two-proton decay of the 6.15 MeV level is *diproton* or *democratic*, the momentum, energy and angular correlation between the two protons have bee studied. The relative momentum correlation of the two decaying protons shows a resonance at $\approx$ 20 MeV/c corresponding to the decay of the unbound $^{2}$He [13] (Fig. 2 top-right panel). Therefore, in order to exclusively select the excitation energy spectrum reconstructed by the $^{16}$O+2p events we have applied, event-by-event, a cut in the relative momentum of the detected protons pair in coincidence with an $^{16}$O. Figure 2 (bottom panel) shows the comparison between the inclusive and exclusive spectra respectively. The described cut on the events clearly enhance the contribution of the

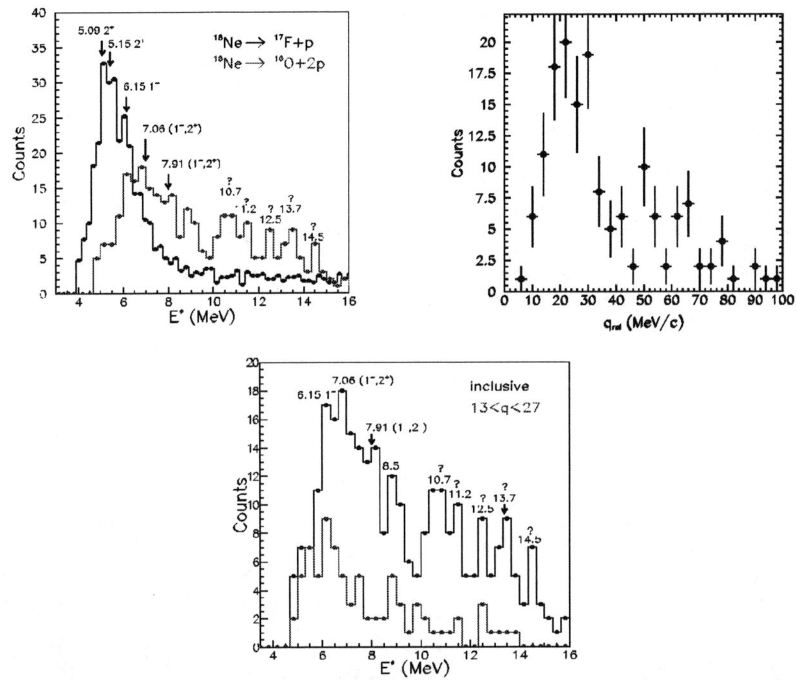


**FIGURE 2.** Top-Left: Comparison of the excitation energy spectra corresponding to the decay of $^{18}$Ne in $^{17}$F+p and $^{16}$O+2p respectively. Top-Right: relative momentum of the two decaying protons of the $^{16}$O+2p channel. Bottom: Inclusive excitation energy spectrum for the $^{18}$Ne decay in $^{16}$O+2p. On the same plot also the corresponding spectrum obtained applying a cut in the relative momentum of the two proton (corresponding to the $^{2}$He resonance) is also shown.

6.15 MeV state with respect to the other states. Figure 3 shows the relative energy and the relative CM angle of the two protons for events selected inside the excitation energy window 5.9≤(E*)≤6.5 MeV. Data are compared to Monte Carlo simulations, filtered for the geometry and detector constrains, assuming $^{2}$He emission (solid line) and three-body decay neglecting final state interaction (dashed line). As shown by the simulations, the two possible mechanism for simultaneous two-proton emission lead to different energy and angular correlation between the two protons, provided the correlations are studied over a large enough angular range [6]. Despite the poor statistic, both the relative energy and angle indicate the presence of events where the two protons are emitted uncorrelated and in the form of a $^{2}$He. To extract the branching ratios the simulation curves, normalized to the data, have been integrated: the $^{2}$He emission is about 40% of the total two-proton decay while the three-body emission is about 60%, in good agreement with the ratio (49%) between the inclusive and selected yields of the excitation energy spectra of figure 2 (bottom panel).

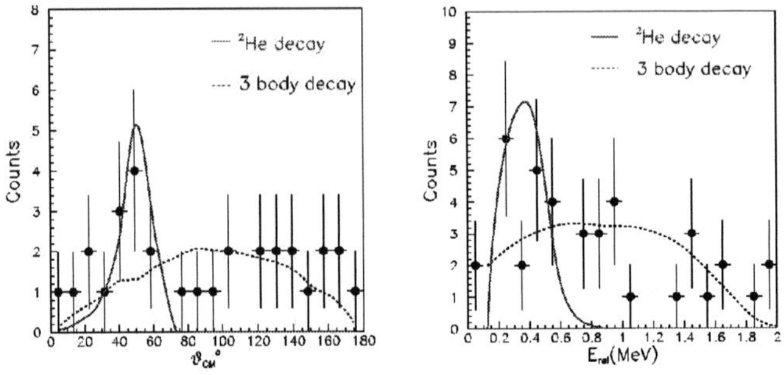


**FIGURE 3.** CM relative angle (left) and relative energy (right) of the two protons emitted from the 6.15 MeV level, compared to Monte Carlo simulations.

## CONCLUSIONS

Recently, an experiment to search for and study two-protons emission from excited states of $^{18}$Ne was performed at the Laboratori Nazionali del Sud. The beam was produced by $^{20}$Ne fragmentation on a $^{9}$Be target at 45 AMeV at the LNS-FRIBS facility. The $^{18}$Ne levels were populated by Coulomb excitation on a $^{209}$Pb target. Several levels were identified in the excitation energy spectrum built by kinematical reconstruction from the $^{17}$F+p and $^{16}$O+2p fully measured events. The presence of the 6.15 MeV peak in the $^{16}$O+2p energy spectrum confirms the observed two-protons decay of such level. In particular, the relative energy and angle of the two protons, analysed in the excitation energy window $5.9 \leq (E^*) \leq 6.5$ MeV, indicate the presence of both *diproton* (40%) and *democratic* (60%) decay mechanisms. However, for a precise measure of the branching ratio more statistics is needed. Finally, the Coulomb excitation allowed to populate high-lying known and unknown states in $^{18}$Ne. Whether the presence of two protons in the decay of such states is the result of a simultaneous two-protons emission is still under investigation.

## REFERENCES

1. V.I. Goldansky, *Nucl. Phys.* **19**, 482 (1960).
2. D.F. Geesaman et al., *Phys. Rev. C* **15**, 1835 (1977).
3. R.A. Kryger et al., *Phys. Rev. Lett.* **74**, 860 (1995).
4. J. Giovinazzo et al, *Phys. Rev. Lett.* **89**, 102501 (2002).
5. M. I. Mukha et al., *Nature* **439**, 298 (2006).
6. J. Gomez del Campo et al, *Phys. Rev. Lett.* **86**, 43 (2001).
7. G. Raciti et al., in *Proc. of the 10th International Conference on Nuclear Reaction Mechanisms, Varenna* 11 (2003).
8. E. Rapisarda et al., *Eur. Phys. Jour. A* to be published.
9. G. Cardella et al., in *Proc. of the International Conference on Proton Emitting Nuclei*, (2007), to be published.
10. www.nndc.bnl.gov.
11. C.A. Bertulani et al., *Phys. Lett. B* **650**, 233 (2007).
12. B. Blank, *Nucl. Phys. A* **734**, 303 (2004).
13. D.H. Boal, C.K. Gelbke and B.K. Jenning, *Rev. Mod. Phys.* **62**, 3 (1990).

# Laser Spectroscopy at On-Line Isotope Separators

J. Billowes

*School of Physics and Astronomy, Schuster Building, University of Manchester M13 9PL, UK*

**Abstract.** Optical hyperfine structures and isotope shifts of radioactive atoms and ions can be measured at ISOL facilities with high precision and sensitivity using a variety of laser spectroscopic techniques. The analysis of these structures and shifts provides information on some of the most fundamental properties of nuclear ground states and isomeric states, namely the magnetic dipole and electric quadrupole moments, the nuclear spin, and the difference in the nuclear mean square charge radius between isotopes. The status of current experimental programmes at ISOL facilities will be reviewed. The techniques employed include in-source, collinear-beams, atom-trap, and cooler-buncher laser spectroscopy.

Keywords: Charge distribution, Spin, Electromagnetic moments, Fast radioactive beam techniques.
**PACS:** 21.10.Ft, 21.10.Hw, 21.10.Ky, 29.38.Db

## INTRODUCTION

This is a review of current research programmes at on-line isotope separator facilities that use laser spectroscopic techniques to measure properties of the nucleus. I will start with a brief reminder of how these nuclear properties are deduced from the atomic spectroscopy. I will start the experimental review at the ISOLDE facility, CERN where there are a number of programmes running, including the highly-sensitive in-source laser spectroscopy of heavy, neutron-deficient isotopes of lead, bismuth and polonium. Recent collinear-beams experiments at the facility include studies of the copper isotope chain and ground state properties of neutron-rich Mg isotopes in the "island of inversion". I will say a few words about on-line atom traps, although applications are generally for fundamental physics rather than nuclear structure. I will however dwell on the $^6$He and $^8$He charge radius measurements. I finish with a description of recent experiments using the ion-cooler at Jyväskylä and this will bring me full-circle back to ISOLDE where a new ion cooler, ISCOOL, will be installed next month.

## NUCLEAR PROPERTIES FROM LASER SPECTROSCOPY

If one compares the frequency of an optical transition for two even isotopes of the same element, a small part-per-million shift may be detected called the isotope shift [1]. The analysis of this shift can yield the difference in the nuclear mean square

CP1012, *Frontiers in Nuclear Structure, Astrophysics, and Reactions: FINUSTAR 2*
edited by P. Demetriou, R. Julin, and S. V. Harissopulos
© 2008 American Institute of Physics 0-7354-0532-5/08/$23.00

charge (msc) radii between the two isotopes. The msc radius is controlled by a number of nuclear factors such as volume, static and dynamic deformations, and surface diffuseness. If the nucleus has non-zero spin then the atomic states will have a hyperfine structure with splittings comparable to the magnitude of the isotopes shift. Analysis can yield the nuclear spin, the magnetic dipole moment and the spectroscopic quadrupole moment. The isotope shift is found using the centroids of the hyperfine multiplets. Table 1 shows the relative contributions to the isotope shift from the mass shift and the volume shift. The latter contribution contains all the nuclear information - it is directly proportional to the change in the nuclear msc radius.

**TABLE 1.** Approximate magnitudes of the contributions to the isotope shift for a pair of isotopes differing in atomic mass by $\Delta A=2$. This is compared in the last column with the room-temperature Doppler broadening for the optical transition.

| Element | Transition | Normal Shift | Specific Shift | Volume Shift | Doppler Width |
|---|---|---|---|---|---|
| Na (Z=11) | 3s – 3p | 550 MHz | 200 MHz | -10 MHz | 1400 MHz |
| Yb (Z=70) | 6s – 6p | 20 MHz | <20 MHz | -1500 MHz | 500 MHz |

It is seen that for heavy elements the isotpe shift is dominated by the volume shift and for such isotopes useful information is obtained even from Doppler-broadened spectra. For light elements sub-Doppler techniques are necessary and the quality of the nuclear information also depends on a reliable evaluation of the specific mass shift.

The two main factors controlling the change in msc radius between isotopes are the nuclear volume and shape. If a neutron is removed from a spherical nucleus, its effect on the nuclear (charge) volume is easily detected in the isotope shift. An example is the lead isotope chain in figure 1 where the msc radii are approximately linearly dependent on the nucleon number. The change in gradient at the neutron shell closure (N=126) is a feature seen in all isotope chains at the N=28, 50, 82, 126 shells. If the removal of neutrons also leads to a change in nuclear deformation then the effect on the msc radius can be an order of magnitude larger than the volume effect. The light mercury isotopes in figure 1 display the well-known alternation in nuclear shape between the weakly-oblate even isotopes and the normal-prolate odd isotopes.

## THE EXPERIMENTAL PROGRAMME AT ISOLDE

### In-Source Laser Spectroscopy

Measurements on neutron-deficient isotopes of lead [2], bismuth and polonium [3] have been carried out by a Leuven-led collaboration using the RILIS [4] pulsed laser facility. Although this technique suffers from the full Doppler broadening of the 2000°C ion source, it combines laser resonance ionization with nuclear detection to achieve extremely high sensitivity which has allowed measurements in the lead chain to be extended well below the N=104 mid-shell point on the neutron-deficient side. The lightest isotope measured, $^{182}$Pb, has a half-life of 55 ms and gave a signal rate of one ion detected per second. The laser ionization efficiency in these measurements was 1%. The isotopes were produced by bombarding a uranium carbide target with a 1.4 GeV proton beam. The radioactive products effuse and diffuse from the target to a hot tube into which the RILIS laser beams are focused. The atoms are step-wise resonantly excited and ionized by the pulsed lasers. The ions are extracted, accelerated, mass-analyzed and detected through their nuclear decay characteristics (either $\alpha$ or $\beta$-$\gamma$ in the present work). The ion detection rate is monitored as one of the lasers is scanned in frequency to locate the resonances for each isotope.

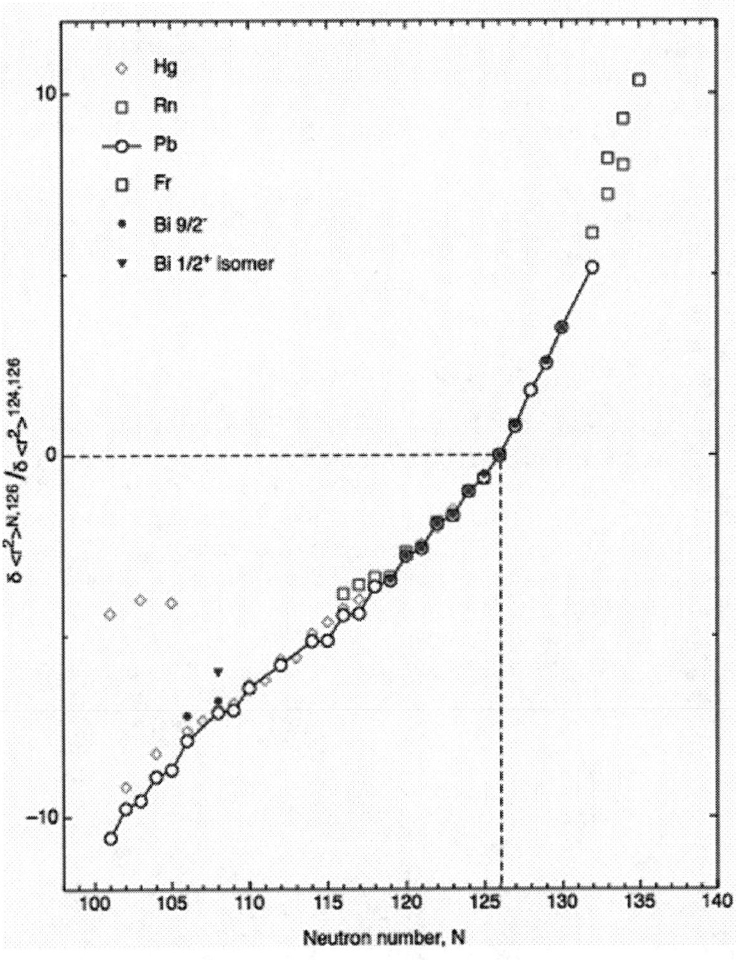


**FIGURE 1.** Isotope shifts for some isotope chains in the lead region. The shifts, which are approximately proportional to the change in mean square charge radii, have been normalized to the N=124,126 shift in each chain in order to compare the features [5].

The relative msc radii for lead isotopes (figure 1) show only a small and smooth deviation above the general downward trend of the heavier isotopes immediately below N=126. The authors [2] conclude that the lead ground states stay essentially spherical in the mid-shell region even though there are known 2p-2h oblate and multi-particle-hole prolate structures at low energy in this region. The behaviour of the lead isotopes is directly compared with that of the light mercury isotopes in figure 1 by normalizing to the N=124,126 isotope shift [5]. It is evident that the lead isotopes fall below even the weak-oblate even mercury isotopes in the mid-shell region, consistent with the conclusions of ref [2].

The same collaboration has made a pilot experiment on $^{189,191,191m}Bi$ isotopes and these results are also shown in figure 1 using the same normalization procedure. Here a slight increase is seen in the msc radii compared to lead (particularly for the $^{191m}Bi$ isomer) which hints at some developing weak deformation. The collaboration's data for the most recent measurements on polonium isotopes with A=193-200, 202, 204 is still under analysis but the indications are that a significant departure to more deformed shape is seen below $^{200}Po$. The spectroscopic quadrupole moments as deduced from hyperfine structures of the odd (I=3/2) Po isotopes are consistent with this observation, although the structures are not fully resolved due to Doppler-broadening and band-width of the frequency-scanned pulsed laser.

## Collinear-Beams Laser Spectroscopy

The collinear-beams method of laser spectroscopy approaches Doppler-free resolution but does not have the extreme sensitivity of the in-source technique. A comparison of the resonance line-shapes for the two methods is shown in figure 2 for the medium-mass nuclide, $^{68}Cu$. The in-source data can still provide useful isotope shift and magnetic moment information, but for improved accuracy and, crucially, for the extraction of the quadrupole moment from the hyperfine splittings the high resolution of the collinear-beams method is required.

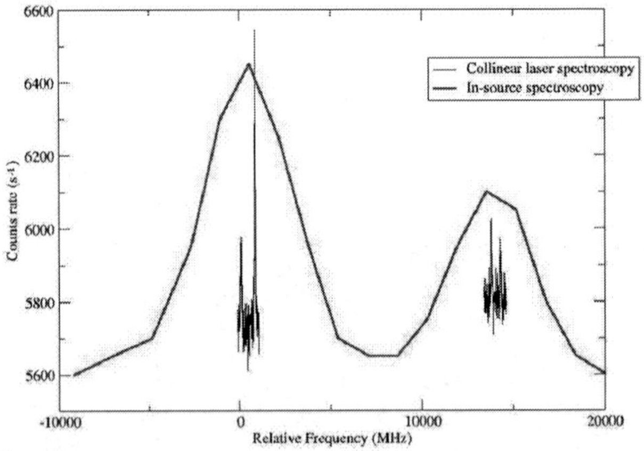


**FIGURE 2.** Comparison of resonance spectra for $^{68}Cu$ [6] obtained with in-source laser spectroscopy and high-resolution collinear-beams spectroscopy.

Measurements on the copper (Z=29) isotopes from A=63 to 72 have recently been carried out in the last two years on the COLLAPS line [7] on the $^2S_{1/2} - {}^2P_{3/2}$ (324 nm) transition in the neutral atom. This isotope chain lies in the middle of the region between Ti (Z=22) and Kr (Z=36) where there has been no high-resolution laser study of any radioisotope. Although the copper data is still under analysis, the neutron-rich isotopes should reveal features of the sub-shell closure at N=40. For example, there appears to be a minimum in the quadrupole moments of the I=3/2 ground states at N=40.

The same beam-line has been used for measurements on the neutron-rich $^{31,33}Mg$ isotopes [8,9] in the so-called island of inversion. The collinear laser was circularly polarized and optically pumped the $Mg^+$ ions on the $D_2$ line ($^2S_{1/2} - {}^2P_{3/2}$) at 280 nm. The hyperfine coupling transfers the polarization to the nucleus and this was detected as a function of laser frequency by a β-asymmetry measurement after the ions were implanted downstream in a cubic crystal. For each isotope the nuclear *g*-factor was deduced from the NMR destruction of the asymmetry signal. A second measurement of the hyperfine splitting of the $^2S_{1/2}$ ionic ground state gives an experimental value for the product $g(I+1/2)$. In this way the nuclear spins for both $^{31,33}Mg$ were firmly established as 1/2 and 3/2 respectively. Furthermore the value of the negative magnetic moment for $^{33}Mg$ is consistent with an odd-neutron occupation of the negative parity 3/2[321] Nilsson orbital. These results were quite unexpected at the time but are clear evidence of the large prolate deformation these isotopes have which allows the *fp* shell orbitals to intrude in the *sd* shell.

## LASER SPECTROSCOPY IN ON-LINE ATOM TRAPS

There are a number of groups using on-line atom traps for fundamental physics studies. These traps use three counter-propagating pairs of orthogonal laser beams to confine and cool atoms in a magnetic trap. The loading of a trap can be very inefficient but once caught, optical measurements are possible on single atoms. The TRINAT trap at TRIUMF has been used to measure β-ν correlations looking for physics beyond the Standard Model and Stony Brook has made a number of studies on the atomic structure of francium in preparation for parity non-conservation studies. Relevant to the topic of this review is the measurement of the nuclear msc radii of $^6He$ [10] and $^8He$ [11] by a group at Argonne National Laboratory. The $^6He$ measurement was a *tour de force* since the trapping lasers were operating from a metastable atomic state and the trapping efficiency was about $10^{-8}$. For such a light system the $^4He - {}^6He$ mass shift is enormous (43,194.772(56) GHz compared to a field shift of -1.430(58) MHz) and indeed the largest contribution to the error on the msc radius comes from the error from the $^6He$ atomic mass used in the mass shift correction. The rms radius of $^6He$ is 2.054(14) fm compared to 1.673 fm for $^4He$. This immediately indicates the displacement of the nuclear charge (of the α-core) from the centre of mass of the nucleus and this is directly related to the correlation in position of the two valence neutrons. The $^8He$ measurement was carried out at GANIL earlier this year [11] and is now under analysis.

## LASER SPECTROSCOPY WITH AN ION-BEAM COOLER

The Manchester-Birmingham-Jyväskylä collaboration has been using the unique combination of an IGISOL ion source and ion beam cooler-buncher to measure light neutron-rich fission fragments around Z=40. The advantages of an IGISOL are its chemical insensitivity (it will provide beams of any element) and its fast extraction time (1 ms). The ion cooler has allowed the group to use a sensitive method of bunched-beam spectroscopy in situations where the ion beam flux may be only a few hundred ions per second. A major problem has been encountered with recent work on the yttrium isotopes [12] where firm nuclear spin determinations have not been possible on the A=98m, 100, 102 isotopes. Without the nuclear spin, neither the static moments nor the isotope shift can be extracted from the measured hyperfine structures. A solution is to measure on a second transition with different lower or upper atomic spin values. Since the $Y^+$ ion ground state is J=0 all upper states accessible by laser will be J=1 and so a measurement from a metastable state is required. This has been achieved by preparing the ion beam in a metastable state by optical pumping the $Y^+$ ions while they are in the ion cooler. Figure 3 shows the signal enhancement when measuring on the 322 nm line from the metastable $^3D_2$ (1045 cm$^{-1}$) state when light at 363 nm illuminates the ions in the cooler. The increase in signal has been just sufficient for the spin determination of $^{100}$Y.

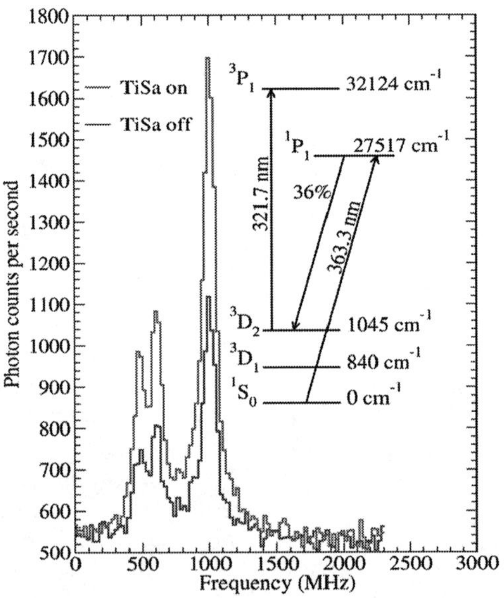


**FIGURE 3.** The spectra show the increase in laser resonance fluorescence signal for the 321.7 nm transition when the yttrium ions in the cooler are pumped on the 363.3 nm transition.

The in-cooler pumping technique opens the way for selecting a measurement transition for its strength or particular hyperfine structure. It will also allow measurements to be made on elements without accessible ground state transitions.

Other applications include the production of spin-polarized nuclear beams by pumping with circularly-polarized laser light.

As a final comment, next month a similar ion cooler, ISCOOL [13], will be installed in the ISOLDE hall and be available for collinear laser spectroscopic measurements on the considerably more intense ion beams provided by that facility.

## REFERENCES


1. J. Billowes and P. Campbell, *J. Phys.* G **21,** 707 (1995).
2. H. De Witte *et al.*, *Phys. Rev. Lett.* **98,** 112502 (2007).
3. T.E. Cocolios *et al.*, To be published.
4. V. Fedosseev *et al.*, *Hypf. Int.* **162,** 15 (2005).
5. B.A. Marsh, Ph.D. Thesis, University of Manchester (unpublished) 2007.
6. K.T. Flanagan, Private Communication.
7. http://is389-proj-collaps.web.cern.ch/is389-proj-collaps/Experiments.htm
8. G. Neyens *et al., Phys. Rev. Lett.* **94,** 022501 (2005).
9. D. Yordanov *et al.*, Accepted by *Phys. Rev. Lett.* (2007).
10. L.-B. Wang *et al.*, *Phys. Rev. Lett.* **93,** 142501 (2004).
11. P. Mueller, Private Communication.
12. B. Cheal *et al., Phys. Lett.* B **645,** 133 (2007).
13. H. Frånberg *et al.*, Proceedings of EMIS 2007, XV$^{th}$ Int. Conf. on Electromagnetic Isotope Separators and Techniques Related to their Application.

# Isomer Tagging with a Dual Multi-Wire Proportional Counter & a Differential Plunger.

D.M. Cullen*, P.J.R Mason*, S.V. Rigby*,†, S. Khan*, A.M. Kishada*, B.J. Varley*, C. Scholey**, T. Grahn**,†, P. Greenlees**, P. Rahkila**, P.M. Jones**, R. Julin**, S. Juutinen**, M. Leino**, A.P. Leppänen**, P. Nieminen**,‡, M. Nyman**, J. Pakarinen**,† and J. Uusitalo**

*School of Physics and Astronomy, University of Manchester, Manchester M13 9PL, UK.
†Oliver Lodge, Lab., University of Liverpool, Liverpool, L69 7ZE, UK.
**Department of Physics, University of Jyväskylä, Jyväskylä, FIN-40014, Finland.
‡Department of Nuclear Physics, Australian National University, Canberra, ACT 0200, Australia.

**Abstract.** This report details the status of an experimental research programme which has studied isomeric states in the mass 130–160 region of the nuclear chart. Several new isomers have been established and characterised near the proton drip line using a recoil isomer tagging technique at the University of Jyväskylä, Finland. The latest experiments have been performed with a modified setup where the standard GREAT focal-plane double-sided silicon-strip detector was changed to a dual multi-wire proportional-counter arrangement. This new setup has improved capability for short-lived isomer studies where large focal-plane rates can be tolerated. The results of key recent experiments for nuclei situated *above* ($^{153}$Yb, $^{152}$Tm) and *below* ($^{144}$Ho, $^{142}$Tb) the N=82 shell gap were presented. These studies have charted the evolution of isomeric states across the neutron shell from K-Isomers at N=74, to shape isomers at N=77 and shell-model isomers at N=82, 83. The excitation energies for some of the lowest-lying excited states in these isomeric nuclei show behaviour which is characteristic of an X(5) symmetry falling midway between the limits expected for pure vibrational and rotational behaviour. The future prospects for studies of these nuclei were discussed using an isomer-tagged differential-plunger setup. This technique will be capable of establishing the deformation of the states above the isomers and will aid in determining whether their behaviour is indeed well described by the X(5) symmetry limit.

**Keywords:** Gamma-ray arrays, RITU, Multi-Wire Proportional counters, Recoil-Isomer tagging, Mass 130–160 proton drip-line region, K-Isomers, Isomeric States, X(5) Symmetry.
**PACS:** PACS: 21.10.Re, 23.20.-g, 23.20.Lv, 27.60.+j.

Many isomeric states are known to exist in the mass 130–160 region of the nuclear chart. Over the past ≈10 years, this collaboration has identified and studied many of these nuclei using recoil-isomer tagging at the University of Jyväskylä [1, 2, 3, 4, 5, 6, 7, 8]. Figure 1 shows the region of interest and the individual nuclei studied to date. The motivation for this work was to chart the evolution of isomeric states from N=73 to N=84 across the N=82 spherical shell gap. The lighter mass 130–140 region nuclei are known to exhibit isomers based on deformed structures. For example, the N=74 nuclei $^{138}$Gd [1] and $^{140}$Dy [4] both show $K^{\pi} = 8^-$ isomers built upon deformed shapes. However, as the neutron number increases to N=77, the nuclear shape becomes less rigid and is known to be susceptible to gamma vibrations [9]. Isomers in these nuclei are thought to occur due to differences in the nuclear shapes for various configurations, e.g. $^{140}$Eu, $^{136}$Pm [3, 2, 6]. A consequence of this gamma-softness is that nuclear chirality has been predicted to occur in some of these nuclei [10, 11]. Furthermore, some of the

CP1012, *Frontiers in Nuclear Structure, Astrophysics, and Reactions: FINUSTAR 2*
edited by P. Demetriou, R. Julin, and S. V. Harissopulos
© 2008 American Institute of Physics 0-7354-0532-5/08/$23.00

lowest-lying excited states built upon isomers show characteristic behaviour which can be classified by the X(5) symmetry limit where the ratios of the excited-state energies lie midway between the limits expected for pure vibrational and rotational behaviour. Our latest proposed experiments will test these ideas in $^{144}$Ho in November 07 with the first isomer-tagged differential-plunger experiment. Finally, as the N=82 shell gap is approached, the nuclear shapes become spherical. In the even-even nuclei beyond the closed shell, several seniority $I^{\pi} = 10^{+}$ shell-model isomers are known to exist in this region built upon shell-model $\pi(h_{11/2})^2$ states [12] and $I^{\pi} = 27/2^{-}$ isomers in the odd mass-nuclei from the coupling of an odd $f_{7/2}$ neutron to the two-quasiproton $\pi(h_{11/2})^2$ configuration [12]. Although the isomeric states are known in some of these nuclei, in many cases little- or no-information is known about the states built upon the isomers.

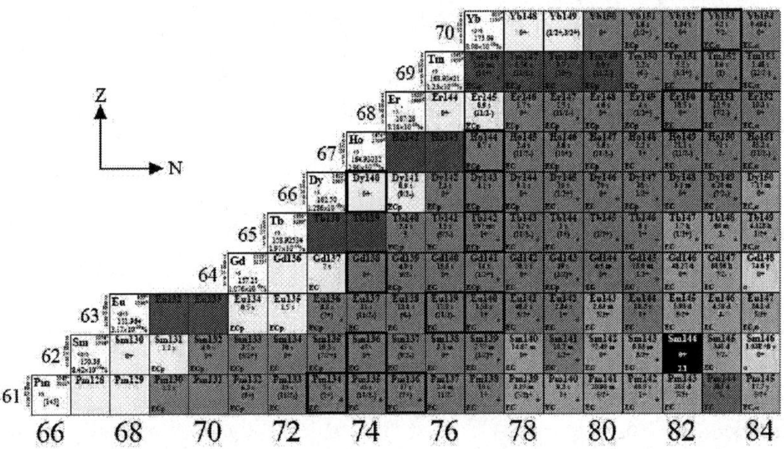


**FIGURE 1.** Partial chart of the nuclei showing the isomeric nuclei studied (black squares) in the mass 130–160 region using recoil-isomer tagging at the University of Jyväskylä.

This paper discusses the results of two new experiments using a modified experimental setup at the University of Jyväskylä. Recoil-isomer tagging has been discussed in several recent references [1, 2, 3, 4, 5, 6, 7, 8]. However, in the present work, the latest experiments have been performed with a modified setup where the standard GREAT [13] focal plane Double-sided Silicon Strip Detectors (DSSDs) were replaced by a dual Multi-Wire Proportional Counter (MWPC) setup which was designed and built at the University of Manchester [7]. This new setup has improved capability for short-lived isomer studies where large focal plane rates can be tolerated. One experiment, below the N=82 shell gap, established a new isomeric state in $^{136}$Pm [6] and also gave new information on an established isomeric state in $^{142}$Tb [2]. Another experiment, above the N=82 shell gap, gave new information [14] on the transitions above known isomeric states in $^{153}$Yb [15] and $^{152}$Tm [16]. Finally, new ideas were discussed to use an isomer-tagged differential-plunger setup to establish the deformation, and thereby aid in the process of assigning the underlying single-particle configurations of these isomers.

**Isomeric States in $^{142}$Tb:** A $^{54}$Fe beam was accelerated onto a 500 $\mu$g/cm$^2$ $^{92}$Mo target at energies of 245 MeV (30 hours at ≈6.4 pnA), 252 MeV (18 hours at ≈8.2 pnA) and

265 MeV (17 hours at ≈7.3 pnA). Prompt $\gamma$ rays were recorded with the JUROGAM-detector array around the target position. The recoiling nuclei passed through the gas-filled separator RITU (Recoil Ion Transport Unit) [17] into the focal plane spectrometer, GREAT [13]. The time-of-flight and energy-loss measurements were made between the two MWPCs in the revised setup. Gamma rays from the de-excitation of isomeric states from the implanted nuclei were subsequently detected in the GREAT Planar and Clover focal-plane detectors. The GREAT triggerless Total Data Readout (TDR) was used, where all data were recorded with a time-stamp generated by a 100 MHz clock. Two-dimensional spectra were constructed from these data using the 'GRAIN' software package [18]. The energies of the prompt JUROGAM $\gamma$ rays were correlated with the energies of the delayed $\gamma$ rays from the planar detector, with various correlation time windows. During this experiment, an attempt was made to raise the beam current to take full advantage of the new setup. However, the beam intensity could not be raised beyond ≈7 pnA, which was similar to the limit for experiments using the DSSDs, because at 7pnA, the prompt reaction rate in the individual JUROGAM detectors exceeded 16 kHz per detector. With the current conventional analogue electronics this caused a high dead-time in the JUROGAM detectors. However, from a comparison with experimental data from the previous experimental DSSD setup [6] at this rate, the new dual-MWPC configuration was more than twice as efficient. The limitation introduced by using conventional electronics will be addressed in future experiments where digital electronics will be used with individual JUROGAM detectors counting at rates of up to 60 kHz [19].

$^{142}$Tb was populated through the $^{92}$Mo($^{54}$Fe,3pn) reaction. The existence of two isomeric states in this nucleus has been established in previous studies [20, 2, 21]. The lifetime of the shorter-lived $^{142m2}$Tb state has been measured as 15(4) $\mu$s [20] and later as 25(1) $\mu$s [21] with a different setup. The recoil-isomer tagging technique has also been previously used to establish prompt transitions above this state [2], however, that prompt level scheme was established based solely on the intensities of the tagged transitions as no coincidence information was available at that time [2]. In the present work, the statistics obtained with the new dual-MWPC setup were sufficient to allow prompt recoil-isomer tagged $\gamma-\gamma$ coincidences to be examined above the isomer. A new level scheme was discussed in the talk which was extended over that discussed in Ref. [2] with the addition of new bands and a reordering of certain other transitions. A full publication of these data is currently in preparation [22]. One significant difference between our present work on $^{142}$Tb and that in the most recent publication [21] concerns the 25(1) $\mu$s half-life measurement of the shorter-lived isomeric state. Figure 2 shows the time difference between the time-stamp of the recoil in the first MWPC and that of the signal in the focal-plane planar detector from the present experiment along with the half-life fit to these data. The half-life was taken as the mean of the fits to several $\gamma$ ray gated time spectra, weighted by its statistical error (≈0.1 $\mu$s). The systematic error in the final result was taken as the standard deviation of the individual measurements yielding a half-life of 20 $\pm$ 1 $\mu$s. This is consistent within 2 standard errors of the original result in Ref. [20], although it is incompatible with the recent result found in Ref. [21].

**Excitation-Energy Systematics:** Figure 3 shows the energies systematics of the excited states for the N=73 - 77 nuclei. For each nucleus shown, the ratio of the 2nd to 1st excited-state energies are plotted which should yield respective values of 2.0 and 3.33

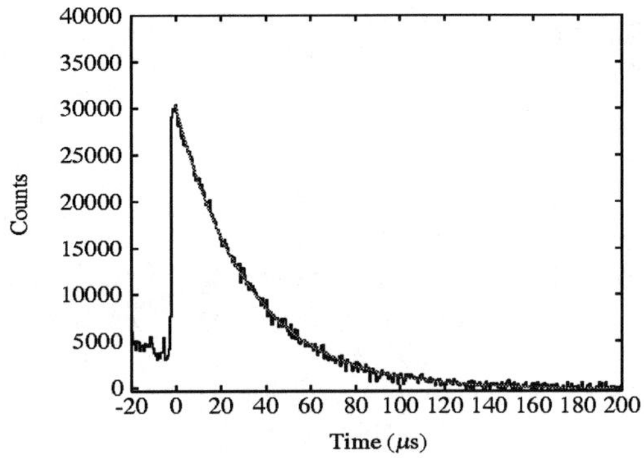


**FIGURE 2.** Sum of the individual time spectra for the 137, 165 and 303 keV delayed $\gamma$ rays in $^{142}$Tb.

for a pure vibrational or a pure rotational nucleus. For the odd-mass N=73 and N=75 isotones, the energy ratios have been calculated for the $19/2^-$ to $15/2^-$ states in the neutron-$h_{11/2}$ bands. For the N=74 even-even isotones, the $4^+$ to $2^+$ ratio for the ground-state band is shown. For the N=77 odd-odd nuclei, the ratios of the $10^+$ to $8^+$ state energies for the $\pi h_{11/2} \otimes \nu h_{11/2}$ bands built upon isomeric states are shown. The data for the $\pi h_{11/2} \otimes \nu h_{11/2}$ band in $^{142}$Tb (Z=65) is from the new level scheme discussed in this isomer-tagging work [22] shows similar behaviour to the known isomer-tagged level scheme for $^{140}$Eu (Z=63) [3]. New data for the isomer-tagged $\pi h_{11/2} \otimes \nu h_{11/2}$ in $^{144}$Ho (Z=65) is currently under analysis [22] and these values are also used in Fig. 3.

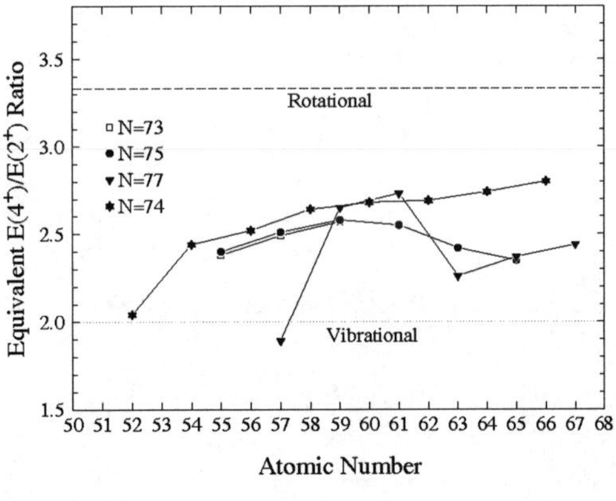


**FIGURE 3.** Excitation energy systematics for the N=73, 74, 75 and 77 isotones, see text for details.

The figure shows that most of these nuclei fall into a region which have energy ratios lying almost midway between that of a pure vibrational- and a pure-rotational nucleus in the so-called X(5) symmetry region. For the N=73, 74 and 75 nuclei with $Z > 54$ the values mostly cluster between 2.3 to 2.8. For the N=75 isotones with $Z > 61$ these values start to show a decrease with increasing atomic number. This trend appears to be more pronounced in the new data points for the odd-odd N=77 $\pi h_{11/2} \otimes \nu h_{11/2}$ bands built upon isomeric states with Z=63, 64 and 65. This interesting feature is not currently understood and further tests are underway for the N=77 nuclei. Key tests will rely on our isomer-tagged differential-plunger lifetime measurements in these nuclei in Nov. 07.

**Isomeric states in $^{153}$Yb and $^{152}$Tm:** In order to populate isomeric states in the heavier-mass region, a beam of $^{64}$Zn was incident on $^{92}$Mo target at 280 MeV. A series of isomeric states were produced in various exit channels populated in this study. By correlating prompt and delayed data across these states, preliminary new information was obtained for the states built upon a 15$\mu s$ isomer in $^{153}$Yb [15] and a 294 *ns* isomer in $^{152}$Tm [16]. The current experimental setup allows delayed $\gamma - \gamma$ coincidences to be examined and is expected to be able to fully establish the delayed level scheme of $^{153}$Yb. (The decay of this isomer was also not fully established in Ref. [15]). The analysis of the new correlated prompt and delayed data is currently underway [14].

The future prospects for this research will allow the detection of isomers at and beyond the proton drip line to reveal valuable information about deformations, single-particle excitation energies and hindrance factors where little or no information exists. The small cross sections for producing these nuclei are offset by the efficiency increases of the recoil-isomer tagging technique. The new two-MWPC detector setup, combined with the use of digital electronics for the JUROGAM detectors, will allow the use of large beam currents & the new Clover array (Summer 08) will increase the prompt $\gamma$-ray efficiency. The first isomer-tagged differential plunger experiments will take place in Nov. 07 to provide the first lifetime, deformation and X(5) symmetry information.

## REFERENCES


1. D.M. Cullen et al., Phys. Rev. C**58** (1998) 846.
2. C. Scholey et al., Phys. Rev. C**63** (2001) 034321.
3. D.M. Cullen et al., Phys. Rev. C**66** (2002) 034308.
4. D.M.Cullen et al., Phys. Lett. **B529** (2002) 42.
5. S.V. Rigby et al., J. Phys. **G. 31** (2005) S1949.
6. S.V. Rigby et al., Submitted to Phys. Rev. C June 07.
7. D.M.Cullen et al., A.I.P. CP831 (2005) 365. "FINUSTAR" Conference Proceedings.
8. D.M.Cullen et al., A.I.P. CP (2007). "PROCON 07" Conference Proceedings.
9. P. Möller *et al.* Phys. Rev. Lett. **97**, 162502 (2006).
10. K. Starosta *et al.*, Phys. Rev. Lett. **86**, 91 (2001) and references therein.
11. D. J. Hartley *et al.*, Phys. Rev. C**64**, 031304 (2001).
12. R.D.Lawson et al., Z. Phys. A**303** (1981) 51.
13. R.D. Page et al., Nucl. Instrum. Meth. **B204** (2003) 634.
14. S. Khan et al., To be submitted to Phys. Rev. C.
15. J.H. McNiell *et al.*, Z. Phys. **A344**, 369 (1993).
16. J.H. McNiell *et al.*, Z. Phys. **A325**, 27 (1986).
17. M. Leino *et al.*, Nucl. Instrum. Meth. **B** 99, 653 (1995).
18. P. Rahkila, "Grain, the analyser", University of Jyväskylä, private communication,
19. P.M. Jones, University of Jyväskylä, private communication.
20. I. Zychor et al., GSI Report No. 89-1, 1989.
21. N.M. Tantawy, Phys. Rev. C.**73**, 024316 (2006).
22. P.J.R. Mason et al., To be submitted to Phys. Rev. C.
23. C. W. Beausang, *et al.* Phys. Rev. C.**36**, 1810 (1987) and references therein.
24. M. A. Riley *et al.*, Phys. Rev. C**47**, R441 (1993).

# The Structure of $^7$He

A.H. Wuosmaa[a], K.E. Rehm[b], J.P. Greene[b], D.J. Henderson[b], R.V.F. Janssens[b], C.L.Jiang[b], L. Jisonna[c], J.C. Lighthall[a], S.T. Marley[a], E.F. Moore[b], R.C. Pardo[b], N. Patel[d], M. Paul[e], D. Peterson[b], S.C. Pieper[b], G. Savard[b], J.P. Schiffer[b], R.E. Segel[c], R.H. Siemssen[f], X. Tang[b], R. B. Wiringa[b]

[a]*Physics Department, Western Michigan University, Kalamazoo MI 49008-5252 USA.*
[b]*Physics Division, Argonne National Laboratory, Argonne IL 60439 USA*
[c]*Physics Department, Northwestern University, Evanston IL 60208 USA*
[d]*Physics Department, Colorado School of Mines, Golden CO 80401 USA*
[e]*Hebrew University, Jerusalem Israe 91904*
[f]*Kernfysich Versneller Instituut, 9747 AA Groningen, The Netherlands*

**Abstract.** We have studied the properties of the unbound isotope of helium $^7$He, using the $^2$H($^8$Li,$^3$He)$^7$He reaction. When combined with prior measurements of the $^2$H($^6$He,*p*)$^7$He reaction the data present a consistent picture for the low-lying excited states of $^7$He, specifically the negative parity sequence $3/2^-$ (ground state) $1/2^-$ (first-excited) and $5/2^-$ (second-excited). The shapes but not the absolute magnitudes of the angular distributions are reproduced by ab-initio theory coupled with a DWBA reaction framework.

**Keywords:** Transfer reactions, unstable beams, spectroscopic factors
**PACS:** 25.60.Je, 21.10.Jx, 21.60.De

## INTRODUCTION

The structure of the unbound isotope of helium $^7$He has attracted significant interest in recent years. Despite it being first observed 40 years ago [1], the precise nature of its properties, including the positions and quantum numbers of the excited states of $^7$He remain uncertain. Much of this uncertainty may be traced to the difficulty of using simple, direct-transfer reactions with stable beams and targets to populate the excited states of $^7$He, and the fact that the excited levels are expected to be broad, overlapping resonances. One attractive feature of this nucleus is that its structure is widely described as a valence neutron interacting with a core consisting of the "Borromean" halo nucleus $^6$He. The interactions between the loosely bound valence neutrons outside the deeply bound alpha-particle core present a challenge for theoretical descriptions of such light systems.

Also, with only seven nucleons, this system presents an ideal laboratory for testing critically the predictions of so-called "ab-initio" models of nuclear structure. These methods, which include the Quantum Monte Carlo (QMC) [2] and No-Core Shell Model (NCSM) [3] approaches, use fundamental two- and three-nucleon interactions to calculate nuclear wave functions which can in turn be used to determine a variety of associated properties. Near stability, these methods have been particularly successful

CP1012, *Frontiers in Nuclear Structure, Astrophysics, and Reactions: FINUSTAR 2*
edited by P. Demetriou, R. Julin, and S. V. Harissopulos
© 2008 American Institute of Physics 0-7354-0532-5/08/$23.00

at reproducing the spins, parities, and excitation energies of states in a large number of light nuclei. Other quantities that probe the nuclear wave functions in more detail may also be extracted from these calculations, including transition matrix elements, capture cross sections, and spectroscopic factors for nucleon transfer [4,5]. Recently, the QMC approach has been remarkably successful in reproducing the charge radii in heavy helium isotopes that have been measure in the laboratory using trapping techniques [6]. While these methods have had considerable success in reproducing these properties for particle-bound, or narrow excitations, a question that remains open is how well they will perform for broad, short-lived states, such as will be found in $^7$He.

The excitation spectrum of $^7$He has been the subject of several theoretical studies, including a shell-model calculation [7], and the above mentioned ab-initio methods [8,9]. These different methods all predict a sequence of negative parity states, with a $3/2^-$ ground state, followed by $1/2^-$, $5/2^-$, and $3/2^-$ levels. Each of these calculations suggests a gap of between 2 and 3 MeV between the ground and $1/2^-$ first-excited levels. The resonance structure of states in $^7$He has also been examined using the Resonating Group Model (RGM) [10] and the Continuum Shell Model (CSM) [11]. The RGM results well reproduced the observed ground-state width of $^7$He, and are also consistent with a broad $1/2^-$ first-excited state near 3 MeV excitation energy. The CSM calculations show an excitation spectrum with resonance positions very close to those predicted by the ab-initio methods, and large widths of 2 MeV or greater for the excited levels.

Over the past decade, the availability of unstable beams has provided new opportunities for the detailed study of excited levels in $^7$He. Early evidence for possible excited states in $^7$He was found from the exotic heavy-ion transfer reaction $^9$Be($^{15}$N,$^{17}$F)$^7$He [12], which suggested a broad excited state near 3 MeV. The neutron pick-up reaction $^1$H($^8$He,*d*)$^7$He has been reported by two groups [13,14]; both measurements also suggest a weakly populated broad excited state near 3 MeV that decays primarily to $2n$+$^4$He, suggesting a dominant decay through the first-excited $2^+$ state in $^6$He. This decay would support an assignment of $5/2^-$ for the state populated in $^7$He. $^7$He has also been studied as the product of the breakup of $^8$He [15] with the very interesting suggestion that the first-excited $1/2^-$ state in $^7$He may actually lie at very low excitation energy, near 600 keV in contrast to the 2-3 MeV suggested by theory. Data from one of the two *(p,d)* studies [14] have been suggested to at least qualitatively support this conjecture. This conclusion was not supported by studies of neutron stripping with the $^2$H($^6$He,*p*)$^7$He reaction [16] which saw evidence for a $1/2^-$ resonance near 2.6 MeV but no $1/2^-$ strength at lower excitation energies. Such a low-lying state was also not seen in a recent report of the $^7$Li(*d*,$^2$He)$^7$He reaction [17] Finally, a study of the $^1$H($^6$He,*n*)$^6$Li($0^+$;T=1) reaction [18] also saw no indication of analog $1/2^-$ strength in $^7$Li at an excitation energy comparable to the low $^7$He excitation energies reported in [15].

In order to address many of these issues, we set out to study $^7$He with a reaction complementary to the other nucleon transfer reactions discussed above, namely the proton pickup reaction $^2$H($^8$Li,$^3$He)$^7$He. Table 1 lists the spectroscopic factors for three different one-nucleon transfer reactions leading to $^7$He: *(d,p)*, (*d*,$^3$He), and *(p,d)*. These highly selective reactions dominantly populate different states in $^7$He, and by comparing results for different reactions we can hope to learn more about the different levels. For example, the *(d,p)* reaction populates only the ground and first-excited states, whereas ($d$,$^3$He) has significant strength only for the ground and second-excited states. Also, the 5/2$^-$ and 1/2$^-$ levels have different decay characteristics; the 1/2$^-$ state strongly prefers decay to the $^6$He ground state, whereas the 5/2$^-$ level decays exclusively to the $^6$He(2$^+$) state. These decays may be identified experimentally due to the fact that the 2$^+$ state in $^6$He is unbound with respect to two-neutron decay, and we may use a comparison between data from the previously obtained ($d$,$p$) reaction described in Ref. [16] and the ($d$,$^3$He) reaction to clarify the situation regarding $^7$He excited states.

**TABLE 1.** Spectroscopic factors for different nucleon transfer reactions populating $^7$He.

| $J^\pi$ | Ex(QMC) | Ex(NCSM) | $S_n(d,p)$ | $S_p(d,^3He)$ | $S_n(p,d)$ | Decay |
|---|---|---|---|---|---|---|
| 3/2$^-$ | 0 | 0 | 0.53 | 0.56 | 3.445 | $^6$He(0$^+$)+$n$ |
| 1/2$^-$ | 2.9(3) | 2.3 | 0.87 | 0.009 | 0.166 | $^6$He(**0**$^+$,2$^+$)+$n$ |
| 5/2$^-$ | 3.3(2) | 3.7 | -- | 0.166 | -- | $^6$He(2$^+$)+$n$ |
| 3/2$^-$ | 3.8(2) | 4.4 | -- | 0.003 | 0.099 | $^6$He(0$^+$,2$^+$)+$n$ |

# EXPERIMENT

The experiment was carried out in inverse kinematics using a $^8$Li beam produced at the "In-Flight" facility at the ATLAS accelerator at Argonne National Laboratory [19]. The $^8$Li beam was produced by bombarding a cryogenically cooled cell pressurized to 1400 mbar with $D_2$ gas with a 70 pnA beam of $^7$Li at an energy of 81 MeV. The $^8$Li nuclei produced from $^2$H($^7$Li,$p$)$^8$Li reactions in the cell were collected using a 6T superconducting solenoid, then sent through a bunching resonator to improve the energy and time resolution of the beam. The secondary $^8$Li beam was then separated from the primary beam using a dipole bending magnet followed by a series of slits. The resulting $^8$Li beam had an intensity of between 0.7-1$\times$10$^5$ particles per second on target, at an energy of 76 MeV. Due to the large difference in mass-to-charge ratio between $^7$Li and $^8$Li, the final beam was essentially free of contamination from the primary beam.

The $^8$Li beam was incident on a 420 μg/cm$^2$ $CD_2$ target. $^3$He nuclei were detected in a set of three segmented annular double-sided silicon strip detectors, covering laboratory angles from 9° to 50°. In coincidence with the $^3$He particles, $^6$He and $^4$He nuclei were detected and identified in a set of silicon ΔE-E telescopes, subtending laboratory angles between 1.3° and 7.2° in θ, and 88% of the full azimuthal range. Following the ΔE-E telescope array, a second gold foil target was used as a monitor target, and the scattering of beam particles from that target was measured in a second ΔE-E telescope at very forward angles as a monitor of the absolute beam intensity. Figure 1 shows a schematic drawing of the detector setup.

In order to verify the proper functioning of the setup, we performed a number of calibration measurements using the stable $^7$Li beam. Also, as all particles of interest are detected in the forward hemisphere, we were concerned about contamination from the $^{12}$C present in the $CD_2$ target. We also carried out high-statistics measurements with $^{12}$C targets using both $^7$Li and $^8$Li beams to assess any backgrounds from the $^{12}$C target content. Figure 2 shows Q-value spectra from the $^7$Li calibration beam, with various combinations of types of particles in the annular detectors and the ΔE-E telescopes. In Figs. 2(a)-(c), the particles in the annular counter are detected in coincidence with an identified $^6He$ (a), $^6Li$ (b), or $^4He$ (c) in the ΔE-E array. The $^3He$ or $^3H$ particles are not distinguished by the silicon detector however the different reaction channels are readily separated by the identified heavy fragment, as well as by the Q value. The effect of subtraction of the $^{12}C$ background for the $^4He$ case is shown as the shaded histogram in Fig. 2(c).

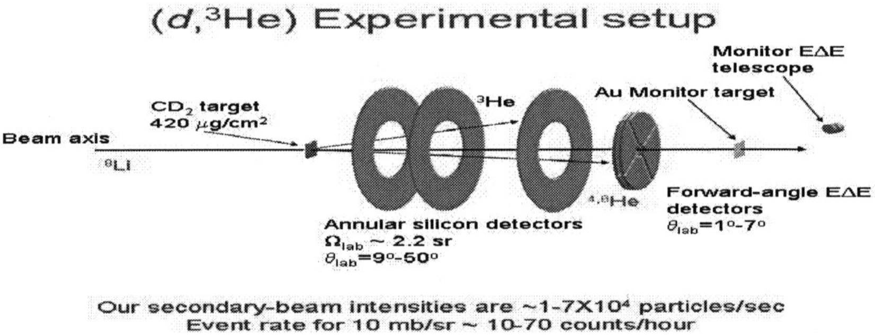


**FIGURE 1.** Schematic diagram of the detector setup.

## RESULTS

Figure 3 shows a comparison between Q-value spectra obtained from the $^2H(^6He,p)^7He$ reaction from Ref. [16], and from the present measurement of $^2H(^8Li,^3He)^7He$. For each reaction, the histograms show coincidences between protons or $^3He$ particles in coincidence with (a),(d) either an identified $^6He$ or $^4He$, (b),(e) an identified $^6He$ only, or (c),(f) an identified $^4He$ only. Upon comparing Fig. 3(a) and 3(d), the spectra appear similar, with a prominent ground-state peak and a broad distributions of counts at higher excitation energy. When requiring the observation of a $^6He$ beam-like fragment alone, however, the differences are clear. The yield at high excitation energy remains for the *(d,p)* case, whereas the spectrum is largely devoid of counts for $^2H(^8Li,^3He)^7He$ at Q< -8 MeV. Conversely, if the particle in the annular counter is required to be in coincidence with a $^4He$ nucleus identified in the forward Δ E-E array, little yield remains at high excitation energy for the *(d,p)* case, whereas all of the high-excitation-energy yield seen in Fig. 3(d) also appears in Fig. 3(f). Evidently the two reactions populate different amplitudes in the final nucleus. Recalling that the $1/2^-$ state decays predominantly to the ground state of $^6He$, while the $5/2^-$ state decays to the unbound $^6He(2^+)$ state, and that the *(d,p)* and $(d,^3He)$ reactions should populate the $1/2^-$, and $5/2^-$ states, respectively, the data shown in Fig. 3 are consistent with a $1/2^-$ resonance at $E_X$=2.6 MeV, and a $5/2^-$ resonance near $E_X$=3 MeV, as expected from the various theoretical calculations.

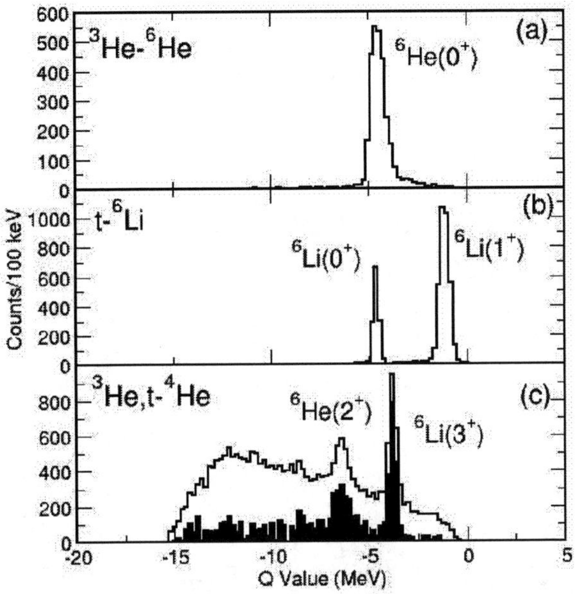


**FIGURE 2.** Q-value spectra obtained with the $^7$Li primary beam. (a) $^3$He-$^6$He coincidences (b) *t*-$^6$Li coincidences (c) $^3$He,*t*-$^4$He coincidences. The shaded histogram in (c) represents the spectrum after subtraction of the $^{12}$C background.

In order to compare the results with the predictions of the QMC approach, we have extracted the angular distribution for the $^2$H($^8$Li,$^3$He)$^7$He ground-state transition, which appears in Fig. 4(a). For comparison, the measured ground-state angular distributions for the $^2$H($^7$Li,$^3$He)$^6$He and $^2$H($^7$Li,*t*)$^6$Li reactions are shown in Fig. 4(b) and 4(c), respectively. In the case of $^2$H($^7$Li,$^3$He)$^6$He, data at the same bombarding energy exist from Ref. [], and are plotted as the square symbols. No adjustment in overall normalization has been made to the present data, and so the magnitude of the cross section for the $^2$H($^7$Li,$^3$He)$^6$He$_{g.s.}$ reaction is in good agreement with the previous measurement, giving confidence in the absolute normalization.

The curves in Fig. 4 represent optical-model calculations where the pickup form factor is taken from overlap functions calculated in the Variational Monte Carlo (VMC) framework as described in [4]. The optical-model potentials for the entrance and exit channels are from Ref. [21] and [22] respectively. For each reaction, the height of the curve is adjusted to match the data. The calculated shapes are in good agreement with the data in each case with the angle dependence expected for an $l$=1 transition. The absolute magnitude of the cross section is, however, not well reproduced in all cases. For the $^7$He data, the data are smaller than the calculation by a factor of 0.3. For the $^6$He case, the situation is not quite as bad with a renormalization factor of 0.7, and the $^6$Li data are in rather good agreement requiring only a 0.85 renormalization between theory and experiment, within the current uncertainty in the absolute normalization.

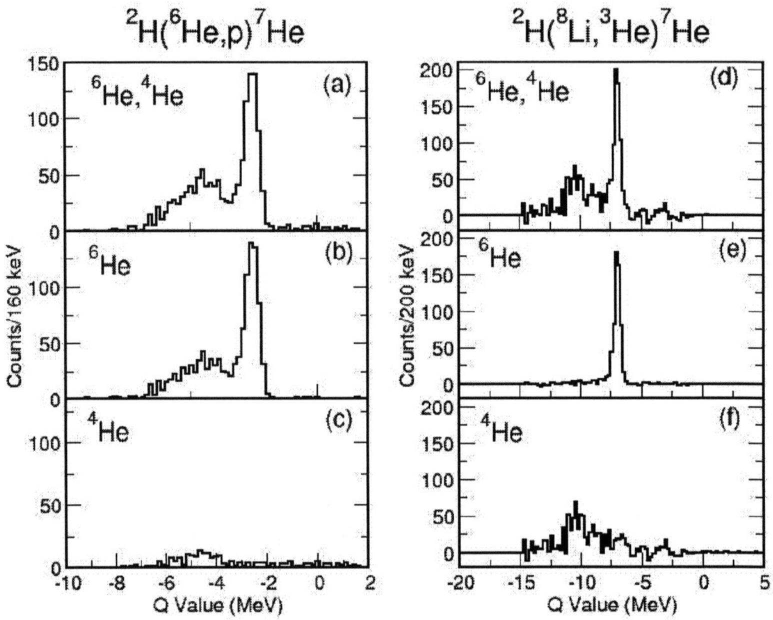


**FIGURE 3.** Comparison between Q-value spectra from $^2H(^6He,p)^7He$ (a-c) and $^2H(^8Li,^3He)^7He$ (d-f).

The reasons for the discrepancies are not yet fully understood, however we can make some speculations. The optical-model parameters used in each case for the exit channels are taken from studies of the $^3He+^7Li$ reaction. Given the diffuse nature of the helium nuclei in the present work, and the known sensitivity to coupling with unbound channels seen in $^6He$ reactions, it is likely that the $^3He+^7Li$ potential is inappropriate for the $^3He+^{6,7}He$ final states. Specifically, the depth and diffuseness of the imaginary well may be substantially larger for $^3He+^{6,7}He$ than for $^3He+^7Li$. The QMC overlap calculations begin from the starting point of the $^8Li$ or $^7Li$ nuclei, and these have been successful in the past for the calculation of neutron-stripping cross sections [4] giving some confidence in that aspect of the calculation. To study these issues further, angular distributions for all of the discrete levels populated in the current work will be compared to the QMC/VMC predictions, in order to deduce any systematic trends.

## CONCLUSIONS

We have studied the $^2H(^8Li,^3He)^7He$ proton pickup reaction in order to learn about properties of excited states in the unbound nucleus $^7He$. Combined with earlier data from the $^2H(^6He,p)^7He$ reaction, the results are consistent with the sequence of negative parity states $3/2^-$, $1/2^-$, $5/2^-$. The excited $1/2^-$ and $3/2^-$ states are broad resonances at excitation energies near 2.6, and 3 MeV, respectively. The sequence and spacing of these levels are consistent with the results of a variety of theoretical calculations for the structure of $^7He$. The ground-state angular distributions obtained from a combination of the QMC/VMC overlap functions with the DWBA reaction framework agree well with the angular-distribution shapes, but for the $^{6,7}He$ final states the absolute magnitude of the cross section is over-predicted. The reason for this discrepancy is presently not known but may be linked to deficiencies in the optical-model potential which does not sufficiently account for the diffuseness or weak binding of $^{6,7}$He.

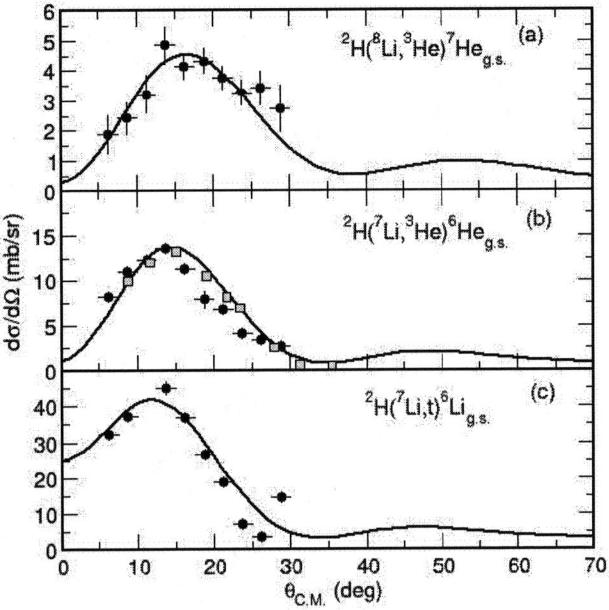


**FIGURE 4.** Angular distributions for the (a) $^2$H($^8$Li,$^3$He)$^7$He$_{g.s.}$, (b) $^2$H($^7$Li,$^3$He)$^6$He$_{g.s.}$, and (c) $^2$H($^7$Li,$t$)$^6$Li$_{g.s.}$ reactions. The horizontal error bars represent the binning width in the center-of-mass angle. The square symbols in (b) represent the data measured by Stokes and Young (Ref. [20]).

## ACKNOWLEDGMENTS

This work was supported by the U. S. Department of Energy, Office of Nuclear Physics, under contract numbers DE-FG02-04ER41320 (WMU), DE-AC02-06CH11357 (ANL), and DE-FG02-98ER4106 (NU).

## REFERENCES


1. R. H. Stokes and P. G. Young, *Phys. Rev. Lett.* **18**, 611 (1967).
2. S. C. Pieper and R. B. Wiringa, *Ann. Rev. Nucl. Part. Sci.* **51**, 53 (2001).
3. P. Navrátil and B. Barrett, *Phys. Rev. C* **54**, 2986 (1996).
4. A. H. Wuosmaa et al., *Phys. Rev. Lett.* **94**, 082502 (2005).
5. Petr Navrátil, *Phys. Rev. C* **70**, 054324 (2004).
6. L. –B Wang et al., *Phys. Rev. Lett.* **93**, 142501 (2004).
7. A.A. Wolters, A.G.M. van Hees, and P.W.M. Glaudemans, *Phys. Rev. C* **42**, 2062 (1990).
8. S. C. Pieper, R. B. Wiringa and J. Carlson, *Phys. Rev. C* **70**, 054325 (2004).
9. P. Navrátil and B. R. Barrett, *Phys. Rev. C* **57**, 3119 (1998).
10. J. Wurzer and H. Hofmann, *Phys. Rev. C* **55**, 688 (1999).

11. A. Voyla and V. Zelevinsky, *Phys. Rev. Lett.* **94**, 052501 (2005).
12. H. G. Bohlen et al, *Phys. Rev. C* **64**, 024312 (2001).
13. A. A. Korsheninnikov et al, *Phys. Rev. Lett.* **82**, 3581 (1999).
14. F. Skaza et al., *Phys. Rev. C* **73**, 044301 (2006).
15. M. Meister et al., *Phys. Rev. Lett.* **88**, 102501 (2002).
16. A. H. Wuosmaa et al., *Phys. Rev. C* **72**, 061301(R) (2005).
17. N. Ryezayeva et al., *Phys. Lett. B* **639**, 623 (2006).
18. G. Rogachev et al., *Phys. Rev. Lett.* **92**, 232502 (2004).
19. B. Harss et al., *Rev. Sci. Instrum.* **71**, 380 (2000).
20. R. H. Stokes and P. G. Young, *Phys. Rev. C* **3**, 984 (1971).
21. J. P. Schiffer et al., *Phys. Rev.* **164**, 1274 (1967).
22. R. L. Dixon and R. D. Edge, *Nucl. Phys. A* **156**, 33 (1970).

# Cluster Model for Near-barrier Fusion Induced by Weakly Bound and Halo Nuclei

C. Beck*, N. Keeley† and A. Diaz-Torres**

*Institut Pluridisciplinaire Hubert Curien, UMR7178, CNRS-IN2P3 et Université Louis Pasteur (Strasbourg I), B.P. 28, F-67037 Strasbourg Cedex 2, France
†DSM/DAPNIA/SPhN CEA Saclay, F-91190 Gif-sur-Yvette, France
**The Australian National University, Canberra ACT 0200, Australia

**Abstract.** The influence on the fusion process of coupling transfer/breakup channels is investigated for the medium weight $^{6,7}$Li+$^{59}$Co systems in the vicinity of the Coulomb barrier. Coupling effects are discussed within a comparison of predictions of the Continuum Discretized Coupled-Channels model. Applications to $^{6}$He+$^{59}$Co induced by the borromean halo nucleus $^{6}$He are also proposed.

**Keywords:** Breakup, fusion, threshold anomaly, weakly bound, halo, di-neutron, cluster model
**PACS:** 25.60.Ge, 25.70.Jj, 25.70.Mn, 25.60.Dz, 24.10.Eq

## INTRODUCTION

In reactions induced by light weakly bound nuclei, the influence on the fusion process of coupling to collective degrees of freedom and to breakup (BU) and transfer channels is a key point for a deeper understanding of few-body systems in quantum dynamics [1, 2, 3, 4, 5, 6]. Due to the very weak binding energies of halo nuclei, such as $^{6}$He, a diffuse cloud of neutrons should lead to enhanced tunneling probabilities below the Coulomb barrier as compared to predictions of one-dimensional barrier penetration models [1, 4]. This was understood in terms of the dynamical processes arising from strong couplings to collective inelastic excitations of the target and projectile [4]. However, in the case of reactions where at least one of the colliding nuclei has a sufficiently low binding energy for BU to become a competitive process, conflicting model predictions and experimental results were reported [4]. Recent experimental results with $^{6,8}$He beams show that the halo of $^{6}$He does not enhance the fusion probability, confirming the prominent role of neutron transfers in $^{6}$He induced fusion reactions [4, 7, 8, 9, 10, 11].

Excitation functions for sub- and near-barrier total - complete (CF) + incomplete (ICF) - fusion cross sections measured for the $^{6,7}$Li+$^{59}$Co reactions [1] when compared to Continuum-Discretized Coupled-Channels (CDCC) calculations [2] indicate a small enhancement of total fusion (TF) for the more weakly bound $^{6}$Li below the Coulomb barrier, with similar cross sections for both reactions at and above the barrier [2]. This result is consistent with BU - although with rather low cross sections even at incident energies larger than the Coulomb barrier [3, 4] - being more competitive for the $^{6}$Li+$^{59}$Co reaction.

In this contribution we present selected CDCC calculations for elastic scattering, TF, transfer (TR), and BU of weakly bound stable ($^{6}$Li considered as a $\alpha$-$d$ cluster) and radioactive ($^{6}$He as a $\alpha$-$2n$ cluster) light projectiles from a medium-mass target ($^{59}$Co).

CP1012, *Frontiers in Nuclear Structure, Astrophysics, and Reactions: FINUSTAR 2*
edited by P. Demetriou, R. Julin, and S. V. Harissopulos
© 2008 American Institute of Physics 0-7354-0532-5/08/$23.00

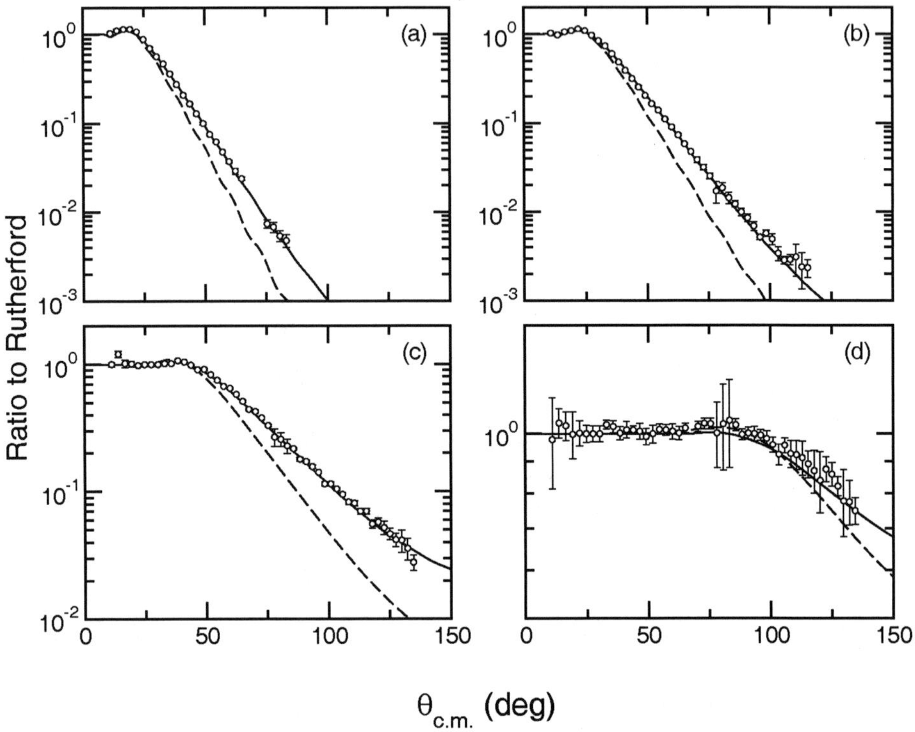


**FIGURE 1.** Elastic scattering for $^6$Li+$^{59}$Co at (a) 30 MeV, (b) 26 MeV, (c) 18 MeV and (d) 12 MeV [3, 4]. The curves correspond to CDCC calculations [5] with (solid lines) or without (dashed lines) couplings with the continuum as discussed in the text.

## FULL CDCC DESCRIPTION OF $^6$LI+$^{59}$CO REACTION

In the present work, detailed CDCC calculations for the interaction of $^6$Li on the medium-mass target $^{59}$Co are applied in order to provide a simultaneous description of elastic scattering, BU, TF as well as transfer. Details of the calculations concerning the breakup space (number of partial waves, resonances energies and widths, maximum continuum energy cutoff, potentials, ...) have been given in previous publications [2, 5]. The CDCC scheme is available in the general coupled channels code FRESCO [2]. Before investigating whether the proposed CDCC formalism can be also applied to halo structures such as the borromean nucleu $^6$He, we present a complete description of the $^6$Li $\rightarrow$ $\alpha$+$d$ cluster as a two-body object. In earlier fusion calculations the imaginary parts of the off-diagonal couplings were neglected, while the diagonal couplings included imaginary parts [2]. We had used short-range imaginary fusion potentials for each fragment separately. This was equivalent to the use of incoming wave boundary conditions applied in previous CCFULL calculations [1]. Here full continuum couplings have been

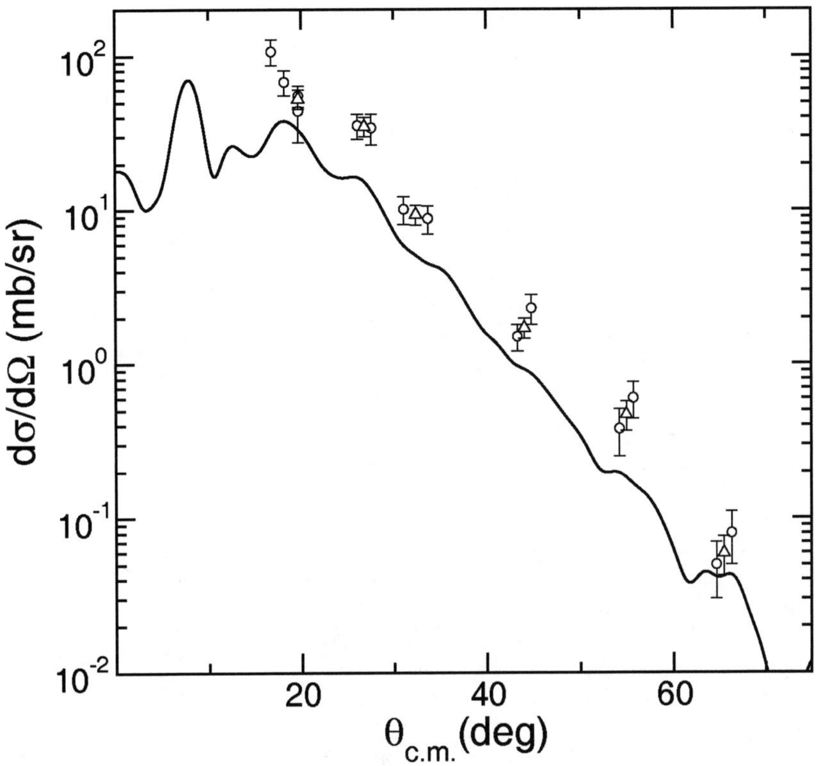


**FIGURE 2.** CDCC calculation for the angular distribution of the $^6$Li $\rightarrow \alpha + d$ sequential breakup via the 2.18 MeV $3^+$ state of $^6$Li compared to data of $^6$Li+$^{59}$Co reaction at 41 MeV.

taken into account so as to reproduce the elastic scattering data [3, 4].

Results of the comparison of the CDCC calculations for the elastic scattering with data of Ref. [3, 4] are shown in Fig. 1 for $^6$Li+$^{59}$Co at four different incident energies. The two different curves are the results of calculations performed with (solid lines) and without (dashed lines) $^6$Li $\rightarrow \alpha + d$ breakup couplings The agreement between the full calculations and data is very good. The effect of the BU channel on elastic scattering is illustrated by the difference between one-channel calculations and full CDCC results.

The total calculated BU cross sections were obtained by integrating contributions from the states in the continuum up to 8 MeV. They are found to be negligible fractions (between 3.7–9.7 %) of the total reaction cross sections and small compared with the TF cross sections (both the data [1] and the two different sets of CDCC calculations [2, 5]). This conclusion can be verified by the angular correlations for the sequential BU of $^6$Li via the 2.18 MeV $3^+$ excited state [5] plotted in Fig. 2 at 41 MeV. Sequential BU via this state is the dominant contribution to the total $^6$Li $\rightarrow \alpha + d$ breakup cross section ($\sigma_{tot} \approx$ 80 mb). Its CDCC cross section ($\sigma_{seq}$ = 22.5 mb) is, however, significantly smaller than the report experimental value ($\sigma_{exp} = 45 \pm 10$ mb) [5].

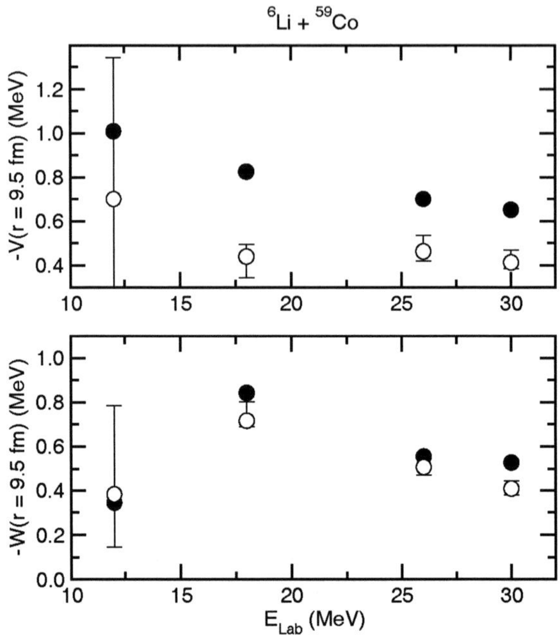


**FIGURE 3.** Energy dependence of the real and imaginary parts of the "bare + DPP" potentials as generated by the CDCC calculations (filled circles) and the best OM fits potentials (open circles) for the $^6$Li+$^{59}$Co system at a radial distance of r = 9.5 fm.

The total reaction cross sections obtained for $^{6,7}$Li+$^{59}$Co either from fits with Optical Model (OM) potentials [5] or CDCC calculations confirm the observed small enhancement of TF cross section for the more weakly bound $^6$Li nucleus at sub-barrier energies [1].

Fig. 3 illustrates how the surface strengths of the "bare plus dynamic polarization (DPP)" potentials, produced by the coupling of the BU channel, are in good agreement with OM potentials. Apparently, they exhibit the energy dependence characteristic of the "threshold anomaly" (TA), i.e. a rise in the strength of the real part as the incident energy is reduced towards the Coulomb barrier accompanied by a drop in that of the imaginary part. However, this conclusion largely rests on the values at 12 MeV incident energy, and as can be seen from the error bars, the potential strength in the nuclear surface is effectively not determined by the data due to its rather poor precision, a very wide range of values giving equally good fits to the data. The spread in values for the other energies, while much less than that at 12 MeV, is still such that we are unable to draw any concrete conclusions concerning the presence or absence of a TA in $^6$Li+$^{59}$Co. Similar observations have been made for $^7$Li+$^{59}$Co and the question of the occurence a "BU threshold anomaly" (BTA) [5] for lithium projectiles on light- and medium-mass targets remains widely open. More precise experimental elastic scattering data at sub- and near-barrier energies will, therefore, be necessary.

## PREDICTIONS FOR $^{6}$HE+$^{59}$CO FUSION AND CONCLUSIONS

Calculations applied to the two-neutron halo nucleus $^{6}$He is much more complicated since $^{6}$He breaks into three fragments ($\alpha$+n+n) instead of two ($\alpha$+d), and the CDCC method for two-nucleon halo nuclei has not yet been implemented in FRESCO [2]. A dineutron model is adopted for the $^{6}$He+$^{59}$Co reaction [3, 4, 5]: i.e. we assume a two-body cluster structure of $^{6}$He = $^{4}$He+$^{2}$n with an $\alpha$ particle core coupled to a single particle representing a di-neutron ($^{2}$n) like cluster. Couplings to resonant ($2^{+}$, $E_{ex}$ = 0.826 MeV) and non-resonant continuum states (up to f-waves) are included. The fact that the dineutron is not an object with both fixed size and fixed energy (Heisenberg principle) might be a critical point in the present model. Results of the CDCC calculations for TF of the $^{6}$He+$^{59}$Co system were compared to $^{4}$He+$^{59}$Co and $^{6}$Li+$^{59}$Co, used as basis reactions (see Refs.[3, 4, 5]). We observed that calculations with and without breakup give much larger TF cross sections for $^{6}$He compared to $^{4}$He and $^{6}$Li. The inclusion of the couplings to the BU channels notably increases the TF cross section for all energies. The predictions for the $^{59}$Co target somewhat over predict the data published for other medium-mass targets such as $^{64}$Zn [8] and $^{63,65}$Cu [9]. Extended calculations are in progress to quantify the role of 1n- and 2n-transfer channels found to be significant in recent $^{6}$He data [7, 8, 9, 10, 11].

The CDCC method [2] can be used to provide the almost complete theoretical description of all competing processes (elastic scattering, TF, TR and BU) in a consistent way. In this contribution we have shown that the $^{6}$Li+$^{59}$Co reaction can be fairly well understood in this framework although CDCC does not separate CF from ICF. CDCC results for the $^{6}$He+$^{59}$Co fusion process are also briefly discussed. A complete understanding of the reaction dynamics involving couplings to the BU and the neutron transfer channels will need high-intensity radioactive ion beams to permit measurements at deep sub-barrier energies and precise measurements of elastic scattering (search for TA and BTA) and yields leading to TR channels and to BU itself. The application of four-body (required for an accurate $\alpha$-n-n description of $^{6}$He) CDCC models still under current development [6] will then be highly desirable. It would be interesting to see what difference this more accurate model would have on the BU coupling effect on TF if applied to a fusion calculation in a similar manner to the calculations presented in this contribution.

## REFERENCES

1. C. Beck *et al.*, Phys. Rev. C **67**, 054602 (2003).
2. A. Diaz-Torres, I. J. Thompson, and C. Beck, Phys. Rev. C **68**, 044607 (2003).
3. C. Beck *et al.*, AIP Conferences Proceedings Vol. **853**, 384 (2006); arXiv:nucl-th/**0605029** (2006).
4. C. Beck, Nucl. Phys. A **787**, 251 (2007); and references therein.
5. C. Beck, N. Keeley, and A. Diaz-Torres, Phys. Rev. C **75**, 054605 (2007).
6. A. M. Moro *et al.*, Phys. Rev. C **75**, 064607 (2007).
7. R. Raabe *et al.*, Nature **431**, 823 (2004).
8. A. Di Pietro *et al.*, Phys. Rev. C **69**, 044613 (2004).
9. A. Navin *et al.*, Phys. Rev. C **70**, 044601 (2004).
10. P. A. De Young *et al.*, Phys. Rev. C **71**, 051601(R) (2005).
11. Yu. E. Penionzhkevich *et al.*, Eur. Phys. Jour. A **31**, 185 (2007).

# Some Questions and Answers in Barrier Distributions

E.Piasecki[1,4], M.Kowalczyk[1,2], J.Jastrzębski[1], T.Krogulski[3], K.Piasecki[2], K.Rusek[4], Ł.Świderski[4], S.Khlebnikov[5], M.Mutterer[6], W.H.Trzaska[7], M.Sillanpää[7], S.Smirnov[8], G.Tiourin[5], S.Dmitriev[9], E.Kozulin[9], A.Ogloblin[9], N. Rowley[10]

*[1]Heavy Ion Laboratory, Warsaw University, PL-02-093 Warsaw, Poland*
*[2]Institute of Exp. Physics, Warsaw University, PL-00-681 Warsaw, Poland*
*[3]University in Białystok, PL- 15-097 Białystok, Poland*
*[4]Sołtan Institute for Nuclear Studies, PL-00-681 Warsaw, Poland*
*[5]Khloplin Radium Institute, St. Petersburg*
*[6]Technical University Darmstadt, Darmstadt, Germany*
*[7]University of Jyväskylä, Jyväskylä, Finland*
*[8]JINR, Dubna, Russia*
*[9]Kurchatov Institute, Moscow, Russia*
*[10] IPHC, Strasbourg, France*

**Abstract.** We report here preliminary results of experiments concerning the influence of "weak" channels on fusion and quasielastic barrier distributions. Some results of measurements of these distributions in the $^{20}Ne + ^{208}Pb$ system and of transfer probabilities for $^{20}Ne + ^{nat}Ni$, $^{90,92}Zr$, $^{118}Sn$ and $^{208}Pb$ backscattering at near-barrier energies will be presented.

**Keywords:** Coupled channels; Coulomb barrier distribution; Quasi-elastic scattering; Transfer.
**PACS:** 25.70.Bc; 25.70.Hi; 25.70.Ji;

## INTRODUCTION

One of the most important near-barrier reactions is fusion. It turns out that a connection between nuclear reaction mechanism and structure of the interacting nuclei exists and manifests itself in strong enhancement of fusion cross-sections at sub-barrier energies. This can be understood as a result of the interplay between various reaction channels: elastic and inelastic scattering, transfer reactions, break-up and fusion. It has been demonstrated experimentally in many systems that the fusion barrier between two nuclei does not have a unique value but rather a weighted distribution $D_{fus}$ of heights, which can be determined directly from fusion excitation function measurements [1]. The best theoretical description of this phenomenon can be made in the framework of the coupled-channels method, according to which the distribution is a manifestation of the strong couplings to different reaction channels, in particular to strong collective excitations of the participating nuclei. In some cases the distribution turns out to be markedly structured and gives a fingerprint of the couplings involved [2,3].

It has been shown both theoretically and experimentally that the difficult fusion measurements can be replaced by much simpler quasi-elastic scattering measurements

CP1012, *Frontiers in Nuclear Structure, Astrophysics, and Reactions: FINUSTAR 2*
edited by P. Demetriou, R. Julin, and S. V. Harissopulos
© 2008 American Institute of Physics 0-7354-0532-5/08/$23.00

at backwards angles, giving rise to the barrier distribution $D_{qe}$, and there exist many experimental data confirming the basic equivalence of these two methods [4,5]. The latter consists of determining the excitation function of quasi-elastic scattering of projectile-like nuclei at large angles. The quasi-elastic cross section is the sum of elastic, inelastic and transfer channels, with no need to individually identify the particular channels involved. The barrier distribution is obtained directly from the data as the first derivative of the ratio the excitation function and the Rutherford cross section.

The advantages of this method are discussed in [2,6]. Experimentally, the measurements boil down to the counting of number of projectile-like nuclei (in some excitation-energy window) registered at backward and forward angles, the latter being used as a measure of the Rutherford scattering.

## OUR PREVIOUS RESULTS: QUESTIONS

Our programme of measurements (performed at the Warsaw Cyclotron) concentrated on the $^{20}$Ne projectile, since this nucleus has spectacularly large deformation parameters: $\beta_2$ = 0.46, $\beta_3$ = 0.39, $\beta_4$ = 0.27 [7-9]. According to calculations performed using the coupled-channels code CCQEL [10], this projectile, used in conjunction with a relatively inert target nucleus, should give rise to a strongly structured barrier distribution.

The results of our measurements were perplexing: the barrier distributions for $^{20}$Ne + $^{112,116,118}$Sn turned out to be smooth [11]. The natural suspicion is that the smoothing is due to couplings not taken into account in our calculations; for example, those connected with transfer channels. These channels are often assumed to be weak and not affecting the distribution dramatically. Nothing was known about the influence of $\alpha$-particle transfer (particularly strong for this system) on the barrier distributions, but our measurements with the $^{22}$Ne beam (where this reaction was much weaker than in the $^{20}$Ne case) eliminated this suspect from our investigations [12].

On the other hand, the measurements (performed simultaneously) for the $^{20}$Ne + $^{nat}$Ni system, resulted in a clearly structured distribution [13] in very good agreement with calculations based on the coupled-channels method.

A possible hypothesis is that the smoothing is due to the neutron transfer (pick-up) channel, which for the Ni target is expected to be much weaker than for Sn. This expectation relies on the Rehm transfer cross-section systematics [14], for which the effective Q-value [15] is the main factor influencing the transfer probability. While for both cases the Q-values are negative, the expected neutron transfer cross sections are smaller by a factor of ~3 for $^{58}$Ni than for $^{118}$Sn.

To check our hypothesis experimentally, we performed, under the same experimental conditions, measurements for the $^{20}$Ne + $^{90,92}$Zr systems. According to the Rehm systematics, for the $^{90}$Zr target, the neutron-transfer probability should be only slightly larger than for Ni, while being considerably smaller than for the $^{118}$Sn target. In the other case, for the $^{92}$Zr target, the n-transfer probability should be similar to that for the $^{118}$Sn target. Thus, if our hypothesis is correct, we should expect a structured barrier distribution for the $^{90}$Zr target, while in the case of $^{92}$Zr we expect to have its structure considerably smoothed out. On the other hand the structure of both Zr nuclei is so similar that, if one excludes the transfer channels, according to the coupled-channels calculations one should not expect any difference between the two distributions.

In agreement with our expectations, the barrier distributions for the $^{90}$Zr and $^{92}$Zr targets turned out to be different [17]: the former is structured, while the latter is structureless and wider. In the case of the $^{90}$Zr target the observed structure is similar to that calculated using the CCQEL code, while the larger width and the lack of structure in the case of the $^{92}$Zr target is apparently due to a large transfer coupling strength, which cannot be taken into account by the calculations, at least using the present code. This points to the same reason for the disagreement between experiment and theory observed in the case of Ne + Sn system: that is, it is the transfer channels, which smooth out the expected structure.

However, one can ask whether the Rehm's systematics, which this reasoning relies on, is in this case sufficiently precise. Moreover, we still do not know whether the weak channels are equally effective in smoothing the barrier distributions determined from fusion and quasi-elastic scattering. To answer these questions we performed two experiments. In the first one we measured simultaneously both $D_{qe}$ and $D_{fus}$ distributions for the system $^{20}$Ne + $^{208}$Pb. In the second we measured the transfer probability in backward scattering of $^{20}$Ne ions on the $^{nat}$Ni, $^{90,92}$Zr, $^{118}$Sn and $^{208}$Pb. We report here some preliminary results of these experiments.

## NEW EXPERIMENTS: RESULTS AND DISCUSSION

Basically, the method and experimental set-up was similar to that described in Ref. [6]. That is, we measured quasi-elastic large-angle scattering and (this was new in respect to our previous measurements) fusion/fission excitation functions using semiconductor detectors placed at 120°, 130°, 145° and 155° (with aperture 2°) in the laboratory system and four "Rutherford" semiconductor detectors placed symmetrically at 35° (collimated to ±0.1°) with respect to the beam.

The $^{20}$Ne beam, with an intensity of a ~20 pnA, from the K-130 Jyvaskyla cyclotron (the acceleration of noble gas ions, such as Ne, cannot be performed with the tandem accelerators usually used in this kind of study) bombarded 150 μg/cm$^2$ target of $^{208}$Pb (enriched to 99%) on a 60 μg/cm$^2$ $Al_2O_3$ backing. To enable measurements with small energy steps, we used nickel foils as energy degraders. Energy calibration was performed using a precise pulse generator and a $^{226}$Ra α-particle source (the estimated pulse-height defect of the semiconductor detectors was negligibly small).

The energy resolution was continuously monitored during the experiment using the energy spectra measured in the forward detectors, and turned out not to exceed 0.8 MeV [FWHM], much smaller than the expected ~4 MeV structure width. About half of the resolution came from the characteristics of the beam, the rest was due to the straggling in the degrader and the target. Other effects, for example detector geometry, had very little influence on the energy resolution. In addition to checking the resolution, the stability of the electronics and detectors was continuously monitored using a precise pulse generator and the peak of elastic scattering.

To distinguish between fusion/fission and the backscattering quasielastic events, we identified the mass of registered ions using the TOF technique, with the start/stop generated by the semiconductor detector and cyclotron RF signals, respectively. The data analysis was performed in a similar way to that described in ref. [6]. From the kinetic-energy spectra, assuming two-body kinematics, we calculated the Q-value spectra for the forward and backward detectors. Then the number of counts in the wide integration limits for the forward and backward detectors were obtained. Next, the $E_{eff}$ (which takes account of the "angle-dependent" centrifugal energy [17]) was calculated and the ratios $\sigma_{qe}/\sigma_R$ as well as $\sigma_{fiss}/\sigma_R$ (being a measure of $\sigma_{fus}$) were constructed. The barrier distributions were determined using the numerical finite-difference method. Compilation of our results concerning the interaction of $^{20}Ne$ with various targets is shown in Fig. 1.

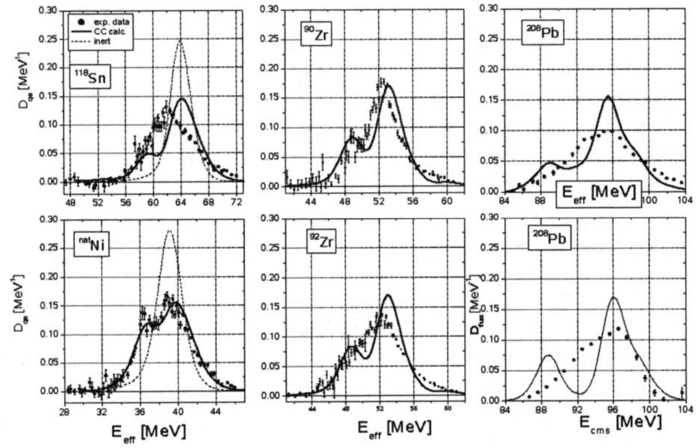


**FIGURE 1.** Barrier height distributions obtained by this collaboration for $^{20}Ne$ + various targets, compared with the coupled channels calculations. The dotted narrow distributions are results of disregarding all couplings (nuclei treated as inert). The lower right panel shows the fusion barrier distribution $D_{fus}$, other panels present quasielastic barrier distributions $D_{qe}$.

In the calculations, although it was checked that the shape of $D_{qe}$ is determined almost entirely by the properties of $^{20}Ne$, we also took into account the vibrational excitations of the targets. The coupling scheme included the states $0^+$, $2^+$, $4^+$, $6^+$ in the $^{20}Ne$ rotational band. The convergence of the results, with the number of states included, was checked. In addition we took into account the strong octupole-phonon state in the projectile ($E^* = 5.62$ MeV, $\beta_3 = 0.39$).

As can be seen, in the case of $^{208}Pb$ target, the experimental barrier distributions, quasielastic and fusion, are very similar but both structureless, in clear disagreement with the coupled channels calculations. We expected such a result since according to the Rehm systematics the transfer probability in $^{20}Ne$ + $^{208}Pb$ should be similar to this in $^{20}Ne$ + $^{118}Sn$. But are this systematics sufficiently reliable?

To check this, we identified the mass of ions scattered to ~150° from the targets shown in Fig.1. The energy was measured by an array of ten 20x20 mm$^2$ semiconductor detectors, which together with the Multichannel Plate Detector gavc us time resolution of ~280 ps. For the TOF base of 50 cm this gave us a mass resolution better than 1 amu. The beam energy was close to the Coulomb barrier (54.4, 63.4, 63.0, 76.3 and 95.6 MeV for the $^{nat}Ni$, $^{90,92}Zr$, $^{118}Sn$ and $^{208}Pb$ in the laboratory reference system, respectively).

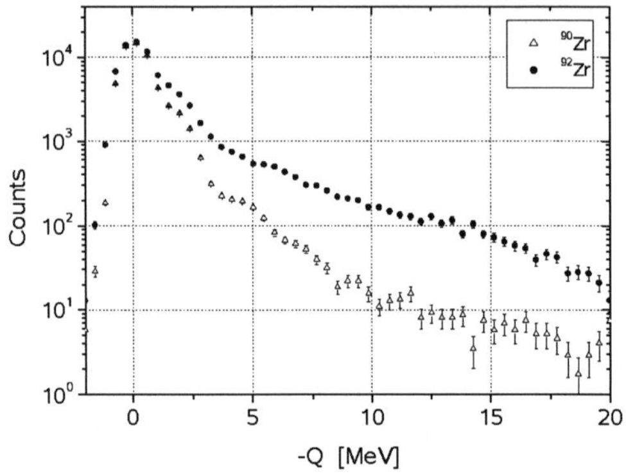


**FIGURE 2.** Excitation-energy spectra of the $^{20}$Ne + $^{90,92}$Zr systems generated in non-transfer backscattering to 150° at near-barrier energy.

For Zr targets experimental data are in agreement with systematics: the transfer probability for $^{92}$Zr is larger by factor of almost 2 in comparison with the $^{90}$Zr (and the Ni) targets. Moreover, the inelastic excitations are clearly stronger in the former case (see Fig. 2). Whether this is related to the smoother barrier distribution in the former case - remains to be seen.

## ACKNOWLEDGMENTS

We are grateful to the University of Jyväskylä Cyclotron staff for the excellent beam they provided. This work was funded in part by Grant No N202 152 31/2796 and supported by the cooperation agreement (03-110) between the IN2P3 (France) and the Polish Laboratories.

## REFERENCES

1. N. Rowley et al., Phys. Lett. B **254**, 25 (1991).
2. M. Dasgupta et al., Annu. Rev. Nucl. Part. Sci. **48**, 401 (1998).
3. A. M. Stefanini et al., Phys. Rev. Lett. **75**, 864 (1995) .
4. H. Timmers et al., Nucl. Phys. A, **584**, 190 (1995).
5. L.F.Canto et al., Phys. Repts. **424**, 1 (2006).
6. E.Piasecki et al**.,** Phys. Rev. **C65,** 054611 (2002).
7. S. Raman et.al., At. Data and Nucl. Data Tables **78,** 1 (2001).
8. R.H.Spear, At. Data and Nucl. Data Tables **42,** 55 (1998).
9. G.Blanpied et al., Phys. Rev. **C38,** 2180 (1988).
10. K.Hagino, N.Rowley, unpublished.
11. L.Świderski et al., Int. Journ. of Modern Phys. **E 13,** 315 (2004).
12. E.Piasecki et al., Phys. Lett. B **615,** 55 (2005).
13. L.Świderski et al., Int. Journ. of Modern Phys. **E 14,** 341 (2005).
14. K.E.Rehm et al., Phys. Rev. **C42,** 2497 (1990).
15. R.Bass, "Nuclear Reactions with Heavy Ions", Berlin, Springer-Verlag, 1980, p. 152.
16. E.Piasecki et al., Int. Journ. of Modern Phys. **E 14,** 341 (2005).
17. H.Timmers et al., Nucl. Phys. **A633,** 421 (1998).

# Spin-dipole excitations studied with tensor polarised deuteron beams

H.J. Wörtche

*Kernfysisch Versneller Instituut (KVI)*
*University of Groningen*

**Abstract.** Taking advantage of the spin, isospin and $J^\pi$ selectivity of cross sections and analysing powers of the tensor polarised $^{12}C(\vec{d},^{2}He)$ reaction, the $J^\pi = 0^-, 1^-, 2^-$ components of the spin-dipole resonance in $^{12}B$ have been investigated. The experiment was performed at the Big-Bite Spectrometer of the KVI taking advantage of purely tensor polarised deuteron beams provided by the AGOR cyclotron. The unique energy resolution achieved at KVI allowed determination of analysing powers for well resolved transitions up to 5 MeV excitation energy. The analysing powers of states with known spin and parity provided the means for the first-time for verification of model-independent predictions of analysing powers at extreme forward angles. In the continuum region, the analysing powers allowed identification of the $J^\pi$ components of the IVSGDR, including a $0^-$ component located at $E_x = 9.3$ MeV.

**Keywords:** charge exchange reactions, giant resonances
**PACS:** 23.40Hc, 24.30Gd, 25.40Kv, 27.20.+n

## INTRODUCTION

The spin and the isospin structure of nuclear excitations is a decisive piece of information to be investigated and quantified in experimental nuclear-structure studies. A natural way of inducing isovector ($\Delta I = 1$) nuclear transitions is provided by charge exchange reactions. Recent years have seen tremendous experimental progress in this field and the present sensitivity and the achieved resolutions offer excellent testing grounds for theoretical model calculations and provide highly desired input for adjacent fields like astrophysics and neutrino physics. In respect to the latter, the double-beta decay community is requesting a determination of relevant nuclear transition-matrix elements with improved precision which implies a $J^\pi$ selective determination of Gamow-Teller (GT) transitions for higher-order multipoles ( see e.g. [1] and references therein). In this paper we present results of experimental investigations of the $\Delta S = \Delta L = 1$ spin-dipole resonance (SDR) in $^{12}B$ [2]. Due the finite orbital angular momentum transfer the SDR splits into three components having $J^\pi = 0^-, 1^-, 2^-$ when the target nucleus is in a $J^\pi = 0^+$ ground state. Besides the location of SDR, the identification of the $J^\pi$ components has gained special attention. This is not the least due to the requirements imposed by the theoretical interpretation of $0\nu2\beta$-decay measurements where the $J^\pi$ components of the SDR provide significant contributions to the nuclear transition-matrix element and experimental information on the decomposition of the SDR provides important calibration for theory. A special challenge is posed by the identification of the $J^\pi = 0^-$ component, which is the most elusive component and which is not accessible to electromagnetic probes. The collective $J^\pi = 0^-$ component of the SDR is considered to act as mediator

CP1012, *Frontiers in Nuclear Structure, Astrophysics, and Reactions: FINUSTAR 2*
edited by P. Demetriou, R. Julin, and S. V. Harissopulos
© 2008 American Institute of Physics 0-7354-0532-5/08/$23.00

for parity mixing in complex nuclear states and the excitation energy is sensitive to the treatment of pion exchange in the nuclear potential.

## EXPERIMENTAL SETUP

The superconducting cyclotron AGOR at the Kernfysisch Versneller Instituut (KVI) provides a spectrum of light and heavy ion beams at low and medium energies, including vector and tensor polarized proton and deuteron beams. KVI offers a variety of facilities dedicated for nuclear instrumentation applications like irradiation of materials and basic research like few-body physics and fundamental physics. For the latter, significant instrumentation is presently in the phase of commissioning and will eventually constitute a Trapped Radioactive Isotope $\mu$icro Laboratory for Fundamental Physics (TRI$\mu$P).

For nuclear-structure research KVI provides the large acceptance Big-Bite Spectrometer (BBS) [3] equipped with a focal-plane detection system, a focal-plane proton polarimeter and real-time data processing system (ESN system) [4]. In the past years, the BBS has seen a lively experimental physics program driven by international collaborations. Main emphasis has been to use polarised and non-polarised hadronic probes for the spin and isospin selective excitation of collective and non-collective nuclear modes. In addition, the selectivity of measurements has further increased by detecting the charged particle decay, neutron decay and $\gamma$ decay of nuclear excited states [5, 6, 7, 8].

The selectivity to spin and isospin degrees of freedom in nuclear excitation by means of hadronic probes is due to the different energy dependence of the various terms in the nucleon-nucleon interaction [9]. The selectivity can further be enhanced by including spin degrees in the observables of the scattering process, i.e. by measuring cross-section asymmetries and polarisation transfer in the scattering of polarised particle beams. In this context, the $(d,^2He)$ probe plays an extraordinary role, whereas $^2He$ refers to the fact, that the two protons in the exit channel are detected in a relative $^1S_0$ state. Due to the dominant $^3S_1$ character of the deuteron in the entrance channel, the probe is equivalent an $(\vec{n},\vec{p})$ measurement and provides highly selective excitation of GT transitions. In recent years it has been demonstrated, that due to the limited acceptance of the BBS, which effectively limits the relative energy of the two protons detected in coincidence, and the rate capabilities of the ESN system, $(d,^2He)$ measurements are feasible at extreme forward angles even for heavy nuclei and yield virtually background free spectra with energy resolutions of order 100 keV for the excitation energy in the remaining nucleus [10, 11]. In the present experiment, a purely tensor-polarised deuteron beam provided by the polarised ion source POLIS [12] was accelerated to 171 MeV by the superconducting cyclotron AGOR. The beam impinged on natural carbon targets of 9.4 mg/cm$^2$ (flexible graphite, 99.8 %) and 23 mg/cm$^2$ (rigid graphite, 99.95 %), respectively. The beam polarisation changed in intervals of 5 minutes between the unpolarised state and two purely tensor-polarised states with average values of $p_{ZZ}$ = $-1.24$ and $p_{ZZ}$ = 0.57. The BBS was set at central scattering angles $\theta_{BBS} = 0°$ and 3°. The beam line to the spectrometer was set up for dispersion-matched transport to achieve optimum energy resolution. Neutrons emitted at backward angles with respect to the beam direction (110°-160°) were detected in 32 liquid-scintillator detectors from the EDEN detection system [13]. The neutron energies were determined by time-of-flight

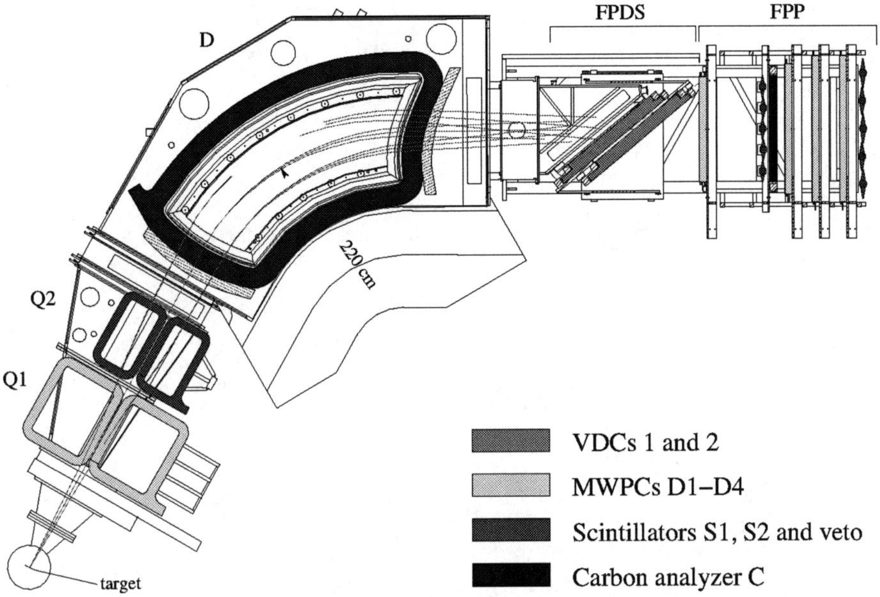


**FIGURE 1.** The Big-Bite Spectrometer in conjunction with the EuroSuperNova collaboration (ESN) detector system. Tracking of light ions is accomplished by means of Vertical Drift Chambers (VDS) along the BBS focal-plane (Focal-Plane Detection System, FPDS) and with Multi-Wire Proportional Chambers (MWPCs) in the focal-plane polarimeter (FPP).

(TOF) measurements using digital techniques to separate events induced by neutrons and $\gamma$'s in the EDEN system. A special scattering chamber with a thin wall of 2 mm stainless steel was used to reduce re-scattering of neutrons. Typical beam current values were 0.1 to 0.5 nA, again depending on the target thickness. The primary trigger rate was around 20-30 kHz in the ESN detector system, yielding a two-proton coincidence rate of 2 Hz and a (triple) neutron coincidence rate of 0.02-0.06 Hz, depending on the target thickness. All measurements were performed with a dead-time of 25 to 30%.

## RESULTS AND DISCUSSION

Figure 2 shows the double-differential cross sections for the $^{12}$C($\vec{\mathrm{d}}$,$^{2}$He)$^{12}$B reaction in the angular bin $\theta_{lab}$ = 0° - 1.7° for the different tensor-polarisation states of the beam. Due to the very small residual background, cross sections have been extracted for all states up to $E_x = 3$ MeV by placing markers and summing up the counts. The decomposition of the cross section above 3.0 MeV was performed by fitting appropriate functions (for details we refer to Ref. [2]).

The DWBA cross-section and analysing-power calculations for the different transitions have been performed using the code ACCBA of Okamura [14], which is specialised for the (d,$^{2}$He) reaction. The optical-model parameters and further input pa-

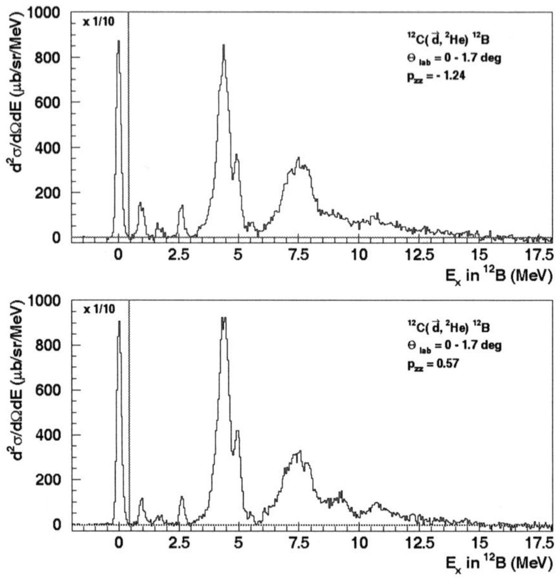


**FIGURE 2.** Double-differential cross sections for the $^{12}C(\vec{d},^{2}He)^{12}B$ reaction in the angular bin $\theta_{lab}$ = 0° - 1.7° for the different tensor-polarisation states of the beam. The background of quasi-elastic contributions has been subtracted (for details we refer to Ref. [2]).

rameters were taken from literature [15]. In all calculations, the $D$-wave contribution of the incident deuteron and the outgoing $^2$He was not taken into account, because its effect is exceedingly small at the forward angles of interest [14].

The measured and calculated cross sections and tensor-analysing powers for well-resolved states up to $E_x$ = 5 MeV are compared in Fig.3 for positive-parity transitions and Fig.4 for negative-parity transitions. For the first time, predictions for the tensor-analysing powers $A_{yy}$ and $A_{zz}$ based on symmetry arguments for natural-parity states using the (d,$^2$He) reaction could be compared with experimental data. From Figs.3 and 4 one can deduce, that for the natural-parity states at $E_x$ = 0.96 and $E_x$ = 2.62 MeV the tensor-analysing powers $A_{zz}$ and $A_{yy}$ nicely match the predictions, i.e. $A_{zz} = 1$ and $A_{yy} = -0.5$ at $\theta = 0°$. This indicates that the analysing powers of tensor polarised (d,$^2$He) provides a sensitive tool to identify natural-parity states at extreme forward scattering angles.

Concerning the decomposition and the localisation of the $J^\pi$ components of the IVSGDR the analysing powers yield the following result. Large negative analysing powers $A_{zz} \simeq -1$, the model-independent indication for the presence of the most elusive $0^-$ component, have been measured for the most forward angle spectra at $E_x$ = 4.0 and 9.3 MeV in agreement with shell-model calculations [16, 17, 18]. The evidence for a concentration of $0^-$ strength at $E_x$ = 9.3 MeV ($\Gamma$ = 330 keV) further confirms the experimental results of Okamura [19]. Evidence for the $J^\pi$ = $1^-$ component, which corresponds to a pure transverse transition, has been found in the high-energy part of

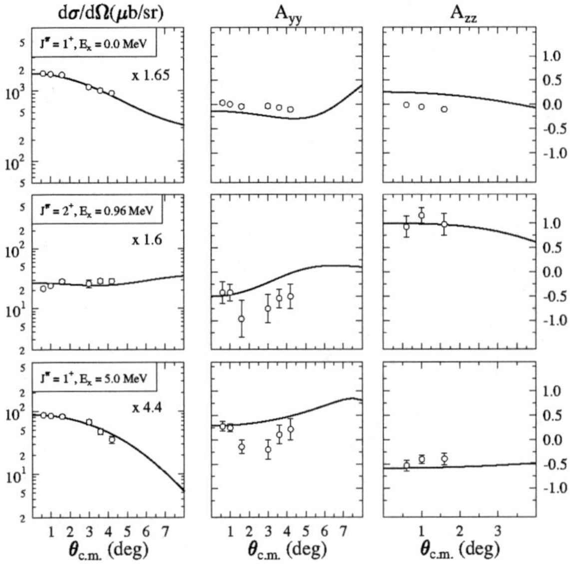


**FIGURE 3.** Measured and calculated $^{12}C(\vec{d},^{2}He)^{12}B$ differential cross sections and tensor-analysing powers for some positive-parity low-lying excited states. The scaling factors of the DWBA cross sections are indicated.

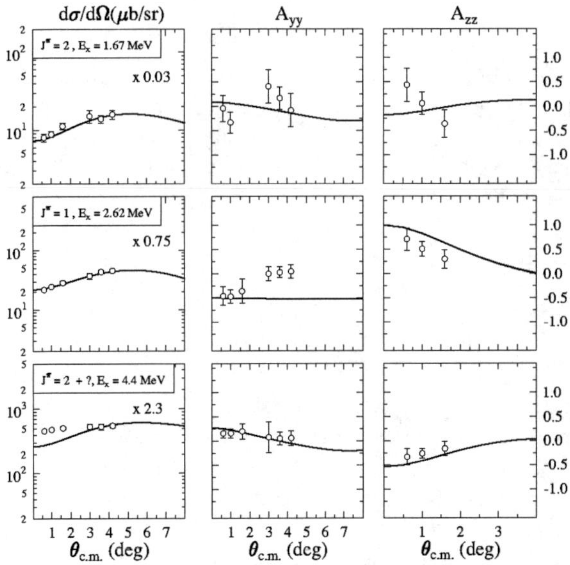


**FIGURE 4.** Same as Fig. 3, but for negative-parity low-lying states. The $A_{yy}$ point at $\theta_{cm}$ = 1.6° for the first $2^-$ state is outside the boundaries of the graph. Its value is -2.11 ± 1.27.

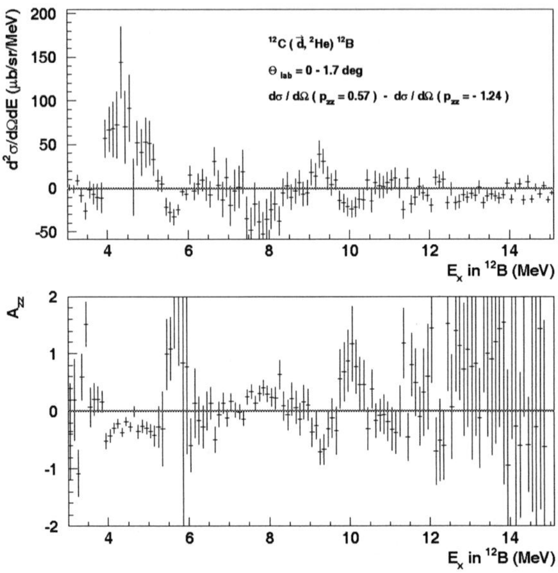


**FIGURE 5.** Upper panel: difference between double-differential cross sections at $\theta_{lab} = 0°\text{-}1.7°$ for the two beam polarisations as a function of excitation energy in $^{12}$B. Lower panel: corresponding $A_{zz}$ spectrum [2].

the bump at $E_x$ = 7.5 MeV. One state at $E_x$ = 7.85 MeV seems to contain most of the strength. This is again in accordance with the shell-model calculation referred to above and in agreement with the conclusions of Inomata regarding the mirror structure in $^{12}$N [20]. A large concentration of positive $A_{zz}$ is located at $E_x$ = 10.05(8) MeV, for which a width of $\Gamma$ = 470 keV could be deduced from the data but no conclusive identification of the $1^-$ character could be achieved. The $J^\pi = 2^-$ component is localised in two energy regions, in the bump at $E_x$ = 4.5 MeV and in the low-energy part of the bump at $E_x$ = 7.5 MeV. Again, a good agreement with shell-model predictions is found.

## CONCLUSIONS

The unique energy resolution achievable at KVI allowed determination of tensor-analysing powers for well-resolved transitions up to 5 MeV excitation energy for the $^{12}$C(d,$^2$He)$^{12}$B reaction . The analysing powers of states with known spin and parity provided the means for the first-time for verification of model-independent predictions of analysing powers at extreme forward angles. In the continuum region, the analysing powers allowed identification of the $J^\pi$ components of the IVSGDR, including a $0^-$ component located at $E_x = 9.3$ MeV. The results provide the stimulating basis to extend the analysing-power measurements to heavier nuclei and investigate the feasibility of the IVSGDR decomposition in complex nuclei, e.g., of relevance for $0\nu2\beta$-decay measurements.

## ACKNOWLEDGMENTS

This work was performed as part of the research program of the *Stichting voor Fundamenteel Onderzoek der Materie (FOM)* with financial support from the *Nederlandse Organisatie voor Wetenschappelijk Onderzoek (NWO)*.

## REFERENCES

1. *Summary of the workshop on: Nuclear matrix elements for neutrinoless double-beta decay*, May 23-24, 2005, Institute for Part. Phys. Phenomenology, University of Durham, UK, edited by K. Zuber, 200.
2. M.A. de Huu et al., *Phys. Lett.* **B 649**, 35 (2007).
3. A.M. van den Berg, *Nucl. Instr. and Meth.* **B 99**, 637 (1995).
4. H.J. Wörtche *Nucl. Phys.* **A 687**, 321c (2001).
5. M. Hunyadi et al., *Nucl. Phys.* **A 731**, 49 (2004).
6. M. Hunyadi et al., *Phys. Rev.* **C 75**, 014606 (2007).
7. D. Savran, M. Babilon, A.M. van den Berg, M.N. Harakeh, J. Hasper, H.J. Wörtche and A. Zilges *Phys. Rev. Lett.* **97**, 172502 (2006).
8. D. Savran, A.M. van den Berg, M.N. Harakeh, K. Ramspeck, H.J. Wörtche and A. Zilges, *Nucl. Inst. and Meth. in Phys. Res.* **A 564**, 267 (2006).
9. W.G. Love and M.A. Franey, *Phys. Rev.* **C 24**, 1073 (1981).
10. S. Rakers et al., *Nucl. Instr. and Meth.* **A 481**, 253 (2002).
11. S. Rakers et al., *Phys. Rev.* **C 71**, 054313 (2005).
12. H.R. Kremers and A.G. Drentje, *Polarized Gas Targets and Polarized Beams*, Vol. 421, AIP Conf. Proc., edited by R.J. Holt and M.A. Miller (AIP, New York, 1997), p. 507.
13. H. Laurent, H. Lefort, D. Beaumel, Y. Blumenfeld, S. Fortier, S. Galès, J. Guillot, J.C. Roynette, P. Volkov, and S. Brandenburg, *Nucl. Instr. and Meth.*, **A 326**, 517 (1992).
14. H. Okamura, *Phys. Rev.* **C 60**, 064602 (1999).
15. S. Rakers et al., *Phys. Rev.* **C 65**, 044323 (2002); and references therein.
16. F.P. Brady, T.D. Ford, G.A. Needham, J.L. Romero, D.S. Sorenson, C.M. Castaned, J.L. Drummond, B. McEachern E.L. Hjort, and N.S.P. King, *Phys. Rev. C* **43**, 2284 (1991).
17. N. Olsson et al., *Nucl. Phys.* **A559**, 368 (1993).
18. T. Suzuki and H. Sagawa, *Nucl. Phys.* **A637**, 547 (1998).
19. H. Okamura, T. Uesaka, K. Suda, H. Kumasaka, R. Suzuki, A. Tamii, N. Sakamoto, and H. Sakai, *Phys. Rev. C* **66**, 054602 (2002).
20. T. Inomata, et al., *Phys. Rev. C* **57**, 3153 (1998).

# Pygmy Dipole Strength and Neutron Skins in Exotic Nuclei

A. Klimkiewicz*,†, N. Paar**, P. Adrich*,†, M. Fallot*, K. Boretzky*, T. Aumann*, D. Cortina-Gil‡, U. Datta Pramanik*, Th. W. Elze§, H. Emling*, H. Geissel*, M. Hellström*, K.L. Jones*, J.V. Kratz¶, R. Kulessa†, C. Nociforo¶, R. Palit§, H. Simon*, G. Surówka†, K. Sümmerer*, D. Vretenar** and W. Waluś†

*Gesellschaft für Schwerionenforschung, D-64291 Darmstadt, Germany
†Instytut Fizyki, Uniwersytet Jagielloński, PL-30-059 Kraków, Poland
**Physics Department, Faculty of Science, Univers. of Zagreb, Bijenicka 32, 10000 Zagreb, Croatia
‡Universidad de Santiago de Compostela, 15706, Santiago de Compostela, Spain
§Institut für Kernphysik, Johann Wolfgang Goethe-Universität, D-60486 Frankfurt/M., Germany
¶Institut für Kernchemie, Johannes Gutenberg-Universität, D-55099 Mainz, Germany

**Abstract.** Dipole strength distributions were determined for the neutron-rich nuclei $^{129-132}$Sn and $^{133,134}$Sb from electromagnetic excitation in an experiment using the FRS-LAND setup. For all nuclei, a sizeable fraction of "pygmy" dipole strength at excitation energies well below the giant dipole resonance was observed. The integrated low-lying dipole strength of the nuclei with low neutron separation energies can be compared to results for stable nuclei (e.g. N=82 isotopes) determined for the energy regime of 5-9 MeV. A clear increase of the dipole strength with increasing asymmetry of the nuclei is observed. Comparing the ratio of the low-lying dipole over the giant dipole strength to recent relativistic mean field calculations, values for the parameters $a_4$ and $p_o$ of the symmetry energy and for the neutron skin thickness are derived. Averaged over $^{130}$Sn and $^{132}$Sn we extract $\overline{a}_4 = 31.8 \pm 1.3$ MeV and $\overline{p}_0 = 2.2 \pm 0.5$ MeV/fm$^3$. The neutron skin sizes are determined to $R_n$-$R_p = 0.23 \pm 0.03$ fm and $0.24 \pm 0.03$ fm for $^{130}$Sn and $^{132}$Sn, respectively. For $^{208}$Pb a neutron skin thickness of $R_n$-$R_p = 0.18 \pm 0.035$ fm follows, when applying the same method and using earlier published experimental findings on the dipole strength.

**Keywords:** $^{130,132}$Sn, Pygmy Dipole Resonance, Giant Dipole Resonance, Electromagnetic excitation, Neutron Skin, Symmetry Energy
**PACS:** 21.65.Cd, 21.65.Ef, 24.30.Cz, 25.60.-t, 25.70.De, 27.60.+j

## INTRODUCTION

A new collective dipole mode, the so called pygmy dipole resonance (PDR), was predicted to evolve in medium-heavy nuclei at the beginning of the 1990s [1, 2]. This concentration of dipole strength is energetically located below the dominant giant dipole resonance (GDR) and close to the particle-separation threshold. It is related to structural changes in nuclei with large neutron excess, resulting in the formation of a neutron skin. A redistribution of dipole strength towards lower energies could be observed in light nuclei [3]. According to microscopic calculations [4, 5], this effect originates from non-collective single-particle transitions. In contrast, for medium-heavy nuclei, recent random-phase approximation (RPA) calculations [5, 6, 7] support the picture of a resonant dipole oscillation of a neutron or proton skin against the isospin-symmetric core. In

CP1012, *Frontiers in Nuclear Structure, Astrophysics, and Reactions: FINUSTAR 2*
edited by P. Demetriou, R. Julin, and S. V. Harissopulos
© 2008 American Institute of Physics 0-7354-0532-5/08/$23.00

this collective picture, the low-lying dipole strength is directly related to the neutron skin thickness. The neutron skin thickness, the difference between the matter and the charge radii, is of fundamental interest, since its evolution is driven by the density dependence of the nuclear symmetry energy S($\rho$) and thus the equation of state in proton-neutron asymmetric matter. It has been suggested (e.g.[8, 9] that, from a precise determination of the neutron skin thickness in only one nucleus (e.g. in $^{208}$Pb), a valuable constraint on S($\rho$) can be deduced. An accurate and model-independent measurement on the neutron radius of $^{208}$Pb (to within 0.05 fm) is planned via parity-violating electron scattering [10].

In a recent paper [11] the question was raised whether the experimentally observed pygmy dipole (E1) strength could play a role equivalent to the neutron skin thickness in constraining S($\rho$), since both quantities are of the same origin.

Up to now, only little experimental evidence for the PDR in medium- heavy nuclei exists. The dipole responses of stable N=82 isotones and of $^{208}$Pb investigated in real-photon-absorption measurements below the particle threshold exhibit a concentration of E1 strength of less than 1% of the TRK sum rule below $\approx$8 MeV [12, 13, 14]. Today, the availability of high-energy radioactive beams enables a systematic study of the dipole response of nuclei as a function of isospin with much larger proton-neutron asymmetries than provided by stable isotopes.

Using Coulomb dissociation of exotic secondary beam particles at kinetic energies of about 500 $A$ MeV and the LAND complete- kinematics reaction setup at GSI, the dipole response has been studied for the tin isotopic chain $^{129-132}$Sn and for $^{133,134}$Sb. In this paper, we present a first attempt to constrain the parameters of the nuclear symmetry energy S($\rho$) by comparing the experimental findings for $^{130}$Sn and $^{132}$Sn to recent calculations within the framework of the relativistic Hartree-Bogoliubov model (RHB) and the relativistic quasi-particle random phase approximation (RQRPA). Values for the neutron skin thickness in these two neutron-rich Sn isotopes and for $^{208}$Pb are derived.

## EXPERIMENTAL FINDINGS ON LOW-LYING STRENGTH IN NEUTRON-RICH TIN ISOTOPES

The electromagnetic excitation of unstable tin isotopes, including $^{132}$Sn and neighboring nuclei, was measured in inverse kinematics at beam energies of about 500 $A$ MeV using the FRagment Separator FRS and the reaction setup LAND at GSI [15]. The measured four-momenta of all decay products, i.e. heavy fragment, neutron(s), and $\gamma$-rays, allow to determine the excitation energies E* of the projectiles applying the invariant-mass method. The validity of this method, being limited to excitation energies above the neutron threshold, was proven for stable heavy ions and is explained in detail in [16]. Besides the dominant giant dipole resonance (GDR) peak, the excitation-energy spectra for all measured nuclei exhibit additional strength at lower excitation energies. On the left-hand side of Fig.1, low-lying dipole-strength spectra are presented after subtraction of the GDR and the isovector and isoscalar giant-quadrupole-resonance components. The remaining strength amounts to about 5% of the Thomas-Reiche Kuhn (TRK) sum rule. For the nuclei with lower neutron-separation thresholds an extension of the low-lying dipole strength to lower energies is observed. On the right-hand side of Fig.1 we

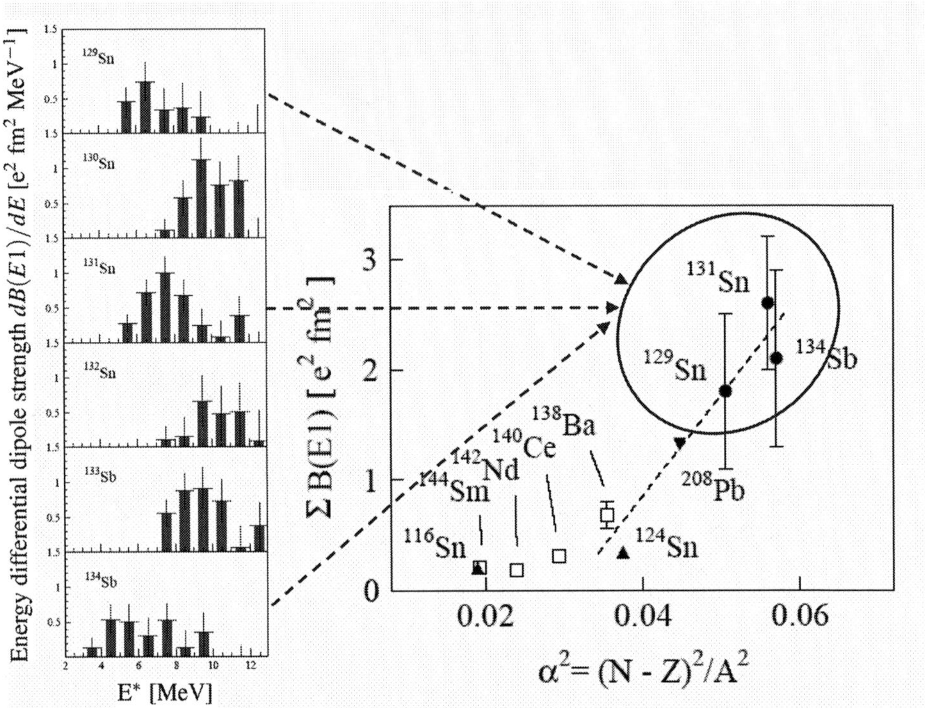


**FIGURE 1.** Left-hand side: Differential dipole- strength distribution $dB(E1)/dE$ obtained for unstable Sn and Sb isotopes (the arrows indicate the neutron separation energies). Right-hand side: Experimental results for the integrated B(E1) strength as a function of $((N-Z)/A)^2$, reflecting the proton-neutron asymmetry of the nuclei. Data are displayed for stable N=82 isotopes [13], for $^{116,124}$Sn [12] and for $^{208}$Pb [14]. The B(E1) strength for $^{129,131}$Sn and $^{134}$Sb (this experiment) has been integrated from 5 to 9 MeV, for stable nuclei up to the neutron threshold.

compare for the first time the results from experiments using real-photon absorption on stable nuclei to the results presented here for exotic nuclei, covering the same excitation-energy range from 5-9 MeV. A clear increase of the low-lying dipole strength with increasing neutron-proton asymmetry is observed.

## LOW-LYING STRENGTH AND THE NEUTRON SKIN

Using the experimental findings on the low-lying dipole strength in $^{130}$Sn and $^{132}$Sn of 7(3)% and 4(3)% of the TRK sum rule, respectively, [15] we deduce values for the parameters of the symmetry energy [17]. By comparing to a systematic series of relativistic mean-field calculations within the framework of RHB+RQRPA [7, 18], we find an average value for the symmetry energy in pure neutron matter of $\overline{a}_4 = 31.8 \pm 1.3$ MeV from the $^{130,132}$Sn analysis. For the symmetry- energy pressure we obtain a value of $\overline{p}_0 = 2.2 \pm 0.5$ MeV/fm$^3$. This value is strongly correlated, however, with $\overline{a}_4$ within this framework. Neutron-skin thicknesses of $R_n$-$R_p = 0.23 \pm 0.03$ fm

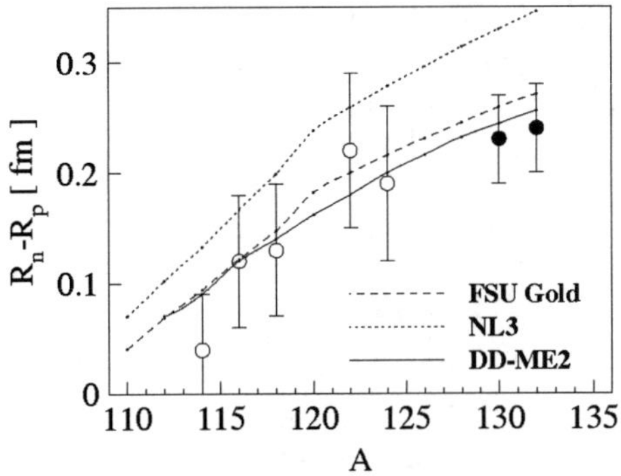


**FIGURE 2.** The neutron skin thickness for $^{130}$Sn and $^{132}$Sn (filled circles), determined in this work is compared to data for stable isotopes (open circles) from [20]. The error bars for the unstable Sn isotopes represent the statistical errors only. Results from different relativistic mean-field calculations [11, 21] are displayed as lines. From [17].

and of $0.24 \pm 0.03$ fm follow for $^{130}$Sn and $^{132}$Sn, respectively, when employing the averaged $\overline{a}_4$ parameter in the calculations. From Fig.2, it is evident that the neutron skin thicknesses of the neutron-rich Sn isotopes follow the trend of those determined for stable Sn nuclei from measured cross sections of the isovector spin-dipole resonances (SDR) [20].

Similar relativistic mean-field calculations were performed for $^{208}$Pb. Taking into account also data on the low-lying dipole strength [14] and the GDR strength [19] in $^{208}$Pb, the resulting $\overline{a}_4$ value is compatible with the one obtained for the $^{130,132}$Sn isotopes. For the neutron skin thickness we determine a value of $R_n$-$R_p = 0.18 \pm 0.035$ fm, which is consistent with the value of 0.17 fm reported for $^{208}$Pb from a new analysis of proton elastic-scattering data in Ref.[22].

## CONCLUSIONS AND FUTURE PERSPECTIVES

From an experiment with the FRS-LAND setup, using electromagnetic excitation of the neutron-rich projectiles impinging on a Pb target, dipole-strength distributions were determined for $^{129-132}$Sn and $^{133,134}$Sb. A substantial amount of dipole strength was found close to the neutron threshold at energies below the GDR region, in contrast to experimental findings in stable nuclei. For the nuclei with lower neutron separation energies, an extension of this low-lying dipole strength down to the neutron threshold is observed. Since the experimental data on pygmy dipole strength distributions exhibit a correlation with the degree of neutron-proton asymmetry in nuclei, values for the neutron skin thickness as well as for the parameters of the symmetry energy term of the equation of state can be extracted by comparison to recent calculations within the RHB+RQRPA framework . The determination of the neutron skin thickness via the relative strength of the PDR relative to the GDR mode leads to accuracies compatible with the one envisaged from a very ambitious parity-violating electron scattering experiment on $^{208}$Pb at Jefferson Laboratory [10]. We emphasize, however, that further experimental and theoretical investigations are required in order to confirm the reliability of our method.

Recently, an experiment exploring the dipole response above the neutron separation threshold in neutron-rich Ni isotopes, including $^{72}$Ni, was performed using the FRS-LAND setup [23]; the data are presently being analyzed. A complementary $(\gamma,\gamma\prime)$ experiment in $^{68}$Ni was performed at similar beam energies using the RISING setup at the FRS at GSI [24, 25]. From the combination of these two complementary datasets, an insight into the decay characteristics and thus the nature of the PDR mode is expected.

A first experiment studying the dipole strength in a proton-rich nucleus, namely $^{32}$Ar, is planned at GSI in 2008. [26].

## ACKNOWLEDGMENTS

This work was supported by the German Federal Minister for Education and Research (BMBF) under Contract No. 06MZ174, by the Polish Committee of Scientific Research under Contract No. 1 P03B 001 27, by the European Commission within the $5^{th}$ Framework Programme through Contract No. HPRI-CT-1999-50010 and by the Helmholtz Association of German Research Centres through the Research Programme 'Physics of Hadrons and Nuclei'.

## REFERENCES

1. Y. Suzuki, K. Ikeda, and H. Sato, Prog. Theor. Phys. **83**, 180 (1990).
2. P. V. Isacker, M. A. Nagarajan, and D. D. Warner, Phys. Rev. C **45**, R13 (1992).
3. A. Leistenschneider et al., Phys. Rev. Lett. **86**, 5442 (2001).
4. G. Colò and P.F. Bortignon, Nucl. Phys. A **696**, 427 (2001).
5. D. Vretenar, N. Paar, P. Ring, and G. A. Lalazissis, Nucl. Phys. A **692**, 496 (2001).
6. N. Tsoneva, H. Lenske, and C. Stoyanov, Phys. Lett. B **586**, 213 (2004).
7. N. Paar, P. Ring, T. Nikšić, and D. Vretenar, Phys. Rev. C **67**, 34312 (2003).
8. B.A. Brown, Phys. Rev. Lett. **85**, 5298 (2000).
9. R.J. Furnstahl, Nucl. Phys. A **706**, 85 (2002).
10. C.J. Horowitz, S.J. Pollock, P.A. Souder, R. Michaels, Phys. Rev. C **63**, 025501 (2001).
11. J. Piekarewicz, Phys. Rev. C **73**, 044325 (2006).
12. K. Govaert et al., Phys. Rev. C **57**, 2229 (1998).
13. A. Zilges et al., Phys. Lett. B **542**, 43 (2002).
14. N. Ryezayeva et al., Phys. Rev. Lett. **89**, 272502 (2002).
15. P. Adrich et al., Phys. Rev. Lett. **95**, 132501 (2005).
16. K. Boretzky et al., Phys. Rev. C **68**, 024317 (2003).
17. A. Klimkiewicz et al., submitted to Phys. Rev 2007.
18. D. Vretenar, A.V. Afanasjev, G.A. Lalazissis, P. Ring, Phys. Rep. **409**, 101 (2005).
19. A. Vessière et al., Nucl. Phys. A **159**, 561 (1970),
20. A. Krasznahorkay et al., Phys. Rev. Lett. **82**, 3216 (1999)
21. N. Paar et al., Phys. Lett. B **606**, 288 (2005).
22. S. Karataglidis et al., Phys. Rev. C **65**, 044306 (2002).
23. P. Adrich et al., Experiment S287: The giant dipole resonance and low-lying E1 strength in Ni isotope, performed in 2005.
24. A. Bracco et al., Acta Phys. Pol. **38**, 1229 (2007).
25. S. Leoni, proceedings to this conference.
26. K. Boretzky et al., Experiment S327 accepted by GSI PAC April 2006, http://www.gsi.de/documents/DOC-2006-Mar-126-1.pdf.

# Neutron skins in spherical and deformed nuclei from Skyrme Hartree-Fock calculations

P. Sarriguren*, M.K. Gaidarov†,*, E. Moya de Guerra** and A.N. Antonov†

*Instituto de Estructura de la Materia, CSIC, Serrano 123, E-28006 Madrid, Spain
†Institute for Nuclear Research and Nuclear Energy, Bulgarian Academy of Sciences, Sofia 1784, Bulgaria
**Departamento de Fisica Atomica, Molecular y Nuclear, Facultad de Ciencias Fisicas, Universidad Complutense de Madrid, E-28040 Madrid, Spain

**Abstract.**
The development of neutron skins as a function of the neutron number is investigated within a self-consistent framework based on deformed Hartree-Fock calculations with density dependent Skyrme forces and pairing correlations in BCS approximation. We study several isotopic chains and consider all the experimentally observed isotopes from neutron-deficient to neutron-rich. We discuss various possibilities to define the neutron skin thickness based on the differences between neutron and proton radii as well as definitions based on comparison of the tails of the neutron and proton density distributions. We finally discuss the effects of deformation on the neutron skin based on the example of Kr isotopes. We study the neutron skins in different directions for oblate and prolate shapes and discuss it in terms of the quadrupole components of the density distributions.

**Keywords:** Neutron skins, Deformed nuclei, Selfconsistent calculations
**PACS:** 21.10.Gv, 21.60.Jz

## INTRODUCTION

The detailed study of the properties of unstable nuclei far from the stability line has been one of the main goals of nuclear physics in the last years. In heavy nuclear systems, due to availability of far more neutrons than protons, the formation of pronounced neutron skins is possible. To get information on the neutron skin thickness one needs data obtained with probes having different sensitivities to the proton and neutron distributions. The methods for extracting the neutron skin thickness mostly include hadron scattering, antiprotonic atoms, parity violating electron scattering, as well as giant dipole resonance and spin-dipole resonance methods.

On the theoretical side, calculations of nuclear charge and matter radii of exotic nuclei are usually made in the framework of mean-field approaches with pairing correlations. Theoretical identification of skin structures in neutron-rich weakly bound nuclei, however, is still a matter of discussion. In Ref. [1] a definition of the neutron skin and its appearance were explored in terms of some proposed criteria which deal with the proton and neutron density tails. The Helm model has been applied in Ref. [2] to analyze neutron skins in terms of form factors and different definitions were proposed for neutron-proton radii differences.

In this work we study the spread of the results obtained for the neutron skin by using different definitions proposed previously within a given microscopic nuclear structure

CP1012, *Frontiers in Nuclear Structure, Astrophysics, and Reactions: FINUSTAR 2*
edited by P. Demetriou, R. Julin, and S. V. Harissopulos
© 2008 American Institute of Physics 0-7354-0532-5/08/$23.00

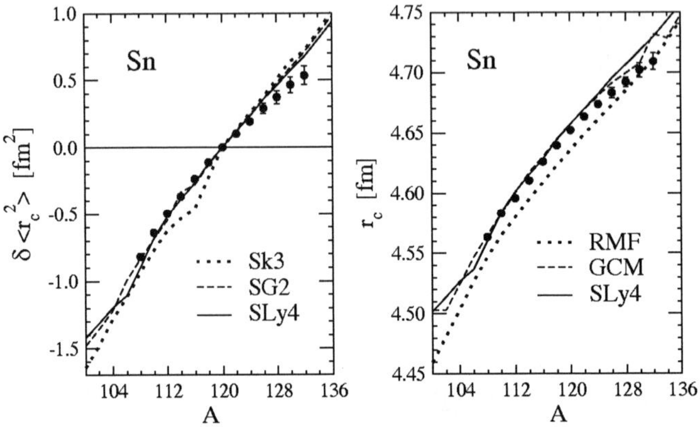


**FIGURE 1.** Left panel: Theoretical (with different Skyrme forces) and experimental isotope shifts $\delta\langle r_c^2\rangle$ of tin isotopes relative to $^{120}$Sn. Right panel: Charge rms radii $r_c$ of tin isotopes.The SLy4 result is compared with the results from RMF calculations [3] and GCM [4]. Experimental data are from [5];

model, the deformed self-consistent mean-field Skyrme HF+BCS method with SLy4 force. We also study, on the example of Kr isotopes, whether the emergence of a skin is influenced by the nuclear shape, an issue that has not been sufficiently studied yet.

## RESULTS AND DISCUSSION

We start by showing our results for the rms radii of the charge distributions, comparing them to the experimental information available [5]. We also compare our results with different theoretical calculations: Relativistic Mean Field (RMF) calculations with NL3 parametrization [3] and nonrelativistic GCM calculations [4] performed on top of HFB states obtained with the D1S Gogny effective interaction.

We show on the left panel of Fig. 1 our results for the mean squared charge radii differences in Sn isotopes obtained from three different Skyrme forces, SLy4, SG2 and Sk3. We compare them to experiment, taking the radius of $^{120}$Sn as the reference. On the right panel we compare our SLy4 results for the charge radii with the other theoretical approaches mentioned above. The general purpose of Fig. 1 is firstly to show that different Skyrme forces do not differ much in their predictions of charge rms radii and secondly, to show that our results with SLy4 are comparable to other theoretical predictions including approaches that go beyond the mean-field approximation, as well as relativistic approaches. Therefore, the approach used here is a good starting point to make predictions for other quantities such as neutron-proton radii differences, where the experimental information is scarce and it is not as accurate as in the case of charge radii. In Fig. 2 we plot the differences between the rms of neutrons and protons $\Delta r_{np} = r_n - r_p$ for Sn isotopes and compare them to RMF results and to experimental data.

The thickness of a neutron skin in nuclei may be defined in different ways. One of these possibilities is to define it as the difference between the rms radius of neutrons and that of protons, as we have plotted in Fig. 2. Similarly, it can be defined as the

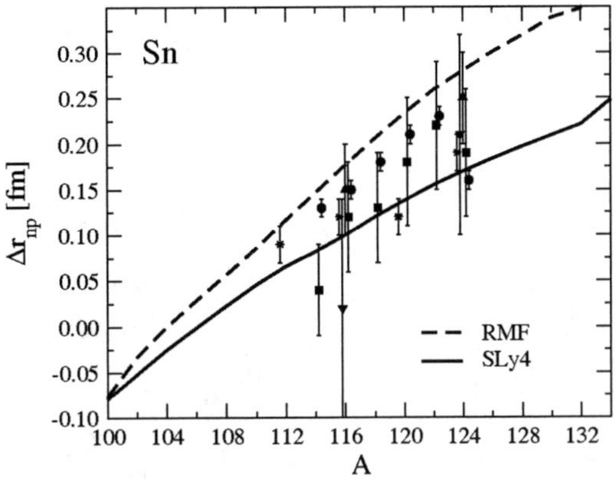


**FIGURE 2.** Difference between neutron and proton rms radii $\Delta r_{np}$ of Sn isotopes calculated with SLy4 force. The RMF calculation results are from Ref. [3]. The experimental data for Sn isotopes measured in $(p,p)$ reaction (open stars) [6], antiproton atoms (full stars) [7], giant dipole resonance method (full circles) [8] and spin dipole resonance method (full and open squares) [9] are also shown.

difference between the neutron and proton radii of the equivalent uniform spheres $R_{hs} = \sqrt{5/3} < r^2 >^{1/2}$. Alternatively, it can be defined as the difference between the neutron and proton diffraction radii $R_d$ or Helm radii $R_{\text{Helm}}$. All of these quantities have already been discussed and used in the past as possible ways to quantify the skin thickness (see for example [2]).

On the other hand, the skin thickness can be also defined in terms of some criteria that the neutron and proton densities must fulfill. In Ref. [1] the neutron skin thickness is defined as the difference between two radii, $R_1$ and $R_2$. $R_1$ is the radius at which the ratio of the neutron density to the proton density is equal to some given value (4 in [1]). $R_2$ is the radius at which the neutron density becomes smaller than some percentage of the density at the center of the nucleus (1 % in [1]). These criteria could have been chosen in rather different manners. Thus, the absolute sizes of the skin thickness do not have a very precise meaning. Nevertheless, these values are useful to judge how the nucleon skins develop as the number of nucleons change. Indeed, we have also considered the case where the first criterion for the inner radius $R_1$ is changed from the above criterion to the radius at which the proton density becomes smaller than 1% of the latter at the center. When we use the conditions in Ref. [1], we call it criterion (a). When we use the alternative condition for $R_1$, we call it criterion (b).

We show in Fig. 3 the results obtained for the neutron skin thickness in Sn isotopes according to the different definitions discussed above. The left panel contains the results for definitions involving directly the difference between neutron and proton radii, either the equivalent hard spheres radii $\Delta R_{\text{hs}}$ corresponding to the rms radii, the diffraction radii $\Delta R_{\text{d}}$, and the Helm radii $\Delta R_{\text{Helm}}$. The skin thickness predicted by the difference of the very simple diffraction radii is in general smaller than the thickness predicted by the other two more involved options that are very similar in this range of masses. The right panel contains the neutron skin thickness defined according to the criteria on the density distributions (a) (solid line) and (b) (dashed line). It is observed that in the region of

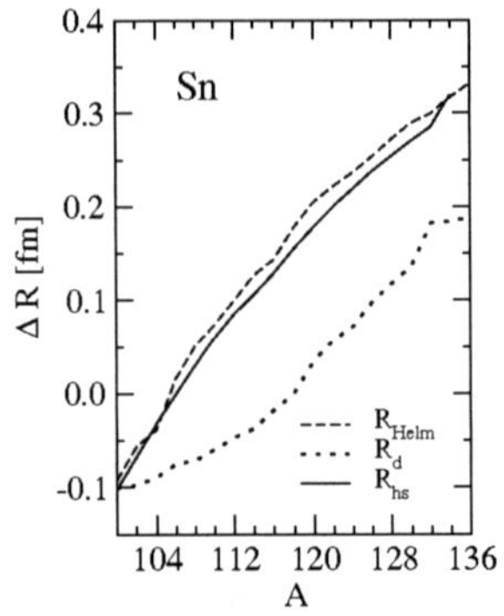


**FIGURE 3.** Neutron skin thicknesses for tin isotopes. Left panel: $\Delta R_{\rm d}$, $\Delta R_{\rm hs}$, and $\Delta R_{\rm Helm}$ as defined in the text. Right panel: corresponding to criterion (a) (solid line) and criterion (b) (dashed line).

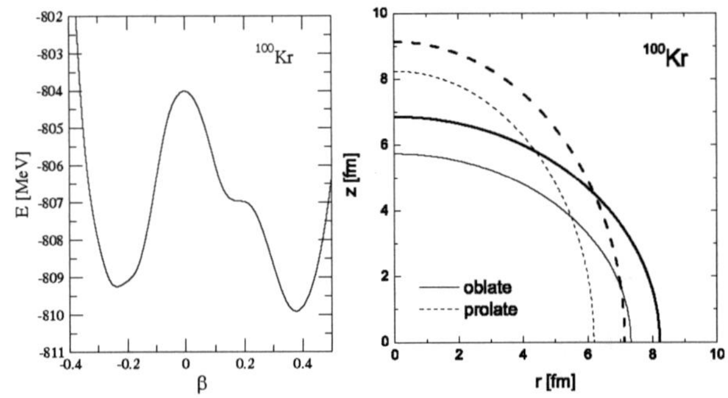


**FIGURE 4.** Left: Binding energies $E$ calculated with the SLy4 force as a function of the quadrupole parameter $\beta$ for $^{100}$Kr. Right: Borders of the neutron skins in $^{100}$Kr shown in $rz$ plane. They correspond to its oblate (solid lines) and prolate (dashed lines) shape.

heavier isotopes, where we can properly talk about a neutron skin formation, criterion (b) somehow establishes a lower limit for the skin thickness. Similar results are obtained in Ni and Kr isotopic chains [10].

When the nucleus is deformed, the thickness of the neutron skin might depend on the direction. It is an interesting and natural question to ask whether the deformed densities give rise to a different skin size in the different directions. It is also interesting to know whether the emergence of the skin may be influenced by the nuclear shape. We study in this work such a dependence on the example of $^{100}$Kr, which is an example of a well deformed nucleus. The results obtained for the binding energy as a function of the quadrupole parameter $\beta$ can be seen in Fig. 4. As we can see, both prolate and oblate shapes produce minima very close in energy.

The profiles of the densities as well as the spatial extensions change with the direction, being more extended in the $z(r)$-direction in the case of prolate (oblate) shapes. A discussion of these features in terms of the quadrupole components of the density distributions can be found in Ref. [10].

It is also worth looking at the points in the $(r,z)$ plane that define the borders of the neutron skin, where the criteria for $R_1$ and $R_2$ are met. Figure 4 shows these points for protons ($R_1$, thin lines) and neutrons ($R_2$, thick lines) and for the two shapes in $^{100}$Kr, prolate (solid) and oblate (dashed). As we can see, the radial extensions are different along the various directions but the neutron skins are similar.

## CONCLUSIONS

In this work we perform a theoretical analysis of nuclear skins, exploring various definitions. The densities are calculated within a given microscopic model, the deformed HF+BCS approach with Skyrme-type density-dependent effective interactions. Different definitions which have been previously proposed to determine the neutron skin thickness, involving both matter radii and tails of nuclear densities, have been tested within this model. We find that all definitions of the neutron skin predict to a different extent the existence of a skin in nuclei far from the stability line. We also find that for a given isotopic chain the increase of the skin with the neutron number in the neutron-rich nuclei exhibits a rather constant slope, which is different depending on the definition of nuclear skin.

In the present work the effects of deformation on the skin formation are studied in Kr isotopes which are well deformed nuclei. Taking as an example $^{100}$Kr isotope, we find that the profiles of the proton and neutron densities, as well as the spatial extensions change with the direction in both oblate and prolate shapes. At the same time, the neutron skin thickness remains almost equal along the different directions perpendicular to the surface.

## ACKNOWLEDGMENTS

This work was partly supported by the Bulgarian National Science Fund under Contracts No. Φ–1416 and Φ–1501, and by Ministerio de Educación y Ciencia (Spain) under contracts No. FIS2005–00640 and SAB2005–0012.

## REFERENCES

1. N. Fukunishi, T. Otsuka, and I. Tanihata, *Phys. Rev. C* **48**, 1648 (1993).
2. S. Mizutori, J. Dobaczewski, G. A. Lalazissis, W. Nazarewicz, and P.-G. Reinhard, *Phys. Rev. C* **61**, 044326 (2000).
3. G. A. Lalazissis, S. Raman, and P. Ring, *At. Data Nucl. Data Tables* **71**, 1 (1999).
4. J. Libert, B. Roussiere, and J. Sauvage, *Nucl. Phys.* **A786**, 47 (2007).
5. C. Piller *et al.*, *Phys. Rev. C* **42**, 182 (1990); F. Le Blanc *et al.*, *Phys. Rev. C* **72**, 034305 (2005).
6. L. Ray *et al.*, *Phys. Rev. C* **19**, 1855 (1979); G. W. Hoffmann *et al.*, *Phys. Rev. Lett.* **47**, 1436 (1981).
7. A. Trzcinska *et al.*, *Phys. Rev. Lett.* **87**, 082501 (2001).
8. A. Krasznahorkay *et al.*, *Nucl. Phys.* **A567**, 521 (1994).
9. A. Krasznahorkay *et al.*, *Phys. Rev. Lett.* **82**, 3216 (1999); *Nucl. Phys.* **A731**, 224 (2004).
10. P. Sarriguren, M.K. Gaidarov, E. Moya de Guerra, and A.N. Antonov, submitted for publication.

# Probing collectivity in the vicinity of neutron deficient Pb nuclei

T. Grahn*, A. Dewald†, M. B. Gómez Hornillos**, P. T. Greenlees‡, A. Görgen§, K. Helariutta¶, J. Jolie†, P. Jones‡, R. Julin‡, S. Juutinen‡, S. Ketelhut‡, R. Krücken||, T. Kröll||, M. Leino‡, J. Ljungvall§, P. Maierbeck||, B. Melon†, M. Nyman‡, R. D. Page*, A. Petts*, Th. Pissulla†, P. Rahkila‡, J. Sarén‡, C. Scholey‡, J. Sorri‡, J. Uusitalo‡, R. Wadsworth†† and M. Zielińska§,‡‡

*Oliver Lodge Laboratory, Department of Physics, University of Liverpool, Liverpool L69 7ZE, UK
†Institut für Kernphysik, Universität zu Köln, Zülpicher Str. 77, 50937 Köln, Germany
**STFC, Daresbury Laboratory, Warrington WA4 4AD, UK
‡Department of Physics, University of Jyväskylä, P.O.Box 35, 40014 Jyväskylä, Finland
§CEA-SACLAY, DSM/DAPNIA/SPhN, F-91191 Gif-Sur-Yvette Cedex, France
¶Laboratory of Radiochemistry, Department of Chemistry, PL 55, FI-00014 Helsinki
||Physik-Department E12, TU München, 85748 Garching, Germany
††Department of Physics, University of York, Heslington, York YO10 5DD, UK
‡‡Heavy Ion Laboratory, Uninversity of Warsaw, Pasteura 5A, 02-093 Warszawa, Poland

**Abstract.** A series of recoil distance Doppler-shift lifetime measurements have been carried out to probe collectivity and configuration mixing of different shapes in the vicinity of neutron mid-shell Pb nuclei. Lifetime measurements of $^{186}$Pb and $^{194}$Po, the first ever utilizing the recoil-decay tagging method, probed the collectivity of coexisting prolate and oblate shapes in this region. Futher lifetime measurements of excited states in $^{180}$Hg, $^{182}$Hg and $^{196}$Po have been carried out.

**PACS:** 21.10.Ky, 21.10.Tg, 27.20.+q, 27.80.+w

## INTRODUCTION

The origin of shape coexistence observed in nuclei near the $Z = 82$ shell closure and close to the proton drip line is one of the current challenges for nuclear structure research. In the neutron mid-shell nucleus $^{186}$Pb, the first two excited states above the spherical ground state have been assigned as $0^+$ states. In the shell model approach these states can be associated with intruder $\pi(2p-2h)$ and $\pi(4p-4h)$ excitations across the $Z = 82$ gap [1]. This picture is supported by hindrance factors derived from the $\alpha$ decays feeding these states [2]. In calculations based on the deformed mean field they appear as oblate and prolate minima intruding down close to the ground state when approaching the neutron mid-shell [3, 4]. A similar picture is seen in recent Hartree-Fock-Bogoliubov (HFB) self-consistent mean-field studies [5, 6]. The yrast line above the spherical ground state in the even mass isotopes $^{182-188}$Pb [7, 8, 9, 10] is formed by a collective band associated with the prolate minimum. Recently, candidates for collective non-yrast bands built on the coexisting oblate minimum have been observed in $^{186}$Pb and $^{188}$Pb [11, 12].

CP1012, *Frontiers in Nuclear Structure, Astrophysics, and Reactions: FINUSTAR 2*
edited by P. Demetriou, R. Julin, and S. V. Harissopulos
© 2008 American Institute of Physics 0-7354-0532-5/08/$23.00

In Po nuclei, just two protons outside the $Z = 82$ shell closure, even more striking behavior of intruder structures has been observed. When approaching the neutron mid-shell, states associated with oblate deformed $\pi(4p-2h)$ configurations cross the yrast line of nearly spherical states reaching the ground state in $^{192}$Po [13, 14]. Furthermore, a sudden change is observed in $^{190}$Po [15] to an yrast band similar to the prolate bands in $^{182-188}$Pb . This picture is also supported by earlier Nilsson-Strutinsky calculations [3].

The mean-field and HFB calculations predict that the degree of deformation associated with the prolate shape of mid-shell Pb and Po nuclei is larger than that for the oblate one. Indeed, the moment of inertia extracted from the level energies is in general much larger for the prolate than for the oblate bands. However, the moment of inertia is not a good measure of deformation and can be influenced by other factors.

In the case of nuclear shape coexistence, fundamental questions are the separation of the energy minima and the degree of mixing of shapes as a function of spin. In addition to predictions of various theoretical calculations [16, 17, 18], simple mixing calculations based on experimental data have been carried out for both mid-shell Pb and Po nuclei [19, 12, 20].

For the conclusive measure of collectivity and the degree of mixing of different shapes, the knowledge of absolute transition probabilities is required. Recently, a study of lifetimes of low-lying yrast states in $^{186}$Pb, $^{188}$Pb and $^{194}$Po have been completed [21] and lifetime measurements in $^{180}$Hg, $^{182}$Hg and $^{196}$Po have been carried out.

## EXPERIMENTS

Heavy nuclei near the neutron mid-shell can be produced in heavy-ion induced fusion evaporation reactions providing high recoil velocities. Therefore, transition probabilities can be obtained in lifetime measurements employing the Recoil Distance Doppler-Shift (RDDS) method. Due to low production cross sections of these nuclei, in-beam detection of $\gamma$ rays of interest requires high selectivity. This can be provided by using tagging techniques with a plunger device combined with an efficient Ge-detector setup and a recoil separator. However, the stopper foil of the conventional plunger device has to be replaced by a degrader foil, allowing the evaporation residues to recoil into the recoil separator.

In the experiments carried out at the Accelerator Laboratory of the University of Jyväskylä (JYFL) the Köln plunger device was combined with the JUROGAM Ge-detector array at the RITU gas-filled separator [22]. The separated recoils were detected at the RITU focal plane by the GREAT spectrometer [23]. Prompt $\gamma$ rays were detected with 15 EUROGAM Phase I Ge detectors [24] of the JUROGAM array, five at 158° and ten at 134° with respect to the beam direction. Signals from all detectors were collected employing a triggerless Total Data Readout data acquisition system [25]. Selection of recoils was made using the energy loss and time-of-flight information provided by the GREAT multiwire proportional counter and the Double Sided Silicon Strip Detectors (DSSSDs). Temporal and spatial correlations of a recoil and its subsequent radioactive decay in the DSSSDs were performed and Recoil-Decay Tagged (RDT) [26] and recoil-gated $\gamma$-ray spectra were constructed.

The spectra of the two JUROGAM Ge-detector rings were both separately analyzed

**TABLE 1.** Experimental details.

| Reaction | $E_{beam}$ [MeV] | $v/c$ | Tagging mode | No of distances | min/max distance [$\mu$m] |
|---|---|---|---|---|---|
| $^{106}$Pd($^{83}$Kr,3n)$^{186}$Pb | 357 | 3.8% | RDT | 11 | 10 - 1600 |
| $^{108}$Pd($^{83}$Kr,3n)$^{188}$Pb | 340 | 3.8% | Recoil gating | 13 | 5 - 3000 |
| $^{114}$Cd($^{83}$Kr,3n)$^{194}$Po | 375 | 3.6% | RDT | 10 | 10 - 600 |
| $^{113}$Cd($^{86}$Kr,3n)$^{196}$Po | 382 | 3.5% | Recoil gating | 13 | 20 - 6000 |
| $^{94}$Mo($^{88}$Sr,2n)$^{180}$Hg | 378 | 4.0% | RDT | 11 | 30 - 1500 |
| $^{96}$Mo($^{88}$Sr,2n)$^{182}$Hg | 371 | 4.0% | Recoil gating | 15 | 15 - 3000 |

by means of Differential Decay Curve Method [27]. Intensities of the fully Doppler-shifted components of the transitions under investigation, recorded at different target-to-degrader distances, were normalized to the sum of the areas of the fully Doppler-shifted and degraded components. The resulting lifetime of each level is an average of the lifetimes extracted from the decay curves measured by the detector rings at 158° and 134°.

## Lifetimes in $^{186}$Pb, $^{188}$Pb and $^{194}$Po

For the first time, the RDT method was employed in RDDS lifetime measurements of $^{186}$Pb and $^{194}$Po. The $\alpha$ decays of $^{186}$Pb ($t_{1/2}$ = 4.8 s, $E_\alpha$ = 6.38 MeV) and $^{194}$Po ($t_{1/2}$ = 390 ms, $E_\alpha$ = 6.99 MeV) were used to tag prompt $\gamma$ rays of interest and singles RDT $\gamma$ ray spectra were collected. Lifetimes of the yrast levels in $^{186}$Pb up to the prolate $8^+$ state and in $^{194}$Po up to the oblate $4^+$ state were extracted [21]. Due to the relatively long $\alpha$ decay half-life and low branching ratio, the recoil-gating method was used in the RDDS measurement of $^{188}$Pb, where the lifetime information was extended up to the prolate $8^+$ yrast state [21]. Experimental details of these measurements are listed in Table 1.

## RDDS measurements of $^{180}$Hg, $^{182}$Hg and $^{196}$Po

The aforementioned RDDS measurements demonstrated the feasibility of tagging methods in lifetime measurements and, hence, enabled RDDS studies of rare species far from stability. Recently, lifetime measurements of low-lying yrast states in $^{180,182}$Hg and $^{196}$Po have been completed at JYFL. Table 1 presents the details of these measurements. The $\alpha$ decay with $E_\alpha$ = 6.20 MeV and $t_{1/2}$ = 2.58 s was used to associate prompt $\gamma$ rays with $^{180}$Hg and singles RDT $\gamma$ ray spectra were recorded. In the RDDS measurements of $^{182}$Hg and $^{196}$Po recoil gating method was used and $\gamma\gamma$ coincidence spectra were collected.

## RESULTS AND DISCUSSION

With the assumption of a rotating, quadrupole-deformed nucleus, absolute values of transition quadrupole moment ($Q_t$) and quadrupole deformation parameter ($\beta_2$) can be extracted from the measured $B(E2)$ values. Fig. 1 shows the preliminary $|Q_t|$ values for the three lowest yrast states in $^{180}$Hg and two lowest yrast states in $^{196}$Po together with those for $^{186,188}$Pb and $^{194}$Po.

The spherical ground state in Pb nuclei [28] is reflected by the low $|Q_t|$ value of the $2^+$ state in $^{186,188}$Pb. The large $|Q_t|$ values of the $6^+$ and $8^+$ states reveal high collectivity of these states and are in agreement with the hypothesis of prolate yrast states. A similar high $|Q_t|$ value is observed for the $4^+$ state in $^{186}$Pb indicating a rather pure prolate structure for the $4^+$ state. However, in $^{188}$Pb the $|Q_t|$ value of the $4^+$ state is significantly lower than that in $^{186}$Pb suggesting a strongly mixed character of the first excited $2^+$ state in $^{188}$Pb. In neutron deficient Hg nuclei, the weakly deformed oblate band based on the ground state is crossed by an intruding prolate band with a very similar moment of inertia to those in Pb nuclei [14]. Preliminary results of the RDDS measurement of $^{180}$Hg reveal that the collectivity of this prolate band is slightly lower than those in $^{186,188}$Pb (see Fig.1). The oblate ground state in $^{180}$Hg is reflected by the lower $|Q_t|$ value of the $2^+$ state.

For $^{194}$Po the collectivity of the $2^+$ and $4^+$ states is much lower than for the prolate states in $^{186,188}$Pb, which is expected for an oblate structure. The similar $|Q_t|$ values for these states indicate their similar intrinsic structure. Hence, the ground state of $^{194}$Po represents predominantly oblate shape. Preliminary results reveal that the situation is somewhat different in $^{196}$Po. The significantly lower $|Q_t|$ value for the $2^+$ state than that for the $4^+$ state suggests the ground state of $^{196}$Po has a large spherical contribution, whereas the $4^+$ state is the pure oblate state.

As a summary, lifetimes of the low-lying yrast states in $^{186,188}$Pb and $^{194}$Po have been measured in order to verify the collectivity of the bands and shed more light on the configuration mixing at low spin. For the $2^+$ state in $^{188}$Pb the $|Q_t|$ value indicates that mixing of the oblate and prolate shapes plays a crucial role. In $^{194}$Po, the contribution of oblate shape in the ground state could be dominant. To elucidate this question, lifetimes in $^{196}$Po, where intruder structures lie higher in energy, have been probed. In addition, lifetimes in $^{180,182}$Hg have been measured in order to study the the collectivity of nuclei near the neutron mid-shell as a function of Z. These pioneering RDDS experiments have demonstrated the feasibility of the RDT technique in lifetime measurements of nuclei with relatively low production cross section.

## ACKNOWLEDGMENTS

This work has been supported through EURONS (European Commission contract no. RII3-CT-2004-506065), the Academy of Finland under the Finnish Centre of Excellence Programme 2006-2011 (Nuclear and Accelerator Based Physics contract 213503) and the BMBF (Germany) under contract No. 06K167. The UK/France (EPSRC/IN2P3) Loan Pool and GAMMAPOOL network are acknowledged for the EUROGAM detectors of JUROGAM.

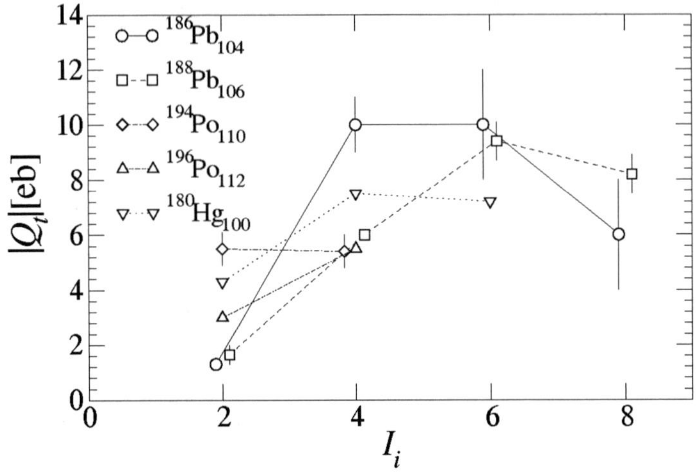


**FIGURE 1.** Preliminary $|Q_t|$ values for $^{180}$Hg and $^{196}$Po. The $|Q_t|$ values for $^{186,188}$Pb and $^{194}$Po are taken from Ref. [21]. Some of the $|Q_t|$values are slightly displaced from their actual $I_i$ values to maintain clarity of presentation.

## REFERENCES


1. C. De Coster, B. Decroix, and K. Heyde, *Phys. Rev. C* **61**, 067306 (2000).
2. A. N. Andreyev *et al.*, *Nature (London)* **405**, 430 (2000).
3. F. R. May, V. V. Paskevich, and S. Frauendorf, *Phys. Lett.* **B 68**, 113 (1977).
4. W. Nazarewicz, *Phys. Lett.* **B305**, 195 (1993).
5. M. Bender, P. Bonche, T. Duguet, and P.-H. Heenen, *Phys. Rev. C* **69**, 064303 (2004).
6. J. L. Egido, L. M. Robledo, and R. R. Rodríguez-Guzmán, *Phys. Rev. Lett.* **93**, 082502 (2004).
7. D. G. Jenkins *et al.*, *Phys. Rev. C* **62**, 021302(R) (2000).
8. J. F. C. Cocks *et al.*, *Eur. Phys. J. A* **3**, 17 (1998).
9. A. M. Baxter *et al.*, *Phys. Rev. C* **48**, R2140 (1993).
10. J. Heese *et al.*, *Phys. Lett.* **B302**, 390 (1993).
11. J. Pakarinen *et al.*, *Phys. Rev. C* **72**, 011304(R) (2005).
12. G. D. Dracoulis *et al.*, *Phys. Rev. C* **69**, 054318 (2004).
13. N. Bijnens *et al.*, *Phys. Rev. Lett.* **75**, 4571 (1995).
14. R. Julin, K. Helariutta, and M. Muikku, *J. Phys G* **27**, R109 (2001).
15. K. Van de Vel *et al.*, *Eur. Phys. J. A* **17**, 167 (2003).
16. V. Hellemans *et al.*, *Phys. Rev. C* **71**, 034308 (2005).
17. A. M. Oros *et al.*, *Nucl. Phys.* **A 645**, 107 (1999).
18. J. Pakarinen *et al.*, *Phys. Rev. C* **75**, 014302 (2007).
19. A. Dewald *et al.*, *Phys. Rev. C* **68**, 034314 (2003).
20. K. Helariutta *et al.*, *Eur. Phys. J. A* **6**, 289 (1999).
21. T. Grahn *et al.*, *Phys. Rev. Lett.* **97**, 062501 (2006).
22. M. Leino *et al.*, *Nucl. Instrum. Methods B* **99**, 653 (1995).
23. R. D. Page *et al.*, *Nucl. Instrum. Methods B* **204**, 634 (2003).
24. C. W. Beausang *et al.*, *Nucl. Instrum. Methods A* **313**, 37 (1992).
25. I. H. Lazarus *et al.*, *IEEE Trans. Nucl. Sci.* **48** (2001).
26. E. S. Paul *et al.*, *Phys. Rev. C* **51**, 78 (1995).
27. A. Dewald, S. Harissopoulos, and P. von Brentano, *Z. Phys. A* **334**, 163 (1989).
28. H. De Witte *et al.*, *Phys. Rev. Lett.* **98**, 112502 (2007).

# Competing Shapes And Alignments In Neutron-Rich Hf Nuclei

P. Chowdhury[a], U.S. Tandel[a], S.K. Tandel[a], D. Cline[b], C.-Y. Wu[b], M.P. Carpenter[c], R.V.F. Janssens[c], T.L. Khoo[c], T. Lauritsen[c], C.J. Lister[c], D. Seweryniak[c] and S.Zhu[c]

[a]*Department of Physics, University of Massachusetts Lowell, Lowell MA 01854*
[b]*Nuclear Structure Research Laboratory, University of Rochester, Rochester NY 14627*
[c]*Argonne National Laboratory, Argonne, IL 60439*

**Abstract.** The talk will focus on spin-dependent competition between oblate and prolate shape minima in the potential energy landscape of $^{180}$Hf (the most neutron-rich stable isotope), mediated via the alignment of valence nucleons. Results of a prompt spectroscopic study, using deep inelastic reactions with Gammasphere and CHICO, bombarding a thin $^{232}$Th target with a $^{180}$Hf beam ~25% above the Coulomb barrier, will be presented. Nucleon alignments in both prolate and oblate minima will be discussed, as well as the favoring of oblate collective rotation at high spins, observed through a mixing with gamma vibrations built on the prolate shape.

**Keywords:** Nuclear shapes, collective rotation, prolate, oblate, Gammasphere, alignment, gamma vibration, neutron-rich, deep inelastic
**PACS:** 21.10.Re, 21.60.Ev, 23.20.Lv, 27.70.+q

## INTRODUCTION

Shape competition as a function of spin continues to be a dominant theme in nuclear structure. The few examples of coexisting and competing oblate and prolate shapes at low spins in deformed nuclei involve low potential energy barriers in the triaxial degree of freedom. With increasing angular momentum, for comparable deformations of the competing shapes, the lower moment of inertia of oblate collective rotation makes this mode less energetically favored than prolate collective rotation. Consequently, there are no clear examples of collective oblate rotation becoming energetically favored over collective prolate rotation at high spins in a well-deformed, prolate, gamma-rigid nucleus.

Nucleon alignments, interestingly, play an important role in this competition. For example, in $^{180}$Hf, which is a textbook example of a well-deformed, axially symmetric, gamma-rigid, *prolate* nucleus, there is a long-standing prediction of a "giant backbend" at high spins (~26 ) [1]. This is the result of a continuous lowering of an *oblate* minimum with increasing spin, which eventually becomes yrast following alignments of particles in the oblate minimum. Our earlier prompt studies of $^{180}$Hf, originally motivated by the physics of K-isomers, revealed a significantly delayed alignment of the first pair of nucleons in the prolate minimum [2,3]. These provided dual motivation for extending the prompt spectroscopy of $^{180}$Hf to higher spins.

CP1012, *Frontiers in Nuclear Structure, Astrophysics, and Reactions: FINUSTAR 2*
edited by P. Demetriou, R. Julin, and S. V. Harissopulos
© 2008 American Institute of Physics 0-7354-0532-5/08/$23.00

# EXPERIMENT

Since $^{180}$Hf is the most neutron-rich stable isotope of hafnium, there is no stable beam-target combination that could be used to populate this nucleus at high spins via fusion-evaporation reactions. The present experiment, therefore, was performed using deep inelastic reactions, with a 1300-MeV $^{180}$Hf beam incident on a 700 μg/cm$^2$ $^{232}$Th target (~25% above the Coulomb barrier). Data were recorded with an event trigger where at least two gamma rays were detected in 100 Compton-suppressed germanium detectors of the Gammasphere array, and both binary reaction partners (target-like and projectile-like fragments) were detected in coincidence by the heavy-ion detector CHICO [4]. This allows the spectroscopy of nuclei that are excited with both inelastic and transfer reaction mechanisms. The present work reports on inelastic excitations of the $^{180}$Hf beam nuclei, and builds on our earlier work on the same nucleus using a $^{136}$Xe beam on an enriched $^{180}$Hf target [2,3], in which the spins necessary to observe the predicted phenomena of interest had not been reached.

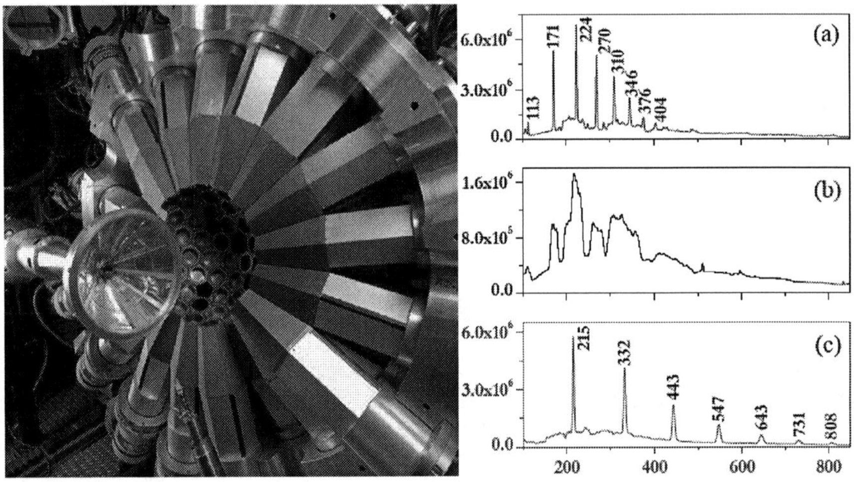


**FIGURE 1.** Experimental set-up and spectra of CHICO with Gammasphere: (a) Doppler correction for Th-like fragments; (b) no Doppler correction; and (c) Doppler correction for Hf-like fragments.

# RESULTS

The low-K bands in $^{180}$Hf were extended to I=20 in the present experiment [5]. A partial level scheme is shown in Fig. 2, which highlights the specific physics issues discussed in this talk. Although the mass of the reaction partner ($^{232}$Th compared to $^{136}$Xe), and the beam energy chosen for the present experiment (~25% above the barrier in the center-of-mass frame compared to ~15% in our earlier study) were significantly higher than before, only a modest gain in spin of ~2 was observed for the highest states placed in the level scheme.

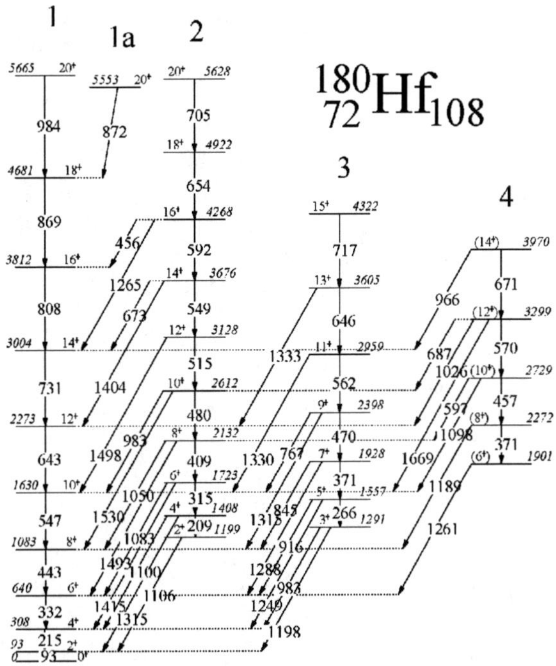


**FIGURE 2.** Partial level scheme of $^{180}$Hf deduced and extended in the present work.

This modest gain, however, is enough to reveal two new $20^+$ states at 5665 and 5553 keV feeding the $18^+$ state of the g-band. These are identified as members of the g-band (Band 1) and s-band (Band 1a), respectively, and locates the first alignment in the prolate yrast structure at a rotational frequency of $\omega$=0.43 MeV. As had been projected and discussed in our previous study, this is considerably higher than the frequency at which the first neutron pair aligns in the lighter even-even isotopes. This "delay" in the first nucleon alignments for N≥108 isotones is also observed for W and Os nuclei (Fig.3), and is partly attributed to an increased gap near the Fermi level, a consequence of a neutron subshell closure at $N$ = 108 [2, 6].

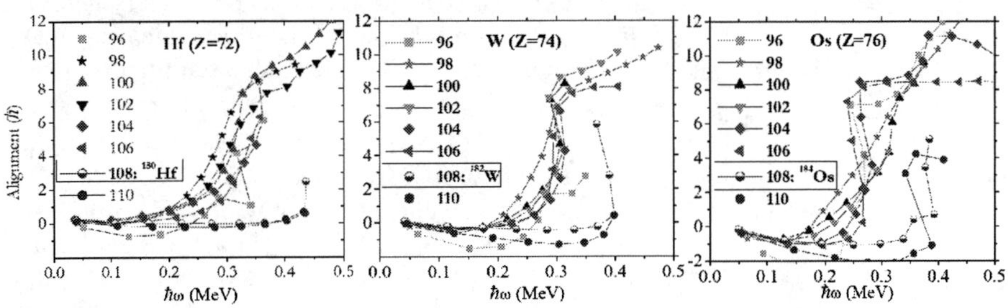


**FIGURE 3.** Systematics of the first nucleon alignment in even-even Hf, W and Os isotopes.

Bands 3 and 2 (below 10 ) are odd- and even-spin signatures of the γ-vibrational band. Their excitation energies parallel the g-band (Fig. 4a) with almost no signature splitting, and the intensity branching ratios of in-band to out-of-band (to the g-band) transitions follow the Alaga rules for gamma vibrations. Above 10 , they diverge, with band 2 energies dipping sharply and reaching Bands 1 and 1a around 20 , at essentially the same spins as where the first alignment in the prolate band occurs. The divergence can be understood as the crossing of the even-spin members of the γ-band with band 4, which is identified from their crossing trajectories in energy-angular momentum space (Fig. 4a) as well as through inter-band transitions in this crossing region.

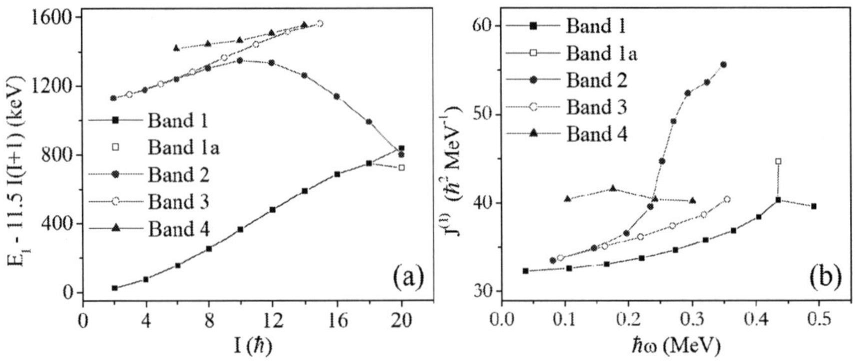


**FIGURE 4.** (a) Relative energies (rigid rotor reference subtracted) and (b) kinematic moments of inertia of the various bands. Band 1 is the prolate g-band and Band 2 (at high spins) the proposed oblate band. Bands 3 and 2 (at low spins) are the odd and even signatures of the γ-vibrational band.

The kinematic moments of inertia for the bands are plotted in Fig. 4b and illustrate the delayed first alignment in the yrast structure. In addition, Band 2, at higher spins and rotational frequencies, exhibits the highest moment of inertia compared to all other bands. This feature, coupled with its strong down-sloping energy trajectory, strongly suggests an oblate character for this band. Simple estimates of the increased moment of inertia for an oblate rotor are consistent with observations.

Quasi-particle level diagrams calculated with a cranked Woods-Saxon potential (Fig.5) for prolate and oblate shape minima, with deformation parameters obtained from Total Routhian Surface calculations, highlight the stronger alignment of quasi-particle trajectories for rotations of an oblate shape. The oblate states at high spin have very small mixing with prolate states, as would be expected from the γ-rigid shape minima of either shape in $^{180}$Hf. The observed energy difference between the prolate and oblate minima is ~1 MeV lower than expected from the theoretical calculations. The unexpected mixing and interaction of the oblate structure with γ vibrations built on the prolate shape may be a reason for the lowering of energies of oblate states, leading to a prolate-oblate crossing at lower spin than predicted earlier for this nucleus. The weak population of the bands makes it difficult to unambiguously verify the proposed oblate character for this band.

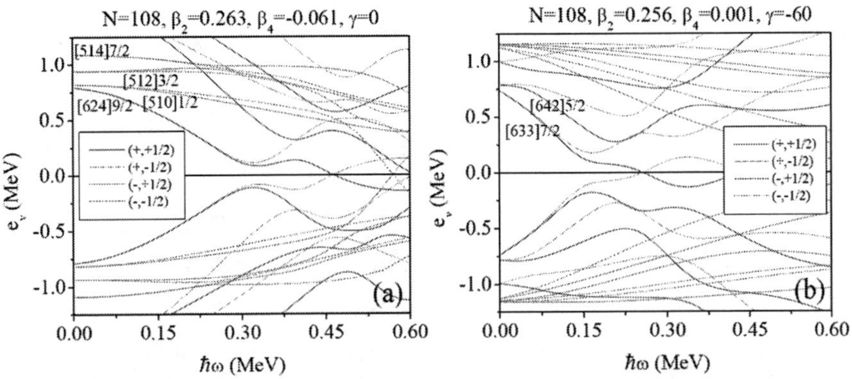


**FIGURE 5.** Quasi-particle levels in $^{180}$Hf from a cranked Woods-Saxon potential, using parameters of the (a) prolate and (b) oblate shape minima that appear in the Total Routhian Surface.

# SUMMARY

In summary, the low-K high-spin structure of $^{180}$Hf, a textbook example of a γ-rigid, strongly deformed prolate rotor at low spins, is extended to I = 20 via multiple bands. The first nucleon alignment in the prolate minimum, as well as the predicted novel physics of the favoring of oblate collective rotation at high spins is observed. The competition is facilitated via strongly-aligning valence neutrons in an oblate minimum, and identified through a mixing of the oblate collective rotation with γ vibrations built on the prolate shape. The prolate-oblate crossing is observed at lower spin than predicted earlier for this nucleus, and may result from the unexpected mixing of the oblate structure with gamma vibrations built on the prolate shape, leading to a lowering of the oblate states. For more neutron-rich isotopes, the crossing is expected to occur at even lower spins.

# ACKNOWLEDGMENTS

This research is supported by the U.S. Department of Energy, Office of Nuclear Physics, under grants DE-FG02-94ER40848 and DE-AC-02-06CH11357, and the National Science Foundation.

# REFERENCES

1. R. R. Hilton and H. J. Mang, Phys. Rev. Lett. **43**, 1979 (1979).
2. E. Ngijoi-Yogo et al., Phys. Rev. **C75**, 034305 (2007).
3. E. Ngijoi-Yogo, "Collective Band Structures in Neutron-rich $^{180,182}$Hf Nuclei", Ph.D. Thesis, University of Massachusetts Lowell, 2004.
4. M.W. Simon et al., Nucl. Inst. Meth. **A452**, 205(2000).
5. U.S. Tandel, "Investigating Shape Evolution and Delayed Nucleon Alignment in the Heaviest Stable Isotope of Hafnium", M.S. Thesis, University of Massachusetts Lowell, 2007.
6. M. P. Carpenter et al., Nucl. Phys. **A513**, 125 (1990).

# Recent progress in theories of reactions off stability

Angela Bonaccorso [1]

*Istituto Nazionale di Fisica Nucleare, Sez. di Pisa and Dipartimento di Fisica, Università di Pisa, Largo Pontecorvo 3, 56127 Pisa, Italy.*

**Abstract.** This talk will review the state of art of nuclear and Coulomb breakup theories and their relation to optical models of elastic scattering of exotic projectiles. The effect of the final state interactions between the breakup particle and the core and target nuclei will be clarified and some typical numerical calculations for the relevant observables will be presented and compared to experimental data. Finally new results will be shown to demonstrate the feasibility of a novel type of experiment involving heavy projectiles far from stability on heavy targets.

**Keywords:** Neutron, proton breakup, projectile fragmentation, unbound and borromean nuclei.
**PACS:** 21.10.Jx, 24.10.-i, 25.60.Gc, 27.30.+t

## Introduction

Nuclear reaction theory has experienced a great revival in the last twenty years following the large increase in quantity as well as quality of experiments with exotic beams. Exotic nuclei are located away from the stability valley and have large differences in the number of neutrons and protons. Their valence particle separation energies $S_n$ are smaller than the average 8 MeV expected in nuclear matter. In the extreme case of halo nuclei such as several beryllium and lithium isotopes ($^{11}$Be,$^{12}$Be,$^{14}$Be,$^{11}$Li), $S_n$ can be even less than 1 MeV. As a consequence as much as 10% of the total reaction cross section is due to just one channel: 1n or 2n breakup. Therefore, *out of necessity* (N. Orr), breakup has been the most studied reaction for very weak beams and the one for which several new models have been developed. In this short review I will first present the mechanisms which lead to breakup and the relative observables that are measured. For each of them a description will follow of the most recent advances in the models used for theoretical calculations.

## Cross section

Most theoretical methods used so far to describe breakup rely on a basic approximation to describe the collision with only the three-body variables of nucleon coordinate, projectile coordinate, and target coordinate. Thus the dynamics is controlled by the three

---

[1] In collaboration with C.A. Bertulani, G. Blanchon, D.M. Brink, F. Carstoiu, A. García-Camacho, A.A. Ibraheem, J. Margueron, N.Vinh Mau.

CP1012, *Frontiers in Nuclear Structure, Astrophysics, and Reactions: FINUSTAR 2*
edited by P. Demetriou, R. Julin, and S. V. Harissopulos
© 2008 American Institute of Physics 0-7354-0532-5/08/$23.00

potentials describing nucleon-core, nucleon-target, and core-target interactions. Often the projectile-target relative motion is treated semiclassically by using a trajectory of the center of the projectile relative to the center of the target $\mathbf{R}(t) = \mathbf{b}_c + vt\hat{\mathbf{z}}$ with constant velocity $v$ in the $z$ direction and impact parameter $\mathbf{b}_c$ in the $xy$ plane. This approximation makes the formalism applicable for incident energies above the Coulomb barrier. Along the semiclassical trajectory the amplitude for a transition from a nucleon state $\psi_i$ bound in the projectile, to a final continuum state $\psi_f$, is given by [1, 2]

$$A_{fi} = \frac{1}{i\hbar}\int_{-\infty}^{\infty} dt \langle \psi_f(\mathbf{r},t)|V(\mathbf{r},\mathbf{R}(t))|\psi_i(\mathbf{r},t)\rangle, \tag{1}$$

where $V$ is the interaction responsible for the transition which will be specified in the following. The probabilities for different processes can be represented in terms of the amplitude as $dP/d\xi = \sum |A_{fi}|^2 \delta(\xi - \xi_f)$ where $\xi$ can be momentum, energy or any other variable for which a differential cross section is measured. Direct one-particle re-arrangment reactions of the peripheral type in presence of strong core-target absorption can be described by an equation like [2]-[5] $d\sigma/d\xi_f = C^2 S \int d\mathbf{b_c} P_{ct}(b_c) dP(b_c)/d\xi_f$, (see Eq. (2.3) of [3]) and $C^2S$ is the spectroscopic factor for the initial single particle state. The core survival probability is defined in terms of a S-matrix function of the core-target distance of closest approach $b_c$. A simple parameterisation is $P_{ct}(b_c) = |S_{ct}|^2 = e^{(-\ln 2 exp[(R_s - b_c)/a])}$. It takes into account the peripheral nature of the reaction and naturally excludes the possibility of large overlaps between projectile and target. The strong absorption radius $R_s \approx 1.4(A_p^{1/3} + A_t^{1/3})$ fm is defined as the distance of closest approach for a trajectory that is 50% absorbed from the elastic channel and a=0.6 fm is a diffuseness parameter. The values of $R_s$ thus obtained agree within a few percent with those of the Kox parameterization [6].

## Breakup mechanisms

Based on the time dependent amplitude Eq.(1) and the classical projectile-target trajectory of relative motion given above, in Ref.[4] we considered the breakup of a halo nucleus like $^{11}$Be consisting of a neutron bound to a $^{10}$Be core in a collision with a target nucleus. The system of the halo nucleus and the target was described by Jacobi coordinates $(\mathbf{R},\mathbf{r})$ where $\mathbf{R}$ is the position of the center of mass of the halo nucleus relative to the target nucleus and $\mathbf{r}$ is the position of the neutron relative to the halo core, and the coordinate $\mathbf{R}$ is assumed to move on a classical path. The Hamiltonian of such a system is $H = T_R + T_r + V_{nc}(\mathbf{r}) + V_{nt}(\beta_2\mathbf{r} + \mathbf{R}) + V_{ct}(\mathbf{R} - \beta_1\mathbf{r})$, with $\beta_1$ and $\beta_2$ the mass ratios of neutron and core, respectively, to that of the projectile. $T_R$ and $T_r$ are the kinetic energy operators associated with the coordinates $\mathbf{R}$ and $\mathbf{r}$ and $V_{cn}$ is a real potential describing the neutron-core final state interaction. $V_{cn}$ was neglected under the hypothesis that the observables measured and calculated did not depend significantly on it. In the "Projectile Fragmentation" section we will discuss cases in which such an interaction dominates instead the measured data. The potential $V_2 = V_{nt} + V_{ct}$ describes the interaction between the projectile and the target. Both $V_{nt}$ and $V_{ct}$ are represented by complex optical potentials. The imaginary part of $V_{nt}$ describes absorption of the neutron by the target to form a compound nucleus. It gives rise to the stripping part [2] of the halo breakup. The imaginary part of $V_{ct}$ describes reactions of the halo core with the target. The potential $V_{ct}$ also includes the Coulomb interaction between the halo core and the target. This part of the interaction is responsible for Coulomb breakup. In fact the Coulomb force does not act directly on the neutron but it affects it only indirectly by causing the recoil of the charged core.

Since the mass ratio $\beta_1$ is small for a halo nucleus with a heavy core ($\beta_1 \approx 0.1$ for $^{11}$Be) the Coulomb potential was approximated by the dipole term in Ref.[4]. Then, making an eikonal approximation for the neutron final state $\psi_f(t) = \exp(i\mathbf{k}\cdot\mathbf{r} - i\varepsilon_{\mathbf{k}}t/\hbar)\exp\left(-\frac{1}{i\hbar}\int_t^\infty V_2(\mathbf{r},t)\,dt\right)$, the amplitude became

$$A_{lm}(\mathbf{k},\mathbf{b}_c) = \frac{1}{i\hbar}\int d^3\mathbf{r}\, dt e^{-i\mathbf{k}\cdot\mathbf{r}+i\omega t} e^{\left(\frac{1}{i\hbar}\int_t^\infty V_2(\mathbf{r},\mathbf{R}(t))dt\right)} V_2(\mathbf{r},\mathbf{R}(t))\,\phi_{lm}(\mathbf{r}) \qquad (2)$$

where $\omega = (\varepsilon_{\mathbf{k}} - \varepsilon_i)/\hbar$ and $\phi_{lm}$ is the radial part of the neutron initial wave function. $\mathbf{k}$ is the neutron-core relative momentum vector in the final state. The corresponding energy is $\varepsilon_{\mathbf{k}}$, while $\varepsilon_i$ is the initial state binding energy. Eq. (2) is appropriate to calculate the coincidence cross section $A_p \rightarrow (A_p - 1) + n$.

The components $V_{nt}$ and $V_{ct}$ of $V_2$ are treated differently because of the long range of the Coulomb interaction. The neutron-target interaction is strong and has a short range. We assumed that the interaction time $\tau$ for this part of the interaction is very short in the sense that $\omega\tau$ is small compared with unity. On the other hand the long range Coulomb interaction between the halo core and the target is weaker and changes more slowly. The way to treat it is discussed in the next subsection.

## Coulomb breakup

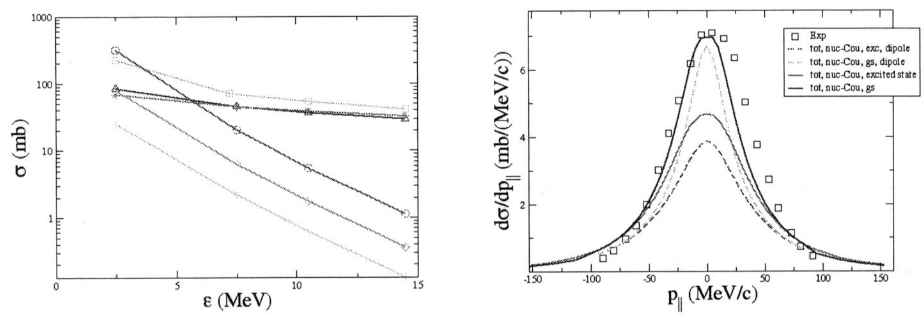


**FIGURE 1.** LHS: *Integrated breakup cross-section for a hypothetical $^{34}$Si beam against Pb at 70 A.MeV as a function of the neutron separation energy. Different initial parameters: circles (squares) are for Coulomb (nuclear) breakup from an initial s-wave; diamonds (triangles) for Coulomb (nuclear) breakup from a d-wave; pluses (stars) for Coulomb (nuclear) breakup from an initial f-wave. Nuclear breakup is the sum of diffractive and stripping contributions [2, 3].* RHS: *Calculated momentum distribution of $^7$Be fragments after proton-removal from $^8$B against Pb at 936 MeV/A. Both dipole and full multipole results are shown for the ground state and first excited state. Calculations according to this section and Ref.[8] where more details can be found. Data are from Ref.[15].*

Introducing the notation $V_c = Z_c Z_t e^2$, $V_v = Z_v Z_t e^2$ and $V_C = (Z_v + Z_c) Z_t e^2$, with $Z_v = 0, 1$ for a neutron and for a proton respectively, the Coulomb potential can be written as $V(\mathbf{r}, \mathbf{R}) = \frac{V_c}{|\mathbf{R} - \beta_1 \mathbf{r}|} + \frac{V_v}{|\mathbf{R} + \beta_2 \mathbf{r}|} - \frac{V_C}{R}$. Here $\beta_1$ can be also the mass ratio of the proton to that of the projectile. In Ref.[7] we have shown that the Coulomb phase $\chi_{eff}(\mathbf{b}_c, \mathbf{r}, k) = \int dt e^{i\omega t} V_C(\mathbf{r}, t)/\hbar$. which is solvable in the dipole approximation, can similarly be calculated with the whole multipole expansion if a screening term is added and subtracted to the potential.

Aiming for an all-order formalism, in [4] it was shown that a possible way to achieve this is to use the sudden approximation, subtract the first order term, which diverges for large impact parameter, and then to add a first order term calculated in time-dependent perturbation theory. The sudden limit ($\omega \to 0$) must be therefore taken in the expression for $\chi_{eff}$, yielding $\chi_{eff}^{sudd} = \frac{2V_C}{\hbar v} \log \frac{b_c}{R_\perp}$. Following a procedure analogous to that of [7], the Coulomb phase for the proton is shown to be [8] $\chi^p = \frac{2}{\hbar v} \left( V_c e^{i\beta_1 \omega z/v} K_0(\omega b_c/v) - V_C K_0(\omega R_\perp/v) + V_v e^{-i\beta_2 \omega z/v} K_0(\omega b_v/v) \right)$. Since $V_C = V_c + V_v$, the phase can be written as $\chi^p = \chi(\beta_1, V_c) + \chi(-\beta_2, V_v)$ where $\chi$ is the $\chi_{eff}$ of the previous section and $\mathbf{b}_v = \mathbf{b}_c + \mathbf{r}_\perp$ is the proton impact parameter with respect to the target. The Coulomb phase is therefore the sum of two terms: one of them describes the recoil of the core whereas the other accounts for the direct proton-target Coulomb interaction. Of course, in the case of the neutron the latter vanishes and the phase reduces to the one derived in [7]. It is easy to see that the expansion of $\chi_{eff}$ to first order in $\mathbf{r}$ yields the well known dipole approximation to the phase: which only differs from the neutron breakup case in the different constant factor, which is now $(V_c\beta_1 - V_v\beta_2)$ instead of $V_c\beta_1$ of Ref. [7]. The probability amplitude can be written as the sum of three contributions. The recoil term,

$$A^{rec} = \int d\mathbf{r} e^{-i\mathbf{k}\cdot\mathbf{r}} \phi_i(\mathbf{r}) \left( e^{i\frac{2V_c}{\hbar v} \log \frac{b_c}{R_\perp}} - 1 - i\frac{2V_c}{\hbar v} \log \frac{b_c}{R_\perp} + i\chi(\beta_1, V_c) \right), \tag{3}$$

where, according to the discussions in [4, 7], the sudden limit has been used in order to include all orders in the interaction. The direct proton Coulomb interaction term $A^{dir}$ which has the same form as Eq.(3) but with the substitution $V_c \to V_v$, $\beta_1 \to -\beta_2$ and $b_c \to b_v$. The nuclear part is given by $A^{nuc} = \int d\mathbf{r} e^{-i\mathbf{k}\cdot\mathbf{r}} \left( e^{i\chi_{nt}(b_v)} - 1 \right) \phi_i(\mathbf{r})$, which is the well known eikonal form of the diffractive nuclear breakup with the nucleon-target phase $\chi_{nt}(\mathbf{b}_v) = \int dz V_{nt}(\mathbf{b}_v, z)/(v\hbar)$. Finally the expression for the differential cross-section is $\frac{d\sigma}{d\mathbf{k}} = \frac{1}{8\pi^3} \frac{mk}{\hbar^2} \int d\mathbf{b}_c |S_{ct}(b_c)|^2 |A^{nuc} + A^{dir} + A^{rec}|^2$. In a number of papers higher order effects [9] and proton breakup have been discussed, among which we recall Refs.[10, 11, 12]. Some works have also addressed the problem of asymmetry in the core parallel momentum distribution after proton knockout [13, 14]. The fact that this asymmetry comes from high order terms can be directly extracted from our formalism. If the Coulomb part of the amplitude is simply expanded to first order in $\chi$, it can be written, in terms of the one-dimensional Fourier transform in $z-$direction of the initial wave function $\tilde{\phi}_i$, as

$$A^{Cou} \simeq \int d\mathbf{r}_\perp e^{-i\mathbf{k}_\perp\cdot\mathbf{r}_\perp} \frac{2}{\hbar v} \left( V_c K_0(\omega b_c/v) \tilde{\phi}_i(\mathbf{r}_\perp, k_z - \beta_1 \omega/v) - V_C K_0(\omega R_\perp/v) \tilde{\phi}_i(\mathbf{r}_\perp, k_z) \right.$$

$$+ \quad V_v K_0(\omega b_v/v)\tilde{\phi}_i(\mathbf{r}_\perp, k_z + \beta_2\omega/v)\big) . \qquad (4)$$

Thus the Coulomb breakup probability amplitude can be regarded as a coherent sum of three terms, each of which contains a shifted $z-$Fourier transform. The shifts are in opposite directions, $\beta_1\omega/v$ and $-\beta_2\omega/v$, but they are not visible directly in the calculated momentum distributions as $\omega$ depends on $k$ itself. Moreover, the $1/v$ factor indicates that the asymmetry decreases as the beam energy increases. In the dipole approximation, however, the amplitude does not contain any asymmetry for the momentum distribution as it involves square modulii of $\tilde{\phi}_i(\mathbf{r}_\perp, k_z)$ separately. Hence we have confirmed analytically that the asymmetry in Coulomb breakup parallel momentum distributions is due to the presence of higher multipole terms, in agreement with earlier works [9, 13, 14]. However, the presence of the nuclear interaction introduces an interference that does depend on the sign of $k_z$ and thus an additional asymmetry to that due to higher multipole terms in the Coulomb interaction.

In Fig. 1LHS we show that the Coulomb break up cross section on heavy targets would be negligible compared to the total nuclear breakup cross section for heavy exotic nuclei in which the valence nucleons are expected to be in $d$ or $f$ orbits and with separation energies of the order of 10 MeV. For those nuclei such novel type of experiments would be therefore possible and more appropriate than the use of light targets for which the strong absorption model for the core-target interaction is not appropriate. Fig. 1RHS compares data [15] for proton breakup to calculations from our new model [8] shortly described above.

## Projectile Fragmentation

We call *projectile fragmentation* the elastic breakup (diffraction dissociation) discussed above, when the observable studied is the neutron-core relative energy spectrum. This kind of observable has been widely measured in relation to the Coulomb breakup on heavy target. Results on light targets have also been presented [16]. These data enlighten the effect of the neutron final state interaction with the core of origin, neglected in the previous sections, while observables like the core energy or momentum distributions enlighten the effect of the neutron final state interaction with the target.

Projectile fragmentation has also been used experimentally with two neutron halo projectiles [17]-[34]. The reaction might proceed by the simultaneous emission of the two neutrons or by successive emissions [17]. The successive emission can be due to a mechanism in which one neutron is stripped by the interaction with the target, as in the one-neutron fragmentation case, while the other is left behind, for example in a resonance state, which then decays. The emission can be expected sequential if the two neutrons are not strongly correlated.

If the two neutrons are strongly correlated they will preferentially be emitted simultaneously. If the neutron which is not detected is stripped while the other suffers an elastic scattering on the target, then in both cases to first order in the interaction the neutron ends-up in a plane wave final state [2]. It can then re-interact with the core which, for example, is going to be $^{10}$Be in the case of the one-neutron halo projectile $^{11}$Be, while it will be $^{12}$Be in the case of the projectile fragmentation of $^{14}$Be, since $^{13}$Be is not bound or $^9$Li in the case of the projectile fragmentation of $^{11}$Li, since in this case $^{10}$Li is not bound. Experiments with a $^{14}$B projectile [19] have also been performed, in which the n-$^{12}$Be relative energy spectra have been reconstructed by coincidence measurements. In such a nucleus the valence neutron is weakly bound, while the valence proton is strongly bound. Thus the neutron will probably be emitted in the first step and then re-scattered by the core minus one proton nucleus. The projectile-target distances at which this kind of mechanism would be relevant are probably not so large to neglect the effect of the neutron-target interaction.

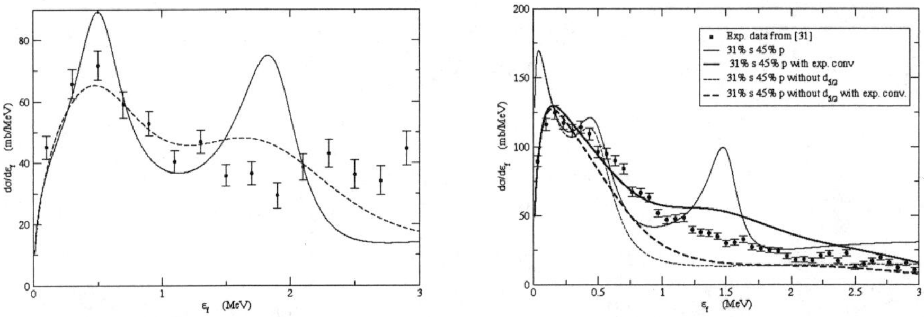


**FIGURE 2.** LHS: *n-$^{12}$Be relative energy spectrum for the reaction $^{14}$Be+$^{12}$C $\rightarrow$ n+$^{12}$Be+X, ($\varepsilon_f$ is the same as $\varepsilon_{\mathbf{k}}$ of the text); the cross section is given by the sum of all transitions from the s initial state with $\varepsilon_i$=-1.85 MeV (solid line). Experimental points from H. Simon et al. [24] for the same reaction at 250 A.MeV. Dashed line is the folding of the calculated spectrum with the experimental resolution curve.* RHS: *n-$^9$Li relative energy spectrum, for the reaction $^{11}$Li+$^{12}$C $\rightarrow$ n+$^9$Li+X at 264 A.MeV. Only the contributions from an s and p initial state with experimental spectroscopic factors [35] $C^2S$= 0.31 and 0.45 respectively and separation energy 0.3 MeV are included. The thin solid curve is the total calculated result. The thick solid curve curve is after convolution with the experimental resolution function. The thin dashed curve is the calculation without the d-resonance while the thick dashed curve is the same calculation after convolution. The symbols with error bars are the experimental points from Ref.[24]. Calculations are normalised to the data.*

To first order the inelastic-like excitations can be described again by the time dependent perturbation amplitude Eq.(1) [1, 2]. In this section also, the potential $V(\mathbf{r},\mathbf{R}(t))$, which is the interaction responsible for the neutron transition, moves past on a constant velocity path as described in the previous sections. The radial part $\phi_i(\mathbf{r})$ of the single particle initial state wave function $\psi_i(\mathbf{r},t)$ is calculated in a potential $V_{WS}(r)$ which is fixed in space. The coordinate system and other details of the calculations can be found in Ref.[20]. In the special case of exotic nuclei the traditional approach to inelastic excitations needs to be modified. For example the final state can be eigenstate of a potential $V_1$ modified with respect to $V_{WS}$ because some other particle is emitted during the reaction process as discussed in the introduction. The final state interaction might also have an imaginary part which would take into account the coupling between a continuum state and an excited core. The first order time dependent perturbation amplitude then reads $A_{fi} = \frac{1}{i\hbar v}\int_{-\infty}^{\infty} dxdydz\ \phi_f^*(x,y,z)\phi_i(x,y,z)e^{iqz}\tilde{V}(x-b_c,y,q)$, where $\tilde{V}(x-b_c,y,q) = \int_{-\infty}^{\infty} dzV(x-b_c,y,z)e^{iqz}$, and we changed variables and put $z' = z - vt$ or $t = (z-z')/v$, $q = \varepsilon_{\mathbf{k}} - \varepsilon_i/\hbar v$.

The target represented by $\tilde{V}$ perturbs the initial bound state wave function and allows the transition to the continuum by transferring some momentum to the neutron. Then it is enough to choose a simplified form of the interaction, such as a delta-function potential $V(r) = v_2\delta(x)\delta(y)\delta(z)$. The value of the strength $v_2 \equiv$ [MeV fm$^3$] used in the calculation is taken equal to the volume integral of the appropriate neutron-target interaction. It is clear that while in the sudden approach the initial and final state overlap is taken in the whole coordinate space, irrespective of the target and of the beam velocity, here the overlap of the initial and final wave functions depends on the core-target impact parameter. The neutron is emitted preferentially on the reaction plane and the z-component, being along the relative velocity axis is boosted by a momentum $q$.

Due of the strong core absorption discussed in the first part of the paper these calculations are also performed using the asymptotic form of the initial and final state wave functions. Introducing the quantization condition according to Ref.[2] the probability spectrum reads

$$\frac{dP_{in}}{d\varepsilon_{\mathbf{k}}} = \frac{2}{\pi}\frac{v_2^2}{\hbar^2 v^2}C_i^2 \frac{m}{\hbar^2 k}\frac{1}{2l_i+1}\Sigma_{m_i,m_f}|1-\bar{S}_{m_i,m_f}|^2|I_{m_i,m_f}|^2. \tag{5}$$

The generalization including spin is given in Appendix B of Ref.[20] and $|I_{m_i,m_f}|^2 = \left|k\frac{i}{2}i^{l_i}\int_{-\infty}^{\infty}dze^{iqz}\gamma h_{l_i}^{(1)}(i\gamma r)Y_{l_i,m_i}(\theta,0)h_{l_f}^{(-)}(kr)Y_{l_f,m_f}(\theta,0)\right|^2$. The quantity $\bar{S} = e^{2i(\delta+\nu)}$ is an off-the-energy-shell S-matrix representing the final state interaction of the neutron with the projectile core. It depends on a phase which is the sum of $\delta$, the free particle n-core phase shift, plus $\nu$ the phase of the matrix element $I$. Some examples of our calculations [20] are shown in Fig. 2 and compared to recent data. See also Ref.[21].

We conclude by mentioning the importance of taking into account the nuclear and Coulomb breakup channels in the optical potentials of projectiles off stability. In Refs.[36, 37] we have shown that this can be done microscopically for the imaginary part of the optical potential. Because of the form factor behaviour such potential must be of the surface type. Its large diffuseness is due to the large decay length of the initial wave function and to the adiabaticity parameter of Coulomb breakup. Finally we refer to two books [38, 39] and recent reviews [40, 41] in which further discussion and bibliography can be found.

## Conclusions and Outlook

The field of Rare Isotopes Studies is very active, growing steadily and rapidly. Some recent achievements in reaction models for neutron and proton breakup from exotic beams have been presented. From the structure point of view, in the search for the dripline position, a very important role is played by the study of nuclei unstable by neutron emission. On the other hand increasing the mass of the projectiles produced we are going to face the problem of envisaging new experiments to study them. These two are among the most important subjects which need to be adressed and further developed in the near future and for which some suggestions have been presented.

# REFERENCES


1. K. Alder and A. Winther, Electromagnetic Excitation, North-Holland, 1975.
2. A. Bonaccorso and D.M. Brink, Phys. Rev. C **38** , 1776 (1988); Phys. Rev. C **43**, 299 (1991); Phys. Rev. C **44**, 1559 (1991).
3. A. Bonaccorso, Phys. Rev. C **60**, 054604 (1999).
4. J. Margueron, A. Bonaccorso and D.M. Brink, Nucl. Phys. **A 703**, 105 (2002);Nucl. Phys. **A 720**, 337 (2003).
5. R. A. Broglia and A. Winther, Heavy Ion Reactions, Benjamin, Reading, Mass, 1981.
6. S. Kox et al., Phys. Rev. C **35**, 1678 (1987).
7. A. García-Camacho, A. Bonaccorso and D.M. Brink, Nucl. Phys. **A 776**, 118 (2006).
8. A. García-Camacho, G. Blanchon, A. Bonaccorso and D.M. Brink, Phys. Rev. C **76**, 014607 (2007).
9. H. Esbensen, G.F. Bertsch and C. Bertulani, Nucl. Phys. **A 581**, 107 (1995).
10. J.H. Kelley et al., Phys. Rev. Lett. **77**, 5020 (1996) and references therein.
11. K. Hencken, G. Bertsch and H. Esbensen, Phys. Rev. C **54**, 3043 (1996).
12. A. Bonaccorso, D. M. Brink and C. A. Bertulani, Phys. Rev. C **69**, 024615 (2004).
13. H. Esbensen and G.F. Bertsch, Nucl. Phys. **A 600** (1996) 37; Phys. Rev. C **59**, 3240 (1999); Nucl. Phys. **A 706**, 383 (2002).
14. B. Davids et al., Phys. Rev. Lett. **81**, 2209 (1998); Phys. Rev. C **63**, 0654806 (2001).
15. D. Cortina-Gil et al., Nucl. Phys. **A 720**, 3 (2003).
16. N. Fukuda et al., Phys. Rev. C **70**, 054606 (2004).
17. F.M. Marqués et al., Phys. Rev. C **64**, 061301(R) (2001); N.Orr, Prog. Theor. Phys. Suppl. 146i, 201 (2003).
18. G.F. Bertsch, K. Hencken and H. Esbensen, Phys. Rev. C **57** , 1366 (1998).
19. J.L. Lecouey, Few Body Syst. **34**, 21 (2004).
20. G. Blanchon, A. Bonaccorso, D.M. Brink, A.García-Camacho and N. Vinh Mau, Nucl. Phys. **A 784**, 49 (2007); G. Blanchon, A. Bonaccorso, D.M. Brink, and N. Vinh Mau. Nucl. Phys. **A 791**, 303 (2007).
21. G. Blanchon, A. Bonaccorso and N. Vinh Mau, Nucl. Phys. **A 739**, 259 (2004).
22. M. Thoennessen et al., Phys. Rev. C **59**, 111 (1999); Phys. Rev. C 60, 027303 (1999); Proceedings the Erice Int. School of Heavy-Ion Physics, $4^{th}$ Course. Eds. R.A. Broglia and P.G. Hansen. World Scientific, Singapore 1998, pag.269.
23. M. Labiche et al. Phys. Rev. Lett. **86**, 600 (2001).
24. H. Simon et al., Nucl. Phys. **A 734**, 323 (2004); Nucl. Phys. **A 791**, 267 (2007). 53.
25. A.A. Korsheninnikov et al., Phys. lett. **B 343**, 53 (1995).
26. G.F. Bertsch and H. Esbensen, Ann. Phys. (N.Y.) **209** , 327 (1991).
27. I.J. Thompson and M.V. Zhukov, Phys. Rev. C **53**, 708 (1996).
28. N. Vinh Mau and J.C. Pacheco, Nucl. Phys. **A 607** , 163 (1996).
29. J.C. Pacheco and N. Vinh Mau, Phys. Rev. C **65**, 044004 (2002).
30. P. Descouvemont, Phys. Lett. **B 331**, 271 (1994); Phys. Rev. C **52**, 704 (1995).
31. A. Adachour, D. Baye and P. Descouvemont, Phys. Lett. **B 356** 445 (1995).
32. D. Baye, Nucl. Phys. A 627, 305 (1997).
33. T. Tarutina, I.J. Thompson, J.A. Tostevin, Nucl. Phys. **A 733**, 53 (2004).
34. M. Labiche, F.M. Marques, O. Sorlin and N. Vinh Mau, Phys. Rev. C **60**, 027303 (1999).
35. S. N. Ershov et al., Phys. Rev. C **70**, 054608 (2004).
36. A. Bonaccorso and F. Carstoiu Nucl. Phys. **A 706**, 322 (2002).
37. A. A. Ibraheem, A. Bonaccorso. Nucl. Phys. **A 748** , 414 (2005).
38. C.A. Bertulani, M.S. Hussein, G. Münzenberg. Physics of Radioactive Beams. Nova Science Publ. 2001.
39. Y. Suzuki, R. G. Lovas, K. Yabana, K. Varga. Structure and Reactions of Light Exotic Nuclei. Taylor and Francis Eds. 2003.
40. L.F. Canto, P.R.S. Gomes, R. Donangelo, M.S. Hussein, Phys. Rep. **424**, 1 (2006).
41. B. Jonson, Phys. Rep. **389**, 1 (2004).

# Recent developments in mean-field theories

Michael Bender

*Université Bordeaux 1; CNRS/IN2P3; Centre d'Etudes Nucléaires de Bordeaux Gradignan, UMR5797, Chemin du Solarium, BP120, F-33175 Gradignan, France*

**Abstract.** This contribution gives an overview over recent efforts to construct a universal energy density functional method for nuclear structure.

**PACS:** 21.10.-k, 21.10.Dr, 21.10.Re, 21.60.Jz

## INTRODUCTION

Self-consistent mean-field (SCMF) models using energy density functionals (EDF) are one of the standard approaches in nuclear structure theory, see Ref. [1] and references given therein. Taking advantage of their relative computational simplicity, that replaces the nuclear $A$-body problem with a system of coupled one-body problems, they are the only microscopic method that at present can be systematically applied on a large scale for heavy nuclei. This contribution will only address those SCMF methods that use the complete model space of occupied particles and universal energy density functionals. These nuclear SCMF approaches have many similarities with Density Functional Theory (DFT), widely used to describe electronic many-body systems [2], but also a number of key differences, which in one way or the other are all rooted in the fact that in nuclear EDF-based approaches the energy functional is constructed from intrinsic densities [3, 4], while the formalism of DFT is based on the observable one.

## GOING BEYOND THE SELF-CONSISTENT MEAN FIELD

### Why going beyond the mean field

There are three mayor motivations to extend self-consistent mean-field models by adding explicitely certain classes of correlations to the density matrices the energy functional is constructed from. As a reference state without correlations (except for the antisymmetry of Fermions) we take a spherically symmetric Slater determinant.

First, a mean-field (or independent (quasi-)particle) state breaks symmetries in the laboratory frame. On the one hand, broken symmetries are a feature of nuclear mean-field models; the symmetry-breaking of the mean-field in the intrinsic frame of the nucleus is exploited to describe the most important $n$-particle-$n$-hole and particle-particle correlations at very moderate computational cost. On the other hand, a broken symmetry mixes excitations related to the symmetry operator into the mean-field state. For example, broken rotational symmetry mixes states with different eigenvalues of $\hat{J}^2$, i.e. the members of a rotational band into a deformed mean-field state. Broken parity mixes states of op-

CP1012, *Frontiers in Nuclear Structure, Astrophysics, and Reactions: FINUSTAR 2*
edited by P. Demetriou, R. Julin, and S. V. Harissopulos
© 2008 American Institute of Physics 0-7354-0532-5/08/$23.00

posite parity, broken translational symmetry (which also cannot be escaped when in the interacting shell model [5] that otherwise works in the laboratory frame) admixes states with finite center-of-mass motion. An independent quasi-particle state of BCS type used to describe pairing is spread over many different particle numbers.

Second, the mean-field approach is ill-defined when the binding energy changes slowly with a collective degree of freedom, i.e. when there are many near-degenerate states with different shell structure. This is a common situation in transitional nuclei.

Third, self-consistent mean-field models mainly provide a tool to describe ground states, while all other microscopic methods available for light systems are routinely used to describe spectroscopic properties, i.e. excited states and the transitions between them. There are certain exceptions: rotational bands can be obtained adding a cranking constraint to a SCMF model [6], or collective harmonic vibrations and giant resonances that can be accessed through the the linear response of the mean field to external perturbations in the (quasi-particle) random phase approximation [7]. An energy density functional based method that is capable of describing different types of excited states on the same footing in all nuclei irrespective of their structure is still missing. The rest of this contribution will sketch recent work into that direction.

## How to go beyond the mean field

Three kinds of correlations have to be distinguished in the methods discussed here. First, *in-medium* correlations are resummed into the effective energy density functional. Second, *static* correlations related to deformation and pairing are described by the symmetry-breaking deformed mean field. Third, *dynamical* correlations describe fluctuations in collective degrees of freedom, for example the order parameters of the broken symmetries. While the last two are clearly separated through the many-body method, one has to take care not to double-count the effect of the explicitely treated correlations when constructing the EDF. The dynamical correlations cannot be described by a single mean field state, they require what one often calls "to go beyond the mean field".

### *Symmetry restoration through projection*

Restoring the symmetries decomposes the mean-field states into states with proper quantum numbers. Eigenstates of the particle number operator $\hat{N}$ with eigenvalue $N_0$ are obtained applying the particle-number projection operator

$$\hat{P}_{N_0} = \frac{1}{2\pi}\int_0^{2\pi} d\phi_N \, e^{i\phi_N(\hat{N}-N_0)} \tag{1}$$

to the SCMF states. Eigenstates of the total angular momentum operator in the laboratory frame $\hat{J}^2$ and its projection on the $z$ axis $\hat{J}_z$ with eigenvalues $\hbar^2 J(J+1)$ and $\hbar M$, respectively, are obtained applying the operator

$$\hat{P}^J_{MK} = \frac{2J+1}{16\pi^2}\int_0^{4\pi} d\alpha \int_0^{\pi} d\beta \, \sin(\beta) \int_0^{2\pi} d\gamma \, \mathscr{D}^{J*}_{MK}(\alpha,\beta,\gamma)\,\hat{R}(\alpha,\beta,\gamma), \tag{2}$$

where $\hat{R} = e^{-i\alpha\hat{J}_z}e^{-i\beta\hat{J}_y}e^{-i\gamma\hat{J}_z}$ is the rotation operator and $\mathscr{D}^J_{MK}$ a Wigner function. $\hat{P}^J_{MK}$ picks the component with angular momentum projection $K$ along the intrinsic $z$ axis. The projected state is then obtained by summing over all $K$ components of the mean-field state $|q\rangle$ with weights $f^J_K$ determined from a variational equation [8, 9]

$$|JMq\rangle = \frac{1}{\mathcal{N}} \sum_{K=-J}^{+J} f^J_K \, \hat{P}^J_{MK} \, |q\rangle \,. \tag{3}$$

Note that $\hat{P}^J_{KM}$ is not a projection operator in the strict mathematical sense [9]. All results discussed below were obtained starting from HF+BCS or HFB states $|q\rangle$ with even proton and neutron numbers, good parity, axial and time-reversal symmetry. This allows for the analytical evaluation of the $\alpha$ and $\gamma$ integration in Eq. (2) at the price of restricting the projected states to positive parity $P = +1$, even integer total angular momentum $J$, and intrinsic angular momentum projection $K = 0$.

## *Variational configuration mixing*

The ambiguities many near-degenerated states with different deformation can be overcome by a variational mixing of these states within the Generator Coordinate Method (GCM). The mixed projected many-body state is set-up as a coherent superposition of projected mean-field states $|JMq\rangle$ with different intrinsic deformations $q$

$$|JMk\rangle = \sum_q f_{Jk}(q) \, |JMq\rangle \,, \tag{4}$$

where $k$ labels the GCM states and $f_{J,k}(q)$ is a weight function determined from

$$\frac{\delta}{\delta f^*_{Jk}} \frac{\langle JMk|\hat{H}|JMk\rangle}{\langle JMk|JMk\rangle} = 0 \,. \tag{5}$$

This leads to the Hill-Wheeler-Griffin equation [10]

$$\sum_{q'} \left[ \langle JMq|\hat{H}|JMq'\rangle - E_k \, \langle JMq|JMq'\rangle \right] \, f_{J,k}(q') = 0 \,, \tag{6}$$

that gives a correlated ground state and a spectrum of excited states. The weight functions $f_{Jk}(q)$ are not orthonormal. A set of orthonormal collective wave functions is obtained from a transformation involving the square root of the norm kernel [8]. Note that projection is a special case of the GCM, where the collective path and weights are determined by the corresponding symmetry group.

Combining projection and variational GCM mixing of states with different mass quadrupole moments allows to calculate transition moments directly in the lab frame for in-band and out-of-band transitions, for example $B(E2)$ values

$$B(E2; J'_{k'} \to J_k) = \frac{e^2}{2J'+1} \sum_{M=-J}^{+J} \sum_{M'=-J'}^{+J'} \sum_{\mu=-2}^{+2} |\langle JMk|\hat{Q}_{2\mu}|J'M'k'\rangle|^2 \tag{7}$$

without the need for effective charges. The $B(E2)$ value scales with mass and angular momentum. A more intuitive measure is the transitional quadrupole deformation obtained from the $B(E2)$ using the static rotor model

$$\beta_2^{(t)}(J'_{k'} \to J_k) = \frac{4\pi}{3R^2A}\sqrt{\frac{B(E2; J_k \to J'_k - 2)}{(J020|J-20)^2 e^2}} \tag{8}$$

with $R = 1.2A^{1/3}$. The method gives also the spectroscopic quadrupole moments, and $E0$ transition moments [11], which we will not discuss here.

Depending on the collective coordinates chosen, other excitation modes and transition moments can be accessed as well. This points to a difficulty, however. The GCM relies on the intuition for the most important collective degree(s) of freedom. The actual choice depends on the mode to be described, for example the quadrupole or octupole moment of the mass density or the monopole moment of the pair density, which then delivers a description of quadrupole, octupole or pairing vibrations, respectively. Examples for such calculations, without angular-momentum projection, can be found in Ref. [12]. Several generator coordinates can be easily combined for multi-dimensional calculations, although this has been rarely done so far; see [13] for an example without projection. For all results shown here, the axial quadrupole moment of the mass density distribution serves as the generator coordinate. Hence, the excited states are either quadrupole vibrational or rotational states. We use the Skyrme energy density functional in our implementation of the method; other groups have developped similar approaches using the Gogny interaction [14] and relativistic energy functionals [15].

## *Example I: Spectroscopy of collective states*

The neutron-deficient Pb isotopes show unique spectroscopic features, which are associated with a spherical ground state, and competing oblate, prolate, and superdeformed structures. The left panel of Fig. 1 shows representative results for the collective excitation spectrum and $E2$ transition moments for $^{188}$Pb obtained with the Skyrme interaction SLy6 and taken from Ref. [16]. The projected GCM delivers collective states that can be associated with a spherical ground state and excited prolate, oblate and superdeformed bands. There is a nice qualitative agreement with experimental data, both for the relative placement and evolution of the coexisting bands, as well as for the few reduced quadrupole transition moments currently known in $^{188}$Pb [17].

The right panel shows results obtained for $^{74}$Kr, also a nucleus situated in a region with near-degenerate oblate and prolate states. Here, the calculations do not even nearly reproduce the relative order of the band heads, a problem that we mainly relate to difficulties of most parameterizations of nuclear energy density functionals to describe the relative distances of single-particle levels in the $fp$ shell [18].

Figure 1 also demonstrates a common problem faced by all current implementations of angular-momentum projected GCM of axial quadrupole-deformed states, namely that the rotational bands are too dilute [11, 14, 15, 16, 18]. As the in-band $B(E2)$ values are much better described than the excitation energies, this hints at a limitation of the variational space that mainly affects the moments of inertia, not so much the shapes.

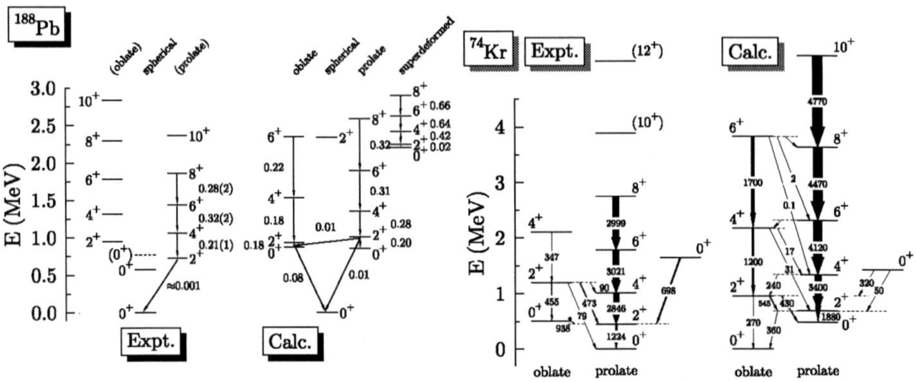


**FIGURE 1.** Left: Experimental and calculated excitation spectra and selected transition quadrupole moments $\beta_2^{(t)}(J'_{k'} \rightarrow J_k)$ for $^{188}$Pb [16]. Right: the same, but using the $B(E2)$ values, for $^{74}$Kr [18].

## *Example II: Mass systematics*

Mass formulas based on self-consistent mean-field models using Skyrme interactions have reached a quality where they compete with the best available microscopic-macroscopic models [19, 20, 21, 22]. A key to this success is to add various correlation energies phenomenologically through correction terms, as the Wigner energy or a rotational correction. There is no correction for vibrations, as it cannot be formulated in terms of a simple expression. The angular-momentum projected GCM allows to consistently calculate the quadrupole correlation energy from symmetry restoration and fluctuations of the quadrupole moment.

Figure 2 shows some representative results obtained with the Skyrme interaction SLy4 taken from Refs. [23, 24]. We compare the mass residuals (i.e. the difference between experimental and calculated masses, where negative values correspond to overbound nuclei) obtained on the mean-field level (upper panel) with those when collective quadrupole correlations are added (lower level), where the same data are drawn for isotopic chains as a function of $N$ (left) and for isotonic chains as a function of $Z$ (right). There is an overall erroneous trend with $A$, already observed in Ref. [25], which can be removed with a perturbative refit of the energy functional [26]. The major effect of the refit is a reduction of the volume energy coefficient by as little as 0.09 MeV.

What cannot be removed easily by a refit are the oscillations that are correlated to magic numbers, and for which shell effects are overestimated. The dynamical quadrupole correlations reduce the oscillation of the mass residuals between the closed shells, but does not remove it completely. It was unexpected that the oscillations of the mass residuals are mainly correlated to neutron magic numbers, not so much proton ones, which hints much more at remaining deficiencies of the energy functional than large missing correlation effects. For the energy, the dynamical correlations mainly play a role in the vicinity of closed shells, which is examplified when looking at energy

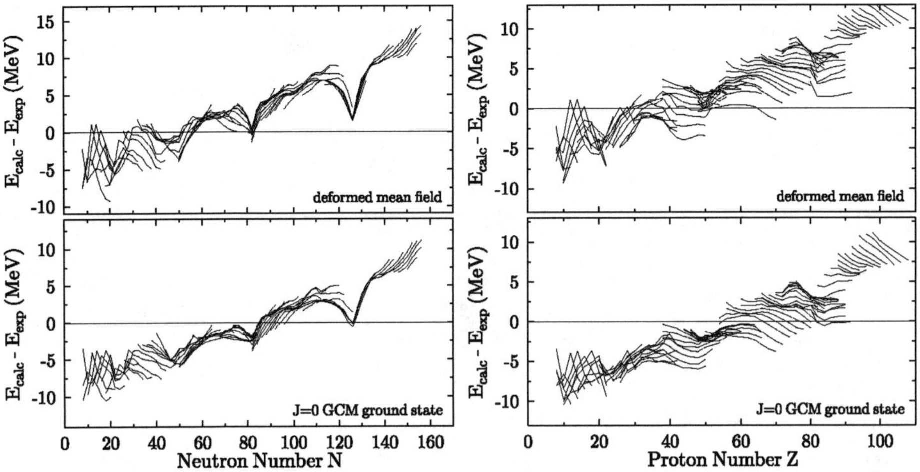


**FIGURE 2.** Mass residuals for mean-field states and $J = 0$ angular-momentum projected quadrupole GCM states drawn for isotopic chains as a function of $N$ and for isotonic chains as a function of $Z$ [23, 24].

differences across the closed shell like the two-proton gap

$$\delta_{2p}(N,Z) = E(N,Z-2) - 2E(N,Z) + E(N,Z+2)\,, \tag{9}$$

see Fig. 3. Spherical mean-field calculations give near-constant $\delta_{2p}$ in accordance with the stable $Z = 50$ and 82 shells. Allowing for deformation and adding correlations substantially reduces $\delta_{2p}$ for mid-shell nuclei, as the non-magic $Z \pm 2$ nuclei might be deformed and gain more dynamical correlation energy than the magic ones with Z, in agreement with experiment. Similar results were obtained in Ref. [27].

## The necessity for improved energy functionals

As we have seen in the few examples discussed above, systematic calculations of mass systematics and the analysis of collective excitation spectra from projected GCM indicate that the single-particle spectra from the currently used self-consistent mean-field models are far from being perfect. This is not too surprising as most popular and sucessful parameterizations are biased on nuclear matter properties, not single-particle spectra. There are problems with all parameterizations of the Skyrme interaction, and similar deficiencies of the Gogny forces, as well as relativistic mean-field Lagrangians, although not always in the same mass regions, none of which can be removed through a modified fit protocol without spoiling other properties of the functionals. This points at deficiencies of the functional form of the energy functionals used in all these approaches.

The comparison between the eigenvalues of a spherical single-particle Hamiltonian and experimental one-nucleon separation energies is problematic, see [1, 28, 29] and references given therein. Still, it serves as an starting point to analyze the impact of new

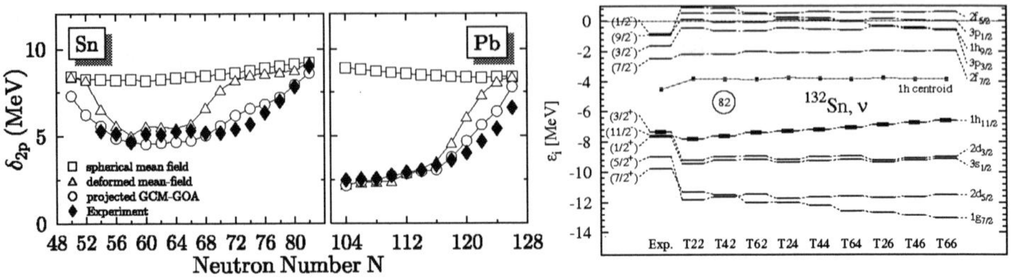


**FIGURE 3.** Left: Two-proton gap $\delta_{2p}$ for Sn and Pb isotopes [23, 24]. Right: eigenvalues of the neutron single-particle Hamiltonian in $^{132}$Sn compared with empirical single-particle energies for Skyrme interactions with systematically varied tensor interaction [29].

terms in the energy functional on the single-particle spectra. Of course more refined studies are necessary as soon as the relevant degrees of freedom have been identified.

A traditionally omitted degree of freedom that gained a lot of interest recently is the tensor interaction [28, 29, 30, 31, 32, 33], that enters the Skyrme energy density functional for even-even nuclei in form of a term bilinear in the spin-orbit current. While the tensor interaction with suitable coupling constants indeed gives a local improvement of various spectroscopic properties, a recent systematic study [29] demonstrated that it is unable to give a global improvement when added to a standard Skyrme energy functional. For example, the free variation of the tensor terms does not lower the $\chi^2$ of standard observables within a standard Skyrme energy functional. Positive values of the coupling constants of proton-neutron and like-particle tensor terms allow for a qualitative description of the evolution of spin-orbit splittings in chains of Ca, Ni, and Sn isotopes; for the same values of the tensor term coupling constants, however, the overall agreement of the single-particle spectra in doubly-magic nuclei is deteriorated, see Fig. 3 for the example of $^{132}$Sn, which can be traced back to features of the single-particle spectra that are not related to the tensor terms. The main conclusion is that the currently used central and spin-orbit parts of the Skyrme energy density functional are not flexible enough to allow for the presence of large tensor terms. In [29] it was also shown that tensor terms should not be added perturbatively to existing parametrizations as done in [31, 32, 33]; a complete refit of the entire parameter set is imperative.

## OUTLOOK

As we have seen, there are many recent developments that significantly improve the performance of mean-field models for standard observables, and open the door for the description of phenomena that were not accessible before. However, it is clear that a lot of work remains to be done regarding (i) the number of simultaneously treated collective degrees of freedom, which will hopefully improve the density of excited levels, (ii) the functional form and fit protocols of the effective energy density functionals used, and (iii), a point only mentioned in passing although very critical, the development of a consistent formal framework for beyond mean-field calculations using energy functionals, as the existing *ad-hoc* formalism, constructed in analogy to the case using a Hamiltonian, has unexpected spurious effects [34, 35, 36]

## ACKNOWLEDGMENTS

Work presented here was performed in collaboration with K. Bennaceur, G. F. Bertsch, P. Bonche, T. Duguet, H. Flocard, P.-H. Heenen, D. Lacroix, T. Lesinski, and J. Meyer.

## REFERENCES


1. M. Bender, P.-H. Heenen, P.-G. Reinhard, *Rev. Mod. Phys.* **75**, 121 (2003).
2. W. Kohn, *Rev. Mod. Phys.* **71**, 1253 (1998) and references therein.
3. J. Engel, *Phys. Rev.* C **75**, 014306 (2007).
4. B. G. Giraud, B. K. Jennings, and B. R. Barrett, preprint arXiv:0707.3099 (2007); B. G. Giraud, preprint arXiv:0707.3901 (2007).
5. E. Caurier, G. Martinez-Pinedo, F. Nowacki, A. Poves, A. P. Zuker, *Rev. Mod. Phys.* **77**, 427 (2005).
6. W. Satuła and R. Wyss, *Rep. Prog. Phys.* **68**, 131 (2005).
7. N. Paar, D. Vretenar, E. Khan, and G. Colò, *Rep. Prog. Phys.* **70**, 691 (2007).
8. C. Y. Wong, *Phys. Rep.* **15**, 283 (1975).
9. P. Ring, P. Schuck, *The Nuclear Many-Body Problem*, Springer Verlag, Heidelberg, 438 (1980).
10. D. L. Hill and J. A. Wheeler, *Phys. Rev.* **89**, 1106 (1953); J. J. Griffin and J. A. Wheeler, *Phys. Rev.* **108**, 311 (1957).
11. M. Bender, H. Flocard, and P.-H. Heenen, *Phys. Rev.* C **68**, 044321 (2003).
12. P.-H. Heenen, A. Valor, M. Bender, P. Bonche, and H. Flocard, *Eur. Phys. J.* **A11**, 393 (2001).
13. J. Meyer, P. Bonche, M. S. Weiss, J. Dobaczewski, H. Flocard, and P.-H. Heenen, *Nucl. Phys.* **A588**, 597 (1995).
14. R. R. Rodriguez-Guzman, J. L. Egido, and L. M. Robledo, *Phys. Rev.* C **62**, 054319 (2002); J. L. Egido and L. M. Robledo, in *Extended Density Functionals in Nuclear Physics*, edited by G. A. Lalazissis et al., Lecture Notes in Physics No. 641 (Springer, Berlin, 2004), p. 269.
15. T. Nikšić, D. Vretenar, and P. Ring, *Phys. Rev.* C **73**, 034308 (2006); *Phys. Rev.* C **74**, 064309 (2006).
16. M. Bender, P. Bonche, T. Duguet, and P.-H. Heenen, *Phys. Rev.* C **69**, 064303 (2004).
17. J. Pakarinen et al., *Phys. Rev.* C **75**, 014302 (2007).
18. M. Bender, P. Bonche, and P.-H. Heenen, *Phys. Rev.* C **74**, 024312 (2006).
19. F. Tondeur, S. Goriely, J. M. Pearson, and M. Onsi, *Phys. Rev.* **62**, 024308 (2000);
20. M. Samyn, S. Goriely, P.-H. Heenen, J. M. Pearson, and F. Tondeur, *Nucl. Phys.* **A700**, 142 (2002).
21. S. Goriely, M. Samyn, M. Bender, and J. M. Pearson, *Phys. Rev.* C **68**, 054325 (2003);
22. M. Samyn, S. Goriely, M. Bender, and J. M. Pearson, *Phys. Rev.* C **70**, 044309 (2004).
23. M. Bender, G. F. Bertsch, and P.-H. Heenen, *Phys. Rev. Lett.* **94**, 102503 (2006).
24. M. Bender, G. F. Bertsch, and P.-H. Heenen, *Phys. Rev.* C **73**, 034322 (2006).
25. M. V. Stoitsov, J. Dobaczewski, W. Nazarewicz, S. Pittel, D. J. Dean, *Phys. Rev.* C **68**, 054312 (2003).
26. G. F. Bertsch, B. Sabbey, and M. Uusnäkki, *Phys. Rev.* C **71**, 054311 (2005)
27. P. Fleischer et al., *Eur. Phys. J.* **A22**, 363 (2004).
28. B. A. Brown, T. Duguet, T. Otsuka, D. Abe, and T. Suzuki, *Phys. Rev.* C **74**, 061303(R) (2006).
29. T. Lesinski, M. Bender, K. Bennaceur, T. Duguet, and J. Meyer, *Phys. Rev.* C **76**, 014312 (2007).
30. T. Otsuka, T. Matsuo, and D. Abe, *Phys. Rev. Lett.* **97**, 162501 (2006).
31. J. Dobaczewski, in *Proceedings of the Third ANL/MSU/JINA/INT RIA Workshop*, edited by T. Duguet et al., World Scientific, Singapore, 2006.
32. G. Colò, H. Sagawa, S. Fracasso, and P. F. Bortignon, *Phys. Lett.* **B646**, 227 (2007).
33. D. M. Brink and F. Stancu, *Phys. Rev.* C **75**, 064311 (2007).
34. J. Dobaczewski, M. V. Stoitsov, W. Nazarewicz, P.-G. Reinhard, preprint arXiv:0708.0441 (2007).
35. D. Lacroix, T. Duguet, and M. Bender, in preparation.
36. M. Bender, T. Duguet, and D. Lacroix, in preparation.

# Study of the tensor correlation in a neutron-rich *sd*-shell region with the charge- and parity-projected Hartree-Fock method

Satoru Sugimoto*, Hiroshi Toki† and Kiyomi Ikeda**

*Kyoto University, Kyoto 606-8502, Japan
†Research Center for Nuclear Physics (RCNP), Osaka University, Osaka 567-0047, Japan
**The Institute of Physical and Chemical Research (RIKEN), Saitama 351-0198, Japan

**Abstract.** We study the effect of the tensor force on nuclear structure with mean-field and beyond-mean-field methods. An important correlation induced by the tensor force is two-particle–two-hole ($2p2h$) correlation, which cannot be treated with a usual mean-filed method. To treat the $2p2h$ tensor correlation, we develop a new framework (charge- and parity-projected Hartree-Fock (CPPHF) method), which is a beyond-mean-field method. In the CPPHF method, we introduce single-particle states with parity and charge mixing. The parity and charge projections are performed on a total wave function before variation. We apply the CPPHF method to oxygen isotopes including neutron-rich ones. The potential energy from the tensor force has the same order of magnitude with that from the LS force and becomes smaller with neutron number, which indicates that excess neutrons do not contribute to the $2p2h$ tensor correlation significantly. We also study the effect of the tensor force on spin-orbit-splitting ($ls$-splitting) in a neutron-rich fluorine isotope $^{23}$F. The tensor force reduces the $ls$-splitting for the proton $d$-orbits by about 3MeV. This effect is important to reproduce the experimental value. We also find that the $2p2h$ tensor correlation does not affect the $ls$-splitting in $^{23}$F.

**Keywords:** tensor force, Hartree-Fock method, spin-orbit-splitting
**PACS:** 21.10.Dr,21.10.Pc,21.60.Jz

## INTRODUCTION

The strong tensor force is characteristic of nuclear many-body systems. The study of nuclear matter using the Brueckner $G$-matrix theory demonstrated that the tensor force gives large attractive energy and plays an important role for the saturation mechanism of nuclear binding energy and density [1]. Exact calculations for light nuclei with realistic nuclear forces actually revealed that the tensor force produces large attractive energy in finite nuclei [2]. Recently, the relation between shell evolution with neutron or proton number with the tensor force has been widely discussed [3]. The effect of the tensor force on a halo structure in $^{11}$Li is also discussed [4].

The tensor force induces two-particle–two-hole ($2p2h$) correlation, which is responsible for large attractive energy in nuclei. The $2p2h$ tensor correlation is a kind of over-shell correlation and cannot be treated in a simple mean-field model. To treat the $2p2h$ tensor correlation we need a framework beyond mean-field. The tensor force is mediated by the pion, which is a pseudoscalar and isovector particle. Considering these characters, we introduce single-particle states with parity and charge mixing [5, 6]. Because a total wave function consisting of the single-particle states with parity and charge mixing,

CP1012, *Frontiers in Nuclear Structure, Astrophysics, and Reactions: FINUSTAR 2*
edited by P. Demetriou, R. Julin, and S. V. Harissopulos
© 2008 American Institute of Physics 0-7354-0532-5/08/$23.00

does not have good parity and a definite charge number, we perform the parity and charge projections on it before variation. By taking a variation with respect to a single-particle state, we obtain a Hartree-Fock-like equation (charge- and parity-projected Hartree-Fock equation) [7, 8]. Then, we call our method the charge- and parity projected Hartree-Fock (CPPHF) method.

The CPPHF method is applied to both non-relativistic [7, 9] and relativistic calculations [8, 10]. In the present contribution, we discuss our result for oxygen isotopes including neutron-rich ones [9]. We also discuss the effect of the tensor force on spin-orbit-splitting ($ls$-splitting) in $^{23}$F, which has a proton around a neutron-rich nucleus $^{22}$O [11].

## STUDY OF OXYGEN ISOTOPES WITH THE CPPHF METHOD

Before showing the result for oxygen isotopes, we discuss the result for $^{16}$O. The spherical symmetry is assumed in the present calculation. We use the modified Volkov force No. 1 (MV1) [12] for the central force and the G3RS force [13] for the LS and tensor forces. The Majorana parameter in the MV1 force is fixeted to 0.6. We multiply the $^3$E central force by 1.025 and the three-body force by 1.25 in the MV1 force, which are determined to reproduce the binding energy and the radius of $^{16}$O in the presence of the tensor force.

In Table 1 we show the results for $^{16}$O with the simple Hartree-Fock (HF) and CPPHF methods. In the HF case, the potential energy from the tensor force is negligible small, because $^{16}$O has LS-closed-shell structure. In the CPPHF case, the potential energy from the tensor force becomes finite. It indicates that the CPPHF method can treat the $2p2h$ tensor correlation, which cannot be treated in the simple HF method. Even if parity mixing is allowed for single-particle states in the HF calculation, we only obtain a result with no parity-mixing. It means that the projection before variation is essential to treat the $2p2h$ tensor correlation in the CPPHF method.

The $2p2h$ tensor correlation not only affects the binding energy but also induces high-momentum component in single-particle wave functions [7, 9]. To see the effect of the high-momentum component, the absolute value of the charge form factor for $^{16}$O as a function of momentum squared is shown in Fig. 1. By comparing with the CPPHF case and the HF case, it is seen that the $2p2h$ tensor correlation enhances the charge form factor in high momentum region. Although the charge form factor becomes close to the experimental data by including the $2p2h$ tensor correlation in the CPPHF method, we

**TABLE 1.** Results for $^{16}$O in the HF and CPPHF methods. $E$ and $T$ are the total energy and the total kinetic energy respectively. $V_{\mathrm{C}}$, $V_{\mathrm{3B}}$, $V_{\mathrm{T}}$, $V_{\mathrm{LS}}$ and $V_{\mathrm{Coul}}$ are the potential energies for the central, three-body, tensor, LS, and Coulomb forces, respectively. Those are give in MeV. $R_{\mathrm{c}}$ is the root-mean-square charge radius in fm.

| | $E$ | $T$ | $V_{\mathrm{C}}+V_{\mathrm{3B}}+V_{\mathrm{Coul}}$ | $V_{\mathrm{T}}$ | $V_{\mathrm{LS}}$ | $R_{\mathrm{c}}$ |
|---|---|---|---|---|---|---|
| HF | -124.1 | 230.0 | -353.2 | 0.0 | -0.9 | 2.73 |
| CPPHF | -127.1 | 237.1 | -351.6 | -11.7 | -1.0 | 2.73 |

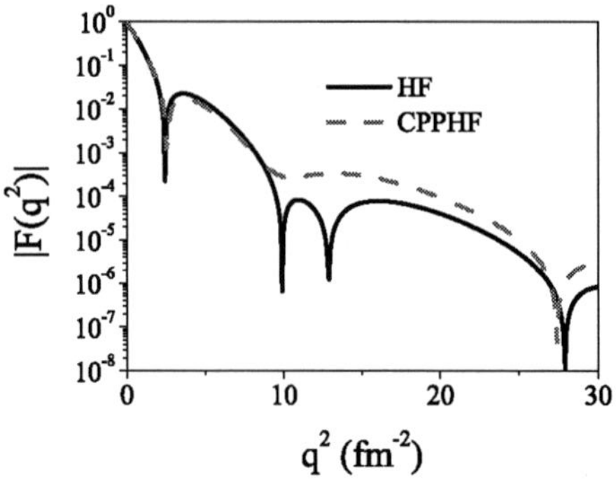


**FIGURE 1.** Absolute values of the charge form factor for $^{16}$O in the Hartree-Fock (HF) and charge- and parity-projected Hartree-Fock (CPPHF) methods as a function of momentum squared ($q^2$). The solid and dashed curves correspond to the HF and CPPHF methods respectively.

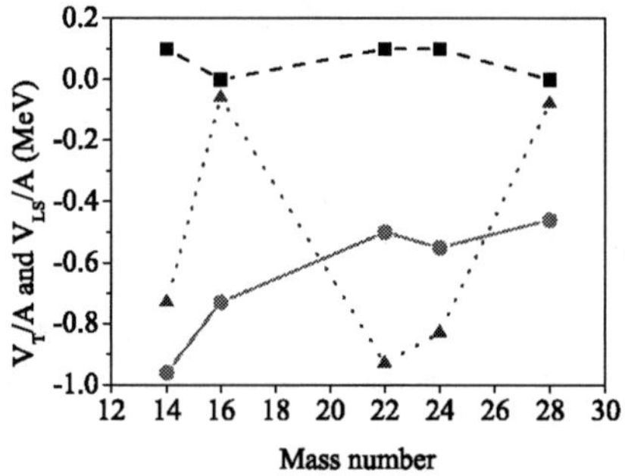


**FIGURE 2.** Potential energies from the tensor and LS forces in oxygen isotopes divided by mass number. The horizontal axis indicates mass number. The circle and triangle symbols correspond to the potential energies from the tensor and LS forces in the CPPHF method. The square symbols correspond to the potential energy from the tensor force in the HF method.

should need a further investigation to compare the present result with the experimental data, taking into account other correlations which cannot be treated in the present method like the short-range correlation and the meson exchange current.

We apply the CPPHF method to closed-shell and close-subshell oxygen isotopes, $^{14}$O, $^{16}$O, $^{22}$O, $^{24}$O, and $^{28}$O, which are assumed to have neutron closed-shell or closed-subshell structure up to the $0p_{3/2}$, $0p_{1/2}$, $0d_{5/2}$, $1s_{1/2}$, and $0d_{3/2}$ orbits respectively. The proton shell is fixed to have the closed-shell structure up to the $0p_{1/2}$ orbit. Because both parities are mixed in a single-particle state in the CPPHF method, we could not label the single-particle orbit as above. But parity mixing is not so large ($\leq 20\%$) and, then, we can approximately label single particle orbits as usual.

In Fig. 2, we show the potential energy per particle from the tensor and LS fores as a function of mass number. The potential energy from the LS force is plotted only for the CPPHF case because that for the HF case is almost the same as the CPPHF case. While the potential energy from the tensor force in the HF case is very small and repulsive, that in the CPPHF case is relatively large and attractive. In fact, the potential energy from the tensor force is comparable to that from the LS force. The potential energy from the LS force becomes negligible small for LS-closed-shell nuclei and large for jj-closed-shell nuclei. The potential energy from the tensor force behaves differently. It becomes smaller with neutron number. Because the $2p2h$ tensor correlation is mainly induced by the $^3$E tensor force and there are no protons in the *sd* shell, excess neutrons around $^{16}$O hardly contribute to the $2p2h$ tensor correlation. It is the reason why the tensor potential energy decreases with neutron number.

## EFFECT OF THE TENSOR FORCE ON *ls*-SPLITTING IN $^{23}$F

It is known that the tensor force or the pion field affects single-particle $ls$-splitting in jj-closed-shell nuclei at the mean field level [3, 14, 15]. It means the tensor force does not influence $ls$-splitting in $^{16}$O but has an effect on $ls$-splitting in $^{22}$O, where the neutron $0d_{5/2}$ orbit is fully occupied. Recently, an experiment to study single-particle states in $^{23}$F was performed [16]. The experiment shows that the $ls$-splitting for the proton $0d$ orbits are reduced by about 1MeV from that in $^{17}$F. To see the effect of the tensor force on the $ls$-splitting in $^{23}$F, we perform the Hartree-Fock calculation with the tensor force and the results are shown in Table 2. Our calculation reveals that the tensor force reduces the $ls$-splitting by about 3MeV. With this reduction, the $ls$-splitting for the proton $0d$ orbits becomes close to the experimental value. This effect is caused by the interaction between a valence proton and excess neutrons around $^{16}$O through the tensor force. Our result indicates that the tensor force actually will change shell structure in neutron-rich nuclei [3].

We also study the $ls$-splitting with the parity-projected Hartree-Fock (PPHF) method to see the effect of the $2p2h$ tensor correlation on the $ls$-splitting. In the PPHF method only the parity-projection is performed. The calculation with the PPHF method shows that the $2p2h$ tensor correlation does not affect the $ls$-splitting in $^{23}$F significantly. It is because excess neutrons around $^{16}$O do not contribute to the $2p2h$ tensor correlation

**TABLE 2.** $ls$-splitting for the proton $0d$-orbits in $^{23}$F. $\Delta(V_{\rm LS})$, $\Delta(V_{\rm T})$, and $\Delta$(others) are the contributions to the $ls$-splitting from the LS force, the tensor force, and the other forces including the kinetic term respectively. Those are given in MeV. The experimental value for $\Delta(0d_{3/2}-0d_{5/2})$ = 4.06MeV. [16]

| | $\Delta(0d_{3/2}-0d_{5/2})$ | $\Delta(V_{\rm LS})$ | $\Delta(V_{\rm T})$ | $\Delta$(others) |
|---|---|---|---|---|
| MV1 with the tensor force | 4.2 | 7.8 | −3.1 | −0.5 |
| MV1 without the tensor force | 7.2 | 8.3 | 0.0 | −1.1 |

as mentioned above. In the PPHF or CPPHF method with spherical symmetry, single-particle states with high orbital angular momenta are not treated. There are a possibility that the single-particle states with high orbital angular momenta affect the $ls$-splitting through the $2p2h$ tensor correlation [17].

## SUMMARY

We studied the effect of the tensor force with mean-field (HF) and beyond mean field (CPPHF) methods. The $2p2h$ tensor correlation induces high-momentum component in single-particle states and the potential energy form the tensor force has the same order of magnitude as that from the LS force in oxygen isotopes. The $2p2h$ tensor correlation energy becomes smaller with neutron number, because excess neutrons around $^{16}$O do not contribute to the $2p2h$ tensor correlation.

We also studied the effect of the tensor force on $ls$-splitting in $^{23}$F. The tensor force reduces the $ls$-splitting for the proton $0d$ orbits by 3MeV. With the reduction, the $ls$-splitting for the proton $0d$ orbits becomes close to the experimental value. We performed the calculation with the PPHF method to see the effect of the $2p2h$ tensor correlation on the $ls$-splitting in $^{23}$F. Our calculation shows that $2p2h$ tensor correlation does not affect the $ls$-splitting largely.

## ACKNOWLEDGMENTS

This work is supported by Grant-in-Aid for the 21st Century COE "Center for Diversity and Universality in Physics" and Grant-in-Aid for Scientific Research 19740140 from the Ministry of Education, Culture, Sports, Science and Technology (MEXT), Japan.

## REFERENCES

1. H.A. Bethe, *Annu. Rev. Nucl. Sci.* **21**, 93 (1971).
2. S. C. Pieper and R.B. Wiringa, *Annu. Rev. Nucl. Part. Sci.*, **51** (2001) 53.
3. T. Otsuka, T. Suzuki, R. Fujimoto, H. Grawe, and Y. Akaishi, *Phys. Rev. Lett.* **95**, 232502 (2005).
4. T. Myo, K. Katō, H. Toki, and K. Ikeda, *Phys. Rev. C*, **76**, 024305 (2007).
5. H. Toki, S. Sugimoto, and K. Ikeda, *Prog. Theor. Phys.* **108** 903 (2002).
6. Y. Ogawa, H. Toki, S. Tamenaga, H. Shen, A. Hosaka, S. Sugimoto, and K. Ikeda, *Prog. Theor. Phys.* **111**, 75 (2004).
7. S. Sugimoto, K. Ikeda, and H. Toki, *Nucl. Phys. A* **740**, 77 (2004).
8. Y. Ogawa, H. Toki, S. Tamenaga, S. Sugimoto, and K. Ikeda, *Phys. Rev C* **73**, 034301 (2006).
9. S. Sugimoto, K. Ikeda, and H. Toki, *Phys. Rev. C* **75**, 014317 (2007).
10. Y. Ogawa, H. Toki, and S. Tamenaga, *Phys. Rev. C* **76**, 014305 (2007).
11. S. Sugimoto, H. Toki, and K. Ikeda, *Phys. Rev. C* (to be published); arXiv:0705.1419.
12. T. Ando, K. Ikeda, and A. Tohsaki-Suzuki, *Prog. Theor. Phys.* **64**, 1608 (1980).
13. R. Tamagaki, *Prog. Theor. Phys.* **39**, 91 (1968).
14. R. M. Tarbutton and K. T. R. Davies, Nucl Phys. A **120**, 1 (1968).
15. M. Löpez-Quelle, N. Van Giai, S. Marcos, and L. N. Savushkin, Phys. Rev. C **61**, 064321 (2000).
16. S. Michimasa *et al.*, Phys. Lett. B **638**, 146 (2006); Nucl. Phys. A **787**, 569c (2007).
17. T. Terasawa, *Prog. Theor. Phys.* **23**, 87 (1960); A. Arima and T. Terasawa, *Prog. Theor. Phys.* **23**, 115 (1960).

# Coulomb Excitation of the N=50 nucleus $^{80}$Zn

J. Van de Walle[a], F. Aksouh[b], F. Ames[c], T. Behrens[d], V. Bildstein[e], A. Blazhev[f], J. Cederkäll[g], E. Clément[h], T.E. Cocolios[a], T. Davinson[i], P. Delahaye[g], J. Eberth[f], A. Ekström[j], D.V. Fedorov[k], V.N. Fedosseev[g], L.M. Fraile[g], S. Franchoo[g], R. Gernhauser[d], G. Georgiev[l], D. Habs[c], K. Heyde[m], G. Huber[n], M. Huyse[a], F. Ibrahim[o], O. Ivanov[a], J. Iwanicki[p], J. Jolie[f], O. Kester[q], U. Köster[r], T. Kröll[d], R. Krücken[d], M. Lauer[s], A.F. Lisetskiy[q], R. Lutter[c], B.A. Marsh[g], P. Mayet[a], O. Niedermaier[s], T. Nilsson[q], M. Pantea[t], O. Perru[o], R. Raabe[a], P. Reiter[f], M. Sawicka[a], H. Scheit[s], G. Schrieder[t], D. Schwalm[s], M.D. Seliverstov[u], T. Sieber[g], G. Sletten[v], N. Smirnova[w], M. Stanoiu[q], I. Stefanescu[a], J.-C. Thomas[x], J.J. Valiente-Dobón[y], P. Van Duppen[a], D. Verney[o], D. Voulot[g], N. Warr[f], D. Weisshaar[f], F. Wenander[g], B.H. Wolf[g] and M. Zielińska[z]

*[a]Instituut voor Kern- en Stralingsfysica, K.U. Leuven, Leuven, Belgium*
*[b]Instituut voor Kern- en Stralingsfysica, K.U. Leuven, Leuven, BelgiumCEA Saclay, DAPNIA/SPhN, Gif-sur-Yvette, France*
*[c]Ludwig-Maximilians-Universität, München, Germany*
*[d]Physik Department E12, Technische Universität München, Garching, Germany*
*[e]Physik Department E12, Technische Universität München, Garching, GermanyMax-Planck-Institut für Kernphysik, Heidelberg, Germany*
*[f]Institut für Kernphysik, Universität Köln, Köln, Germany*
*[g]ISOLDE, CERN, Geneva, Switzerland*
*[h]ISOLDE, CERN, Geneva, SwitzerlandCEA Saclay, DAPNIA/SPhN, Gif-sur-Yvette, France*
*[i]University of Edinburgh, Edinburgh, United Kingdom*
*[j]Physics Department, University of Lund, Lund, Sweden*
*[k]Department of High Energy Physics, Petersburg Nuclear Physics Institute, Gatchina, Russia*
*[l]ISOLDE, CERN, Geneva, SwitzerlandCSNSM, IN2P3-CNRS, Université Paris-Sud, Orsay, France*
*[m]Vakgroep Subatomaire en Stralingsfysica, Universiteit Gent, Gent, Belgium*
*[n]Institut für Physik, Johannes Gutenburg Universität Mainz, Mainz, Germany*
*[o]Institut de Physique Nucléaire, IN2P3-CNRS, Orsay, France*
*[p]Heavy Ion Laboratory, University of Warsaw, Warsaw, Poland*
*[q]Gesellschaft für Schwerionenforschung mbH, Darmstadt, Germany*
*[r]Institut Laue-Langevin, Grenoble, FranceISOLDE, CERN, Geneva, Switzerland*
*[s]Max-Planck-Institut für Kernphysik, Heidelberg, Germany*
*[t]Institut für Kernphysik, Technische Universität Darmstadt, Darmstadt, Germany*
*[u]Institut für Physik, Johannes Gutenburg Universität Mainz, Mainz, GermanyDepartment of High Energy Physics, Petersburg Nuclear Physics Institute, Gatchina, Russia*
*[v]Physics Department, University of Copenhagen, Denmark*
*[w]Vakgroep Subatomaire en Stralingsfysica, Universiteit Gent, Gent, BelgiumCENBG, CNRS/IN2P3, Université Bordeaux, Gradignan cedex, France*
*[x]Instituut voor Kern- en Stralingsfysica, K.U. Leuven, Leuven, BelgiumGANIL, IN2P3-CNRS-CEA, Caen, France*
*[y]Instituto Nazionale di Fisica Nucleare, Laboratori Nazionali di Legnaro, Legnaro, Italy*
*[z]Heavy Ion Laboratory, University of Warsaw, Warsaw, PolandCEA Saclay, DAPNIA/SPhN, Gif-sur-Yvette, France*

CP1012, ***Frontiers in Nuclear Structure, Astrophysics, and Reactions:*** *FINUSTAR 2*
edited by P. Demetriou, R. Julin, and S. V. Harissopulos
© 2008 American Institute of Physics 0-7354-0532-5/08/$23.00

**Abstract.** Neutron rich Zinc isotopes, including the N=50 nucleus $^{80}$Zn, were produced and post-accelerated at the Radioactive Ion Beam (RIB) facility REX-ISOLDE (CERN). Low-energy Coulomb excitation was induced on these isotopes after post-acceleration, yielding B(E2) strengths to the first excited $2^+$ states. For the first time, an excited state in $^{80}$Zn was observed and the $2_1^+$ state in $^{78}$Zn was established. The measured $B(E2,2_1^+ \rightarrow 0_1^+)$ values are compared to two sets of large scale shell model calculations. Both calculations reproduce the observed *B(E2)* systematics for the full Zinc isotopic chain. The results for *N*=50 isotones indicate a good *N*=50 shell closure and a strong Z=28 proton core polarization. The new results serve as benchmarks to establish theoretical models, predicting the nuclear properties of the doubly magic nucleus $^{78}$Ni.

**Keywords:** Nuclear Structure, Radioactive Ion Beams, Coulomb Excitation
**PACS:** 21.10.-k,21.60.Cs,23.20.-g,25.70.De

## INTRODUCTION

The evolution of shell structure in nuclei with unusual neutron-to-proton (N/Z) ratio's has become an intensive field of research in the past years. In this contribution, experimental results are presented around the *Z*=28 and *N*=50 shell gaps, i.e. the region around the astrophysical r-process waiting-point nuclei $^{80}$Zn and $^{78}$Ni.

Recent experiments have investigated the strength of the *Z*=28 proton shell gap. Lifetime measurements of excited states in the stable Ni isotopes (*Z*=28) have indicated the need for inclusion of the $1\pi f_{7/2}$ orbital, below the *Z*=28 gap, in the valence space in order to describe the *B(E2)* trend in these isotopes. The recent RIB measurement of the *B(E2)* value in $^{70}$Ni [1] has indicated an enhanced proton core polarization when neutrons are added in the $1\nu g_{9/2}$ orbital. Both publications highlight the need for proton scattering across the *Z*=28 shell gap in order to describe the *B(E2)* strengths to the first excited $2_1^+$ state.

The *N*=50 shell gap has been investigated in a series of experiments at Legnaro. The level scheme of *N*=50 isotones down to $^{82}_{32}$Ge has been obtained and compared to shell model calculations in [2]. From [2], it was concluded that the $2_1^+$ state in *N*=50 isotones down to *Z*=32 have a dominating proton component in the wave function.

Combining the observations around *Z*=28 and *N*=50, one might expect that the $2_1^+$ state in $^{80}_{30}$Zn has a dominant proton character and possible weakening of the *Z*=28 shell gap should be reflected in a larger then expected *B(E2)* value for this nucleus. With the presence of a strong *N*=50 shell gap, the $2_1^+$ energy in the $_{Z=30}$Zn isotopic chain should increase drastically at *N*=50.

Furthermore, the Zinc isotopic chain exhibits an onset of collectivity around *N*=40, indicated by a smooth decrease of the E($2_1^+$) and an increase of the *B(E2)* values in $^{70-74}$Zn [1, 3, 4]. In order to probe the further development of collectivity in neutron-rich Zinc isotopes up to the *N*=50 closed neutron shell, a program of low-energy Coulomb excitation was started at the Radioactive Ion Beam facility REX-ISOLDE [5]. This technique is a very selective tool to investigate low-lying $2^+$ states and their *B(E2)* value in even-even isotopes because of the dominating E2 excitation cross section and the accurate description of the inelastic scattering process in terms of the electromagnetic interaction.

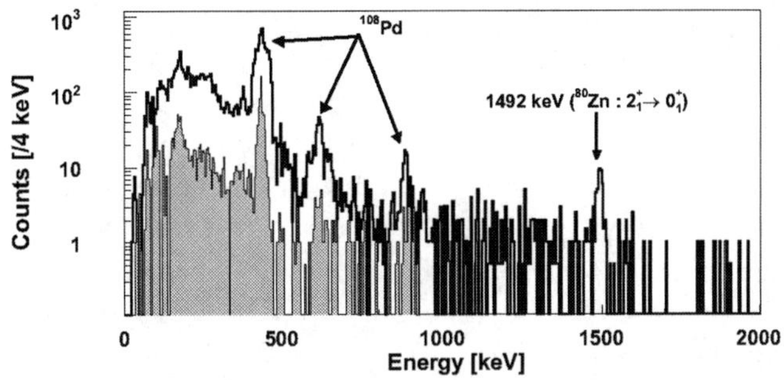


**FIGURE 1.** $\gamma$-spectrum in prompt coincidence with a particle, after random subtraction. The filled grey spectrum is acquired during laser off periods ( 10 h), the black line is the spectrum during laser on periods ( 100 h).

# RESULTS

The neutron-rich Zinc beams were produced at ISOLDE with a standard ISOLDE UCx target and 1.4 GeV protons from the CERN PS Booster. Details on the production method can be found in [6] and [7]. The Zinc ions were post-accelerated by the REX linear accelerator [5] to a final energy of 2.87 MeV/nucleon ($A$=74,78), 2.83 MeV/nucleon ($A$=76) and 2.79 MeV/nucleon ($A$=80). The post-accelerated beam was Coulomb excited on a 2.3 mg/cm$^2$ $^{120}$Sn target ($A$=74,76) and a 2.0 mg/cm$^2$ $^{108}$Pd target ($A$=78,80). The $\gamma$-rays following the de-excitation process were detected by the MINIBALL Germanium detector array [8]. The scattered particles were recorded in a 500 $\mu$m thick double sided silicon strip detector (DSSSD), covering laboratory angles between $\theta$=29°-52° ($A$=74,76,78) and $\theta$=16.4°-52° ($A$=80). Nuclei scattered in these angular ranges stem from collisions where the distance between the nuclear surfaces does not drop below 5 fm, ensuring that the observed excitation is induced by the electromagnetic interaction. Typical average beam intensities at the secondary target were $3.0\times10^5$ ($A$=74), $1.1\times10^5$ ($A$=76), $4.3\times10^3$ ($A$=78) and $3.0\times10^3$ ($A$=80) particles per second. The beam purity was determined by switching the laser ionization periodically on and off. The difference in particle scattering in the DSSSD during laser on and off periods gives the ratio of surface (Ga,Rb) over laser ionized species (Zn). The purity (=Zinc content) of the RIB's was 83(4)% ($A$=74), 73(7)% ($A$=76), 64(13)% ($A$=78) and 43(5)% ($A$=80).

$\beta$-decay activity and room background in random coincidence with elastic scattered particles were subtracted from the prompt coincident $\gamma$-spectra. Comparing $\gamma$-ray spectra accumulated with the lasers on and with the lasers off allowed to discriminate between lines originating from the Coulomb excitation of Zinc, isobaric contaminants and the target. In the A=78 and A=80 experiments, the $2_1^+ \rightarrow 0_1^+$ transitions could be established unambiguously from this comparison of laser-on and laser-off spectra, thereby fixing the $2_1^+$ state in $^{78}$Zn to 730 keV and in $^{80}$Zn to 1492 keV (see Fig. 1 for the $^{80}$Zn laser-on and laser off-spectrum).

The known $0_1^+ \rightarrow 2_1^+$ (E2) excitation cross section of the even-even target nucleus served as normalization to determine the excitation cross section of the Zinc isotopes, which depends on the unknown $B(E2)$ value. The cross sections were deduced using the coupled channels Coulomb excitation code GOSIA [9]. For $^{120}$Sn and $^{108}$Pd nuclei, adopted *B(E2)* values [10, 11, 12] between all states involved in the excitation process were used in the calculation. The calculated de-excitation cross section depends on the re-orientation matrix element ($<2_1^+||E2||2_1^+>$), related to the spectroscopic quadrupole moment of the $2_1^+$ state. In all the current calculations, this matrix element was fixed to 0.0 eb. When fitting the $<0_1^+||E2||2_1^+>$ matrix element, assuming a rotational quadrupole moment for the $2_1^+$ state, the *B(E2)* value increases by 22%(18%,15%) for prolate deformation and decreases by 13%(12%,11%) for oblate deformation for A=74(A=76,A=78). In those cases, the resulting quadrupole moments have a magnitude of 0.40(0.35,0.25) b for A=74(76,78) for both prolate and oblate deformation. Lifetime measurements of the $2_1^+$ state in these isotopes can provide constraints on the *B(E2)* values and would enable us to fit the re-orientation matrix-element.

## DISCUSSION

The resulting *B(E2)* and E($2_1^+$) values are summarized in Fig. 2. Two shell model calculations (SMC) are included in the figures. The first set of SMC (dotted line, SMI) was performed with the shell model code ANTOINE and the realistic effective nucleon-nucleon interaction based on G-matrix theory by M. Hjorth-Jensen [13] with monopole modifications by Nowacki [14]. The model space consisted of proton/neutron $2p_{3/2}$,$1f_{5/2}$,$2p_{1/2}$ and $1g_{9/2}$ orbits around an inert $^{56}$Ni core. High proton/neutron effective charges of 1.9e/0.9e were used to compensate for the large $^{56}$Ni core polarization. The second set of SMC (dashed lines, SMII) were obtained with the JJ4B effective interaction [15] which is an extension of the renormalized G-matrix interaction based on the Bonn-C NN potential (JJ4APN) constructed to reproduce the experimental data for exotic Ni, Cu, Zn, Ge and *N*=50 isotones in the vicinity of $^{78}$Ni. $\pi/\nu$ polarization charges of 1.76e and 0.97e were used. Both SM calculations reproduce the *B(E2)* trend in the Zinc isotopic chain (see Fig. 2). Around *N*=40, the SMII results are sensitive to the position of the excited $0_2^+$ state. A similar problem in the Ge isotopes has been discussed in [16]. The calculated *B(E2)* strength for the *N*=50 isotones is solely due to proton excitations, since neutron excitation across the *N*=50 gap are not included in the valence space. The first set of SMC overestimates the *B(E2)* strength around mid-shell between *Z*=28 and *Z*=40, whereas it reproduces the two *B(E2)* values in $^{82}$Ge and $^{80}$Zn. The use of a high proton effective charge seems to be needed for these two isotopes, which have a limited numbers of valence protons outside the *Z*=28 core (four and two, resp.). This hints to a higher proton core polarization close to *Z*=28. The second set of SMC reproduces the full trend of *B(E2)* values for the *N*=50 isotones down to $^{80}$Zn. Possible proton excitations across the *Z*=28 shell should be incorporated in the empirical set of matrix elements.

In conclusion, these set of Coulomb excitation experiments at low energy on neutron-rich Zinc isotopes have provided *B(E2)* and E($2_1^+$) values up to the *N*=50 nucleus $^{80}$Zn. The E($2_1^+$) systematics of *N*=50 isotones shows a small increase in excitation energy for the $2^+$ state in $^{80}$Zn, with only two protons added to the assumed doubly magic $^{78}$Ni. In comparison with SMC, it can be concluded that the *N*=50 neutron shell gap is still a good shell closure down to *Z*=30. No definite conclusions can be drawn on the strength of the

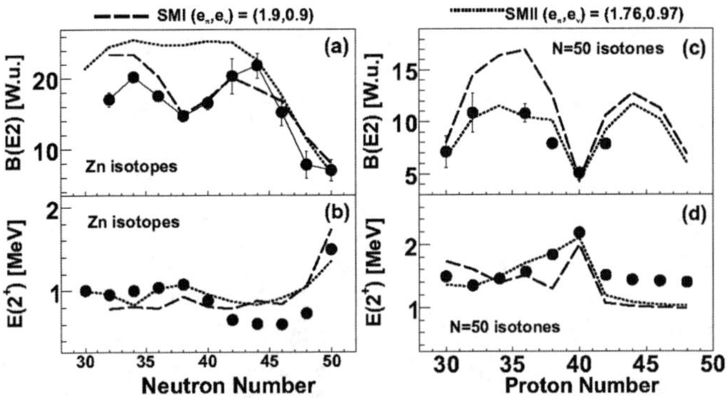


**FIGURE 2.** E(2) and *B(E2)* values for Zn isotopes (a,b) and *N*=50 isotones (c,d). Two sets of SMC are included, labeled by SMI and SMII, see text for details on the calculations.

*Z*=28 shell gap, though, the need for high proton effective charges with few valence protons outside the *Z*=28 shell gap indicate a strong proton core polarization.

## ACKNOWLEDGMENTS

This work was supported by the European Union Sixth Framework through RII3-EURONS (Contract Co. 506065), the German BMBF Grant No. 06KY205I, the BriX-IAP Research Program No. P06/23, NSF Grant No. PHY-0555393, and FWO-Vlaanderen (Belgium).

## REFERENCES

1. O. Perru *et al.*, *Phys. Rev. Lett.* **96**, 232501 (2006).
2. Y. Zhang *et al.*, *Phys. Rev. C* **70**, 024301 (2004).
3. S. Leenhardt *et al.*, *Eur. Phys. J. A* **14**, 1 (2002).
4. J. Van Roosbroeck *et al.*, *Phys. Rev. C* **70**, 054307 (2005).
5. O. Kester *et al.*, *Nucl. Instr. Meth. B* **204**, 20 (2003).
6. U. Köster *et al.*, *AIP Conf. Proc.* **798**, 315 (2005).
7. J. Van de Walle *et al.*, *Phys. Rev. Lett.* **99**, 142501 (2007).
8. J. Eberth *et al.*, *Prog. Part. Nucl. Phys.* **46**, 389 (2001).
9. T. Czosnyka *et al.*, *Bull. Am. Phys. Soc.* **28**, 745 (1983).
10. S. Raman, *At. Data Nucl. Data Tables* **78**, 1–128 (2001).
11. URL `http://www.nndc.bnl.gov/nndc/nudat/`.
12. L. Svensson *et al.*, *Nucl. Phys. A* **584**, 547 (1995).
13. M. Hjorth-Jensen *et al.*, *Phys. Rep.* **261**, 125 (1995).
14. N. Smirnova *et al.*, *Phys. Rev. C* **69**, 044306 (2004).
15. A. Lisetskiy *et al.*, *Phys. Rev. C* **70**, 044314 (2004).
16. M. Hasegawa *et al.*, *Nucl. Phys. A* **789**, 46–54 (2007).

# Sub-Barrier Coulomb Excitation of $^{106,108,110}$Sn

A. Ekström*, J. Cederkäll†,*, C. Fahlander*, M. Hjorth-Jensen**, F. Ames‡, P. A. Butler§, T. Davinson¶, J .Eberth‖, G. Georgiev†, A. Gorgen††, M. Górska‡‡, D. Habs§§, M. Huyse¶¶, O. Ivanov¶¶, J. Iwanicki***, O. Kester‡‡, U. Köster†, B. A. Marsh†††,‡‡‡, P. Reiter‖, H. Scheit§§§, D. Schwalm§§§, S. Siem¶¶¶, I. Stefanescu¶¶, G. M. Tveten¶¶¶, J. Van de Walle¶¶, P. Van Duppen¶¶, D. Voulot‡‡‡, N. Warr‖, D. Weisshaar‖, F. Wenander‡‡‡ and M. Zielinska***

**Physics Department, University of Lund, Box 118, SE-221 00 Lund, Sweden*
*†PH Department, CERN 1211, Geneva 23, Switzerland*
***Physics Department and Center of Mathematics for Applications, University of Oslo, Norway*
*‡TRIUMF, Vancouver, Canada*
*§Oliver Lodge Laboratory, University of Liverpool, United Kingdom*
*¶Department of Physics and Astronomy, University of Edingburg, United Kingdom*
*‖Institute of Nuclear Physics, University of Cologne, Germany*
*††CEA Saclay, DAPNIA/SPhN, gif-sur-Yvette, France*
*‡‡Gesellschaft für Schwerionenforschung, Darmstadt, Germany*
*§§Physics Department, Ludwig-Maximilian University, Munich, Germany*
*¶¶Instituut voor Kern- en Stralingsfysica, K.U. Leuven, Belgium*
****Heavy Ion Laboratory, Warsaw University, Poland*
*†††Department of Physics, University of Manchester, United Kingdom*
*‡‡‡AB Department, CERN 1211, Geneva 23, Switzerland*
*§§§Max-Planck Institute of Nuclear Physics, Heidelberg, Germany*
*¶¶¶Department of Physics, University of Oslo, Norway*

**Abstract.** The reduced transition probabilities between the first excited $2^+$ state and the $0^+$ ground state, $B(E2;0^+ \rightarrow 2^+)$, have been measured in $^{106,108,110}$Sn using sub-barrier Coulomb excitation in inverse kinematics at REX-ISOLDE. The results are, $B(E2;0^+ \rightarrow 2^+) = 0.220(22)$, 0.226(17), and 0.228(32) e$^2$b$^2$, for $^{110}$Sn, $^{108}$Sn, and $^{106}$Sn, respectively. The results for $^{106,108}$Sn are preliminary. De-excitation $\gamma$-rays were detected by the MINIBALL Ge-array. The $B(E2)$ reveals detailed information about the nuclear wave function. A shell model prediction based on an effective CD-Bonn interaction in the $\nu(0g_{7/2}, 2s, 1d, 0h_{11/2})$ model space using $e^{\nu}_{eff} = 1.0$ e follows the experimental values for the neutron rich Sn isotopes, but fails to reproduce the results presented here.

**Keywords:** multipole matrix elements, shell model, Coulomb excitation, nuclei with mass number 90 to 149
**PACS:** 23.20.Js, 21.60.Cs, 25.70.De, 27.60.+j

## INTRODUCTION

The experimental knowledge about the shell structure evolution towards the doubly-magic self-conjugate $^{100}$Sn nucleus is now becoming available through radioactive ion beam (RIB) techniques. The investigation of exotic isotopes can reveal novel effects of the underlying effective nucleon-nucleon interaction. Furthermore, the Sn isotopes span a region between the $N = Z = 50$ and $N = 82, Z = 50$ shell closures making it the

CP1012, *Frontiers in Nuclear Structure, Astrophysics, and Reactions: FINUSTAR 2*
edited by P. Demetriou, R. Julin, and S. V. Harissopulos
© 2008 American Institute of Physics 0-7354-0532-5/08/$23.00

longest isotopic chain available for experiment. This enables a unique study of the shell structure variations as a function of the number of neutrons outside the closed $^{100}$Sn core. The constancy of the energy separation between the first excited $2^+$ state and the $0^+$ ground state in the even-mass Sn isotopes is well explained within the generalized seniority model [1]. Furthermore, according to this theory, non-diagonal matrix elements of the even one-body E2 tensor operator will exhibit a parabolic behaviour as a function of mass number across the Sn isotope chain. Large scale shell-model calculations [2] based on an effective CD-Bonn nucleon-nucleon interaction agree with the generalized seniority model. The adopted experimental $B(E2)$ values on the neutron-rich side of the Sn chain follow the theoretical predictions. The experimental RIB results on $^{106,108,110}$Sn presented here and in [2, 3, 4] are consistent with each other and display a clear discrepancy with theoretical models.

## EXPERIMENTAL TECHNIQUES

Radioactive $^{110}$Sn, $^{108}$Sn, and $^{106}$Sn beams were produced at ISOLDE by bombarding a thick Lanthanum Carbide primary target by 1 GeV protons. Atomic Sn was ionized through a resonant three-step laser scheme and the isotope of interest was separated using the General Purpose Separator of the facility. The selected isotope was subsequently cooled in a Penning-type trap after which the beam was charge bred in the Electron Beam Ion Source. Post-acceleration was performed by the REX linac. The final beam energy was 2.8 MeV/u, well below the Coulomb barrier in order to exclude any excitation caused by direct nuclear-nuclear overlap. The first $2^+$ state was populated through sub-barrier Coulomb excitation against a 2 mg/cm$^2$ thick $^{58}$Ni target. De-excitation $\gamma$ rays were detected by the highly segmented MINIBALL Ge-array. The recoils and ejectiles were detected by a Double Sided Silicon Strip (DSSSD) detector placed 3 cm from the $^{58}$Ni target covering an angular range of about 16°-54°. High segmentation in both $\gamma$-ray and particle detectors allowed for a Doppler correction. This is needed since scattered particled had velocities of $\sim 0.05c$. A side effect of using RIBs is the significant $\gamma$-ray background caused by unstable nuclei deposited in the target chamber. By gating on prompt particle-$\gamma$ coincidences the background was reduced. A further gate in the analysis was to select only those events in which both the ejectile and the recoil was detected in the DSSSD, a so called 2p event. Due to the type of inverse kinematics this implied selecting events where both the ejectile and the recoil was scattered with an angle larger than 24°. In the case when only one particle, a so called 1p event, was detected over 24°, due to e.g. noise or double hits, the missing particle could be reconstructed. The integrated Coulomb excitation peaks presented in this paper come from the total set of $1p+2p$ events. The $\gamma$-ray statistics was increased by $\sim$10% due to add-back between neighbouring Ge-crystals. It should also be mentioned that the beam did not consist of only the selected Sn isotope, but also of surface ionized In. The contamination originate in the cavity immediately after the primary target. It is of primary importance to map out the isobaric contamination over time. In the case of $^{110}$Sn, the beam contamination was not that severe. Due to the high beam intensity of $\sim 10^6$ p/s at the secondary target, the In component could be measured by switching off the laser and register the beam current just before the $^{58}$Ni target. The contamination gradually increase as the proton drip line

**TABLE 1.** Experimental $B(E2)$ values from this work. Note that the values on $^{108,106}$Sn are preliminary. The value on $^{110}$Sn has previously been published in [3]

| Isotope | $^{110}$Sn | $^{108}$Sn | $^{106}$Sn |
|---|---|---|---|
| $B(E2;\uparrow)$ [e$^2$b$^2$] | 0.220(22) | 0.226(17) | 0.228(32) |

is approached. In the $^{106,108}$Sn experiments the laser power was recorded continously in the data stream as well as the laser status. The laser was run in on/off mode for one hour every three hours throughout the experiments. In summary, this yielded Sn fractions of 90.0(14)%, 58(1)%, and 25(1)% for $^{110}$Sn, $^{108}$Sn, and $^{106}$Sn, respectively. It should be pointed out that the intensity of the $^{108}$Sn beam was comparable to the $^{110}$Sn beam, while for $^{106}$Sn it was one order of magnitude lower. This is the primary reason for the larger uncertainty in the $B(E2)$ value for $^{106}$Sn. The transition probabilities were extracted by measuring the $\gamma$-ray yield coming from the $2^+ \rightarrow 0^+$ transition in $^{110,108,106}$Sn and normalizing against the corresponding $\gamma$-ray yield of the equivalent transition in $^{58}$Ni which has a known $B(E2;0^+ \rightarrow 2^+)$ value. Angular integration as well as energy loss in the $^{58}$Ni target was performed by the coupled-channels code GOSIA2.

## RESULTS AND DISCUSSION

The results from this work are presented in Tab. 1 and displayed in Fig. 1. The $B(E2)$

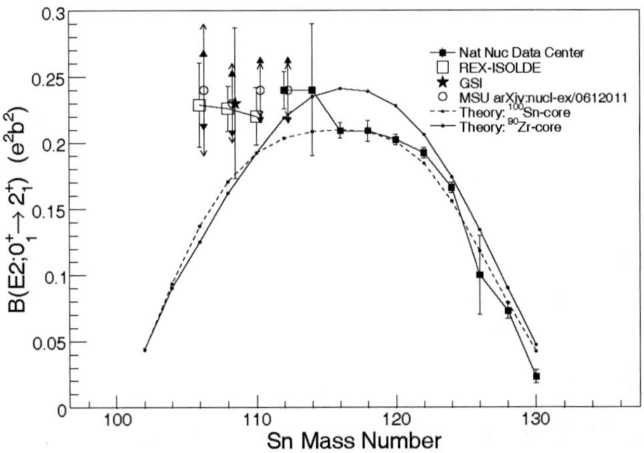


**FIGURE 1.** The known experimental $B(E2\uparrow)$ values expressed in e$^2$b$^2$ across the Sn isotope chain. The results from REX-ISOLDE presented in this paper are marked with empty squares. The value on $^{110}$Sn has been published in [3]. The value on $^{108}$Sn indicated with a star comes from a measurement at GSI [2] using intermediate Coulomb excitation. The empty circles mark the $B(E2)$ values for $^{106,108,110,112}$Sn measured at MSU [4] using intermediate Coulomb excitation. The dotted and solid lines extending over the entire isotope chain indicate the result of shell-model calculations [2] using $^{100}$Sn (dotted) or $^{90}$Zr (solid) as closed shell cores.

values are in disagrement with theoretical predictions based on shell-model calculations using an effective CD-Bonn interaction renormalized with G-matrix theory. The details regarding the calculations presented in Fig. 1 can be found in [2]. It is however worth mentioning that the model space included only $\nu(0g_{7/2}, 2s, 1d, 0h_{11/2})$ with an effective neutron charge $e^{\nu}_{eff} = 1.0$ e. Including proton-neutron excitations from $0g_{9/2}$ was computationally not possible due to the size of the model space. The effect of core-polarization was considered using a seniority truncated model space $\pi(0g1d2s)$ and $\nu(0g_{7/2}1d2s0h_{11/2})$ using $e^{\nu}_{eff} = 0.5$ e and $e^{\pi}_{eff} = 1.5$ e with $^{90}$Zr as a closed core. A Relativistic Quasiparticle Random Phase Approximation done in Ref. [5] is consistent with the results presented in this paper but fails to reproduce the adopted $B(E2)$ values in the mid-shell and neutron-rich region. Inclusion of proton-neutron excitations across the $N = Z = 50$ shell gap will of course increase the collectivity of the $2^+$ state. This is immediately seen in the $^{90}$Zr core calculation in Fig. 1. The observed transition rates reported on here is independent of $A$ indicating that seniority is violated towards $^{100}$Sn and implying stronger renormalization effects on the effective charges. The picture is reversed on the neutron rich side of the isotopic chain, where a good shell closure is reached at $N = 82, Z = 50$ as seen in Fig. 1. Furthermore, this region has been studied experimentally to a greater extent. The number of neutrons in the $(0g_{7/2}1d_{5/2})$ plays an important role for the size of the $\pi(0g_{9/2}) - \nu(0g_{7/2}1d_{5/2})$ energy separation. The same effect is responsible for the shell structure evolution between Zr-Sn as the $\pi(0g_{9/2})$ is filled [6, 7]. We are presently preparing a parallel shell-model calculation in an extended model space which includes proton-neutron excitations from the $0g_{9/2}$ orbit.

## REFERENCES


1. I. Talmi, Nucl. Phys. **A172**,1 (1971).
2. A. Banu et al., Phys. Rev. C **72**, 061305(R) (2005).
3. J. Cederkäll et al., Phys. Rev. Lett. **98**, 172501 (2007).
4. C. Vaman et al., arXiv:nucl-ex/0612011v1. To be published in Phys. Rev. Lett.
5. A. Ansari, Phys. Lett **B623**, 37 (2005).
6. B. L. Cohen, Phys. Rev. **127**, 597 (1962).
7. T. Otsuka et al., Phys. Rev. Lett. **95**, 232502 (2005).

# Study of multi-nucleon transfer reactions with light nuclei

G. Benzoni*, D. Montanari*, A. Bracco *, N. Blasi *, F. Camera *, F.C.L. Crespi *, A. Corsi*, S. Leoni*, B. Million*, R. Nicolini*, O. Wieland*, A. Zalite*, F. Zocca*, F. Azaiez†, S. Franchoo†, I. Stefan†, F. Ibrahim†, D. Verney†, S. Battacharyya**, G. De France**, A. Navin**, M. Rejmund**, L. Corradi‡, G. de Angelis‡, A. Gadea‡, A.M. Stefanini‡, J.J. Valiente-Dobón‡, E. Farnea§, S. Lunardi§, P. Mason§, G. Montagnoli§, F. Scarlassara§, S. Szilner¶,‖ and G. Pollarolo††

*Dipartimento di Fisica and INFN, Sezione di Milano, Milano, Italy.*
†*IPN Orsay, France.*
***GANIL, Caen, France*
‡*LNL, Legnaro, Italy.*
§*Universitá di Padova and INFN sezione di Padova, Padova, Italy.*
¶*LNL, Legnaro, Italy*
‖*RBI, Zagreb, Croatia*
††*Università di Torino, Torino, Italy*

**Abstract.** Multi-nucleon transfer reactions are useful tools to populate exotic nuclei, particularly the neutron-rich ones. In this view, two different experiments have been performed employing a stable ($^{22}$Ne) and a radioactive ($^{24}$Ne) beam, both impinging on a $^{208}$Pb target. The first reaction has been studied using the CLARA-PRISMA-DANTE set-up at Laboratori Nazionali di Legnaro (Legnaro-Italy), while the second reaction was performed at Ganil (Caen-France) employing a SPIRAL radioactive beam of $^{24}$Ne. In this case recoils and coincident $\gamma$ rays were detected with the VAMOS-EXOGAM set-up.

The data show that MNT reactions can selectively populate states of different nature and, therefore, are a good tool to study nuclear structure further away from stability.

**Keywords:** gamma-ray spectroscopy, multinucleon transfer reactions, radioactive beams
**PACS:** 21.10.-k,25.60.-t,27.30.+t

## INTRODUCTION

The spectroscopy of nuclei around mass 20 is particularly interesting since these nuclei are located at the boundaries of the so-called "island of inversion". Important experimental information in this region has been gained using single- and double- step fragmentation [1] and light-particle transfer reactions employing radioactive beams [2, 3].

Multi-nucleon transfer (MNT) and deep-inelastic (DIC) reactions are proved to be useful tools to populate exotic nuclei, particularly the neutron-rich ones. Although these reaction mechanisms have already been extensively exploited in the past years, only recently the availability of efficient spectrometers opens up a variety of new possibilities to study the reaction mechanism itself and to investigate the spectroscopy of medium mass neutron-rich nuclei. In this view, two different experiments were performed employing

CP1012, *Frontiers in Nuclear Structure, Astrophysics, and Reactions: FINUSTAR 2*
edited by P. Demetriou, R. Julin, and S. V. Harissopulos
© 2008 American Institute of Physics 0-7354-0532-5/08/$23.00

a stable ($^{22}$Ne) and a radioactive ($^{24}$Ne) beam.

In this proceeding we present preliminary results from the analysis of both experiments.

## THE EXPERIMENTS

The first reaction has been performed using an intense beam (3-4 pnA) of $^{22}$Ne at 150 MeV, from the ECR source - positive ion injector (PIAVE), impinging on an enriched 700 $\mu$g/cm$^2$ $^{208}$Pb target [4].

The events were measured by means of the CLARA-PRISMA-DANTE set-up. The PRISMA spectrometer [5], placed around the grazing angle for this reaction (estimated to be $\theta_{lab} = 58°$), allows the identification in charge and mass of the reaction products on an event by event basis, through the reconstruction of the individual trajectories. The coupling of PRISMA to the Clover array CLARA [6] allows to detect $\gamma$ rays emitted in coincidence with the recoils. The 25 Clover detectors are placed around the scattering chamber covering the angles between 98° and 170° with a photopeak efficiency of 3% for 1 MeV $\gamma$ rays. The addition of DANTE [7], a multi-detector array formed by a variable number of MCP detectors, enables the kinematic reconstruction of part of the events that are outside the PRISMA acceptance, therefore increasing the statistics and allowing the creation of $\gamma-\gamma$ coincidence matrices. In 5 days of data taking a number of $4\times10^6$ good CLARA-PRISMA coincidences were collected.

The second reaction, performed at Ganil (Caen-France) employed a SPIRAL radioactive beam of $^{24}$Ne (at 190 MeV with an intensity of $1.5 \times 10^5$ pps) also impinging on a thick $^{208}$Pb target [8]. The thickness of the target (10.9 mg/cm$^2$) was chosen such as to guarantee a counting rate high enough to enable basic gamma-ray spectroscopic information to be obtained. The reaction products were detected in the focal plane of the VAMOS spectrometer [9, 10] which was positioned at the calculated grazing angle for this reaction ($\sim$ 35°). The $\gamma$ rays were detected by the EXOGAM array [11, 12], consisting of 11 segmented germanium Clover detectors, 9 of which had Compton suppression shields. A total of 1.4*$10^5$ particle-$\gamma$ coincidences were recorded in 7 days of data taking.

The coupling of a spectrometer to an array of $\gamma$-rays detectors allows the study of reaction mechanism dynamics by using the structural information provided by the $\gamma$ rays. In this way, information on both cross sections and angular momentum population can be deduced from such experiments.

The data show that MNT reactions can selectively populate states of different nature and, therefore, are a good tool to study nuclear structure further away from stability.

## PRELIMINARY EXPERIMENTAL RESULTS

The experiment performed in LNL is the most recent and the data analysis is still in progress. In this proceeding we only present the E-$\Delta$E plot (fig. 1) showing the different nuclear species populated in the reaction. The different contours correspond to distinct elements, as indicated in the figure: the reaction populated nuclei ranging from Mg to C,

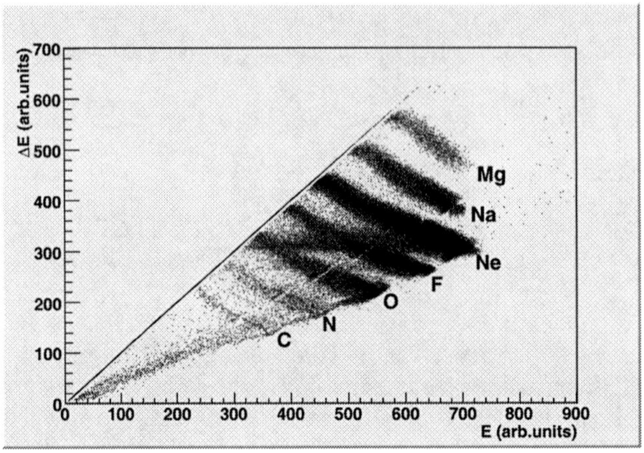


**FIGURE 1.** *E-ΔE plot showing the isotopes populated in the reaction* $^{22}Ne$+$^{208}Pb$.

which correspond to the +2p±xn and -4p±xn channels.

The on-going data analysis will allow a detailed analysis of the $\gamma$ spectra measured in coincidence with every single isotope. Important information on the reaction mechanism can be extracted [13].

The data measured in Ganil are at the moment the most interesting, since we notice a dependence on the population of states of different nature, depending on the scattering angle. This analysis concentrates on $\gamma$ spectra associated with the most intense neon channels (i.e. $^{24}$Ne and $^{25}$Ne).

Software conditions on the Q and $M/Q$ of the populated nuclei allow to separate the contributions coming only from $^{25}Ne^{10+}$. Gates on these nuclei allowed the sorting of isotopically clean $\gamma$ spectra.

In the following we concentrate on $^{25}$Ne which is one of the most intensly populated nuclei with a rich spectrum showing gamma transitions extending up to 3.3 MeV. In particular, the 1.7 MeV $5/2^+ \rightarrow 1/2^+$ transition to the ground state is clearly observed, as the 2.1 MeV $3/2^+ \rightarrow 1/2^+$, 2.35 MeV $7/2^+ \rightarrow 3/2^+$ and 3.3 MeV transition from the direct decay from the $3/2^-$ level to the ground state. There is also a strong peak at 570 keV wich has been identified in $\beta$-decay studies [14], even if possibly misplaced in that work.

The transition at 1.7 MeV is a doublet (1.7 and 1.75 MeV), due to the near degeneracy of the $3/2^+$ and $5/2^+$ states. This degeneracy is also predicted by USD calculations and was confirmed by the recent study of $^{25}$Ne by nucleon knock-out (see reference [16] and references therein).

In our spectrum we clearly distinguish other two transitions (380 keV and 900 keV) which we are at present not able to locate in the level scheme.

The population of the states as function of the $\theta$ angle at target position is also investigated. The gamma spectra were incremented with the requirement that $\theta$ was higher or lower than the estimated grazing angle for this reaction (i.e. $\theta < (>) 34°$). The resulting spectra for $^{25}$Ne are shown in figure 2. Panel a) of figure 2 shows the spectrum

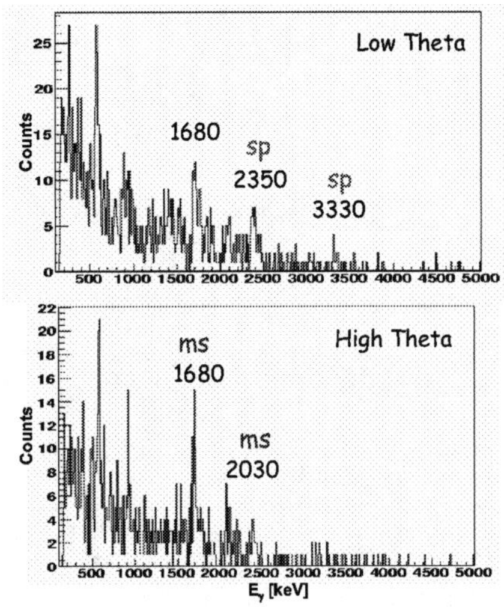


**FIGURE 2.** $\gamma$-ray spectra of $^{25}$Ne requiring the condition that $\theta < 34°$ (panel a)) or $\theta > 34°$ (panel b)).

obtained with the condition $\theta < 34°$ (labelled as *'Low Theta'*) and panel b) $\theta > 34°$ (*'High Theta'*). These spectra clearly show different peaks with those present in the *Low Theta* spectrum absent in the *High Theta* cut. The condition of $\theta$ being $< 34°$ seems to select mainly transitions from states of single particle character, while the condition $\theta > 34°$ selects states of mixed character.

Moreover the doublet around 1.7 MeV also seems to be populated in different proportions by these conditions.

A similar dependence of the population of single particle and collective states as a function of the recoil angle and $\gamma$ multiplicity has been reported by the Surrey group using the CHICO detector coupled to GAMMASPHERE, as described in ref.[17, 18].

More quantitative results will be provided after the response function of the VAMOS spectrometer will be unfolded and its combined dependence on angle and momentum examined.

# CONCLUSION

In this brief report, preliminary results from the first deep inelastic experiment performed using a radioactive $^{24}$Ne beam from SPIRAL have been presented. The presented experiment has been performed with a beam intensity of approximately $1.5*10^5$ pps, leading to a total of $10^4$ $\gamma$-particle coincidences. The data show that, also employing radioactive beams, DIC can selectively populate states of different nature and, therefore, are a good tool to study nuclear structure further away from stability.

This interesting result will be compared by a more recent experiment performed in LNL using a stable $^{22}$Ne beam.

## ACKNOWLEDGMENTS

The GANIL/SPIRAL accelerator crew for providing a particularly stable radioactive $^{24}$Ne beam and the LNL PIAVE-ALPI operators for the delivery of a very intense $^{22}$Ne beam are thanked. We would like to acknowledge the help and support from the local VAMOS-EXOGAM and the CLARA-PRISMA technical staff for making it possible to run very smooth experiments. This work was partially supported by the EU through the EURONS project.

## REFERENCES


1. M. Stanoiu et al., *Phys, Rev.* **C 69**, 034312 (2004).
2. W.N. Catford et al., Eur. Phys. J. **A 25**, S1 250 (2005).
3. A. Obertelli et al., Phys. Rev. C **74**, 064305 (2006).
4. G.Benzoni et al., LNL Report 2006, pp. 39: P.Mason et al., LNL Report 2005, pp. 31.
5. LNL Report 2004, pp.136-146; A. M. Stefanini et al., Nucl. Phys. **A701**, 217c (2002); A. Latina et al., Nucl. Phys. **A 734**, E1 (2004).
6. A. Gadea et al., Eur. Phys. J. **A 20**, 193 (2004).
7. J. J. Valiente-Dobón et al., AIP conference proceeding **853**, 202 (2006); J. J. Valiente-Dobón et al.,LNL Report 2005, pp.175-176.
8. G.Benzoni et al., AIP conference proceeding **853**, 49 (2006).
9. H. Savajols et al, Nucl. Phys. **A 654**, 1027c (1999).
10. GANIL web site http://www.ganil.fr/vamos.
11. S.L. Shepherd et al., Nucl. Inst. Methods Phys. Res. Sect.**A 434**, 373 (1999).
12. GANIL web site http://www.ganil.fr/exogam.
13. P. Mason et al., LNL report 2006, pp.242-243.
14. S.W. Padgett et al., Phys. Rev. **C 72**, 064330 (2005).
15. A.Winther, Nucl. Phy. **A 549**, 203 (1995).
16. J.R. Terry et al., Phys. Lett. **B 640**, 86 (2006).
17. J.J. Valiente Dobon et al., Phys. Rev. **C 69**, 024316 (2004).
18. P.H. Regan et al., Phys. Rev. **C 68**, 044313 (2003).

# Nuclear Astrophysics At LUNA: Status And Perspectives

P.Prati[1], D. Bemmerer[2], R. Bonetti[3], C. Broggini[4], P. Corvisiero[1], F. Confortola[1], E. Conti[4], H. Costantini[1], Z. Elekes[5], A. Formicola[6], Zs. Fülöp[5], G. Gervino[7], A. Guglielmetti[3], Gy. Gyürky[5], C. Gustavino[6], G. Imbriani[8], M. Junker[6], A. Lemut[1], B. Limata[8], M. Marta[2], C. Mazzocchi[3], R. Menegazzo[4], V. Roca[8], C. Rolfs[9], C. Rossi Alvarez[4], E. Somorjai[5], O. Straniero[10], F. Strieder[9], F. Terrasi[11], H.P. Trautvetter[9]
(The LUNA Collaboration)

[1] *Universitá di Genova and INFN Sezione di Genova, Genova, Italy*
[2] *Forschungszentrum Dresden-Rossendorf, Postfach 510119, 01314 Dresden, Germany*
[3] *Istituto di Fisica Generale Applicata, Universit`a di Milano and INFN Sezione di Milano, Italy*
[4] *Istituto Nazionale di Fisica Nucleare (INFN), Sezione di Padova, via Marzolo 8, 35131 Padova, Italy*
[5] *Institute of Nuclear Research (ATOMKI), Debrecen, Hungary*
[6] *INFN, Laboratori Nazionali del Gran Sasso (GRAN SASSO), Assergi (AQ), Italy*
[7] *Dipartimento di Fisica Sperimentale, Universitá di Torino and INFN Sezione di Torino, Torino, Italy*
[8] *Dipartimento di Scienze Fisiche, Universitá di Napoli "Federico II", and INFN Sezione di Napoli, Napoli, Italy*
[9] *Institut für Experimentalphysik III, Ruhr-Universität Bochum, Bochum, Germany*
[10] *Osservatorio Astronomico di Collurania, Teramo, and INFN Sezione di Napoli, Napoli, Italy*
[11] *Seconda Universitá di Napoli, Caserta, and INFN Sezione di Napoli, Napoli, Italy*

**Abstract.** The talk will be mainly devoted to the measurement of the cross section of the $^3He(\alpha,\gamma)^7Be$ reaction which belongs to the p-p chain. Actually, solar neutrino fluxes depend both on astrophysical and on nuclear physics inputs, namely on the cross sections of the reactions responsible for neutrino production inside the Solar core. While the flux of solar $^8B$ neutrinos has been recently measured at Superkamiokande with a 3.5% uncertainty and a precise measurement of 7Be neutrino flux is foreseen in the next future, the predicted fluxes are still affected by larger errors. The largest nuclear physics uncertainty to determine the fluxes of $^8B$ and $^7Be$ neutrinos comes from the $^3He(\alpha,\gamma)7Be$ reaction. The uncertainty on its S-factor is due to an average discrepancy in results obtained using two different experimental approaches: the detection of the delayed $\gamma$ rays from $^7Be$ decay and the measurement of the prompt $\gamma$ emission. The LUNA Collaboration has performed a new high precision experiment with both techniques at the same time. Thanks to the low background conditions of the Gran Sasso LUNA accelerator facility, the cross section has been measured down to 93 keV, the lowest interaction energy ever reached. The S-factors from the two methods do not show any discrepancy within the experimental errors. An extrapolated $S(0) = 0.560 \pm 0.017$ keV barn is obtained. Moreover, branching ratios between the two prompt $\gamma$ -transitions have been measured with 3-8% accuracy. Recently, the LUNA Collaboration has presented to the Scientific Board of Gran Sasso Laboratory its experimental program for the next five years based on the existing 400 kV accelerator facility. A Letter of Intent addressing the possibility to install a 3 MV machine at Gran Sasso has also been submitted to the Board. The LUNA scientific programs with both the 400 kV and 3 MV accelerator facilities will be illustrated in the final part of the talk.

CP1012, *Frontiers in Nuclear Structure, Astrophysics, and Reactions: FINUSTAR 2*
edited by P. Demetriou, R. Julin, and S. V. Harissopulos
© 2008 American Institute of Physics 0-7354-0532-5/08/$23.00

**Keywords:** Nuclear astrophysics, underground accelerator, H-burning, He-burning
**PACS:** 26.20.Cd, 26.20.Fj

## INTRODUCTION

Nuclear reactions that generate energy and synthesize elements take place inside the stars in a relatively narrow energy window: the Gamow peak. The extremely low value of the cross-section inside the Gamow peak has always prevented its measurement in a laboratory at the Earth's surface, where the signal to background ratio is too small because of cosmic ray interactions. In order to explore this new domain of nuclear astrophysics LUNA (Laboratory for Underground Nuclear Astrophysics) started in 1991 its activity by installing a 50 kV electrostatic accelerator underground at Gran Sasso, followed in the year 2000 by a second 400 kV one. The qualifying features of both accelerators are a very small beam energy spread and a very high beam current even at low energy.

Outstanding results have been obtained on the cross-sections of $^3He(^3He,2p)^4He$ [1] and $^2H(p,\gamma)^3He$ [2], both measured within the Gamow peak of the Sun, and of $^{14}N(p,\gamma)^{15}O$, measured down to 70 keV [3,4,5]. With these experiments LUNA has shown that, performing the measurements underground with the typical techniques of low background physics, nuclear cross-sections can be measured down to the energy of the nucleosynthesis inside stars.

In particular, our result on $^3He(^3He,2p)^4He$ down to 16 keV (20 fbarn cross-section with a count rate of 2 events per month) has ruled out the astrophysical solution of the solar neutrino problem based on the existence of a narrow resonance within the Gamow peak of the Sun. The $^{14}N(p,\gamma)^{15}O$ reaction has been studied by LUNA with two different techniques: a germanium detector with solid target at 'high' energy and a BGO summing crystal with a windowless gas target down to 70 keV. Our measured cross-section is about a factor 2 smaller than the previous extrapolation on the NACRE compilation [6]. The astrophysical consequences of such a reduction are significant: the CNO neutrino yield in the Sun is decreased by about a factor two [7,8], the age of the oldest Globular Clusters is increased by 0.7-1 Gyr [9] and the dredge-up of carbon to the surface of Aymptotic Giant Branch (AGB) stars is much more efficient [10].

We have recently completed the analysis of the $^3He(\alpha,\gamma)^7Be$ data [11]. The uncertainty on this cross-section is the main nuclear limitation to the extraction of physics from the $^8B$ and $^7Be$ solar neutrino flux measurements. We have performed a high accuracy study by detecting the emitted prompt $\gamma$-rays and counting the produced $^7Be$ atoms. We obtain an extrapolated S-factor with a 3% error, thus reducing the uncertainty on the predicted $^7Be$ solar neutrino flux from 9.4% to 5.5% (from 12% to 10% for the $^8B$ flux). Nuclear physics does not give anymore a dominant contribution to the total uncertainty of the solar neutrino fluxes [11,12,13]. This result, togheter with the previous determination of the $^{14}N(p,\gamma)^{15}O$ cross section had a clear impact on the BOREXINO experiment at Gran Sasso as quoted in the very recent article [14] describing the first results of BOREXINO.

The 'non solar' phase of LUNA has already started with the running experiment on $^{25}Mg(p,\gamma)^{26}Al$. Two are the reasons making this reaction so relevant: the 1.8 MeV map taken by the satellites which look at the γ-sky and the anomalous meteoritic abundance of $^{26}Mg$. This experiment is scheduled to finish by the end of 2007. The latter experiment will conclude the experimental program so far approved by the Gran Sasso Scientific Committee. A new phase is going to start as described in the following.

## LUNA GOALS FOR THE FUTURE

With a study which lasted about one year we have identified a sample of nuclear reactions of great astrophysical importance which can strongly profit from an underground measurement (Table1).

**TABLE 1.** Candidate reactions for an underground measurement. For each reaction we give the Q-value, the Gamow energy, i.e. the effective mean energy for the fusion reaction in the star, the lowest measured energy and the lowest energy achievable in LUNA. Reaction in italic can be studied at the 400 kV accelerator presently installed in Gran Sasso but they would benefit of a higher energy machine. Reaction written in bold-italic definitely require a beam energy of a few MeV.

| Reaction | Q-value (MeV) | Gamow energy (keV) | Lowest meas. Energy (keV) | LUNA limit estimate (keV) |
|---|---|---|---|---|
| ***$^{12}C(\alpha,\gamma)^{16}O$*** | 7.16 | 300 | 950 | 500 |
| ***$^{22}Ne(\alpha,n)^{25}Mg$*** | -0.47 | 470-700 | 850 | 630 |
| *$^{13}C(\alpha,n)^{16}O$* | 2.21 | 170-250 | 270 | 200 |
| *$^{2}H(\alpha,\gamma)^{6}Li$* | 1.47 | 50-300 | 700 (direct)<br>50 (indirect) | 50 |
| $^{15}N(p,\gamma)^{16}O$ | 12.13 | 10-300 | 130 | 50 |
| $^{17}O(p,\gamma)^{18}F$ | 5.6 | 35-260 | 300 | 65 |
| $^{18}O(p,\gamma)^{19}F$ | 8.0 | 50-200 | 143 | 143 |
| $^{23}Na(p,\gamma)^{24}Mg$ | 11.7 | 100-200 | 240 | 138 |
| $^{22}Ne(p,\gamma)^{23}Na$ | 8.8 | 50-300 | 250 | 68 |
| $^{15}N(a,\gamma)^{19}F$ | 4.01 | 364 | 536 | 364 |

Part of the LUNA scientific program can be performed with the 400 kV accelerator already running underground in the so called LUNA2 laboratory. To study $^{12}C(\alpha,\gamma)^{16}O$, the 'Holy Grail' of nuclear astrophysics, $^{13}C(\alpha,n)^{16}O$ and $^{22}Ne(\alpha,n)$ $^{25}Mg$, the neutron sources inside stars, and other (α,γ) reactions a new accelerator of a few MV voltage installed in a dedicated space underground is required. Such project, which could produce an extremely important step forward in nuclear astrophysics, is the subject of a Letter of Intent recently submitted to the Gran Sasso Scientific Committee.

The first reaction we propose to study is $^2H(\alpha,\gamma)^6Li$, one of the most important Big-Bang nucleosynthesis processes. Existing experimental data reach $E_{cm}$= 400 keV while indirect measurements, strongly discrepant with extrapolations based on the above data, have been recently performed down to about 50 keV. This reaction can well be studied with the present 400 kV machine down to about 50 keV c.m. (150 keV laboratory beam energy).

We now come to thermonuclear reactions inside the stars. The temperature in the stellar interior is determined by the mass of the star. Low-mass stars like our Sun operate mainly on the proton-proton (pp) chain, while more massive stars produce energy through the operation of the Carbon-Nitrogen-Oxygen (CNO) cycles. However, in second generation stars, whose central temperatures are higher than those for the quiescent CNO cycle, additional cycles in the hydrogen burning stage can occur. Because of the higher Coulomb barriers involved in the processes, these higher burning cycles are relatively unimportant as a source of energy, but they are essential for the nucleosynthesis of elements in the mass $A \geq 20$ region. LUNA has already studied in the past the key processes of the pp chain and of the first CNO cycle. We plan now to explore the other cycles by measuring the (p,γ) reactions on $^{15}N$, $^{17}O$, $^{18}N$, $^{22}Ne$ and $^{23}Na$.

Details of the first reaction we plan to measure with the 400 kV accelerator starting from spring 2008 (namely $^2H(\alpha,\gamma)^6Li$ ) and on the more outstanding cases which would benefit of a higher energy machine are given in the following sections.

## $^2H(\alpha,\gamma)^6Li$

Recently, the $^6Li$ isotope has been detected in a number of metal-poor stars [15 and references therein]. These observations are significant because they suggest a $^6Li$ plateau similar to the well-known Spite plateau for $^7Li$ [16]. The observed $^7Li$ abundance [15,17] is a factor of 3 below the value predicted by standard big-bang nucleosynthesis based on the baryon-to-photon ratio from cosmic microwave background observations [18]. The $^6Li$ data, however, are higher by as much as two to three orders of magnitude than the predicted big-bang production. This gives rise to a $^6Li$ problem in big-bang nucleosynthesis, in addition to the well-known $^7Li$ problem.

If the existence of the $^6Li$ plateau is confirmed, a pre-galactic source of $^6Li$ has to be introduced as an explanation. This could either be an extremely enhanced production in big-bang nucleosynthesis, mainly through the $^2H(\alpha,\gamma)^6Li$ reaction, or non-standard physics like the decay of relic gravitinos [19]. It has been shown that $^6Li$ production through the pre-galactic interaction of energetic particles with the interstellar medium cannot explain the level of detected $^6Li$ [20]. Most scenarios of pre-galactic $^6Li$ production, except for an enhanced $^2H(\alpha,\gamma)^6Li$ cross-section, produce not only $^6Li$ but also $^7Li$, therefore worsening the $^7Li$ problem. A sensitivity study of nuclear reaction rates for big-bang nucleosynthesis has shown that the nuclear uncertainty in the predicted $^6Li$ abundance is dominated by the uncertainty in the $^2H(\alpha,\gamma)^6Li$ cross-section [21]. The reaction is dominated by d-wave capture to the first excited state in $^6Li$ and produces single γ-rays from ~1.6 to 2.2 MeV. The Gamow peak ranges from

50 to 300 keV in the centre of mass. Because of its electric quadrupole nature the cross-section is predicted to be extremely small, few pbarn to few tens of pbarn in the energy range 50-100 keV. The cross-section of the $^{2}H(\alpha,\gamma)^{6}Li$ reaction has been previously measured, in the energy range of big-bang nucleosynthesis, in a Coulomb dissociation study at Karlsruhe [22]. Preliminary results from GSI [23] indicate, with large uncertainties, a lower cross-section at similar energies (Figure 1). Theoretical calculations [e.g. 24] yield values that are systematically lower than the data [22,23]. The NACRE compilation consequently cites an uncertainty of a factor 3 in the rate at big-bang nucleosynthesis temperatures.

The Q-values of $^{2}H(\alpha,\gamma)^{6}Li$ and $^{3}He(\alpha,\gamma)^{7}Be$ reactions are similar: 1.474 and 1.586 MeV, respectively. At big-bang energies, only capture to the ground state in $^{6}Li$ is energetically possible. Consequently, the observed $\gamma$-ray spectra from the LUNA $^{3}He(\alpha,\gamma)^{7}Be$ study have been used to evaluate the feasibility of a $^{2}H(\alpha,\gamma)^{6}Li$ measurement.

The measurement could be performed with the presently available 400 keV accelerator in the lab energy range 150-400 keV corresponding to c.m. values of 50-130 keV (the corresponding rate of emitted photons at the lowest energy is 4 $h^{-1}$ with 100 $\mu$A alpha beam and $10^{18}$ atoms/$cm^2$ target thickness). However, a higher energy (3-3.5 MV) accelerator could allow reaching the 0.711 keV resonance and overlap the existing data.

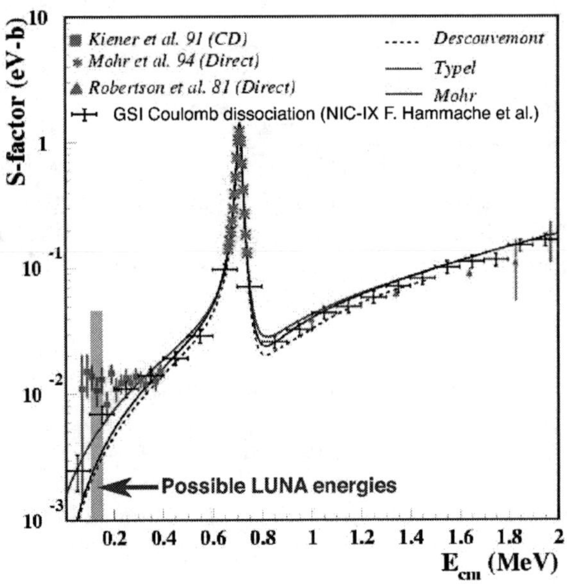


**FIGURE** 1. Existing results on the astrophysical factor of $^{2}H(\alpha,\gamma)^{6}Li$

In conclusion, this reaction could profit of the same set-up used for the $^{3}He(\alpha,\gamma)^{7}Be$ reaction [11] and, more important, is a clear example of the advantages offered by an underground location such at Gran Sasso.

## $^{12}C(\alpha,\gamma)^{16}O$

The measurement of the $^{12}C(\alpha,\gamma)^{16}O$ reaction (Q=7161.91 keV) should be done in a nearly $4\pi$ geometry with an angle-segmented crystal ball detector. The measurement of angular distributions is necessary to separate the E1 and E2 components of both ground-state and cascade transitions. The efficiency of such a detector has been preliminarly studied with Monte-Carlo techniques. For the expected reaction rate calculation, a conservative value of 50 % for the summing peak efficiency has been adopted, although preliminary calculations show that its value would be as high as 90 %, assuming a BGO crystal ball segmented like a truncated icosahedron polyhedron. With this geometry the detector has 11 effective angles with an average detection efficiency per angle of about 5 %. However, we consider a conservative value of 2.5 %.

Typical solid $^{12}C$-enriched targets are those implanted in high purity gold. Preliminary estimates of the expected counting rate were carried on assuming conservative values for the target areal density ($2.0\cdot10^{18}$ atoms/cm$^2$), summing peak detection efficiency (50 %), single angle detection efficiency (2.5 %) and beam intensity (500 μA). For the cross-section values, recent R-matrix fits were assumed. The summing peak, located at Q+$E_{cm}$, is above the natural laboratory background, taking full advantage of the high suppression factor for the cosmic ray background in detectors at Gran Sasso. The laboratory background counting rate is expected to be about 20 counts/day in the $4\pi$ detector summing peak and about 1 count/day in a single-angle detector, without shielding. The reaction rate is estimated to be 20 counts/day for the summing peak at $E_{cm}$=533.6 keV and 1 count/day for the single angle detector at the same energy. The high cosmic ray suppression of Gran Sasso reduces the $\gamma$ ray background of 4 orders of magnitude. This gives a high improvement in the measurement of this reaction at low energy. Under these conditions a new study of the reaction at Gran Sasso could reach about 500 and 600 keV for the total S-factor and the E1/E2 components, respectively.

## Neutron detectors

The measurement of $^{13}C(\alpha,n)^{16}O$ (Q=2215.6 keV) and $^{22}Ne(\alpha,n)^{25}Mg$ (Q= –478 keV) should be done with a nearly $4\pi$ geometry neutron detector. One possibility is to use an existing neutron detector available to the collaboration [25,26], in the following named "Stuttgart detector". This detector is approximately a 1m·1m cylindrical polyethylene moderator containing 12 $^3$He proportional counter chambers. The chambers are arranged into two rings at radii optimized for the neutron energy of interest. Several layers of passive shielding materials (paraffin wax, polyethylene, boron and cadmium) could be arranged around the detector. Typical values for efficiency of the Stuttgart detector range from 10 % to 32 %. For the expected reaction rates calculation, the conservative value of 10 % for the efficiency has been adopted.

Another possibility is to develop a new detection setup based on the traditional approach above described and liquid scintillators chambers which have already been tested at Gran Sasso [27]. Each scintillator chamber is a cylindrical cell containing one litre of scintillator [27]. Each cell is composed of a white Teflon tube 500 mm long, 50 mm internal diameter, 1.5 mm thick, which acts as a diffusing coating. It is inserted into a light-tight stainless-steel tube of 59 mm internal diameter, 1.5 mm thick filled with the scintillator (BC501A, produced by Bicron Corp.): this is ideal for γ-neutron discrimination because of its different pulse shape response. The exact efficiency of the detection setup will be studied during the project development with Monte-Carlo techniques.

## $^{13}C(\alpha,n)^{16}O$

For a reinvestigation of $^{13}C(\alpha,n)^{16}O$, targets enriched in $^{13}C$ are required. The ideal solution is given by solid targets, with typical $^{13}C$ areal density ranging from $1\cdot10^{18}$ to $10\cdot10^{18}$ atoms/cm$^2$.

Preliminary estimates of the expected counting rate were done assuming conservative values for the target areal density ($1.0\cdot10^{18}$ atoms/cm$^2$, corresponding to about 38 keV energy loss), detection efficiency (10 %) and beam intensity (100 μA). The laboratory background counting rate in the detector is expected to be about 20 counts/day without shielding against laboratory neutrons and about 1 count/day with shielding. The reaction rate is expected to be ~20 counts/day at $E_{cm}$=235.8 keV and 1 count/day at $E_{cm}$=205.2 keV. The neutron background suppression of Gran Sasso is 3 orders of magnitude. This gives a high improvement in the measurement of this reaction at low energy which could reach, at LUNA, the upper edge of the Gamow peak (about 200 keV c.m.).

## $^{22}Ne(\alpha,n)^{25}Mg$

In this case the target would be a windowless gas target with recirculation system, since enriched $^{22}Ne$ gas is rather expensive. A typical $^{22}Ne$ gas pressure is 5 mbar corresponding to a gas density of $1.25\cdot10^{17}$ atoms/cm$^3$ (at 293 K). Assuming a gas target length of 10 cm (extended gas target), the resulting Ne areal density is $1.25\cdot10^{18}$ atoms/cm$^2$.

Also in this case, preliminary estimates of the expected counting rate were done considering conservative values for the target areal density ($1.25\cdot10^{18}$ atoms/cm$^2$, corresponding to about 56 keV energy loss), detection efficiency (10 %) and beam intensity (100 μA). The reaction rate is expected to be 20 counts/day at $E_{cm}$=618.3 keV and 1 count/day at $E_{cm}$=553.1 keV. The reaction rate when including the 552 keV resonance is expected to be about 20 counts/day assuming the upper-limit value of 60 neV for the resonance strength [25]. Also in this case we would have a large improvement in an underground measurement at low energy possibly reaching with good statistics the interaction energy of about 600 keV.

## ACKNOWLEDGMENTS

This paper is dedicated to the memory of Prof. Roberto Bonetti, nuclear physicist, spokesperson of the LUNA Collaboration, teacher and friend.

## REFERENCES

1. R.Bonetti et al., Phys. Rev. Lett. **82**, 5205 (1999).
2. C.Casella et al., Nucl. Phys. **A 706**, 203 (2002).
3. A. Formicola et al., Phys. Lett. **B 591**, 61 (2004) .
4. G. Imbriani et al., Eur. Phys. J **A 25**, 45 (2005).
5. A. Lemut et al., Phys. Lett. B **634**, 483 (2006) .
6. C. Angulo et al., Nucl. Phys. **A 656**, 3 (1999).
7. S. Degli Innocenti et al., Phys. Lett. **B 590**, 13 (2004).
8. J.N. Bahcall et al., Phy. Rev. Lett **92**, 121301 (2004).
9. G. Imbriani et al., Astronomy and Astrophysics **420**, 625 (2004).
10. F. Herwig et al., Astrophys. J. **613**, L73 (2004).
11. F. Confortola et al., Phys. Rev. **C75**, 65803 (2007).
12. D. Bemmerer et al., Phys. Rev. Lett **97**, 122502 (2006).
13. Gy. Gyurky et al, Phys. Rev. **C75**, 35805 (2007).
14. C. Arpesella et al., Phys. Lett **B**, in press, arXiv:0708.2251v1.
15. M. Asplund et al., Astrophys. J. **644**, 229 (2006).
16. F. Spite and M. Spite, Astron. Astrophys. **115**, 357 (1982).
17. P. Bonifacio et al., Astron. Astrophys. **390**, 91 (2002).
18. A. Coc et al., Astrophys. J. **600**, 544 (2004).
19. K. Jedamzik, Phys. Rev. D 74, 10309 (2006).
20. N. Prantzos, Astron. Astrophys. **448**, 665 (2006).
21. E. Vangioni-Flam et al., Astron. Astrophys. **360**, 15 (2000).
22. J. Kiener et al., Phys. Rev. **C 44**, 2195 (1991).
23. F. Hammache et al., PoS(NIC-IX)013 (2006).
24. A. Kharbach and P. Descouvemont, Phys. Rev. **C 58**, 1066 (1998).
25. M. Jaeger *et al.*, Phys. Rev. Lett. **87**, 202501 (2001).
26. S. Harissopulos *et al.*, Phys. Rev. **C72**, 62801 (2005).
27. F. Arneodo *et al.*, Nuovo Cimento **A 112**, 819 (1999) .

# Nuclear Matter Equations of State and the Neutron Stars

M. Urbanec*, E. Běták[†,*] and Z. Stuchlík*

*Institute of Physics, Fac. of Philos. and Sci., Silesian Univ., 74601 Opava, Czech Republic
[†]Institute of Physics, Slovak Acad. Sci., 84511 Bratislava, Slovakia

**Abstract.** The equations of state (EoS) of relativistic asymmetric nuclear matter are obtainable from assumed form of the interaction Lagrangian. They are one of important inputs to describe the neutron stars. The structure of the neutron stars, i.e. the density of matter and the pressure as functions of radial distance starting from their values at the center of a star, is straightforwardly dependent on EoS. Similarly, a limitation on the total mass of the neutron star can be obtained therefrom. Thus, EoS and the underlying nucleon interactions can be tested also by the means of astronomical observations.

**Keywords:** Asymmetric nuclear matter, Neutron stars, Quantum hadrodynamics, Relativistic Brueckner-Hartree-Fock, Equations of state
**PACS:** 26.60.+c; 05.70.Ce; 21.60.-n; 97.60.Jd

## INTRODUCTION

A wide spectrum of different equations of state (EoS) of nuclear matter and their applications to astrophysical problems has been reported in literature (see, e.g., [1–5]). Some of the EoS collections (even though not all of them up-to-date already) give an amazing rich overview either in general or with emphasis to a specific aim. These EoS yield (nearly) the same properties close to the standard nuclear density ($\rho_N \approx 0.16$ nucleon/fm$^3$ $\approx 2.7 \times 10^{14}$ g/cm$^3$), but one has to rely more on underlying principles than on possible experimental verification far off this value. Therefore, additional tests as those confronting the neutron star parameters given by different EoS with parameters implied by multi-resonant models of quasiperiodic oscillations observed in some neutron-star binary systems could be of high importance [6–8].

One of straightforward EoS applications is to model neutron stars and to derive their properties, which can be (sometimes) compared to the astronomical data. We do this here for asymmetric nuclear matter within the relativistic mean-field approach.

## ASYMMETRIC NUCLEAR MATTER IN RELATIVISTIC MEAN-FIELD APPROACH

We follow here the relativistic Brueckner-Hartree-Fock mean field (see [1, 9–11] for underlying theories), which easily allows to consider different neutron-proton composition of the matter, and also the inclusion of non-nucleonic degrees of freedom. The EoS obtained in this way have been an input for the calculation of neutron star properties (mass, radius and the central pressure).

CP1012, *Frontiers in Nuclear Structure, Astrophysics, and Reactions: FINUSTAR 2*
edited by P. Demetriou, R. Julin, and S. V. Harissopulos
© 2008 American Institute of Physics 0-7354-0532-5/08/$23.00

We benefit from the results of Kotulič Bunta and Gmuca [12, 13]. The Lagrangian density used was

$$\begin{aligned}
\mathscr{L}(\psi,\sigma,\omega,\rho,\delta) &= \bar{\psi}[\gamma_\mu(i\partial^\mu - g_\omega\omega^\mu - (M - g_\sigma\sigma)]\psi \\
&+ \frac{1}{2}(\partial_\mu\sigma\partial^\mu\sigma - m_\sigma{}^2\sigma^2) - \frac{1}{4}\omega_{\mu\nu}\omega^{\mu\nu} + \frac{1}{2}m_\omega{}^2\omega_\mu\omega^\mu \\
&- \frac{1}{3}b_\sigma M(g_\sigma\sigma)^3 - \frac{1}{4}c_\sigma(g_\sigma\sigma)^4 + \frac{1}{4}c_\omega(g_\omega{}^2\omega_\mu\omega^\mu)^2 \\
&+ \frac{1}{2}(\partial_\mu\delta\partial_\mu\delta - m_\delta{}^2\delta^2) + \frac{1}{2}m_\rho{}^2\rho_\mu\rho^\mu - \frac{1}{4}\rho_{\mu\nu}\rho^{\mu\nu} \\
&+ \frac{1}{2}\Lambda_V(g_\rho{}^2\rho_\mu\rho^\mu)(g_\omega{}^2\omega_\mu\omega^\mu) - g_\rho\rho_\mu\bar{\psi}\gamma^\mu\tau\psi + g_\delta\delta\bar{\psi}\tau\psi, \quad (1)
\end{aligned}$$

i.e. we consider the nucleon field and $\sigma$, $\omega$, $\rho$ and $\delta$ mesons, and also the vector cross-interaction. This Lagrangian is the starting point to obtain the nuclear matter EoS. The real mean-field calculations of [14] have been parametrized by Kotulič Bunta and Gmuca [12, 13], and we employ their parameterization here. These EoS, calculated for different proton-neutron compositions ranging from pure neutron matter up to the proton fraction of 0.18, have been used as an input to construct the properties of neutron stars.

The EoS of Kotulič Bunta and Gmuca [12, 13] with cross-interaction term, which have been found to be a good description of asymmetric nuclear matter, are easily expressed up to about $4 \times \rho_N$, where the complicated algorithm starts to yield (mathematical) instabilities. For a comparison, we have done also the calculations without the cross interaction — these were found feasible up to very high nuclear densities.

## NEUTRON STAR CONSTRUCTION FROM EOS

The nuclear EoS have been the dominant input for the calculations. For lower densities, namely $\rho \leq 10^{14}\mathrm{g/cm^3}$, the EoS used are the following:
Feynman-Metropolis-Teller EoS for $7.9 \leq \rho \leq 10^4\,\mathrm{g/cm^3}$ where matter consists of $e^-$ and $^{56}_{26}Fe$ [15];
Baym-Pethick-Sutherland EoS for $10^4 \leq \rho \leq 4.3 \times 10^{11}\,\mathrm{g/cm^3}$ with Coulomb lattice energy corrections [16];
Baym-Bethe-Pethick EoS for $4.3 \times 10^{11} \leq \rho \leq 10^{14}\,\mathrm{g/cm^3}$: here, $e^-$, neutrons and equilibrium nuclei calculated using the compressible liquid drop model are considered [17].

We use the Schwarzschild metric with $c = G = 1$. In this notation, the hydrostatic equilibrium is given by the Tolman-Oppenheimer-Volkov equation (TOV) [18, 19]

$$\frac{\mathrm{d}p}{\mathrm{d}r} = -(\rho + p)\frac{m(r) + 4\pi p}{r(r - 2m(r))}, \quad (2)$$

where

$$m(r) = 4\pi \int_0^r \rho r_1{}^2 \mathrm{d}r_1 \quad (3)$$

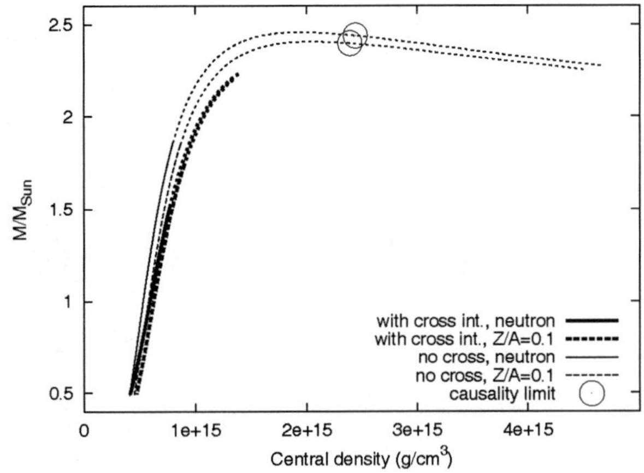


**FIGURE 1.** Predicted mass of the neutron star as a function of the central density for two types of the EoS, namely with and without cross-interaction term (heavy and thin curves, respectively). The calculations depicted are for pure neutron matter (full lines) and for proton fraction of 0.1 (dashed). The portions of the curves where the neutron star radius is a decreasing function of its central density are drawn as dotted lines only. For calculations without the cross-interaction, also the causal limits are marked with circles.

is the mass inside the sphere of radius $r$. Integration of TOV starting from given central density $\rho_c$ uses the EoS and finally yields the radius $R$ and the mass $M = m(R)$ of the neutron star.

The resulting masses are shown [1] in Fig. 1 and the corresponding radii in Fig. 2. We have checked also possible violation of causality at high densities ($v_{\text{sound}} = c\sqrt{\partial p/\partial\varepsilon} \geq c$), and the corresponding limits [2] are denoted by circles.

The maximal mass of the neutron star comes to be about 2.46 $M_\odot$ for pure neutron matter, 2.41 $M_\odot$ for the proton fraction 0.1, and 2.38 $M_\odot$ for the highest considered proton fraction (i.e., 0.18) in the case without the cross interaction, and probably slightly less in the case with the cross interaction [3]. If we expect the neutron star radii to grow with increasing central density, the mass limits are from 1.5 to 1.8 $M_\odot$. Our results do not contradict to recent compilations of the deduced radii of observed and analyzed compact objects (see e.g. the compilations in Refs. [20, 21]). Therein, several dozens of neutron stars and/or similar objects have their masses reported. A great majority of them is in very close vicinity of 1.4 $M_\odot$, and only very few are significantly above (the heaviest two reported ones are of $M = (2.44 \pm 0.27)\ M_\odot$ and $M = (2.1 \pm 0.2)\ M_\odot$, but the interpretation of observations is not without questions).

---

[1] For clarity, only two of ten calculated curves are plotted here, namely that corresponding to pure neutron matter and the one for the proton fraction of 0.1.

[2] For calculations with the vector cross interaction included, we have not reached the causal limit within the density range used.

[3] The radii of neutron stars and also their masses come out to be slightly lower with the vector cross interaction included than they are when one neglects it.

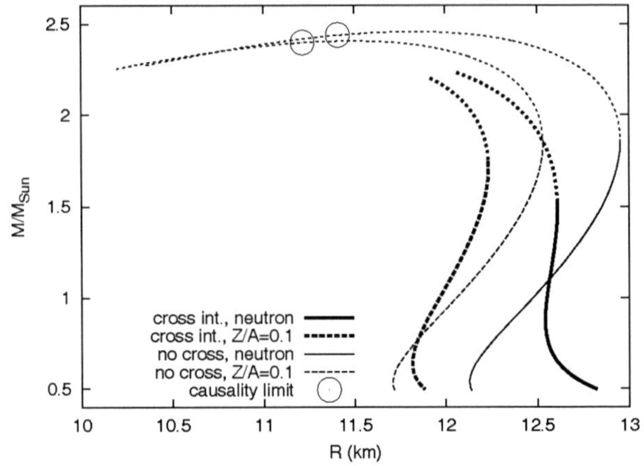


**FIGURE 2.** The same as in the preceding Figure, but the radius-mass dependence.

The neutron star radii are not so easily verifiable. Our calculations range from 11.5 km to 13 km, with minor differences between the calculations with and without the cross interactions, but more sensitive to the fraction of protons inside the neutron star (more protons indicate more compact stars).

Our calculations of the neutron star mass and radius are also in reasonable agreement with the recent results and/or collections reported in Refs. [2, 5, 11, 21-23] [4].

## CONCLUSIONS

We have employed the parametrized form of the relativistic mean-field EoS for asymmetric nuclear matter with vector cross interaction. The proton fraction was changed from 0 (pure neutron matter) to 0.18. Using TOV equation, we constructed model neutron stars. Their maximal allowed radii are between 11.5 and 13 km and the masses below 2.46 $M_{\odot}$, with both the radius and the mass somewhat lowered by the presence of the vector cross interaction when compared to the calculations without it. The deduced masses do not contradict to the observed ones and they are in line with results published by other authors using different approaches. We believe that more detailed tests based on the fitting of observational data of quasi-periodic oscillations in low-mass X-ray systems measurements with the rotating neutron star models based on the Hartle-Thorne metric reflecting mass, spin and the quadrupole moment of the neutron star could bring a new information on the validity of EoS [7, 8].

---

[4] We simply give here recent mass and radii calculations, but do not restrict this comparison to those obtained within relativistic mean-field approach.

## ACKNOWLEDGMENTS

The authors are grateful to P. Bakala, P. Čermák and G. Török for discussions and help with numerical algorithms, to J. Kotulič Bunta and Š. Gmuca for the availability of their computer codes and to F. Weber for sending some of his collected data. The work has been supported by the Czech grant MSM 4781305903 and by grants LC 06014 and VEGA 2/7115/27.

## REFERENCES


1. F. Weber, *Pulsars as Astrophysical Laboratories for Nuclear and Particle Physics*, Inst. Phys. Publishing, Bristol, 1999.
2. J. Rikovska Stone, J. Miller, R. Koncewicz, P. Stevenson, and M. Strayer, *Phys. Rev. C* **68**, 034324 (2003).
3. *International Nuclear Physics Conference, Tokyo, June 3–8, 2007*, 2007, URL `http://www.inpc2007.jp`.
4. K. Oyamatsu, and K. Iida, "Equation of state of nuclear matter and nuclei in laboratories and in neutron-star crusts," in [3], contribution H4-4.
5. F. Weber, R. Negreiros, and P. Rosenfeld, Neutron star properties and the equation of state of superdense matter, San Diego State Univ. (2007), arXiv: 0705.2708v2, May 2007.
6. Z. Stuchlík, G. Török, and P. Slaný, "Multi-resonant models of quasi-periodic oscillations," in [24].
7. M. Urbanec, Z. Stuchlík, G. Török, P. Bakala, and P. Čermák, "Neutron star equations of state limited by QPO observations," in [24].
8. Z. Stuchlík, P. Čermák, G. Török, M. Urbanec, and P. Bakala, "Genetic Selection of Neutron Star Structure Matching the X-ray Observations," in *11th World Multi-Conference on Systemics, Cybernetics and Informatics: WMSCI 2007, July 8–11, 2007 Orlando, Florida, USA*, edited by N. Callaos, Q. Lesso, C. Zinn, and O. Goriachkin, Internat. Inst. of Inf. and Syst., Orlando, 2007, pp. 41–44.
9. J. D. Walecka, *Theoretical Nuclear and Subnuclear Physics, 2nd Edition*, Imperial College and World Scientific, London and Singapore, 2004, ISBN 981-238-898-2.
10. F. de Jong, and H. Lenske, *Phys. Rev. C* **57**, 3099–3107 (1998).
11. P. Krastev, and F. Sammarruca, *Phys. Rev. C* **74**, 025808 (2006).
12. J. Kotulič Bunta, and Š. Gmuca, *Phys. Rev. C* **68**, 054318 (2003).
13. J. Kotulič Bunta, *Asymmetric nuclear matter in relativistic mean-field approach with hyperons and vector cross interactions*, Ph.D. thesis, Comenius University, Bratislava (2003).
14. H. Huber, F. Weber, and M. Weigel, *Phys. Rev. C* **51**, 1790–1799 (1995).
15. R. Feynman, N. Metropolis, and E. Teller, *Physical Review* **75**, 1561–1573 (1949).
16. G. Baym, C. Pethick, and P. Sutherland, *Astrophys. J* **170**, 299–317 (1971).
17. G. Baym, H. Bethe, and C. Pethick, *Nucl. Phys. A* **175**, 225–271 (1971).
18. R. Tolman, *Phys. Rev.* **55**, 364–372 (1939).
19. J. Oppenheimer, and G. Volkoff, *Phys. Rev.* **55**, 374–381 (1939).
20. H. A. Bethe, G. Brown, and C.-H. Lee, *Phys. Rep.* **442**, 5–22 (2007).
21. J. Lattimer, and M. Prakash, *Phys. Rep.* **442**, 109–165 (2007).
22. J. Rikovska Stone, P. Guichon, H. Matevosyan, and A. Thomas, *Nucl. Phys. A* **792**, 341–369 (2007), preprint nucl-th/0611030v1 (9 Nov. 2006).
23. H. Kanzawa, K. Oyamatsu, K. Sumiyoshi, and M. Takano, "Variational calculation for the equation of state of nuclear matter," in [3], poster QT-211.
24. *18th Internat. Conf. on General Relativity and Gravitation, Sydney, July 8–13, 2007*, 2007.

# Nucleosynthesis in jets from rotating magnetized stars during core collapse

Shin-ichirou Fujimoto*, Nobuya Nishimura† and Masa-aki Hashimoto†

*Department of Electronic Control, Kumamoto National College of Technology, Kumamoto 861-1102, Japan
† Department of Physics, School of Sciences, Kyushu University, Fukuoka 810-8560, Japan

**Abstract.** We investigate nucleosynthesis inside magnetically driven jets ejected from rotating, magnetized massive stars, or collapsars, based on long-term, magnetohydrodynamic simulations of the core collapse of six collapsars with the various distributions of magnetic fields and angular momentum before the collapse. We follow the evolution of the abundances of about 4000 nuclides from the collapse phase to the ejection phase and through the jet generation phase using a large nuclear reaction network. We find that the *r*-process successfully operates in the jets from three collapsars, so that U and Th are synthesized abundantly, even when the collapsar have a relatively small magnetic field ($10^{10}$ G) and a moderately rotating core before the collapse. The abundance patterns inside the jets are similar to that of the *r*-elements in the solar system.

**Keywords:** Accretion, accretion disks, nuclear reactions, nucleosynthesis, abundances, stars: supernovae: general, MHD methods: numerical, gamma rays: bursts
**PACS:** 26.30

## INTRODUCTION

During collapse of a star more massive than 35-40$M_\odot$, the stellar core is considered to promptly collapse to a black hole. When the star has sufficiently high angular momentum before the collapse, an accretion disk is formed around the hole and jets are shown to be launched from the inner region of the disk through magnetic processes [1, 2] and neutrino heating. Gamma-ray bursts (GRBs) are expected to be driven by the jets. This scenario of GRBs is referred to as the collapsar model [3]. Assisted by the accumulating observations that imply an association between GRBs and the deaths of massive stars. this model seems to be most promising.

For accretion rates greater than $0.1 M_\odot \mathrm{s}^{-1}$, the accretion disk is so dense and hot that nuclear burning is expected to proceed efficiently. In fact, an innermost region of the disk related to GRBs becomes neutron-rich through electron capture on protons [4, 5, 6]. Nucleosynthesis inside the outflows from the neutron-enriched disk has been investigated with steady, one-dimensional models of the disk and the outflows [5, 7, 8]. Not only neutron-rich nuclei [5] but also *p*-nuclei [7, 8] are shown to be produced inside the outflows. This co-production of *r*- and *p*-elements in the outflows is confirmed with more elaborate calculation of the nucleosynthesis in jets from a collapsar [9], based on two-dimensional magnetohydrodynamic (MHD) simulation [2]; The calculation, however, is performed only for the collapsar with strong magnetic fields ($10^{12}$ G) and a rapidly rotating core whose angular velocity is $10\,\mathrm{rad\,s}^{-1}$, or R12 model in Fujimoto et al. [2, 9]. In the present study, we calculate the chemical composition of jets from six

CP1012, *Frontiers in Nuclear Structure, Astrophysics, and Reactions: FINUSTAR 2*
edited by P. Demetriou, R. Julin, and S. V. Harissopulos
© 2008 American Institute of Physics 0-7354-0532-5/08/$23.00

collapsars, which are shown to eject the magnetically driven jets [2], to investigate the dependence of jet composition on the rotation and magnetic field of the collapsars.

In §2 we briefly describe a numerical code for MHD calculation of the collapsars, or rapidly rotating, massive stars, the initial conditions of the collapsars before the core collapse, and properties of jets from the collapsars. In §3, we present Lagrangian evolution of ejecta through the jets and a large nuclear reaction network to follow abundance evolution of the jets. It is also shown in §3 that $r$-process operates inside the jets from three collapsars, even for that with magnetic fields of $10^{10}$ G and moderately rotating core, whose angular velocity is $2.5\,\mathrm{rad\,s^{-1}}$.

## MHD SIMULATIONS OF COLLAPSARS

We have carried out Newtonian MHD calculations of the collapse of a rotating massive star of $40M_\odot$ [2]. The core of the star is assumed to be collapsed to a black hole promptly. Calculations are performed over the region other than the central region ($\leq$ 50km) of the star. Fluid is freely absorbed through the inner boundary of 50km, which mimics a surface of the black hole with the mass of $M$. The mass is initially set to be that of the central region of the progenitor $\leq$ 50km ($0.001M_\odot$) and is continuously increased by the mass of infalling gas to the region during the calculation.

The numerical code for the MHD calculations employed in this paper is based on the ZEUS-2D code [10]. We have extended the code to include a realistic EOS [11] based on the relativistic mean field theory [12]. For lower density regime ($\rho < 10^5$g/cc), where no data is available in the EOS table with the Shen EOS, we use another EOS, which includes contributions from an ideal gas of nuclei, radiation, and electrons and positrons with arbitrary degrees of degeneracy [13]. We carefully connect two EOS at $\rho = 10^5$g/cc for physical quantities to vary continuous in density at a given temperature [2].The use of the realistic equation of state is important for nucleosynthesis. This is because it enable us to evaluate temperatures of fluid preciously.

We consider neutrino cooling through electron-positron pair capture on nuclei, electron-positron pair annihilation, and nucleon-nucleon bremsstrahlung. The total neutrino cooling rate is evaluated with a simplified neutrino transfer model based on the two-stream approximation [14], with which we can treat the optically thin and thick regime on neutrino reaction, approximately. We ignore resistive heating, whose properties are highly uncertain, not as in Proga et al. [1]. We note that the change in the electron fraction is ignored in the MHD calculations (or $\mathrm{D}Y_e/\mathrm{D}t = 0$).

Spherical coordinates, $(r, \theta, \phi)$ are used in our simulations and the computational domain is extended over 50km $\leq r \leq$ 10000km and $0 \leq \theta \leq \pi/2$ and covered with 200($r$) $\times$ 24($\theta$) meshes, with which we can resolve a fastest growing mode of MRI for models, whose magnetic fields are greater than $10^{10}$G before the core collapse. We assume the fluid is axisymmetric and the mirror symmetry on the equatorial plane. We mimic strong gravity around the black hole in terms of the pseudo-Newtonian potential [15].

We set the initial profiles of the density, temperature and electron fraction to those of the spherical model of a $40M_\odot$ massive star before the collapse [16]. The radial and azimuthal velocities are set to be zero initially, and increase due to the collapse induced by the central hole and self-gravity. The computational domain is extended from the iron core to an inner oxygen layer. The boundaries of the silicon layers between the iron core and the oxygen layers are located at about 1800 km ($1.88M_\odot$) and 3900km ($2.4M_\odot$), respectively. We adopt an analytical form of the angular velocity $\Omega(r)$ of the star before the collapse:

$$\Omega(r) = \Omega_0 \frac{R_0^2}{r^2 + R_0^2}, \tag{1}$$

as in previous work on Supernovae (SNe) [11]. Here $\Omega_0$ and $R_0$ are parameters of our model. We consider three sets of $(\Omega_0, R_0)$ = (10rad/s, 1000km) (cases with rapidly rotating core, or a R model) as in [9], (2.5rad/s, 1000km) (cases with moderately rotating core, or a M model), and, (0.5rad/s, 5000km) (cases with slowly rotating core, or a S model).

For the sets of $\Omega_0$ and $R_0$, the maximum specific angular momentum is about $10^{17}\,\mathrm{cm^2/s}$, which is comparable to that of the Keplerian motion at 50km around a black hole of $3M_\odot$. Therefore the centrifugal force can be larger than the gravitational force of the central black hole and the formation of disk like structure is expected near the hole.

Initial magnetic field is assumed to be uniform, parallel to the rotational axis, and weak elsewhere ($\beta = 8\pi P/B^2 \gg 1$). In the present paper, we consider two distributions with an initial magnetic field, $B_0 = 10^{10}$ and $10^{12}$ G. It should be noted that the magnetic pressure is much smaller than the other pressure before the collapse even if $B_0$ is equal to $10^{12}$ G. In brief, we consider six collapsars in the present study, or R10, M10, S10, R12, M12, and S12.

We briefly describe results of our MHD calculation [2]. We find that the jets can be magnetically driven from the central region of the star along the rotational axis; After material reaches to the black hole with high angular momentum of $\sim 10^{17}\mathrm{cm^2 s^{-1}}$, a disk is formed inside a surface of weak shock, which is appeared near the hole due to the centrifugal force and propagates outward slowly. The magnetic fields, which are dominated by the toroidal component, are chiefly amplified due to the wrapping of the field inside the disk and propagate to the polar region along the inner boundary near the black hole through the Alfvén wave. Eventually, the jets can be driven by the tangled-up magnetic fields at the polar region near the hole. The velocities of the jets are at most 10% of the velocity of light.

## NUCLEOSYNTHESIS IN COLLAPSARS

In order to calculate chemical composition of material inside jets, we need Lagrangian evolution of physical quantities, such as density, temperature, and, velocity of the material. We adopt a tracer particle method [17] to calculate the Lagrangian evolution of the physical quantities from the Eulerian evolution obtained from our MHD calculations of the collapsars. Particles are initially placed from an iron core to an inner O-rich layer. The numbers of the particles in a layer are weighted to the mass in the layer. The total numbers of the particles, $N_p$, are set to be 50000, 1000, 2000, 1000, 2000, and 5000 for R10, M10, S10, R12, M12, and S12, respectively, with which we can follow ejecta through the jets appropriately. The numbers of the particles in a layer are weighted to the mass of the layer. We find that 21, 105, 18, 59, 214, and 113 particles can be ejected via the jets for R10, M10, S10, R12, M12, and S12, respectively. Hereafter, we refer to the particles through the jets as *jet particles*.

Some jet particles can attain to high temperatures greater than $9 \times 10^9$K near the black hole, where material is in nuclear statistical equilibrium (NSE). The abundances of the material in NSE can be expressed with simple analytical expressions [18], which are specified by $\rho$, $T$, and the electron fraction. The electron fraction changes through electron and positron capture on nuclei. It should be emphasized that changes in the electron fraction through neutrino interactions can be ignored because neutrino luminosity is much lower than that from a proto neutron star in core collapse SNe [2, 9].

In the relatively cool regime of $T < 9 \times 10^9$K, NSE breaks and the chemical composition is calculated with a nuclear reaction network. The network includes about 4000 nuclei from neutron and proton up to Fermium, whose atomic number, $Z$, is 100 [see NETWORK B of Table 1 in 19]. The networks contain reactions such as two and three body reactions, various decay channels, and electron-positron capture. The experimental masses and reaction rates are adopted if available or theoretical masses and rates are taken into account otherwise [see 2, 19, for detailed nuclear data]. Moreover, spontaneous and $\beta$-delayed fission is taken into accounts in the network. We note that empirical formula [eq.(5) in 20], in which the distribution of fission yields is asymmetric, is adopted about decay products through the fission.

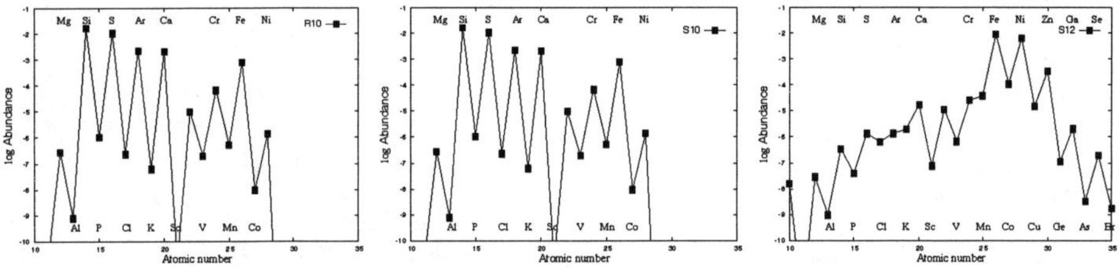


**FIGURE 1.** Composition of jets from collapsars with relatively low densities and temperatures, or R10 (left panel), S10 (center panel), and S12 (right panel).

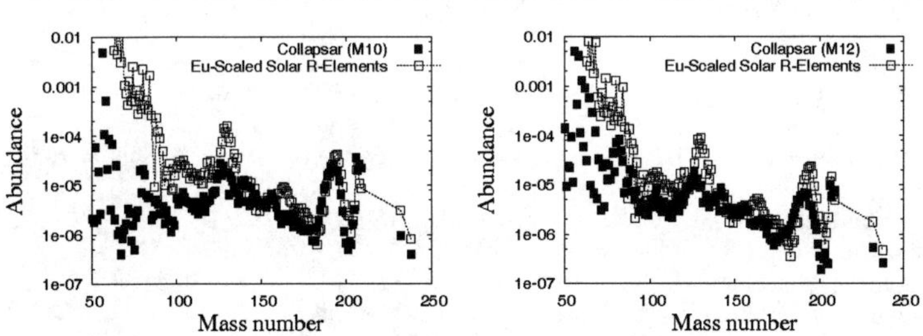


**FIGURE 2.** Composition of jets from collapsars with a moderately rotating core, or M10 (left panel) and M12 (right panel) as well as Eu-scaled $r$-nuclei in the solar system [21].

Figure 1 shows abundances of jets with respect to atomic numbers for R10, S10, and S12. The jets have relatively low densities and temperatures, so that For $T_{\rm max} < 5 \times 10^9$K, the material in the jets are abundant in $\alpha$-elements, such as $^{28}$Si, $^{32}$S, $^{36}$Ar and $^{40}$Ca due to incomplete Si-burning or explosive O-burning. On the other hand, for

$T_{\max} > 5 \times 10^9$K, they undergo complete Si burning, which leads the synthesis of a large fraction of $^{56}$Ni. The compositions of jets for R10 and S10 are therefore similar to those of ejecta from Fe and Si-rich layers of Type II SNe. The jets for S12 is abundant in iron-group elements. This is due to slightly neutron-rich ejecta, whose electron fractions are 0.46 -0.49.

On the other hand, jets from M10 and M12 have higher densities ($> 10^9 \mathrm{g/cm^3}$) and temperatures ($> 10$ K), so that material in the jets become neutron-rich via electron capture on protons. Consequently, neutron capture reactions proceed in the jets to synthesize U and Th abundantly (Figure 2), as in the jets from R12 [9]. The compos ion of the jets is similar to Eu-scaled $r$-nuclei in the solar system for mass numbers, $A > 80-100$. It should be emphasized that $r$-process successfully operates even in the jets from M10, in which magnitude of magnetic fields is relatively small ($10^{10}$ G) and rotation of the core is moderate ($2.5 \mathrm{rad\,s^{-1}}$) before the core collapse.

## ACKNOWLEDGMENTS

This work was supported in part by the Japan Society for Grants-in-Aid for the Scientific Research from the Ministry of Education, Science and Culture of Japan (No.17540267)

## REFERENCES


1. D. Proga, A.I. MacFadyen, P.J. Armitage and M.C. Begelman, Astroph. J. **599**, 5 (2003).
2. S. Fujimoto, K. Kotake, S. Yamada, M. Hashimoto, and K. Sato, Astroph. J. **644**, 1040 (2006).
3. S.E. Woosley, Astroph. J. **405**, 273 (1993).
4. J. Pruet, S.E. Woosley, and R.D. Hoffman, Astroph. J. **586**, 1254 (2003a).
5. S. Fujimoto, M. Hashimoto, K. Arai, and R. Matsuba, Astroph. J. **614**, 817 (2004).
6. S. Fujimoto, M. Hashimoto, K. Arai, and R. Matsuba, in Origin of Matter and Evolution of Galaxies, ed. M. Terasawa et al. (Singapore: World Sci.), 344, (2005).
7. J. Pruet, T.A. Thompson, and R.D. Hoffman, Astroph. J. **606**, 1006 (2004).
8. S. Fujimoto, M. Hashimoto, K. Arai, and R. Matsuba, Nucl. Phys. **A 758**, 47 (2005)
9. S. Fujimoto, M. Hashimoto, K. Kotake, and S. Yamada, Astroph. J. **656**, 382 (2007).
10. J.M. Stone, and M.L. Norman, Astroph. J. Supp. **80**, 791 (1992).
11. K. Kotake, H. Sawai, S. Yamada, and K. Sato, Astroph. J. **608**, 391 (2004).
12. H. Shen, H. Toki, K. Oyamatsu, and K. Sumiyoshi, Nucl. Phys. **A 637**, 435 (1998).
13. S.I. Blinnikov, N.V. Dunina-Barkovskaya, and D.K. Nadyozhin, Astroph. J. Supp. **106**, 171 (1996).
14. T. Di Matteo, R. Perna, and R. Narayan, Astroph. J. **579**, 706 (2002).
15. B. Paczyńsky, and P.J. Wiita, Astr. Astroph. **88**, 23 (1980).
16. M. Hashimoto, Prog. Theor. Phys. **94**, 663 (1995).
17. S. Nagataki, M. Hashimoto, K. Sato, and S. Yamada, Astroph. J. **486**, 1026 (1997).
18. D.D. Clayton, *Principles of Stellar Evolution and Nucleosynthesis* (Newyork: MacGraw-Hill), 1968.
19. S. Nishimura, K. Kotake, M. Hashimoto, S. Yamada, N. Nishimura, S. Fujimoto, and K. Sato, Astroph. J. **642**, 410 (2006).
20. T. Kodama, and K. Takahashi, Nucl. Phys. **A 239**, 489 (1975).
21. E. Anders, and N. Grevesse, Geochim. Cosmochim. Acta **53**, 197 (1989).

# Asymptotic Normalization Coefficients from the $^{15}N(^3He,d)^{16}O$ Reaction and the Astrophysical Factor for the $^{15}N(p,\gamma)^{16}O$ Capture


V. Burjan*, P. Bém*, S. Cherubini†, C. A. Gagliardi**, Z. Hons*, M. La Cognata†, V. Kroha*, J. Mrázek*, A. M. Mukhamedzhanov**, J. Novák*, Š. Piskoř*, R. G. Pizzone†, A. Plunkett**, S. Romano†, E. Šimečková*, C. Spitaleri†, L. Trache**, R. E. Tribble**, A. Tumino† and J. Vincour*

*Nuclear Physics Institute, Czech Academy of Sciences, 250 68 Řež near Prague, Czech Republic
†Università di Catania and INFN Laboratori Nazionali del Sud, Catania, Italy
**Cyclotron Institute, Texas A&M University, College Station, TX 77843



**Abstract.** The angular distributions of the $^{15}N(^3He,d)^{16}O$ reaction were measured with the aim to determine the direct capture rate of the astrophysical reaction $^{15}N(p,\gamma)^{16}O$ by deducing asymptotic normalization coefficients (ANC). The $^{15}N(p,\gamma)^{16}O$ reaction is a part of the CNO cycle having importance in the nucleosynthesis of the N and O isotopes. The measurement was carried out on the cyclotron U120M of NPI CAS at the energy 25.74 MeV of $^3He$ ions in a gas chamber containing the high purity $^{15}N$ isotope. The preliminary results of corresponding spectroscopic factors and ANC's were used for the estimation of the S-factor for the direct capture $^{15}N(p,\gamma)^{16}O$. Using the widths of the resonances $E_R$ = 312 and 962 keV, the total S-factor was determined within the framework of the R-matrix approach.



**Keywords:** direct capture, resonance capture, optical model, CNO cycles
**PACS:** 21.10.Jx, 25.55.Hp, 26.20.-f, 27.2.+n


## INTRODUCTION

The $^{15}N(p,\gamma)^{16}O$ reaction is an important nuclear astrophysical reaction because it initiates a leak from the first CN cycle to other CNO cycles. The loss of catalytic material via the $^{15}N(p,\gamma)^{16}O$ reaction, although a very rare, generates a new sequence of reactions $^{16}O(p,\gamma)^{17}F(e^+\nu_e)^{17}O(p,\alpha)^{14}N$ which restores the material of the CN cycle or generates another path $^{16}O(p,\gamma)^{17}F(e^+\nu_e)^{17}O(p,\gamma)^{18}F(e^+\nu_e)^{18}O(p,\gamma)^{19}F(p,\gamma)^{20}Ne$ which leads to the formation of heavier elements. At stellar environment the $^{15}N(p,\gamma)^{16}O$ reaction proceeds at very low energies, where it is dominated by the resonant capture to the ground state through the first two $J^\pi = 1^-$ $s$ wave resonances at $E_R = 312$ and 964 keV, where $E$ is the resonance energy in the center of mass (c. m.) system [1, 2]. So far the experimental data have been measured directly in [1] at proton energies only higher than 155 keV and earlier in [2] at proton energies higher than 220 keV. It has been found in [1] that to fit the experimental astrophysical $S$ factor at energies below $E_{R2} = 964$ keV one has to take into account the interference of the two $1^-$ resonant capture amplitudes with the nonresonant (direct) capture to the ground state of $^{16}O$. To fit


CP1012, *Frontiers in Nuclear Structure, Astrophysics, and Reactions: FINUSTAR 2*
edited by P. Demetriou, R. Julin, and S. V. Harissopulos


© 2008 American Institute of Physics 0-7354-0532-5/08/$23.00

the data the spectroscopic factor of the configuration $^{15}N-p$ in the ground state of $^{16}O$ was varied arbitrarily in [1]. In the $R$ matrix amplitude for the radiative capture reaction is given by the sum of the resonant and nonresonant (direct capture) amplitudes. The nonresonant capture amplitude is determined by the radial integral with the low integration limit $r_0$ [3]. The channel radius $r_0$ in the $R$ matrix method divides the internal and external regions. Meanwhile, in [1] the nonresonant matrix element has been taken as the integral from $r=0$ to $\infty$. It results in double counting of the internal region $r \leq r_0$ because the contribution from this region has been included into the resonant amplitude. In a strict approach, as we have mentioned, the nonresonant matrix element is contributed by the external region $r > r_0$ where the bound state wave function describing the final state can be approximated by its asymptotic term. Hence the nonresonant capture amplitude in the $R$ matrix approach is proportional to the proton asymptotic normalization coefficient (ANC). Taking into account the importance of the nonresonant capture for determination of the low-energy astrophysical factor for $^{15}N(p,\gamma)^{16}O$, we determined the ANCs for the proton removal from the ground state and 7 excited states of $^{16}O$. In previous studies [4] we have shown that the $(^3He,d)$ proton transfer reaction at incident energies of around 10 MeV/A is a reliable tool to determine ANCs. Here we report a measurement of ANCs for eight levels in $^{16}O$ by the $^{15}N(^3He,d)^{16}O$ reaction.

The $^{15}N(^3He,d)^{16}O$ reaction measured at beam energy $E_{^3He}$ = 11 MeV was already the subject of studies of W. Bohne et al. [5, 6] with an emphasis on shell-model properties of the $^{16}O$ nucleus. H. W. Fulbright et al. [7] measured this reaction at higher energies $E_{^3He}$ = 16.0 and 24.9 MeV also with the attention to the structure of $^{16}O$. We performed our measurement at $^3He$ energy of 25.74 MeV on the $^{15}N$ gas target in contrast to previous studies which were performed with solid targets. Angular distributions were measured with high precision at energy which was sufficient to achieve peripheral character of this reaction and which allowed to apply the method of asymptotic normalization coefficients.

## EXPERIMENTAL SET UP AND RESULTS

The experiment was performed with a momentum analyzed 25.74 MeV $^3He$ beam from the cyclotron U-120M of the Nuclear Physics Institute of CAS incident on a nitrogen gas target. The target gas chamber contained the $^{15}N$ isotope of high purity. Windows of the gas chamber were covered by 3.05 $\mu$m thick havar foils. The working pressure was kept at level of 205 - 208 mbar and was monitored by a gas control system, which registered the pressure and temperature of $^{15}N$ gas inside the chamber. Reaction products were registered by a pair of $\Delta$E-E telescopes consisting of 250-$\mu$m and 3-mm thick Si(Li) surface barrier detectors. Both telescopes were equipped with a pair of collimating slits of 2 x 3 mm$^2$ dimensions. The near and far slits were located 80 and 160 mm, respectively, from the center of the target cell. The effective target thickness, depending on the reaction angles, was determined for the given geometry by a Monte Carlo simulation. One telescope was fixed at the angle of $20^o$ as a monitor and the second one was adjusted at different laboratory angles in the range between $6.5^o$ and $60^o$. All measured data including the charge from the Faraday cup were collected in an on-line computer for later processing.

**TABLE 1.** Fit of optical model parameters in input channel $^3He + ^{15}N$

| Pot | V [MeV] | r [fm] | a [fm] | W [MeV] | $r_w$ [fm] | $a_w$ [fm] | $W_d$ [MeV] | $r_d$ [fm] | $a_d$ [fm] | Ref. |
|---|---|---|---|---|---|---|---|---|---|---|
| seed T | 116.4 | 1.15 | 0.764 | - | - | - | 18.9 | 1.288 | 0.800 | [9] |
| fit FT | 125.2 | 1.15 | 0.6870 | - | - | - | 11.3 | 1.4041 | 0.7469 | |
| seed VII | 183.5 | 1.15 | 0.659 | 7.9 | 2.142 | 0.695 | - | - | - | [10] |
| fit FVII | 202.5 | 1.15 | 0.6074 | 11.4 | 1.7842 | 0.7918 | - | - | - | |

The estimation of uncertainties of measured data included several sources. The uncertainties due to the effective gas target thickness and detector solid angle were about 8%. The charge of the beam was measured with the uncertainty about 3%. These uncertainties together with statistical uncertainties which were mostly less than 5% for more intensive transitions at maxima of angular distributions thus give the total uncertainty for forward angles less than 10% in most of cases.

Eight deuteron groups were identified from the $^{15}N(^3He,d)^{16}O$ reaction below the particle emission threshold of $^{16}O$.

## OPTICAL MODEL POTENTIALS AND DWBA CALCULATIONS

The analysis of angular distributions was carried out by means of DWBA calculations with optical model parameter sets, which were in case of the input channel deduced from the fit of our elastic scattering data. The search of the optical model parameters in the input channel was done by the computer code ECIS79 [8]. We looked for the parameters which describe best the angular distribution of the elastic scattering of the $^3He$ particles on the $^{15}N$ nucleus at energy 25.74 MeV. We have taken 2 sets of optical model parameters as seed parameters to include potentials with surface absorption [9] and with volume absorption [10] (see Table 1). We did not use seed potentials with the spin-orbit coupling. When fitting the optical model parameters ($V$, $a$, $W_d$, $r_d$, $a_d$ or $W$, $r_w$ and $a_w$) we minimized the $\chi^2$-function using the uncertainties of the elastic scattering differential cross section. Resulting optical model parameters from both seed sets are given in Table 1 and denoted as FT (fit of Trost seed calculated from the global formula [9]) and FVII (fit of Vernotte family II seed taken from $^{18}O$ data [10]) sets.

Optical model parameters used for the output channel are given in Table 2. They were taken from the work of Fulbright [7] or calculated from global formulas derived by Daehnick et al. [11].

Angular distributions of deuterons from the reaction $^{15}N(^3He,d)^{16}O$ corresponding to the first four bound states in $^{16}O$ calculated with different combinations of the optical model parameters are shown in Fig. 1.

Experimental angular distributions were then used for the extraction of proton ANCs for different bound states of $^{16}O$ by comparison with distorted-wave Born approximation

**TABLE 2.** Parameters of optical model parameters in output channel $^{3}\mathrm{d} + {}^{16}\mathrm{O}$

| Pot | V [MeV] | r [fm] | a [fm] | $W_d$ [MeV] | $r_d$ [fm] | $a_d$ [fm] | $W_{so}$ [MeV] | $r_{so}$ [fm] | $a_{so}$ [fm] | $r_c$ [fm] | Ref. |
|---|---|---|---|---|---|---|---|---|---|---|---|
| F | 92 | 1.053 | 0.771 | 8.18 | 1.361 | 0.772 | - | - | - | 1.3 | [7] |
| D | 82.88 | 1.17 | 0.7478 | 11.74 | 1.325 | 0.5178 | 6.4 | 1.070 | 0.660 | 1.3 | [11] |

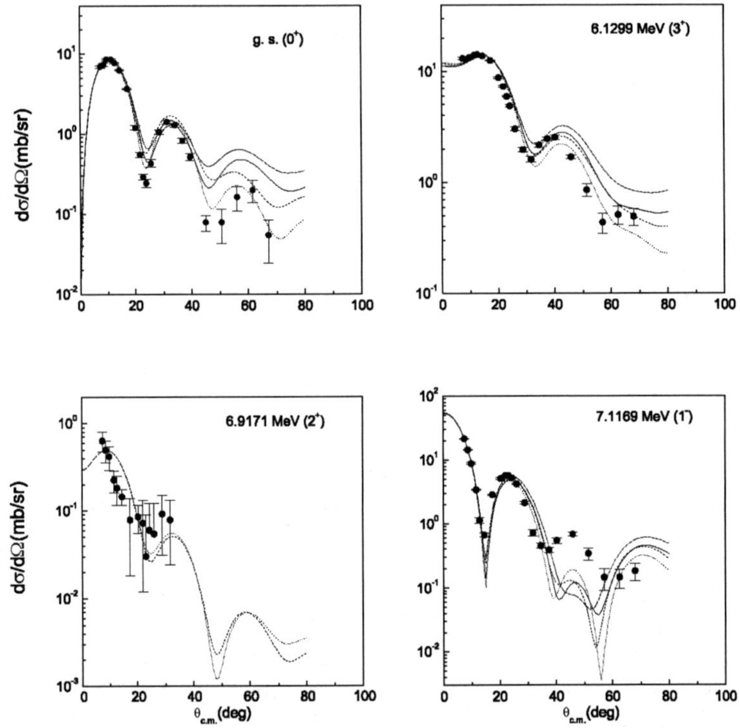


**FIGURE 1.** Angular distributions from the $^{15}\mathrm{N}(^{3}\mathrm{He},\mathrm{d})^{16}\mathrm{O}$ reaction for the transitions leading to the ground, 6.1299 MeV, 6.9171 MeV and 7.1169 MeV states in $^{16}$O. DWBA calculations were made with optical model parameter sets given in Tables 1 (input channel) and 2 (output channel): FVII-D, solid curve; FT-F, dashed curve; FVII-F, dotted curve and FT-D, dash-dotted curve.

**TABLE 3.** Spectroscopic factors $S_{lj}$ and ANCs from the $^{15}N(^{3}He,d)^{16}O$ reaction

| State | $j_{tr}$ | $b_{lj}$ | $S_{lj}$ | ANC |
|---|---|---|---|---|
| g. s. $(0^+)$ | 0.5 | 9.579 | 4.19 | $192.229 \pm 25.6$ |
| 6.1299 $(3^-)$ | 2.5 | 3.026 | 0.77 | $3.5252 \pm 0.423$ |
| 6.9171 $(2^+)$ | 1.5 | 4.754 | 0.018 | $0.2034 \pm 0.059$ |
| | 5.2 | 0.799 | 0.2 | $0.0639 \pm 0.019$ |
| 7.1169 $(1^-)$ | 1.5 | 2.193 | 0.4 | $0.962 \pm 0.12$ |
| 8.8719 $(2^-)$ | 2.5 | 1.654 | 0.6 | $0.821 \pm 0.115$ |
| 9.8445 $(2^+)$ | 1.5 | 2.875 | 0.011 | $0.046 \pm 0.010$ |
| 10.957 $(0^-)$ | 0.5 | 5.278 | 0.98 | $13.648 \pm 1.91$ |
| 11.080 $(3^+)$ | 2.5 | 0.142 | 0.1 | $0.0010 \pm 0.0002$ |

calculations. Using these ANCs (Table 3) and proton and $\alpha$ resonance widths determined from the $R$ matrix fit to another astrophysical reaction $^{15}N(p,\alpha)^{12}C$ [2], we calculated the astrophysical factor for $^{15}N(p,\gamma)^{16}O$ in the $R$ matrix approach [3]. At low energy tail, where the direct capture plays a significant role, the astrophysical S(0)-factor is about $38 \pm 4$ keV.b. We thus found out that the direct capture contribution S(0) = $64 \pm 6$ keV.b in the paper [1] was significantly overestimated.

This work was supported by ME 902(2007) project NSF and MŠMT, ČR, LC07050 center, grant GAČR 202/05/0302 and OTKA T49245.

## REFERENCES


1. C. Rolfs and W. S. Rodney, *Nucl. Phys.* **A235**, 450 (1974).
2. D. F. Hebbard, *Nucl. Phys.* **15**, 289 (1960).
3. F. C. Barker and T. Kajino, *Aust. J. Phys.* **44**, 369 (1991).
4. C. A. Gagliardi, R. E. Tribble, A. Azhari, H. L. Clark, Y.-W. Lui, A. M. Mukhamedzhanov, A. Sattarov, L. Trache, V. Burjan, J. Cejpek, V. Kroha, Š. Piskoř, J. Vincour, *Phys. Rev.* **C59**, 1149 (1999).
5. W. Bohne, H. Homeyer, H. Lettau, H. Morgenstern, J. Scheer, F. Sichelschmidt, *Nucl. Phys.* **A128**, 537 (1969).
6. W. Bohne, H. Homeyer, H. Lettau, H. Morgenstern, J. Scheer, F. Sichelschmidt, *Nucl. Phys.* **A160**, 257 (1971).
7. H. W. Fulbright, J. A. Robins, M. Blann, D. G. Fleming, H. S. Plendl, *Phys. Rev.* **184**, 1068 (1969).
8. J. Raynal, code ECIS79 (unpublished)
9. H. J. Trost, P. Lezoch, U. Strohbusch, *Nucl. Phys.* **A462**, 333 (1987).
10. J. Vernotte, G. Berrier-Ronsin, J. Kalifa, R. Tamisier, *Nucl. Phys.* **A390**, 285 (1982).
11. W. W. Daehnick, J. D. Childs, Z. Vrcelj, *Phys. Rev.* **C21**, 2253 (1980).

# Charge densities and the electron scattering form factors of exotic nuclei

S. Karataglidis* and K. Amos†

*Department of Physics and Electronics, Rhodes University, P.O. Box 94, Grahamstown, 6140, South Africa
†School of Physics, University of Melbourne, Victoria, 3010, Australia

**Abstract.** Elastic and inelastic electron scattering form factors for several neutron-rich exotic nuclei are presented. The results have been obtained using large space no-core shell models. While the elastic scattering form factors are insensitive to the details of the neutron density, it is found that inelastic scattering may be influenced by extensive neutron distributions.

**Keywords:** Electron scattering, nuclear structure, shell model, exotic nuclei
**PACS:** PACS numbers

## INTRODUCTION

There is renewed interest in the observables for the scattering of electrons from exotic nuclei with the construction of the electron-ion collider at GSI [1] and of the SCRIT (Self-Containing Radioactive Ion Targets) project at RIKEN [2]. The facilities will be able to measure the charge densities of neutron-rich nuclei. As the charge is concentrated in the core of such systems, particularly in the case of halos, the experiments will be able to probe the core of the systems directly. Through appropriate kinematic selection, particularly at the collider, transverse as well as longitudinal form factors may be obtained. Taken together with analyses of nucleon-nucleus (*NA*) scattering data, a complete map of the densities of exotic nuclei may be possible.

Electron scattering has been previously considered by Antonov *et al.* [3] for both light and heavy nuclei. For the light mass systems, densities from large space shell models [4, 5] were used to obtain the elastic longitudinal form factors. For the He isotopes they found variations of the charge densities, and so likewise for the form factors, for $^4$He and $^6$He but not a significant change in the form factor between $^6$He and $^8$He. They also found that the proton density extends far with increasing neutron number. However, no conclusion was made as to the role of the halo in $^6$He. That has been the subject of the studies by Bertulani [6]. He used densities obtained from a potential model to obtain the form factors for $^6$He and $^{11}$Li and found that there was no significant contribution to the form factors from the neutron halo. However, for the proton-halo nucleus $^8$B he found that the proton halo did affect the elastic scattering form factor, and significantly. That is to be expected given the large extension of the charge density over that of a skin such a halo creates.

We consider herein the longitudinal form factors for elastic scattering from the He and Li isotopes. For inelastic scattering we also consider the transverse electric form factor to the first excited state in $^6$He as the transverse electric form factor is dependent

CP1012, *Frontiers in Nuclear Structure, Astrophysics, and Reactions: FINUSTAR 2*
edited by P. Demetriou, R. Julin, and S. V. Harissopulos
© 2008 American Institute of Physics 0-7354-0532-5/08/$23.00

on the charge density via Siegert's theorm. As a comparison to the longitudinal form factors, we also consider the longitudinal elastic scattering form factor for the proton halo nucleus $^8$B.

The halo in $^6$He was established unambigously in the analysis of inelastic scattering data of $^6$He from hydrogen [7]. Therein, scattering to the $2_1^+$ (1.8 MeV) state in $^6$He, which is surface peaked, clearly showed the effect of the halo on the inelastic scattering differential cross section, when analysed with a microscopic optical potential obtained by using realistic (shell model) wave functions for $^6$He. As a result, we also present results for the inelastic longitudinal and transverse $E2$ form factors for the scattering to the $2^+$ state in $^6$He to see if there is an effect from the large extension of the neutron density.

## ELECTRON SCATTERING FORM FACTORS

The microscopic model of electron scattering specified by deForest and Walecka [8] and by Karataglidis *et al.* [9], based on the shell model, has been used to obtain the form factors presented herein. Using the notation of the latter, the form factors between nuclear states $J_i$ and $J_f$ involving angular momentum transfer $J$ are expressed as

$$\left|F_J^\eta(q)\right|^2 = \frac{1}{2J_i+1}\left(\frac{4\pi}{Z^2}\right)\left|\langle J_f \| T_J^\eta(q) \| J_i\rangle\right|^2 , \tag{1}$$

where $\eta$ specifies the type, i.e. longitudinal, transverse electric, or transverse magnetic. Assuming one-body operators, the reduced matrix elements may be expressed in the form,

$$\langle J_f \| T_J^\eta(q) \| J_i\rangle = \frac{1}{\sqrt{2J+1}}\mathrm{Tr}(SM), \tag{2}$$

where $S$ is the matrix of one-body transition densities, $S_{j_1 j_2 J}$, defined as

$$S_{j_1 j_2 J} = \left\langle J_f \left\| \left[a_{j_2}^\dagger \times \tilde{a}_{j_1}\right]^J \right\| J_i \right\rangle , \tag{3}$$

and $M$ denotes the matrix elements of the one-body longitudinal or transverse electromagnetic operators for each allowed single-particle transition. Bare operators are used for the results presented, and explicit meson-exchange-current (MEC) effects are ignored. However, MEC have been incorporated implicitly in the transverse electric form factors in the long-wavelength limit by using Siegert's theorem [10]. That serves to introduce an explicit dependence on the charge density into the transverse electric form factor through the use of the continuity equation.

Within the framework of the shell model, the halo has been specified by using Woods-Saxon (WS) single-particle (SP) wave functions with the binding energies of the valence (halo) nucleons set to the appropriate single-nucleon separation energies. Doing so allows for the extension of the nucleon density consistent with there being a halo [5], if the nucleus is very loosely bound such that a halo exists. If the separation energy is large then the results obtained using WS functions are similar to those for the standard shell model results using oscillator (HO) SP functions, and give rise to skins.

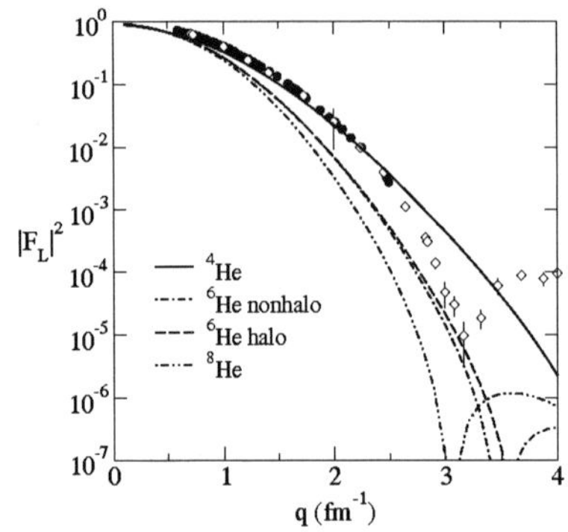


**FIGURE 1.** Elastic electron scattering form factors for $^4$He, $^6$He, and $^8$He.

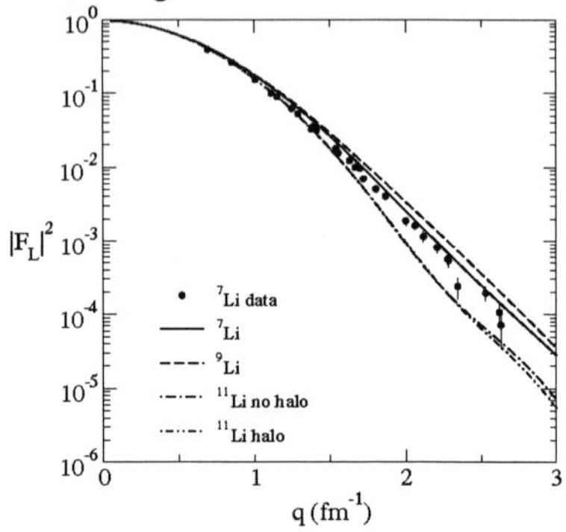


**FIGURE 2.** The longitudinal elastic scattering form factors for $^7$Li, $^9$Li, and $^{11}$Li.

Hence, we denote results obtained using WS and HO SP functions as halo and non-halo respectively. Such has been used previously to identify halos in both $^6$He [7] and $^{11}$Li [5], and confirm the neutron skin characteristics of both $^8$He and $^9$Li [5].

## RESULTS

The elastic scattering form factors for the He isotopes are shown in Fig. 1. For the results presented therein, the shell model wave functions for $^{4,6,8}$He were obtained from a complete $(0+2+4)\hbar\omega$ shell model using the $G$ matrix interaction of Zheng *et al.* [12]. The halo in $^6$He was specified by using WS wave functions with the SP energies of the $0p$-shell neutrons set to the single neutron separation energy of 1.8 MeV [5]. An oscillator parameter of 1.8 fm was used to specify the HO SP wave functions. Such a model was used to explain the elastic and inelastic $^6$He+hydrogen scattering data establishing the halo in $^6$He [7].

The comparison of the $^4$He form factor data of McCarthy *et al.* with the results of the shell model calculation is quite good up to 2.5 fm$^{-1}$. This is consistent with the predicted charge radius of 1.71 fm as compared to the measured value of $1.671 \pm 0.014$ fm [13]. For $^6$He and $^8$He the additional neutrons pull the charge density out and so the form factors decrease with momentum transfer. Note that there is little difference between the results of the halo and non-halo form factor calculations for $^6$He indicating that the elastic scattering form factor is not dependent on the detailed properties of the neutron halo.

The longitudinal elastic scattering form factors for $^7$Li, $^9$Li and $^{11}$Li are displayed in Fig. 2. The data shown are those for the scattering from $^7$Li [14, 15]. The shell model wave functions used were those published previously (see Refs. [16, 5]). The level of agreement between the results for the $^7$Li form factor and the data is quite good. The form factor for $^9$Li is not substantially different from that of $^7$Li. But a noticeable change is observed in the form factor for $^{11}$Li. It decreases faster than the other two with increasing momentum transfer. As with $^6$He, this extension in the charge density does

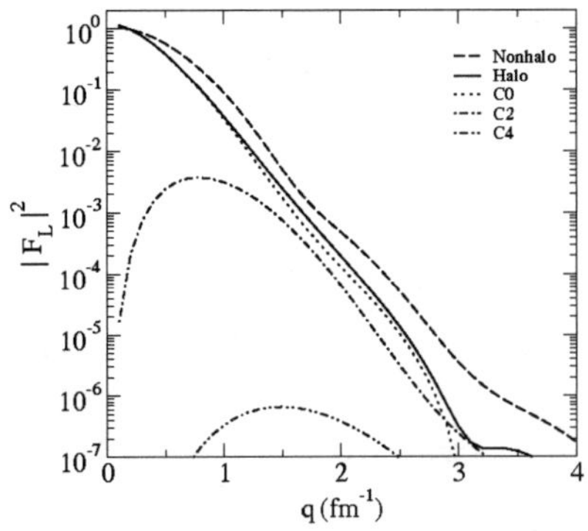


**FIGURE 3.** Longitudinal elastic scattering form factor for $^8$B.

not come about with the specification of the halo; rather it is due only to the coupling of the four extra neutrons to the $^7$Li core. A similar conclusion has been reached by Bertulani [6].

As a comparison to the elastic scattering from the neutron-rich nuclei, we consider the longitudinal elastic scattering form factor from the proton halo nucleus $^8$B. The wave functions for $^8$B were also obtained from a complete $(0+2+4)\hbar\omega$ shell model using the Zheng interaction. The single proton separation energy is 137 keV [17] and the halo is specified by using WS functions for the $0p$-shell protons with that binding energy. The non-halo specification uses the same set of WS functions as for $^8$He. The prediction for the longitudinal elastic scattering form factor is shown in Fig. 3. This form factor is dominated by the $C0$ component below 1 fm$^{-1}$ while the $C2$ contribution becomes significant above that value. The $C4$ component is negligible. The effect of the proton halo, in this case, is evident with the decrease in the form factor with momentum transfer reflecting the more extensive charge distribution.

The halo was established in $^6$He as a result of the analyses of elastic and inelastic scattering data [7]. The inelastic scattering data were to the $2_1^+$ (1.8 MeV) state in $^6$He. As the $E2$ transition density is surface-peaked, it provides a more direct measure of the surface of the nucleus than elastic scattering, which probes the volume. Fig. 4 displays the longitudinal and transverse inelastic $E2$ form factors to the $2_1^+$ state. The longitudinal form factor shows a significant change due to the halo above 1 fm$^{-1}$, where the form factor decreases faster than the non-halo result. This illustrates the extension of the density, consistent with the result of the analysis of the differential cross section from proton scattering [7]. Note that below 1 fm$^{-1}$ the two results are the same; there is no effect on the $B(E2)$ value due to the halo. The transverse form factor shows a similar change. The effect there is due to the dramatic decrease in the neutron component when the halo is specified. As both form factors are much weaker compared to the elastic scattering form factor, the effect of the halo becomes significant. This is especially so for the transverse form factor.

Elastic and inelastic form factors have been presented for the scattering of electrons from light exotic nuclei. For the neutron-rich nuclei, the effect of the halo is not sig-

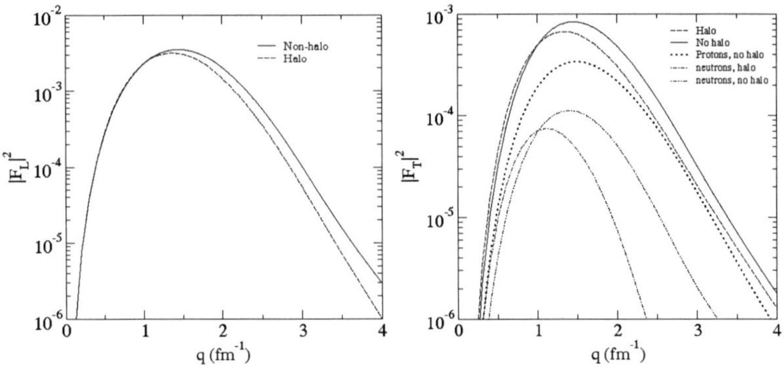


**FIGURE 4.** Longitudinal and transverse $E2$ inelastic scattering form factors to the $2_1^+$ state in $^6$He.

nificant in the elastic scattering form factors. The converse is true for proton-halos: the effect of the proton halo is most significant. By comparison, the inelastic $E2$ form factors for $^6$He do show the effect of the neutron halo, as the neutron component of the form factors become comparable to the proton component in the inelastic scattering. It is hoped that with the new facilities form factors, both longitudinal and tranverse, will be measured. Together with analysis of nucleon-scattering data, a complete map of the densities of exotic systems will be possible.

## REFERENCES


1. *Technical Proposal for the Design, Construction, Commissioning, and Operation of the ELISe setup*, spokesperson Haik Simon, GSI Internal Report, Dec. 2005.
2. T. Suda, and M. Wakasugi, *Prog. Part. Nucl. Phys.* **55**, 417 (2005).
3. A.N. Antonov, D.N. Kadrev, M.K. Gaidarov, E.M. de Guerra, P. Sarriguren, J.M. Udias, V.K. Lukyanov, E. Zemlyanaya, and G.Z. Krumova, *Phys. Rev. C* **72**, 044307 (2005).
4. S. Karataglidis, P.J. Dortmans, K. Amos, and C. Bennhold, *Phys. Rev. Lett.* **79**, 1447 (1997).
5. S. Karataglidis, P.J. Dortmans, K. Amos, and C. Bennhold, *Phys. Rev. C* **61**, 024319 (2000).
6. C.A. Bertulani, *J. Phys. G* **34**, 315 (2007).
7. A. Lagoyannis et al., *Phys. Lett.* **B518**, 27 (2001).
8. T. deForest, and J.D. Walecka, *Adv. Phys.* **15**, 1 (1966).
9. S. Karataglidis, P. Halse, and K. Amos, *Phys. Rev. C* **51**, 2494 (1995).
10. J.L. Friar, and W.C. Haxton, *Phys. Rev. C*, **31** 2027 (1985).
11. J.S. McCarthy, I. Sick, and R.R. Whitney, *Phys. Rev. C* **15**, 1396 (1977).
12. D.C. Zheng, B.R. Barrett, J.P. Vary, W.C. Haxton, and C.-L. Song, *Phys. Rev. C* **52**, 2498 (1995).
13. C.R. Ottermann, G. Köbschall, K. Maurer, K. Rörich, C. Schmitt, and V.H. Walter, *Nucl. Phys.* **A436**, 688 (1985).
14. L.R. Suelzle, M.R. Yearian, and H. Crannell, *Phys. Rev.* **162**, 992 (1967).
15. J. Lichtenstadt, J. Alster, M.A. Moinester, J. Dubach, R.S. Hicks, G.A. Peterson, and S. Kowlaski, *Phys. Lett.* **B219**, 394 (1989).
16. S. Karataglidis, B.A. Brown, P.J. Dortmans, and K. Amos, *Phys. Rev. C* **55**, 2826 (1997).
17. D.R. Tilley, J.H. Kelley, J.L. Godwin, D.J. Millener, J.E. Purcell, C.G. Sheu, and H.R. Weller, *Nucl. Phys.* **A745**, 155 (2004).

# Near-barrier scattering of $^6$He and $^{11}$Be

L. Acosta[1], D. Escrig[4], D. Galaviz[4], A.M. Sánchez-Benítez[2], C. Angulo[2], M.A.G. Álvarez[3], M.V. Andrés[3], M.J.G. Borge[4], E. Casarejos[2], J.M. Espino[3], J.E. García-Ramos[1], J. Gómez-Camacho[3], I. Martel[1], A.M. Moro[3], I. Mukha[3], F. Pérez-Bernal[1], D. Rodríguez[1], K. Rusek[5], and O. Tengblad[4]

[1] *Departamento de Física Aplicada, Universidad de Huelva, E-21071 Huelva, Spain*
[2] *Centre de Recherches du Cyclotron-UCL, B-1348 Louvain-la-Neuve, Belgium*
[3] *Departamento de FAMN, Universidad de Sevilla, E-41080 Sevilla, Spain*
[4] *Instituto de Estructura de la Materia, CSIC, E-28006 Madrid, Spain*
[5] *The Andrzej Soltan Institute for Nuclear Studies, 00-681 Warsaw, Poland*

**Abstract.** Experiments on halo nuclei at energies around the Coulomb barrier can provide good quality data on elastic, inelastic, breakup and nucleon transfer reactions. This information turns out to be very important to understand the structure and dynamics of these exotic nuclei. In this work we present relevant results obtained from recent scattering experiments of $^{11}$Be and $^6$He on heavy targets at energies around the Coulomb barrier. Our data exhibit strong absorption effects in the elastic channel. The role played by transfer and breakup channels, and the coupling to the continuum, are studied using CDCC and DWBA calculations.

**Keywords:** Nuclear reactions; Halo nuclei; Dipole polarizability, Transfer to continuum, Breakup reactions.
**PACS:** 24.10.Eq; 24.10.Ht; 25.50+g; 25.60.Bx; 25.60.Je

## INTRODUCTION

Over the last decade there has been a notorious development of high intensity beams of loosely bound radioactive nuclei, on both neutron and proton rich sides of the nuclear chart. Among these extraordinary specimens are those that exhibit the so called "neutron halo": a rather inert core plus one or two barely bound extra neutrons, forming an extended neutron distribution. The nucleus $^{11}$Be exhibits a one-neutron halo structure, whereas the nucleus $^6$He is a two-neutron Borromean halo nucleus: i.e., none of the individual subsystems n-n or $\alpha$-n forms a bound state.

Some important properties of these nuclei are the very low binding energy, the extended halo structure, and the fact that they should be easily polarizable: i.e., during the collision process the system will be distorted due to the different interactions of the target with the halo and the core. These particular features affect the scattering of halo nuclei on heavy targets, even at collision energies below the Coulomb barrier. Therefore nuclear reactions at energies around the barrier exhibit some common features, like the strong absorption found in the elastic channel or the large cross sections reported for nuclear dissociation [1].

CP1012, ***Frontiers in Nuclear Structure, Astrophysics, and Reactions:*** *FINUSTAR 2*
edited by P. Demetriou, R. Julin, and S. V. Harissopulos
© 2008 American Institute of Physics 0-7354-0532-5/08/$23.00

During the previous years there has been an intense research on near-barrier reactions induced by light halo nuclei on heavy targets [2]. In this paper we discuss the results obtained in recent experiments performed with $^{6}$He beams at the Centre de Recherche du Cyclotron at Louvain-la-Neuve (Belgium) [3, 4], and with $^{11}$Be beams at the REX-ISOLDE facility at CERN (Switzerland) [5]. Theoretical calculations are presented and the relevant reaction mechanisms are discussed on the light of the analysis of these data.

## SCATTERING OF $^{6}$He on $^{208}$Pb

High intensity $^{6}$He beams in a wide range of energies are currently produced at the radioactive beam facility of the Centre de Recherche du Cyclotron (CRC). In a first set of measurements [3] the $^{6}$He beam was produced at laboratory energies of 14, 16, 18 and 22 MeV and the scattering on a $^{208}$Pb target was measured at (5°-65°) and (135°-170°) laboratory angular ranges. In a second experiment [4] we measured the scattering at 22 MeV but covering a wider angular region (5°-165°). In these experiments the LEDA array [6], the DINEX telescope array [7] and a set of two double-sided silicon-strip detector telescopes [8] were used. A detailed description of the experimental methods and analysis can be found in [9, 10]. From these experiments we were able to determine accurate values for the elastic and inclusive break-up cross sections over a wide range of angles and several collision energies.

In Fig. 1.a it is shown the angular distribution of the elastic cross section at E=22 MeV, which is close to the Coulomb barrier for this system. The data exhibits a strong reduction of the elastic flux, extending up to very small scattering angles, so that the rainbow structure typical of the scattering of stable nuclei, is completly absent. This effect can be found at collision energies as low as 16 MeV. Further investigations using Optical model (OM) calculations [10] reveal that in order to reproduce the data, large values of imaginary diffuseness (~1.7 fm) are needed. Simple semiclassical estimates [11] show that flux removal can reach distances as large as 20 fm. These effects are clear indications of the presence of strong reaction mechanisms still active in situations in which the nuclei stay well separated. This is what we call long range absorption, and it is associated to the scattering of halo nuclei.

The relevant reaction mechanisms producing long range absorption might arise from the loosely bound structure of $^{6}$He that favors the coupling to the positive energy continuum. This effect has been investigated by Continuum Discretized Coupled Channel (CDCC) calculations, and the results are shown in Fig. 1.a. The solid line is the result of a CDCC calculation similar to that reported in [12], which includes the $^{4}$He+2n continuum states of the projectile [13], and reproduces the general trend of the data. The dashed line is the result of the one-channel calculation, which now exhibits the typical heavy ion rainbow and is unable to describe the data. An important part of these couplings originate from the distortion induced in $^{6}$He by the intense dipole interaction with the heavy target. However our studies reveal that only a fraction of this absorption is actually due to the Coulomb dipole polarizability [9, 14], the rest arising from other reaction mechanisms.

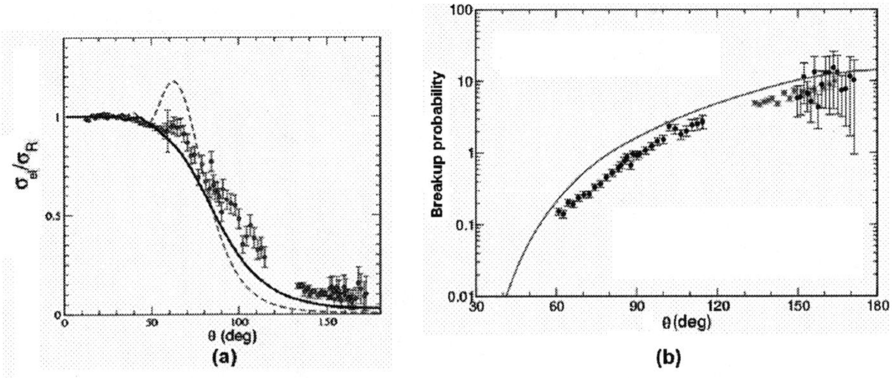


**FIGURE 1.** a) Ratio of cross sections elastic/Rutheford versus the scattering angle (lab) $^6$He on $^{208}$Pb at E= 22 MeV (preliminary). b) Angular distribution of the "breakup probability" of $^6$He in $^4$He at 22 MeV (preliminary). Data are from PH189 and PH215 experiments. See text for details.

A deeper insight into the dynamics of this system can be achieved from the study of the strong alpha yields produced at near-barrier collisions. These are known to exhaust most of the total reaction cross sections [15, 16]. In Fig. 1.b we present the "breakup probability" of $^6$He on $^4$He at 22 MeV, defined as the ratio of the $^4$He fragments to the $^6$He elastic channel, as a function of the scattering angle. It is noticeable the large alpha production at backward angles, which is about ten times stronger than the elastic channel. This reaction channel remains very strong even at deep sub-barrier energies where we still find about 10% of the elastic channel at 14 MeV [10].

Our experimental results can be well described in terms of the *transfer to continuum* (TC) mechanism. In this process the two neutrons are transferred to continuum states of the lead target with very low relative kinetic energy, so that the emerging alpha gets an outgoing energy similar to that of the elastic events. Transfer calculations have been performed in a DWBA approach, including bound and unbound states of the $^{208}$Pb target and using a di-neutron model for $^6$He [12]. Preliminary results of a TC calculation are plotted with a solid line in Fig. 1.b. The calculations based on a TC mechanism show a reasonable agreement with the data, and with the energy distribution measured at backward angles (see ref. [10]). Nevertheless direct breakup and other reaction channels, like one-neutron transfer, might play a more important role at smaller angles. A detailed study of the alpha energy distributions from PH215 will shed some light to this problem. More refined TC calculations using a modified di-neutron model [17] should improve the theoretical description.

## SCATTERING OF $^{11}$Be on $^{120}$Sn

The $^{11}$Be nucleus is an interesting case to study the dynamic of nuclear haloes at near-barrier energies. The weakly bound $^{11}$Be nucleus ($S_n$ = 503 keV) has a relatively long half-life (13.8 s) suitable to be studied at REX-ISOLDE. The only bound excited state lies at 320 keV ($I=½^-$) with a strong coupling to the ground state ($I=½^+$) by E1 transitions [18]. Scattering experiments at near-barrier energies have been performed to disentangle the relative role played by different processes like transfer and fusion mechanisms [19, 20, 21, 22].

We have studied the scattering of $^{11}$Be on $^{120}$Sn at 32 MeV at REX-ISOLDE [4]. We have used an array of six DSSSD telescopes [8], arranged in a close hexagonal configuration [2]. Full details of the experimental setup and data analysis will be given elsewhere [13].

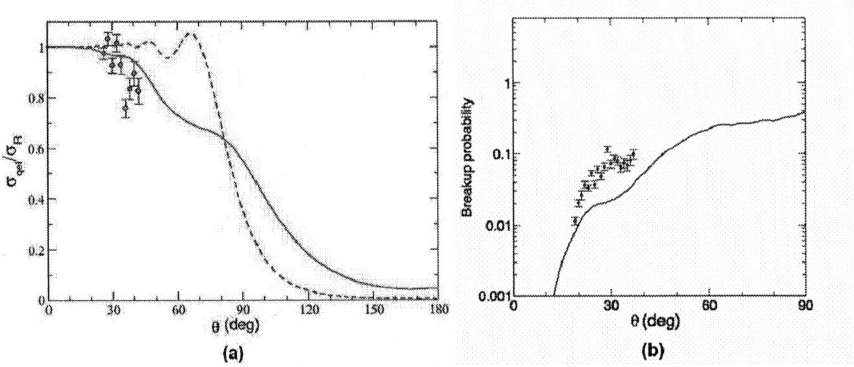


**FIGURE 2.** a) Ratio of cross sections quasi-elastic/Rutheford versus the scattering angle (lab) for $^{11}$Be on $^{120}$Sn at 32 MeV (preliminary). b) Angular distribution of the "breakup probability" of $^{11}$Be in $^{10}$Be at 32 MeV. See text for details.

Due to relatively low beam intensity ($\leq 10^5$ pps) we had to use a thick (3.5 mg/cm$^2$) $^{120}$Sn target. This choice limited our energy resolution to about 500 keV, spoiling the possibility of resolving elastic and inelastic channels. However the resolution was good enough to separate $^{10}$Be break-up from elastic channel using the ΔE-E technique. Although the statistics achieved limited the analysis to angles below 40°, the results obtained are promising. A preliminary angular distribution for the quasi-elastic cross section is shown in Fig. 2.a.

In the forward angular region we should expect the dominance of the direct break-up channel. In order to allow a comparison with the experimental results, we have performed preliminary CDCC calculations including the coupling to the ½- excited state and the $^{10}$Be + n continuum. The results are plotted in Fig. 2.a with a solid line. The dashed line excludes the effects of coupling to continuum states and exhibits a typical rainbow structure. Our calculations predict a strong reduction in the elastic cross sections due to the effect of coupling to the continuum, which is consistent with the general trend of the data. In Fig. 2.b we present the angular distribution of the break-up probability at 32 MeV, normalized to the quasi-elastic channel. The calculation predicts lower cross sections than the measurements. This is a signature of existing reaction mechanisms not included in our calculations. In particular, neutron transfer and core excitation could play a relevant role. Part of this puzzle will be solved by looking to the energy distribution of the break-up events. This work is actually in progress.

## SUMMARY AND CONCLUSIONS

In this work we discuss recent experimental results on reactions of $^{6}$He and $^{11}$Be on heavy targets at near-barrier energies. The elastic cross sections show a strong absorption even at deep sub-barrier energies, suggesting the presence of long range reaction mechanisms. This feature is only partially due to the dipole part of the nuclear interaction. Other reaction processes coupled through unbound continuum states of projectile and target are very important. The analysis of the energy and angular distributions of reaction fragments suggest that "transfer to the continuum" is the dominant reaction mechanism at backward angles. More and high quality data on cross sections for elastic, inelastic and transfer channels would be needed to fully understand the near-barrier scattering of these weakly bound nuclei.

## ACKNOWLEDGEMENTS

This work has been partially supported by the Spanish MCyT under the contracts FPA2005-02379, FPA2005-04660, FPA2006-13807-C02-02 and the Consolider Project CE-SCD2007-0042.

## REFERENCES

1. J.F. Liang, C. Signorini, Int. J. Mod. Phys. **E**, 1121 (2005).
2. L.F. Canto et al. , Physics Reports **424**, 1 (2006).
3. J Gómez-Camacho et al. ," *Exploring the dynamics of low energy 6He elastic scattering on heavy targets)*, Proposal to the Scientific Advisory Committee of the Centre de Recherche du Cyclotron at Louvain-la Neuve (Belgium), January 2002.
4. I. Martel et al., "*Study of the elastic scattering of $^{6}$He at energies around the Coulomb barrier"*, Proposal to the Scientific Advisory Committee of the Centre de Recherche du Cyclotron at Louvain-la Neuve (Belgium), January 2004.
5. M.V. Andrés el al., "Exploring halo effects in scattering of $^{11}$Be halo on a heavy target at REX-ISOLDE", Proposal to the ISOLDE –Neutron Time of flight Committee, CERN-INTC/2006-010, 2006.
6. T. Davinson et al., Nucl. Instr. and Meth. **A 454**, 350 (2000).
7. A. M. Sánchez-Benítez, et al., J. Phys. (London) **G 31**, S1953 (2005).
8. O. Tengblad et al., Nucl. Inst. Meth. **A 525**, 458 (2004).
9. A.M. Sánchez-Benítez, et al., Nucl. Phys. A, in press.
10. D. Escrig et al., Nucl. Phys. **A 792**, 2 (2007).
11. A.M. Sánchez-Benítez et al., Acta Phys. Pol. **B 37**, 1 (2006).
12. K. Rusek et al., Phys. Rev. **C 72**, 037603 (2005).
13. L. Acosta et. al., in preparation.
14. M.V. Andres, J. Gomez-Camacho and M.A.Nagarajan, Nucl. Phys. **A579**, 273 (1994).
15. E.F. Aguilera et al., Phys.Rev.Lett. **84**, 5058 (2000).
16. A. Di Pietro et al., Europhys. Lett. **64**, 309 (2003).
17. A Moro, K. Rusek, M.J. Arias, J. Gómez-Camacho, M. Rodríguez-Gallardo, nucl-th/0703005, Phys. Rev. C, in press.
18. D.J. Millener et al., Phys. Rev. **C 28**, 497 (1983).
19. C. Signorini, et al., Eur. Phys. J. **A 2**, 227 (1998).
20. V. Fekou-Youmbi, et al., Nucl. Instrum. Methods Phys. Res. **A 473**, 490 (1999).
21. C. Signorini et al., Nuclear Physics **A 735**, 329 (2004) .
22. M. Mazzocco et al., Eur. Phys. J. **A 28**, 295 (2006).

# POSTER CONTRIBUTIONS

# Triaxial deformations in the proton emitters $^{161}$Re and $^{185}$Bi

P. Arumugam*, E. Maglione† and L. S. Ferreira*

*Centro de Física das Interacções Fundamentais, and Departmento de Física, Instituto Superior Técnico, Avenida Rovisco Pais, P1049-001 Lisbon, Portugal
†Dipartimento di Fisica "G. Galilei", Via Marzolo 8, I-35131 Padova, Italy and Istituto Nazionale di Fisica Nucleare, Padova, Italy

**Abstract.** The nonadiabatic quasiparticle method which has been recently generalized by us to include the triaxial degree of freedom, is used to perform exact calculations for proton emission from triaxial nuclei. This formalism for triaxial proton emitters takes fully into account the pairing residual interaction and the coupling to the daughter nucleus states. With this versatile formalism, we discuss the sensitivity of our results to the triaxial deformation. We apply our method to study proton emission from the nuclei $^{161}$Re and $^{185}$Bi for which the half-lives have been measured experimentally.

**Keywords:** Proton emission, Dripline nuclei, Triaxial deformation
**PACS:** 23.50.+z, 21.10.Tg, 27.60.+q

## INTRODUCTION

The study of proton emission is well recognized as a sensitive probe for understanding the structural properties of nuclei near the proton dripline. An interesting but yet to be understood phenomenon in the studies of deformed proton emitters, is the role played by triaxial deformation in the decay process. The properties of proton emitting nuclei with axial deformation is well explained [1, 2] within the nonadiabatic quasiparticle approach [3] with a consistent interpretation of all experimental data available. As a next major development in these studies, the role of triaxial deformation on proton emission from some of these exotic nuclei, where such deformations are expected to occur, is yet to be analyzed carefully. All the previous works [4, 5, 6] with approximate treatments, reveal that triaxiality influences the decay properties. Recently [7] we have developed a formalism based on the nonadiabatic quasiparticle method for proton emitters having triaxial deformation, which allows us to do exact calculations. This description overcomes many shortcomings of previous approaches and could elucidate the role of triaxial deformation. In the present work we discuss our results from the exact calculations for the nuclei the nuclei $^{161}$Re and $^{185}$Bi.

## DETAILS OF CALCULATIONS

We follow a particle plus rotor description of an even-$N$ odd-$Z$ nucleus where the odd proton (particle) moves in the triaxially deformed potential of the daughter nucleus (rotor). Thus the total Hamiltonian of this system is written as the intrinsic Hamiltonian

CP1012, *Frontiers in Nuclear Structure, Astrophysics, and Reactions: FINUSTAR 2*
edited by P. Demetriou, R. Julin, and S. V. Harissopulos
© 2008 American Institute of Physics 0-7354-0532-5/08/$23.00

of the valence proton plus a collective part representing the core, $H = H_{in} + H_{col}$, where $H_{in}$ is the triaxial Nilsson Hamiltonian including a deformed spin-orbit term and the residual pairing interaction. Assuming a potential of the Wood-Saxon type, the nuclear potential is evaluated in a similar way as in [8] but extending the set of deformation parameters $\hat{\beta}$ to $(\beta_2, \beta_4, \gamma)$[7]. With this potential we evaluate the single-particle energies and wave functions corresponding to $H_{in}$. Using the single-particle energies we calculate the quasi-particle energies within the BCS approach employing a constant gap approximation where the gap is obtained from experimental binding energies. This takes care of residual pairing interaction and allow us to treat the Coriolis coupling in a system of quasiparticles. In this case the yrast levels appear naturally as the lowest in energy, as it should be, and the coherence in the phases of the various components entering the calculation of the decay widths provides the correct branching ratios [3]. If we consider the Coriolis mixing between particle states, picking up the resultant yrast state becomes heuristic, since it is not the lowest in energy, and many possible values of the total angular momentum are allowed [6]. The decay width is obtained from the overlap of initial (parent) and final wave function which is a coupling between the daughter wave function and the outgoing proton one. This approach to study the decay properties is similar to the one given in Ref. [9] for nucleon resonances in deformed nuclei. More details of the formalism is given in Ref. [7].

In the above-mentioned way, we have calculated the decay and structural properties of $^{161}$Re and $^{185}$Bi using the "universal" set of parameters for the Woods-Saxon potential.

## RESULTS

Our results for the nuclei $^{161}$Re and $^{185}$Bi, where the triaxial deformations are expected to occur, are shown in Table 1. In the table we have presented the ratio between experimental and theoretical proton emission half-lives for different sets of deformation parameters. To enable comparison with previously reported [6] approximate calculations, we have shown our calculated half-lives with particle states assuming decay from the state that has $\Omega^\pi = \frac{1}{2}^+$ at $\gamma = 0°$ (originating from the spherical level $3s_{1/2}$. Our exact calculations yielded entirely different results which are closer to the experimental values. From our results we observe that the half-lives are not very sensitive to the $\gamma$ deformation and the half-lives increase with $\gamma$, while in Ref. [6] the trend is opposite.

In the extreme right column of Table 1, we have shown our results with Coriolis mixing between quasiparticles and hence taking care of the nonadiabatic quasiparticle effects. These results indicate a weak dependence of half-lives on $\gamma$ deformation, except for $^{161}$Re with $\beta_2 = 0.3$ where many changes in the single-particle spectrum happens [7]. In the case of $^{161}$Re, our results have better conformity with experiment with the deformation $\beta_2 = 0.2$ suggesting a strong $\gamma$ deformation in this nucleus. For the nucleus $^{185}$Bi, the first observation from our results is a more weaker dependance on triaxial deformation. After performing calculations at different $\beta_2$ deformations we found that our results are closer to experiment only for small $\beta_2$ values, where the role of $\gamma$ is expected to be negligible. These observations rule out the role of triaxial deformation in explaining proton emission from $^{185}$Bi. We observed that these results are quite sensitive to the pairing gap chosen.

**TABLE 1.** The ratio $T/T_{exp}$ between the calculated and experimental half-lives in $^{161}$Re and $^{185}$Bi at different deformations. The adiabatic limit assumes decay from the state that has $\Omega^{\pi} = \frac{1}{2}^{+}$ at $\gamma = 0°$ (originating from the spherical level $3s_{1/2}$).

| Nucleus | $\beta_2$ | $\gamma$ (deg) | $T/T_{exp}$ Ref. [6] Adiabatic | $T/T_{exp}$ Present work Adiabatic | $T/T_{exp}$ Present work Nonadiabatic |
|---|---|---|---|---|---|
| $^{161}$Re | 0.1 | 0.0 | 39.7 | 0.64 | 0.565 |
| | 0.1 | 26.3 | 4.6 | 1.55 | 0.616 |
| | 0.2 | 0.0 | 39.0 | 0.62 | 0.589 |
| | 0.2 | 26.3 | 7.0 | 1.89 | 0.919 |
| | 0.3 | 0.0 | 40.2 | 0.73 | 1.149 |
| | 0.3 | 26.3 | 15.4 | 2.59 | 5.297 |
| $^{185}$Bi | 0.156 | 0.0 | 11.4 | 0.12 | 0.122 |
| | 0.156 | 22.1 | 3.8 | 0.13 | 0.116 |

## CONCLUSIONS

Our exact calculations with a complete formalism for studying triaxially deformed proton emitters reveal that the role of triaxial deformation in the cases of proton emission from $^{161}$Re and $^{185}$Bi is weak in comparison to previous claims. We also observe that the residual pairing effects can be significant. In case of $^{161}$Re experimental data is explained with a strong triaxial deformation and we could assign an angular momentum $\frac{1}{2}^{+}$ to the emitting state.

## ACKNOWLEDGMENTS

This work was supported by the Fundação para a Ciência e a Tecnologia (Portugal), Project: PTDC/FIS/68340/2006. One of the authors (PA) thanks the Fundação para a Ciência e a Tecnologia (Portugal) for support through the grant SFRH/BPD/26642/2006.

## REFERENCES

1. E. Maglione, L. S. Ferreira, and R. J. Liotta, Phys. Rev. Lett. **81**, 538 (1998).
2. L. S. Ferreira and E. Maglione, Phys. Rev. Lett. **86**, 1721 (2001).
3. G. Fiorin, E. Maglione, and L. S. Ferreira, Phys. Rev. C **67**, 054302 (2003).
4. A. T. Kruppa and W. Nazarewicz, Phys. Rev. C **69**, 054311 (2004).
5. C. N. Davids and H. Esbensen, Phys. Rev. C **69**, 034314 (2004).
6. D. S. Delion, R. Wyss, D. Karlgren, and R. J. Liotta, Phys. Rev. C **70**, 061301R (2004).
7. P. Arumugam, E. Maglione, and L. S. Ferreira, Phys. Rev. C (In Press).
8. L. S. Ferreira, E. Maglione, and D. E. P. Fernandes, Phys. Rev. C **65**, 024323 (2002).
9. L. S. Ferreira, E. Maglione and R. J. Liotta, , Phys. Rev. Lett., **78**, 1640 (1997).

# Systematics Studies of (n,α) reaction cross sections at 14.5 MeV neutrons energy

M.Belgaid[a], F.Kadem[a] and A.Amokrane[a]

[a] *Faculté de physique USTHB, BP 32 EL-Alia, BEZ Alger Algeria.*

**Abstract.** A new semi-empirical formula for the calculation of the (n,α) cross section at 14.5 MeV neutron energy is obtained. It is based on the pre-equilibrium exciton and evaporation models and uses the Droplet model of Myers and Swiatecki to express the reaction energy Q(n, α). The systematics behavior of the different terms of the Droplet model involved in Q(n,α) was checked individually before choosing the pertinent terms and setting up the formula. Fitting this formula to the existing cross section data on 120 nuclei with $40 \leq A \leq 209$, the adjustable parameters have been determined and the systematics of the (n,α) reaction have been studied. The predictions of this formula are compared with those of the existing formulae and with the experimental data. The formula with five parameters is found to give a better fit to the data than the previous comparable formulae

**Keywords:** 14 MeV neutrons, systematics cross sections, pre-equilibrium exciton model, evaporation model.
**PACS:** 24.60.Dr , 24.60.Gv

## FORMALISM AND MODEL PARAMETERS

On the basis of Weisskopf-Ewing evaporation model the cross sections of the emitted particle at the equilibrium process is given by

$$\sigma_{nb}^{eq} = \sigma_{CN}^{a} \frac{F_b}{\sum F_{b'}}$$

With

$$F_b = \frac{2m_b}{\hbar^2}(2I_b+1) \int_0^{E-S_b} \sigma_{inv}^{b} E_b \rho(U)\, dE_b$$

The energy spectrum of the pre-equilibrium emission is given as

$$\frac{d\sigma_{nb}^{pre}}{dE_b} = \sigma_{CN}^{n} \frac{(2S_b+1) m_b \sigma_{inv}^{b} E_b}{\pi^2 \hbar^3 gE} \sum_{n=3} \frac{\omega(n-1,U)}{\omega(n,E)} \tau_n$$

The expression of the exciton state density used here is given by

$$\omega(n,E) = \frac{g^n E^{n-1}}{p!h!(p+h-1)!}$$

CP1012, *Frontiers in Nuclear Structure, Astrophysics, and Reactions: FINUSTAR 2*
edited by P. Demetriou, R. Julin, and S. V. Harissopulos
© 2008 American Institute of Physics 0-7354-0532-5/08/$23.00

The integration leads to :

$$\sigma_{nb}^{pre} = \sigma_{CN}^{n} \frac{(2S_b+1) m_b \pi R^2 (E_n + Q_{nb} - V_b)^3}{\pi^2 \hbar^3 g E^2} 4\tau$$

The cross sections due to the equilibrium and pre-equilibrium emission can be given now as

$$\sigma_{nb} = \sigma_{CN}^{n} \left( \beta_1 \exp(\frac{Q_{nb} - V_b}{T}) + \beta_2 (E_n + Q_{nb} - V_b)^3 \right)$$

By using the Droplet model of Myers and Swiatecki $Q_{n,\alpha}$ can be expressed as $Q_{n,\alpha} = d_1S_1 + d_2S_2 + d_3S_3 + d_4B_1 + d_5B_2 + d_6B_3 +$ terms with relatively small contributions ignored;
$S_1$=(N-Z+0.5)/A: the usual symmetry term found; $S_2$=(N-Z+0.5)/A4=3: the surface symmetry term; $S_3$ =[(N-Z)4-(N-Z+1)4]/A3: the symmetry anharmonicity term; $B_1$=(Z-1)/A1/3: the usual Coulomb term; $B_2$=(Z-1)/A: the Coulomb diffuseness term; $B_3$=(Z-1)A1/3: the Coulomb volume redistribution term.

## THE SYSTEMATIC PARAMETERS

We checked the behaviour of these terms against the ratio $\sigma_{n\alpha}/\sigma_R$ of 120 nuclei where the experimental values of $\sigma_{n\alpha}$ cross sections with $39 \leq A \leq 209$ were taken from EXFOR data compilations.

Then, with the dependence of $\sigma_{n\alpha}$ on $V_\alpha/T$ and $V_\alpha$ neglected and in order to develop the systematic for the (n,α) reaction cross section at 14.5 MeV the following expressions were obtained by fitting the experimental data :

$$\sigma_{n\alpha} = 9.3(1 + A^{1/3})^2 \exp(-170 S_1^2 + -4.510^{-3} B_3) \ for \ Z \leq 50$$
$$\sigma_{n\alpha} = (1 + A^{1/3})^2 (0.5 - 220 S_3^2)^3 \ for \ Z > 50$$

**TABLE 1.** Comparison of chi-squares value $\chi^2$ obtained from eq.(7) to those of the literature .

| Formula | Zone | Σ | $\chi^2$ |
|---|---|---|---|
| This work | Z ≤ 50 | 253 | 3.78 |
| eq. (7) | Z > 50 | 120 | 2.50 |
| | Total | 373 | 3.20 |
| Konobeyev | Z ≤ 50 | 258.37 | 3.91 |
| [6] | Z > 50 | 134.23 | 2.80 |
| | Total | 392.60 | 3.47 |
| Forrest | Z ≤ 50 | 274.2 | 4.15 |
| [5] | Z > 50 | 166.1 | 3.53 |
| | Total | 440.3 | 3.93 |

The ratio of the $\sigma_{exp}$ to $\sigma_{cal}$ for the different nuclei is shown in Fig. 1. This ratio is rather smooth apart from a structure around mass 100. This mass region does not correspond to the presence of any shell. One possibility of this structure is that the nuclei in this mass region may have some quadrupole deformation. Because of this deformation, the Coulomb barrier should be lower leading to a higher tunnelling probability for α-particles and, hence, to higher values of ratio $\sigma_{exp}$ to $\sigma_{cal}$ for the different nuclei that were considered.

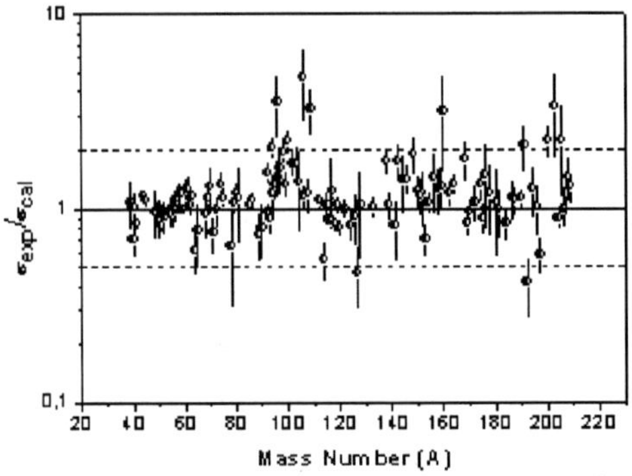


**FIGURE 1.** Ratios of the experimental cross sections to the cross sections calculated with eq.(7) .

# CONCLUSION

This work attempts to derive a new phenomenological formula to systematize (n, α) cross section values. For our case an important improvement in description of the (n,α) cross section has been obtained. This improvement is due to the simultaneous introduction of the pre-equilibrium process and the effective reaction energy deduced from the droplet model of Myers and Swiatecki. The new formula has been tested for 120 nuclei with $40 \leq A \leq 209$.

# REFERENCES


1. W. Hauser and H. Feshbach , Phys. Rev.**87**,366 (1952).
2. K. Kikuchi and M. Kawai, *Nuclear Matter and Nuclear Reactions,* Amsterdam, North-Holland Publishing Company (1968).
3. M. Belgaid, A. Tassadit, F. Kadem and A. Amokrane, Nucl. Instr. Meth **B 239**, 303-313 (2005).
4. M. Belgaid, M. Asghar, Nucl. Instr. Meth **B 142,** 463-472 (1998).
5. R.A.Forrest, AERER **12419**, Harwell, UK (1986).
6. A.Yu. Konobeyev, Yu.A. Korovin, Nucl. Instr. Meth **B 103**, 15 (1995).

# Mirrored Fragmentation Reactions – A New Technique for Probing Isospin Symmetry in Exotic Nuclei

J.R. Brown[a], M.A. Bentley[a], M.J. Taylor[a], P. Aldrich[b], D. Bazin[b], J.M.Cook[b], C.A. Diget[b], A. Gade[b], T. Glasmacher[b], S.M. McDaniel[b], B. Pritychenko[c], A. Ratkiewicz[b], K. Siwek[b] and D. Weisshaar[b]

[a]*University of York, Heslington, York, UK. YO10 5DD*
[b]*Michigan State Univ, Natl Supercond Cyclotron Lab, E Lansing, MI 48824 USA*
[c]*Brookhaven Natl Lab, Natl Nucl Data Ctr, Upton, NY 11973 USA*

**Abstract.** Gamma decays have been observed for the first time in the $T_z$=-$^3/_2$ nucleus $^{53}$Ni. This represents the first gamma-spectroscopy of a $T_z$=-$^3/_2$ nucleus heavier than *A=33*. The nucleus was produced via a two-step fragmentation process, along with its mirror $^{53}$Mn. Differences in excitation energy between isobaric analogue states have been calculated and a preliminary interpretation attempted; shell model calculations are required to further understand these results. This work represents the first study of isobaric analogue states via mirrored fragmentation reactions and demonstrates the power of this new technique.

**Keywords:** $^{53}$Ni, gamma-ray spectroscopy, mirror energy differences, isobaric analogue states, mirrored fragmentation reactions.
**PACS:** 27.40. +z, 21.10.Sf, 23.20.Lv

Recent studies of mirror energy differences (MED) in the $f_{7/2}$ shell have been shown to be very sensitive to many subtle nuclear structure effects [1]. These include radial shape changes [2], single particle effects – such as the electromagnetic spin-orbit interaction - [3], particle alignments [4] and the "*J=2* anomaly" [5]. This last effect refers to recent observations in $T_z$=±$^1/_2$ mirror pairs where particle alignment effects dominate the MED. This realignment of a pair of protons from *J=0* to *J=6* should lead to a continuous reduction in the two-body Coulomb matrix elements (CME) but an increase is observed at *J=2* when these CME are extracted from experimental data on energy differences. This effect has now been observed right across the $f_{7/2}$ shell through studies of the MED of $T_z$=±$^1/_2$, A=47, 49 and 53 [5] mirror pairs and the $^{42}$Ca/$^{42}$Ti pair, although the origin is as yet unclear. To date, studies of MED have been restricted to $T_z$=±$^1/_2$ and $T_z$=±*1* mirror pairs. In fact, the heaviest $T_z$=±$^3/_2$ nucleus on which gamma-ray spectroscopy has been performed is $^{33}$Ar [6]. The natural extension to these studies is to continue out to $T_z$=±$^3/_2$ mirror systems in the $f_{7/2}$ shell. Studies in this region have traditionally been performed using fusion-evaporation reactions, however the cross-sections for such reactions become prohibitively small as increasingly proton rich systems are sought. To overcome this, a test experiment was carried out to produce $T_z$=±$^3/_2$ mirror pairs via mirrored fragmentation reactions.

The experiment was performed at the NSCL facility at MSU. A secondary beam of $^{56}$Ni was fragmented on a Be target to produce the isotopes of interest. These were

CP1012, *Frontiers in Nuclear Structure, Astrophysics, and Reactions: FINUSTAR 2*
edited by P. Demetriou, R. Julin, and S. V. Harissopulos
© 2008 American Institute of Physics 0-7354-0532-5/08/$23.00

identified from their time-of-flight and energy loss characteristics and coincident prompt gamma-rays were detected with the SeGA detector array. Two settings were used on the S800 spectrometer to focus on the $T_z=-^3/_2$ and $T_z=+^3/_2$ isotopes. The $^{53}$Ni and $^{53}$Mn spectra produced from these settings are shown in Fig. 1.

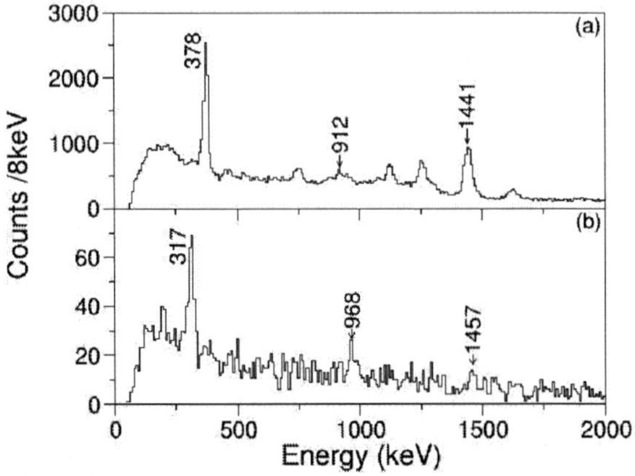


**FIGURE 1.** (a) $^{53}$Mn spectrum produced in this experiment. All peaks can be identified with known transitions in $^{53}$Mn. Peaks of interest are labeled. (b) New $^{53}$Ni spectrum produced in this work. The mirror transitions of the labeled peaks in the $^{53}$Mn spectrum are apparent.

As mirrored reactions were used to produce the isotopes of interest (*3p* and *3n* removal respectively) the identical (i.e. analogue) states are expected to be populated in both mirror nuclei. As a result of this, we can build an initial tentative level scheme based purely on mirror symmetry. The new $^{53}$Ni level scheme is shown in Fig. 2, along with the isobaric analogue states and mirrored transitions in $^{53}$Mn.

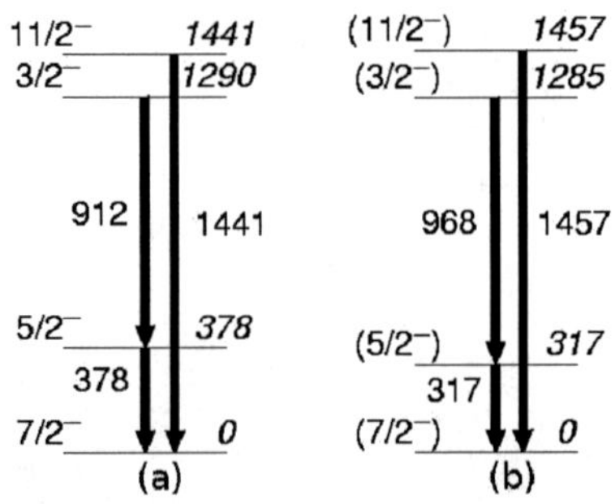


**FIGURE 2.** (a) Relevant levels and transitions in $^{53}$Mn. (b) New $^{53}$Ni level scheme, with spins and parities assigned purely on the basis of mirror symmetry.

As the spins and parities of these levels have been assigned purely on the basis of mirror symmetry, shell model calculations will be performed in order to help confirm these and aid with the interpretation. The MED calculated from these states is plotted in Fig. 3.

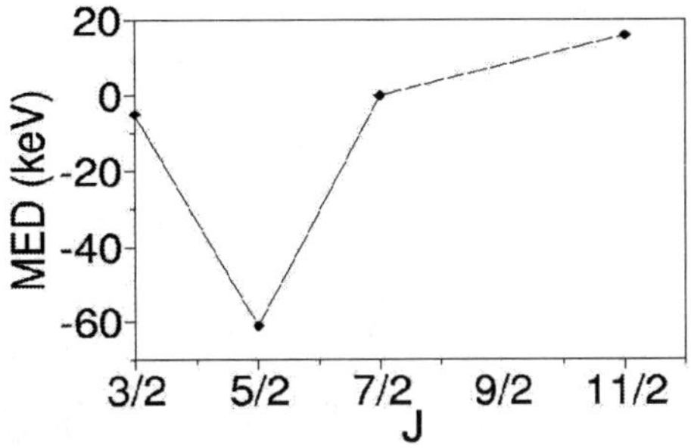


**FIGURE 3.** Mirror energy differences calculated as $MED(J) = E_x(^{53}Ni)_{J,T_z} - E_x(^{53}Mn)_{J,T_z}$.

Interpretation of this MED curve is still at a very preliminary stage; nevertheless some initial observations can be made. We can begin by naively assuming that the $^{53}$Ni (or $^{53}$Mn) ground state is formed from three $f_{7/2}$ neutron (proton) holes in a $^{56}$Ni core. Excited states can be formed by re-coupling two of these holes from $J=0$ to $J=2$, and coupling the remaining neutron (proton) hole in various ways to this. These configurations can produce all of the observed excited states. The re-coupling of two protons from $J=0$ to $J=2$ should lead to a reduction in their Coulomb energy, as their spatial overlap must reduce due to the Pauli Exclusion Principle. No corresponding reduction in Coulomb energy will occur for the neutron re-coupling. One might expect this to result in similar negative energy differences in all three states; alternatively the "$J=2$ anomaly" could come into play causing a reversal of this effect, i.e. three similar positive MED values. Neither of these effects are observed.

An alternative interpretation could involve some configuration mixing from other nearby orbitals and the effect of the electromagnetic spin-orbit (EMSO) interaction. This idea is particularly compelling given the very negative MED value of the $J^{\pi}={}^{5}/_{2}^{-}$ state. If this state were produced by the excitation of a single neutron (or proton, in the case of $^{53}$Mn) to the $f_{5/2}$ orbital - with all other particles coupled to spin zero - this would lead to an MED of ~250 MeV as a result of the EMSO interaction. Therefore, configuration mixing of ~20% would result in the MED observed. This interpretation is further supported by the well known weak shell closure at $Z$ (or $N$) =28, however explanations of the $J^{\pi}={}^{3}/_{2}^{-}$ and ${}^{11}/_{2}^{-}$ states are harder to find. Given these considerations, it is clear that shell model calculations are required to further interpret these new and interesting results.

In summary, gamma decays from excited states in the $T_z=-{}^{3}/_{2}$ nucleus $^{53}$Ni have been observed for the first time, and compared with isobaric analogue states in $^{53}$Mn. This work represents the first use of mirrored fragmentation reactions for the study of isobaric analogue states and the power of this new technique has been demonstrated.

## REFERENCES

1. M.A. Bentley and S.M. Lenzi, Progr. Part. Nucl. Phys. **59**, 497 (2007).
2. S.M. Lenzi et. al., Phys. Rev. Lett. **87.** 122501 (2001).
3. J. Ekman et. al., Phys. Rev. Lett. **92,** 132502 (2004).
4. C.D. O'Leary et. al., Phys. Rev. Lett. **79.** 4349 (1997).
5. S.J. Williams et al, Phys. Rev. C **68,** 011301(R) (2003).
6. R.R.C. Clement et al., Phys. Rev. Lett. **92,** 172502 (2004).

# The isovector imaginary neutron potential: a key ingredient for the r-process nucleosynthesis

S. Goriely* and J.-P. Delaroche†

*Institut d'Astronomie et d'Astrophysique, Université Libre de Bruxelles, Campus de la Plaine, CP 226, 1050 Brussels, Belgium
†DPTA/Service de Physique Nucléaire, CEA/DAM Ile de France, BP 12, 91680 Bruyères-le-Châtel, France

**Abstract.** The isovector contribution to the imaginary component of the microscopic optical model potential is adjusted on experimental neutron strength function data. Within the Brückner–Hartree–Fock approximation of Jeukenne-Lejeune-Mahaux, it is shown that experimental data favours a strong isovector component that can have a drastic impact on the radiative neutron capture cross section for neutron-rich nuclei. If confirmed, this result strongly inhibits the resonant capture by exotic nuclei, so that the traditional r-process picture of the fast neutron captures during the nucleosynthesis r-process needs to be revisited in depth.

**Keywords:** stellar nucleosynthesis – nuclear reactions – optical potential
**PACS:** 26.30.+k,24.10.Ht, 24.60.Dr, 25.40.Lw

The radiative neutron captures by exotic nuclei are known to be of fundamental importance in the rapid neutron-capture process (or r-process) invoked to explain the origin of approximately one half of the nuclides heavier than iron observed in nature (see [1] for a review). The neutron capture rates are commonly evaluated within the framework of the statistical model of Hauser-Feshbach, in which the the nucleon-nucleus optical model potential (OMP) of [2] is used. This so-called JLM potential derived from a Reid's hard core nucleon–nucleon interaction by applying the Brückner–Hartree–Fock approximation to nuclear matter has recently been updated by [3, 4] who empirically renormalized the energy dependence of the potential depth to reproduce scattering and reaction observables for spherical and quasi-spherical nuclei between $^{40}$Ca and $^{209}$Bi in a large energy range from the keV region up to 200 MeV. In the JLMB approach, the renormalization factors are rather well constrained by experimental data, except the $\lambda_{W_1}$ factor affecting the isovector imaginary component in the low-energy regime. The major constraint imposed on the OMP isovector component comes from the quasi-elastic $(p,n)$ scattering data as well as the angle-integrated quasi-elastic $(p,n)$ cross sections to the isobaric analog states at energies above some 20 MeV. For lower energies, the $\lambda_{W_1}$ factor was extrapolated from the confident region around 20 MeV to a constant value of approximately 1.5. Due to the lack of scattering data in the keV region, the low-energy extrapolation of the $\lambda_{W_1}$ factor remains essentially unconstrained. Since the low-energy part of the factor $\lambda_{W_1}$ strongly affects the absorption characteristics by exotic neutron rich nuclei, the corresponding neutron capture at astrophysically relevant energies (i.e. typically about 100 keV for incident neutrons) remains poorly described.

For this reason, we have searched for additional experimental information that could possibly constrain the isovector component of the imaginary potential, i.e. the $\lambda_{W_1}$ factor,

CP1012, *Frontiers in Nuclear Structure, Astrophysics, and Reactions: FINUSTAR 2*
edited by P. Demetriou, R. Julin, and S. V. Harissopulos
© 2008 American Institute of Physics 0-7354-0532-5/08/$23.00

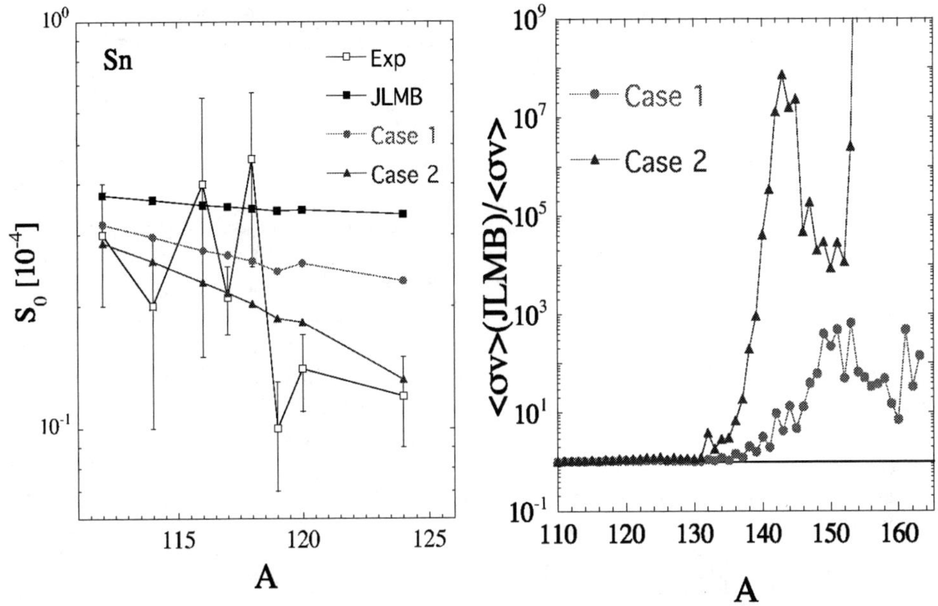


**FIGURE 1.** Left panel: Comparison between the experimental S-wave strength function ($S_0$) for spherical and quasi-spherical Sn isotopes [5] with the values obtained with the JLMB OMP and the modified imaginary potentials corresponding to Cases 1 and 2 (see text). Right panel: Ratio of the radiative neutron capture rates of Sn isotopes at a temperature of $T = 10^9$ K obtained with the JLMB potential to the one obtained with the modified imaginary potentials corresponding to Cases 1 and 2.

at low energies. In this respect, the S-wave neutron strength function ($S_0$) experimentally determined at energies ranging between 1 and 100 keV provide an extremely valuable set of constraints, which is particularly sensitive to the isospin dependence of the OMP. Interestingly, experimental $S_0$ data is available for the long spherical or quasi-spherical isotopic chains of Sn isotopes as given in Fig. 1 which reveals that the isospin trend is not properly described by the JLMB renormalization. This low-energy data is particularly sensitive to the adopted value of the $\lambda_{W_1}$ factor. To study this sensitivity, we consider here two cases corresponding to two modified renormalizations ot the JLM imaginary potential. Case 1 corresponds to a modified value of $\lambda_{W_1}$ at energies below 1 MeV which is larger by 30% with respect to the JLMB value and Case 2 by about 50%. The impact of these modified OMPs on the S-wave neutron strength function of the Sn isotopes is shown in Fig. 1. Case 1 already presents a relatively small slope, in contrast to the JLMB case. Cases 2 predicts a strong slope in agreement with experimental data.

On the basis of these OMPs, the Maxwellian-averaged reaction rates of astrophysics interest have been estimated within the statistical model of Hauser-Feshbach. As observed in Fig. 1 (right panel) for the Sn isotopic chain, the experimental constraints introduced in Case2 has a drastic impact on the reaction rate. At large neutron excesses, the enhanced $\lambda_{W_1}$ factor strongly reduces the imaginary component, i.e. the neutron absorption channel, and consequently the radiative neutron capture cross section. In particular, it can be seen that, for Sn isotopes with $N \gtrsim 135$, the rates obtained in Case 2 rapidly drop, leading to a totally insignificant resonant neutron capture. Case 1 gives a less drastic decrease of the rate, though a factor of 100 to 1000 with respect to JLMB is found. According to Fig. 1, Case 2 better reproduces the slope seen in the experimental $S_0$ values with increasing neutron excess and could therefore be expected to provide a more reliable input for the calculation of reaction rates. Similar results are obtained for other isotopic chains, as shown in [6].

The enhancement of the isovector imaginary potential has a drastic impact on the resonant capture by exotic neutron-rich nuclei. It should be kept in mind that when dealing with nucleosynthesis applications, only rates estimated within the Hauser-Feshbach approach have ever been considered. The newly-derived OMP constrained on S-wave neutron strength functions leads to radically smaller rates by neutron-rich nuclei, so that the traditional picture of a possible fast r-process neutron capture up to the exotic neutron-rich region needs to be revisited. Our new calculation of radiative neutron capture rates, if confirmed, could have far-reaching astrophysics implications that can be summarized as follows, *(i)* the r-process nuclear flow can only follow a relatively narrow path characterized by short $\beta$ half-lives and large neutron capture rates; *(ii)* the $(n,\gamma)-(\gamma,n)$ equilibrium is in general not reached in the $S_n \lesssim 3.5$ MeV region of the nuclear chart; *(iii)* the $A \simeq 120$ and $A \simeq 180$ underabundances obtained under given conditions within the canonical r-process model can be resolved since no equilibrium is achieved; *(iv)* the exotic r-process paths close to the neutron drip line cannot be reached (except at extremely large neutron densities such as in decompressing neutron star matter [1]); *(v)* the so-called "universality" of the r-process abundances (as inferred from spectroscopic observation of ultra-metal-poor stars) could possibly be explained by the rapid drop of the neutron capture rates at increasing neutron excesses (which constrains the r-process flow to remain in the narrow region of the nuclear chart); *(vi)* the r-abundance distribution depends critically on the estimated $\beta$-decay and $(n,\gamma)$ rates (the direct neutron capture by exotic nuclei might give the dominant contribution and needs to be evaluated carefully); and finally *(vii)* all r-process abundance calculations (including cosmochronometry) performed so far on the basis of Hauser-Feshbach reaction rates need to be revisited.

However, before drawing any firm conclusion, the present calculations need to be further constrained by additional theoretical and experimental works. More accurate calculations of the direct capture are also required to estimate to what extent neutron-rich nuclei can potentially capture neutrons under r-process conditions. For astrophysics applications, it is consequently of crucial importance to devote more theoretical as well as experimental effort to better understand the neutron capture by exotic nuclei.

## REFERENCES


1. M. Arnould, S. Goriely, K. Takahashi, Phys. Rep. **450**, 97 (2007).
2. J.P. Jeukenne, A. Lejeune, C. Mahaux, Phys. Rev. **C 16**, 80 (1977).
3. E. Bauge, J.P. Delaroche, M. Girod, Phys. Rev. **C 58**, 1118 (1998).
4. E. Bauge, J.P. Delaroche, M. Girod, Phys. Rev. **C 63**, 024607 (2001).
5. *Handbook for calculations of nuclear reaction data, RIPL-2*, IAEA-TECDOC-1506 (2006); www-nds.iaea.org/RIPL-2/.
6. S. Goriely, J.P. Delaroche, Phys. Lett. B (2007), in press.

# Latest results of HFB mass formulas

S. Goriely* and J.M. Pearson†

*IAA, ULB, Campus de la Plaine, CP 226, 1050 Brussels, Belgium
†Département de Physique, Université de Montréal, Montréal (Québec), H3C 3J7 Canada

**Abstract.** The latest developments made in deriving accurate mass predictions within the microscopic Skyrme Hartree-Fock-Bogolyubov (HFB) approach are described. These include an improved description of infinite nuclear and neutron matter, a reduced pairing strength and the inclusion of a phenomenological vibrational term. A special attention in the determination of the latest effective forces is paid not only to the nuclear ground state and nuclear matter properties, but also to nuclear level densities and fission barriers.

The interplay between the Coulomb and the strong interaction is also analyzed, the associated Coulomb correlations being simulated by suppressing the Coulomb exchange energy. With our latest Skyrme force, BSk14, being refitted in this way, the Coulomb correlation effect is found to reduce the rms deviation with respect to experimental masses and to affect the nuclear mass predictions close to the neutron drip lines significantly.

**Keywords:** stellar nucleosynthesis – atomic masses – fission barriers – nuclear level density
**PACS:** 26.30.+k,24.60.Dr,21.10.Dr, 21.10.Ma

In the last few years it has become possible to construct complete mass tables by the Hartree-Fock (HF) method (see [1] and references therein), with the parameters of the underlying forces being fitted to essentially all of the available mass data. These HF calculations are based on conventional Skyrme forces. A $\delta$-function pairing force acting between like nucleons is also included in the full Bogoliubov framework (HFB), as well as a phenomenological Wigner term. The models HFB-1 to HFB-7 are characterized by different treatments of the pairing force and of the effective nucleon mass, while in the case of HFB-8 and HFB-9 we project with the PLN (projected Lipkin-Nogami) method states of good particle number. However, despite the differences between them, models HFB-2 to HFB-8 all give comparably good fits to the mass data: the rms deviation with respect to the 2149 measured masses of nuclei with $N$ and $Z \geq 8$ given in the 2003 Atomic Mass Evaluation [2] is in every case less than 0.7 MeV, the best being HFB-8, for which the rms deviation is 0.635 MeV.

Since the publication of HFB-8, we have generated several new mass models, but the direction of our work has shifted. Rather than seek ever better fits to the mass data our concern has been more with the construction of a *universal* force for all the various astrophysical applications, and to this end we have been imposing on our mass models extra physical constraints: i) The energy-density curve of neutron matter was fitted, a requirement that is relevant not only to neutron-star applications, but also to the reliability of finite-nucleus extrapolations out towards the neutron drip line. The properties of neutron matter cannot, of course, be measured in the laboratory, but they can be calculated with a high degree of reliability from realistic two- and three-nucleon forces; we took as the basis for our mass fit the zero-temperature results of Friedman and Pandharipande (FP) [3]. For all mass models from HFB-9 to HFB-14, an excellent fit to

CP1012, *Frontiers in Nuclear Structure, Astrophysics, and Reactions: FINUSTAR 2*
edited by P. Demetriou, R. Julin, and S. V. Harissopulos
© 2008 American Institute of Physics 0-7354-0532-5/08/$23.00

the FP curve was achieved simply by imposing a nuclear-matter symmetry coefficient of $J = 30$ MeV.

ii) The pairing force for the HFB-10 to HFB-14 models was considerably weakened with respect to that of all our previous mass models, for which the pairing was excessively strong, in the sense that despite their good mass fits the calculated spectral pairing gaps were much bigger than the experimental even-odd mass differences. The pairing strength plays a decisive role for a reliable predictions of nuclear level densities in particular [4]. The strong-pairing BSk9 and weak-pairing BSk13 effective interactions (corresponding to HFB-9 and HFB-13 mass model, respectively) have been used to generate all nuclear structure properties needed to estimate the nuclear level density within the combinatorial approach [4]. The predicted s-wave neutron spacings at the neutron separation energy $S_n$ are compared in Fig. 1 with experimental data [5]. There is a clear disagreement when using the BSk9 nuclear input, deviations up to a factor of 53 being found. In particular, in the actinide region, the nuclear level densities are largely underestimated due to the significant pairing effect at $S_n$. A much more satisfactory result is obtained with the low-pairing BSk13 effective force. This comparison reflects the well-known remarkable sensitivity of nuclear level densities to the pairing strength at low energies and the importance of considering this quantity when fitting the effective force.

iii) A more general form of collective (rotational plus vibrational) correction was introduced and fitted to experimental energies of shape isomers and fission barriers. All forces prior to HFB-14 (which only consider a rotational correction based on the cranking model) turn out to poorly predict the $Z < 92$ primary barriers. Fitting a model to nuclear masses does not explore all the regions of deformation space that are relevant to barriers. However, it is possible, through adjustment of a phenomenological vibrational term that we have included for the first time in our HFB-14 model [1] to obtain drastically improved barriers without any deterioration of the mass fit. HFB-14 barriers are compared with experimental data in Fig. 1. In particular, the rms deviation for the 52 primary barriers of nuclei with $88 \leq Z \leq 96$, which are always less than 9 MeV high, is as low as 0.67 MeV. A similar accuracy is obtained (0.65 MeV) for the secondary barriers, good values for which are necessary for a reliable calculation of fission probabilities.

Despite the imposition of these conditions, the quality of the mass fit was only slightly worse with HFB-9 to HFB-14 than with HFB-8, the rms deviation rising to acceptable values in the range 0.720 to 0.755 MeV for the same data set.

Finally, the interplay between the Coulomb and strong interactions was shown to lead to an enhancement of the Coulomb energy in the nuclear surface [6] that could solve the Nolen-Schiffer anomaly, i.e the systematic reduction in the estimated binding energy differences between mirror nuclei with respect to experiment. This Coulomb correlation effect could in fact significantly affect the nuclear mass predictions close to the neutron drip line. To analyse its impact, the BSk14 Skyrme force (corresponding to HFB-14 mass model) was refitted excluding the contribution from the Coulomb exchange energy, since, as shown by [6] in an excellent approximation the Coulomb correlation energy cancels the Coulomb exchange energy. The final force leads to an rms error on all the 2149 experimental masses [2] of 0.678 MeV (to be compared with 0.729 MeV for HFB-14). The resulting HFB-14cc masses can differ by more than 10 MeV from the HFB-14 masses close to the neutron drip line, as shown in Fig. 2. This effect is actually relatively

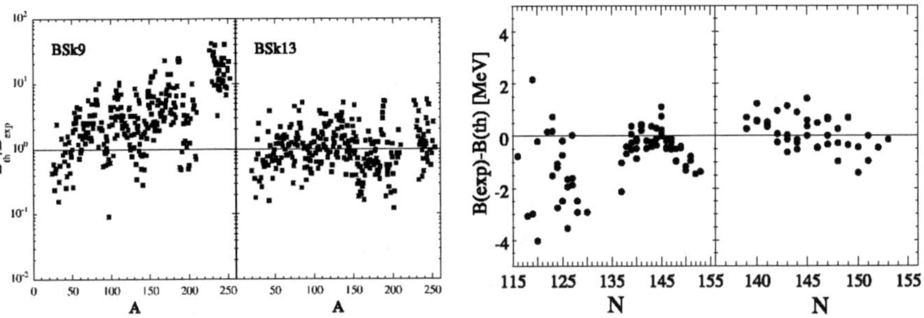


**FIGURE 1.** Left panel: Ratio of theoretical ($D_{th}$) HFB plus combinatorial model based on the BSk9 (left) or BSk13 (right) Skyrme force to experimental ($D_{exp}$) s-wave neutron resonance spacings for the 295 nuclei compiled in [5]. Right panel: Deviations between model HFB-14 and experiment for primary barriers (left) and secondary barriers (right).

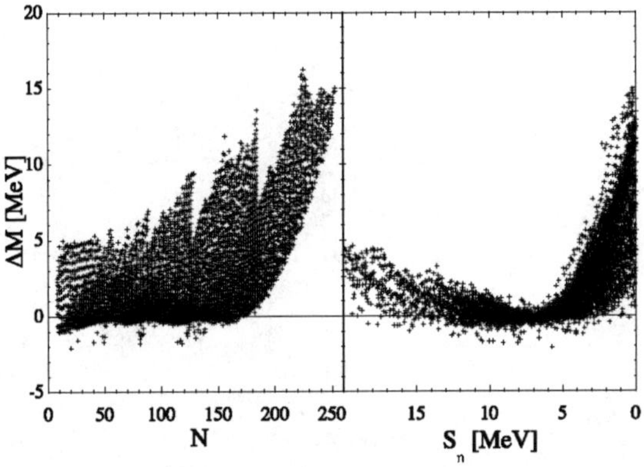


**FIGURE 2.** Deviations between HFB-14 and HFB14cc masses as a function of the neutron number (left panel) and neutron separation energy $S_n$ (right panel).

large and will need to be further scrutinized.

# REFERENCES


1. S. Goriely, M. Samyn and J. M. Pearson, Phys. Rev. **C 75**, 064312 (2007).
2. G. Audi, A.H. Wapstra and C. Thibault, Nucl. Phys. **A 729**, 337 (2003).
3. B. Friedman and V. R. Pandharipande, Nucl. Phys. **A 361**, 502 (1981).
4. S. Hilaire and S. Goriely, Nucl. Phys. **A 779**, 63 (2006).
5. *Handbook for calculations of nuclear reaction data, RIPL-2*, IAEA-TECDOC-1506 (2006); www-nds.iaea.org/RIPL-2/.
6. A. Bulgac A. and V.R. Shaginyan, Phys. Lett. **B 469**, 1 (1999).

# Drip-line To Drip-line Microscopic Nuclear Level Densities

S. Hilaire* and S. Goriely†

* *CEA/DAM Ile-de-France, Service de Physique Nucléaire,*
*BP 12, F-91680 Bruyères-le-Châtel, FRANCE*
† *Institut d'Astronomie et d'Astrophysique, Université libre de Bruxelles,*
*Campus de la plaine CP 226,B-1050, Brussels, BELGIUM*

**Abstract.** New energy-, spin- and parity-dependent nuclear level densities based on the microscopic combinatorial model are proposed for practical applications. The combinatorial model includes a detailed microscopic calculation of the intrinsic state density and of the rotational enhancement factor, but a phenomenological treatment of the vibrational effect. The vibrational description has been improved using a boson partition function to construct phonons' state densities which are then folded in with the incoherent particle-hole densities. The present model is shown to predict the experimental s-wave neutron resonance spacings with a degree of accuracy comparable to that of the best global models available.

**Kewords:** Level densities, Nuclear energy levels, Hartree-Fock.
**PACS:** 21.10.Ma, 21.10-k, 21.60.Jz

The knowledge of nuclear level densities (NLDs) has been a matter of interest and study for years going back at least to 1936 with Bethe's pioneering work[1]. Since then, more or less sophisticated methods have been developed to reproduce the available experimental data [2,3]. For applications such as astrophysics or accelerator driven systems, which require a large number of data far away from the experimentally known regions, extrapolations as reliable and accurate as possible have to be done. If analytical expressions for NLDs are used, the parameters on which they depend have to be extrapolated without having any way to check the quality of these extrapolations. Therefore, one should preferentially use microscopic or semi-microscopic global predictions, provided they compete with more phenomenological highly-parameterized formulae in the reproduction of experimental data. The microscopic character indeed ensures the required reliability, while the searched accuracy is guaranteed by the ability to reproduce experimental data as well as the best global phenomenological approaches.

Many global microscopic approaches have been developed in the past but they are almost never used for practical applications, either because of their lack of accuracy or because they do not offer the same flexibility as the analytical expressions which can be easily adjusted. Therefore, we have performed large scale calculation of NLDs for nearly 8500 nuclei in the framework of the combinatorial method described in [3] which has proven its predictive power [2,4]. One of the main advantages of our approach is to provide not only the NLDs tabulated as a function of the excitation energy, but also the spin and parity distributions without any statistical assumption.

CP1012, *Frontiers in Nuclear Structure, Astrophysics, and Reactions: FINUSTAR 2*
edited by P. Demetriou, R. Julin, and S. V. Harissopulos
© 2008 American Institute of Physics 0-7354-0532-5/08/$23.00

As precisely described in ref.[2], our method consists in using the single-particle level schemes obtained from constrained axially symmetric Hartree-Fock-Bogoliubov method to construct incoherent particle-hole (ph) state densities as functions of the excitation energy, the spin projection (on the intrinsic symmetry axis of the nucleus) and the parity. Once these incoherent ph state densities are determined, we then account for collective effects, as well as for their washing out with increasing excitation energy[1], to deduce total level densities. In ref.[2], we made the choice of multiplying the level densities by the phenomenological enhancement factor described in refs.[2,5] after accounting for rotational motion if necessary (i.e. for deformed nuclei). The resulting nuclear level densities were found to reproduce very well the available experimental data (i.e. both the cumulated low energy discrete level histograms and the s-wave resonances mean spacings $D_0$ at the neutron binding energy). However, we knew that the phenomenological treatment of vibrational effects had to be replaced by a sounder treatment, and feedback from fission cross sections calculations [6] also suggested that issue.

Consequently, we have modified the vibrational description using a boson partition function as in ref.[3] to construct phonons' state densities which are then folded in with the incoherent ph densities. To account for the damping of vibrational effects, we restrict the number of coupled phonons up to a given number $N_{max}$ and, on top of that, only fold in the multiphonon's state densities obtained by coupling $N_{vib}$ phonons, (where $N_{vib} \leq N_{max}$) with the ph state densities which have a total number $N_{ph}$ of particles plus holes such that $N_{ph} + N_{vib} = N_{max}$. The rotational enhancement factor is then taken into account by constructing rotational bands on purely vibrational or ph band-head states as well as on band-head states resulting from the coupling of ph and phonons' states. In practice, it means that supplementary low energy phonon states which were not considered in ref.[2] are now included.

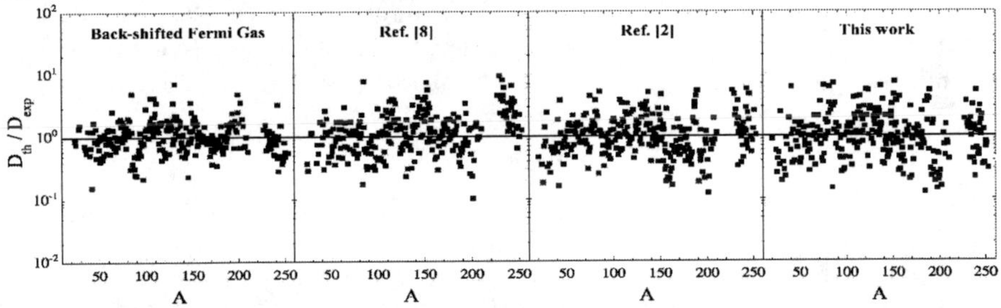


**FIGURE 1.** Ratio of theoretical ($D_{th}$) and experimental ($D_{exp}$) s-wave neutron resonance spacings for the 295 nuclei compiled in ref. [9] as functions of the mass number.

As can be observed on Fig.1, our new level densities reproduce the $D_0$ values as well as those of ref.[2], with the advantage of dealing more correctly with the vibrational effects. Globally, the predictions remain within a factor of 2 with no deviations larger than a factor of 8 being found. In the same figure, are also shown the predictions obtained with an accurate global Back Shifted Fermi Gas-type formula [7] as well as those of ref. [8] which clearly shows that the present approach is comparable to the predictions obtained with other global approaches.

[1] When the excitation energy increases, deformed nuclei tend to become spherical. This shape transition is described as in ref.[2,7], by introducing a phenomenological damping function between the spherical and deformed expressions of the level densities.

In Fig.2, the present vibrational enhancement factor ($K_{vib}$) is compared to the analytical temperature-dependent expression of ref.[2]. The microscopic character of our new treatment results in clearly oscillating structures compared to the smooth behaviour of the temperature-dependent expression. The damping of $K_{vib}$ is clearly observed and occurs for much lower energies than in ref.[2]. Finally, for nuclei having many low energy vibrational levels (as $^{238}$U for instance), our new treatment departs from unity at much lower energy than the analytical expression.

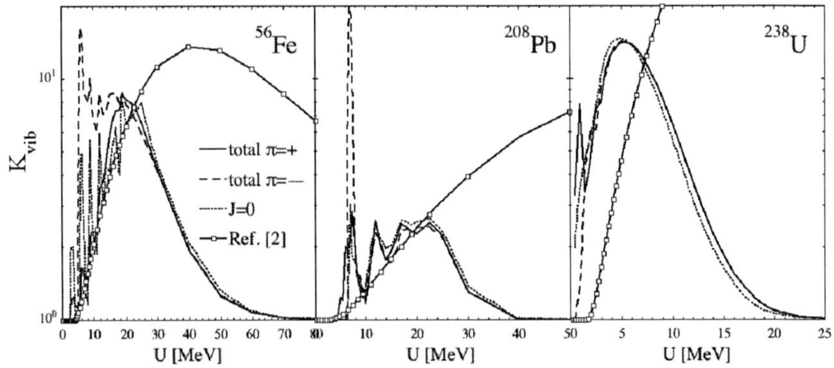


**FIGURE 2.** Vibrational enhancement factor calculated using our new treatment for positive and negative parities as well as for the spin J=0 compared to that of ref.[2].

In summary, tabulated level densities have been determined for nearly 8500 nuclei as function of energy, spin and parity and can be downloaded at the web address http://www-astro.ulb.ac.be/Html/nld_comb.htm (making use of the phenomenological vibrational correction). They have the same level of accuracy as the usually employed global analytical expressions with the advantage of being (i) more reliable thanks to their microscopic grounds and (ii) spin- and parity-dependent.

## REFERENCES


1. H.A. Bethe, Phys. Rev. **50**, 332 (1936).
2. S. Hilaire and S. Goriely, Nucl. Phys. **A 779**, 63 (2006).
3. S. Hilaire, J.P. Delaroche and M. Girod., Eur. Phys. J. **A 12**, 169 (2001) and references therein.
1. S. Goriely in *Proceedings of the International Conference on Frontiers in Nuclear Structure, Astrophysics and Reactions*, edited by S. Harissopulos, P. Demetriou and R. Julin, American Institute of Physics Conference Proceedings 831, NY, 2006, p. 8.
5. A.V. Ignatyuk, IAEA report, TECDOC-1034 (1998).
6. M.Sin et al. to be published in *Proceedings of the International Conference on Nuclear Data for Science and Technology*, EDP-NICE 2007.
7. S. Goriely, J. Nucl. Sci. Technol. Suppl. **2**, 536 (2002).
8. P. Demetriou and S. Goriely, Nucl. Phys. **A 695**, 95 (2001).
9. *Hanbook for calculations of nuclear reaction data, Reference Input Parameter Library-2,* IAEA TECDOC-1506 (2006).

# The AMEDEE Nuclear Structure Database

S. Hilaire and M. Girod

*CEA/DAM Ile-de-France, Service de Physique Nucléaire,*
*BP 12, F-91680 Bruyères-le-Châtel, FRANCE*

**Abstract.** The increasing need for nuclear data far from the valley of stability requires information on nuclei which cannot be accessed experimentally or for which almost no experimental data is known. Consequently, the use of microscopic approaches to predict properties of such poorly known nuclei is necessary as a first step to improve our understanding of nuclear reaction on exotic nuclei. Within this context, large scale axial mean field calculations from proton to neutron drip-lines have been performed using the Hartree-Fock-Bogoliubov method based on the D1S Gogny nucleon-nucleon effective interaction. Nearly 7000 nuclei have been studied under the axial symmetry hypothesis and several properties are now available for the nuclear scientific community on an Internet web site for every individual nucleus.

**Keywords:** Binding energies and masses, Nuclear structure models and methods, Gogny Force.
**PACS:** 21.10.Dr, 21.60-n, 21.60.Jz, 21.30-x

The microscopic Hartree-Fock-Bogoliubov (HFB) method using the Gogny effective nucleon-nucleon interaction D1S [1,2] has shown, for nearly 30 years, its predictive power in various aspects of nuclear structure [3-13]. Thanks to the high computer power available nowadays, constrained axially symmetric HFB+D1S calculations have been performed for nuclei from Carbon (Z=6) up to Darmstadtium (Z=110) for even-even, odd mass and odd-odd nuclei. In total, nearly 7000 nuclei have been considered from proton to neutron drip-lines, and a large amount of information has been put online at the web address *http://www-phynu.cea.fr/HFB-Gogny_eng.htm*.

For each nucleus, we have used the constrained axially symmetric HFB+D1S method to determine potential energy surfaces (PES) as function of the axial deformation parameter $\beta$ defined by

$$\beta = \frac{1}{AR^2}\sqrt{\frac{5\pi}{9}}\, q_{20} \,,$$

where $R=1.2A^{1/3}\,fm$ is the nucleus radius and $q_{20}$ is the quadrupole moment (expressed in $fm^2$). Our deformation parameter $\beta$ ranges between -0.6 and 1.3 in order to locate mass regions where nuclei might exhibit super-deformed or/and hyper-deformed states. For odd mass and odd-odd nuclei, the blocking technique without breaking time reversal symmetry has been used in order to determine also the spin of the nucleus ground state. When looking at the PES of each nucleus, we can locate the minimum of the PES which is the closest to sphericity ($\beta$ =0). In most of the cases, this minimum is also the lowest point of the PES, and thus gives us the nucleus ground state energy $E_{HFB}(Z,N)$. Only for very few nuclei is the lowest point of the PES not the closest to sphericity. However such nuclei are very neutron rich and probably completely unstable with respect to the fission process.

CP1012, *Frontiers in Nuclear Structure, Astrophysics, and Reactions: FINUSTAR 2*
edited by P. Demetriou, R. Julin, and S. V. Harissopulos
© 2008 American Institute of Physics 0-7354-0532-5/08/$23.00

The nuclei considered in the database are those for which $E_{HFB}(Z,N) < E_{HFB}(Z-2,N)$ for the proton drip-line and $E_{HFB}(Z,N) < E_{HFB}(Z,N-2)$ for the neutron drip-line. Of course, we have tried other criteria to defined the drip-lines such as looking if the chemical potential remains negative or if the $Z^{th}$ or $N^{th}$ level is bound or not, but the differences are usually not significant and are certainly not worth a deep discussion in the present paper. It is worth mentioning that our drip-lines definition does not account for the fact that nuclei may fission instantaneously because of a too weak fission barrier.

On the web site, a large amount of information is displayed graphically or can be downloaded in a tabulated format for each nucleus as function of $\beta$. It is also possible to display a given isotopic or isotonic chain for any of the results available for a single nucleus.

Global qualitative as well as quantitative properties can be extracted from our large scale approach, such as the ground state deformation, the position of the proton and neutron drip lines, the regions that may exhibit super or hyper-deformation or the deviation of the predicted binding energies from experimental data. These properties are discussed in details in ref.[14] and we will just compare here, on fig.(1) theoretical and experimental masses, by plotting the differences between them as function of the neutron number N.

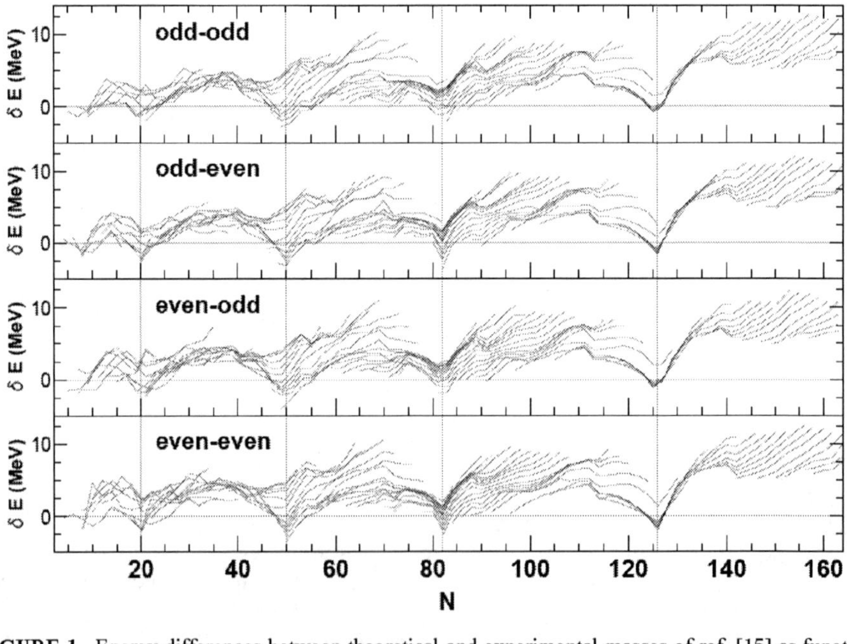


**FIGURE 1.** Energy differences between theoretical and experimental masses of ref. [15] as functions of the neutron number. In each panel, the lines connect an isotopic chain.

As can be observed, the global behaviour of $\delta E$ does not depend on the odd-odd, odd or even-even nature of the nuclei, even if the calculations are more rigorous for even-even nuclei. Moreover, arches are clearly appearing with minima located around magic neutron numbers where nuclei are also overbound. We however know that correlations beyond mean field will somehow wash out these arches. Finally, and more annoying is the increase of $\delta E$ with increasing neutron number N. This tendency, particularly clear for the heaviest nuclei, results from the fact that the neutron matter energy dependence of D1S has not been precisely investigated when the force was adjusted. This problem has been solved [16] by re-adjusting the Gogny force parameters and extensive calculations are presently underway.

To summarize, a database containing structure predictions for almost 7000 nuclei, based on the axially symmetric hypothesis, is available online for the scientific community. The content of the database is going to increase to include results of interest for nuclear reaction calculations such as the nuclear radial matter densities required to construct semi-microscopic optical potentials [17], nuclei beyond Z=110 of particular interest for the study of super-heavy elements, or results obtained from beyond mean field predictions such as low energy collective levels which can be rather well described when compared with experimental data [13].

## REFERENCES


1. J. Decharge and D. Gogny, Phys. Rev. **C 21**, 1568 (1980).
2. J.F. Berger, M. Girod and D. Gogny, Comp. Phys. Comm. **63**, 365 (1991).
3. J.P. Delaroche et al., Nucl. Phys. **A 771**, 103 (2006).
4. M. Kleban et al., Phys. Rev. **C 65**, 024309 (2002).
5. S. Hilaire, J.P. Delaroche and M. Girod, Eur. Phys. J. **A 12**, 169 (2001).
6. S. Peru, M. Girod and J.F. Berger, Eur. Phys. J. **A 9**, 35 (2000).
7. E. Bauge, J.P. Delaroche and M. Girod, Nucl. Phys. **A 654**, 829c (1999).
8. J. Libert, M. Girod and J.P. Delaroche, Phys. Rev. **C 60**, 054301 (1999).
9. X.H. Phan et al, *Phys. Rev.* **C 38**, 1173 (1988); Erratum Phys.Rev. **C 39**, 1645 (1989).
10. J.F. Berger, M. Girod and D. Gogny, Nucl. Phys. **A 428**, 23c (1984).
11. M. Girod et al, Phys. Rev. **C 38**, 1519 (1988).
12. M. Girod et al, Phys. Rev. Lett. **62**, 2452 (1989).
13. G.F. Bertsch et al., Phys. Rev. Lett. **99**, 032502 (2007).
14. S. Hilaire and M. Girod, Eur. Phys. J. **A 33**, 237 (2007).
15. A.H. Wapstra, G. Audi and C. Thibault, Nucl. Phys. **A 729**, 129 (2003).
16. F. Chappert and M. Girod, Int. J. Mod. Phys. E **15**, 339 (2006).
17. E. Bauge, J.P. Delaroche and M. Girod, Phys. Rev. C **63**, 024607 (2001).

# Quasi-elastic Barrier Distribution of the $^{18}O + ^{64}Zn$ System

J. F. P. Huiza[1], E. Crema[1], D. S. Monteiro[1], J. M. B. Shorto[1], R. F. Simões[1], P. R. S. Gomes[2]

[1] *Departamento de Física Nuclear, Instituto de Física da Universidade de São Paulo, SP, Brazil.*
[2] *Instituto de Física, Universidade Federal Fluminense, RJ, Brazil.*

**Abstract.** We present the recent experimental data of the quasielastic excitation function that was measured at $\theta_{lab}$=161° for the $^{18}O$+ $^{64}Zn$ system and its correspondent quasielastic barrier distributions has been derived. These data were taken at energies near and below of the Coulomb barrier. We compared the data with coupled channel calculations, and the results show a strong coupling effect of the quadrupole and octupole states of the $^{64}Zn$ and the state $2_1^+$ of the $^{18}O$.

**Keywords:** quasielastic reaction. Theoretical Coupled Channel Calculations.
**PACS:** 25.70.Bc, 24.10.Eq.

In a collision of two heavy nuclei their internal structure can be affected leading to intrinsic excitation, transfer of nucleons or particles between the nuclei. A powerful tool used to explain these reaction mechanisms is the quasielastic barrier distribution (QEBD) that is extracted from the experimental quasielastic excitation function (QEF) [1-4]. The QEBD allows observing channel coupling effects on the reaction mechanism at energies around the Coulomb barrier.

In this work we present the first precise measurement of the QEF of the $^{18}O$+ $^{64}Zn$ system at energies near and below of the Coulomb barrier, from Elab=30.0 MeV to 45.0 MeV, using steps of 1 MeV in the laboratory system. The $^{18}O$ beam was provided by the São Paulo Pelletron Accelerator facility. The experiment was realized inside a scattering chamber which contained a self-supporting target of 40 μg/cm$^2$ of $^{64}Zn$ isotope, and for distinguish the scattered nuclei at backward angles was placed at $\theta_{LAB}$ = 161° a gas proportional counter, E -ΔE, where ΔE was supplied by the energy loss in the gas. Behind the gas detector, a silicon surface barrier detector determines the residual energy; it was also utilized other three silicon detectors placed at forward angles (at ±30° and +45°) for normalization purposes. The outcome after an electronics-informatics and statistical treatment allows to obtain the QEF and QEBD, this last with 1% error.

We compared the experimental data of QEF and QEBD with coupled channel calculations (CCC) using CCQUEL code [5]. We present the first results of these comparisons which included only inelastic states in the coupling matrix. For incorporate the real nuclear potential in the optical potential of the CCQUEL input, we performed a Woods-Saxon fit to the parameter-free double folding São Paulo Potential (SPP) around the barrier radius (±2 fm) [3, 6], getting the following parameter for the real nuclear potential: $V_0$=133.0 MeV, $r_0$ = 1.06 fm and $a_0$=0.69 fm. A very short-

CP1012, *Frontiers in Nuclear Structure, Astrophysics, and Reactions: FINUSTAR 2*
edited by P. Demetriou, R. Julin, and S. V. Harissopulos
© 2008 American Institute of Physics 0-7354-0532-5/08/$23.00

range imaginary potential was used in order to avoid interference of its absorption with inelastic channel couplings. The Woods-Saxon parameters used were: $W_i$=50.0 MeV, $r_i$ = 0.6 fm and $a_i$=0.4 fm.

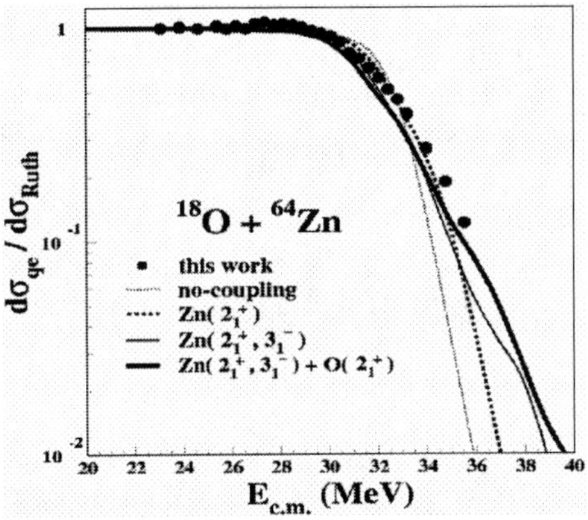


FIGURE 1. Experimental quasielastic excitation function and the theoretical predictions obtained by coupling inelastic states of the target and projectile.

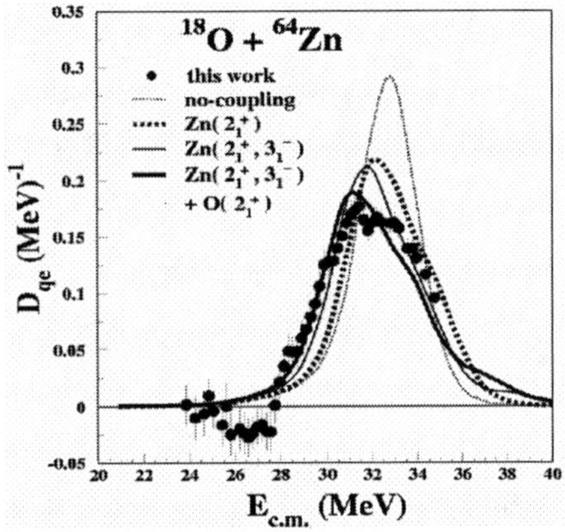


FIGURE 2. Experimental quasielastic barrier distribution, and theoretical predictions obtained by coupling inelastic states of the target and projectile.

In CCC were tested several excited states of the target and projectile [7-9]. Among all excited states, the more relevant are the two vibrational states $2_1^+$ and $3_1^-$ of $^{64}$Zn, and the $2_1^+$ vibrational state of $^{18}$O. The following nuclear parameters were employed: a) for the collective states of the target, the $2_1^+$ (E*=0.992 MeV and deformation parameter $\beta_2$=0.311(14), using a reduced radius $r_0$=1.06 fm), and the $3_1^-$ excited state (E*=2.998 MeV and deformation parameter $\beta_3$=0.333(28) ); b) for the projectile state, the $2_1^+$ excited state (E*=1.982 MeV and deformation parameter $\beta_2$=0.455(10)). The Fig. 1 shows the contributions of the inelastic states in the QEF. The no-coupling calculation is represented by the dotted line, which does not describe the experimental data. The coupling of the $2_1^+$ quadrupole state of the target (the thick dashed line in Fig. 1), diminishes the cross section in the region around of Coulomb barrier (32 MeV), and increases the cross section at higher energies. However, a better agreement of the calculation with data at this energy region is obtained when we include the $3_1^-$ vibrational state of the target, as shown by the thin solid line. This effect is also observed when the $2_1^+$ quadrupole state of the projectile is included in the coupling matrix, as shown by the thick solid line in Fig. 1. The Figure 2 shows that the effect of each channel in the reaction mechanism, at energies near and below the Coulomb barrier, is easier observed in the QEBD. The dotted line in Fig. 2 shows that the simple barrier produced by the bare potential is far from the experimental barrier distribution. The coupling of the $2_1^+$ quadrupole state of the target produces a lowering of the barrier of about 25%, and an energy shift of 0.7 MeV compared to the no-coupling calculation. On the other hand, the inclusion of the $3_1^-$ octupole state of the target has less lowering effect of the distribution maximum (3%), but a similar energy shift of the barrier height (0.5 MeV). Finally, the $2_1^+$ quadrupole state of the projectile has a strong influence on the barrier distribution, lowering the maximum by 12% and reducing the barrier height by approximately 0.4 MeV, compared to the result of the coupling of the quadrupole and octupole states of the target. In conclusion, a good agreement of the CCC with QEBD was obtained, mainly below the Coulomb barrier, for which the excitation of the quadrupole state of the projectile played a very important role. However, at energies above the barrier the agreement is not yet perfect. In order to improve the theoretical description of the reaction dynamic of this system, we are now using the code FRESCO [10] and running a more complete CCC by including several other inelastic channels and the transfer channels.

## CONCLUSIONS

In this work we present the first experimental measurement of a precise quasielastic excitation function and its correspondent quasielastic barrier distribution for the $^{18}$O + $^{64}$Zn system. We compared these experimental data with Coupled Channel Calculation, including the following inelastic channels: quadrupole and octupole states of $^{64}$Zn and the quadrupole of $^{18}$O. With only these three channels a good description of the data was obtained. In this theoretical calculation was used the parameter-free double folding São Paulo Potential as the real nuclear potential between the projectile and the target. It was showed that effects of internal structures are clearly seen in the quasielastic barrier distribution.

## ACKNOWLEDGMENTS

This work was partially supported by the Fundação de Amparo à Pesquisa do Estado de São Paulo (FAPESP), and the Conselho Nacional de Desenvolvimento Cientifico e Tecnológico (CNPQ) and the Fundação de Amparo à Pesquisa do Estado do Rio de Janeiro (FAPERJ).

## REFERENCES

1. R.F. Simões, D.S. Monteiro, J.M.B. Shorto, E. Crema, Phys Lett. **B 527**, 187 (2002).
2. H. Timmers et al., Nucl. Phys. **A 584**, 190 (1995) ; Nucl. Phys. **A 633**, 421 (1998).
3. J. F. P. Huiza, E. Crema, D.S. Monteiro, J.M.B. Shorto, R.F. Simões, N. Added and P.R.S. Gomes, Phys. Rev. **C 75**, 064601/1-7 (2007).
4. D.S. Monteiro, R.F. Simões, J.M.B. Shorto, E. Crema, Nucl. Phys. **A 725**, 60 (2003).
5. K. Hagino and N. Rowley (unpublished).
6. L. C. Chamon et al., Phys. Rev. **C 66**, 014610/1-13 (2002).
7. Balraj Sing, Nucl. Data Sheets **78**, 395 (1996).
8. S. Raman et al., At. Data Nucl. Data Tables **78**, 1 (2001)
9. T. Kibéti and R. H. Spear, At. Data Nucl. Data Tables **80**, 35 (2002).
10. I. J. Thompson, Comp. Phys. Commun. **7**, 167 (1988).
11. C. R. Morton et al., Phys. Rev. **C 60** 044608/1-11 (1996).
12. J. R. Leigh et al., Phys. Rev. **C 52**, 3151 (1995) .
13. M. Dasgupta , D.J. Hinde, J.R. Leigt and K. Hagino, Nucl. Phys. **A 630**, 78c (1998).

# Systematic analysis of astrophysical S-factors and thermonuclear reaction rates

M. Katsuma

*Institut d'Astronomie et d'Astrophysique, Université Libre de Bruxelles, Campus de la Plaine, 1050 Brussels, Belgium*

**Abstract.** The astrophysical S-factors of the $^{13}C(\alpha,n)^{16}O$, $^{17}O(\alpha,n)^{20}Ne$, $^{21}Ne(\alpha,n)^{24}Mg$ and $^{25}Mg(\alpha,n)^{28}Si$ reactions are analyzed with DWBA. The gross structures of the experimental data are reproduced by the DWBA calculations. The resulting reaction rates are compared with those in the CF88 and NACRE compilations.

**Keywords:** Astrophysical S-factors
**PACS:** 24.50.+g; 29.87.+g

## INTRODUCTION

After the pioneering work (CF88) by W.A. Fowler and his collaborators [1], a few compilations of thermonuclear reaction rates have been constructed in view of the experimental efforts. The NACRE (Nuclear Astrophysics Compilation of REaction rates) [2] of charged-particle induced reactions has been an important step forward in the compilation effort, and it is being updated and extended into a new version of NACRE [3].

In evaluating cross sections, theoretical nuclear reaction models are required to cope with the experimental difficulties at low energies, including the electron screening effects. In order to make the reliable extrapolation of the data, we have been investigating global features of nuclear reactions at astrophysical energies, which are expected to be seen with the direct reaction models, i.e. the potential model and distorted-waves Born approximation (DWBA). If cross sections are assumed to be decomposed into its energy-average and a fluctuating part, the cross sections obtained from these two models correspond to the former, which may give us some clues in predicting the Maxwellian-averaged reaction rates.

We have demonstrated the applicability of our potential model and DWBA to some low-energy nuclear reactions relevant to nuclear astrophysics [3]. We describe here our systematic analysis of the thermonuclear reactions with DWBA [4] by a way of showing the results on four $(\alpha,n)$ reactions.

## S-FACTORS AND REACTION RATES FROM DWBA

It is convenient to deal with the astrophysical S-factor, which is defined by $S(E_{\rm c.m.}) = E_{\rm c.m.}\exp(2\pi\eta)\,\sigma(E_{\rm c.m.})$, where $\eta$ is the Sommerfeld parameter and $E_{\rm c.m.}$ is the center-of-mass energy of the initial channel, since cross sections $\sigma(E_{\rm c.m.})$ become very small at astrophysical energies much lower than the Coulomb barrier.

CP1012, *Frontiers in Nuclear Structure, Astrophysics, and Reactions: FINUSTAR 2*
edited by P. Demetriou, R. Julin, and S. V. Harissopulos
© 2008 American Institute of Physics 0-7354-0532-5/08/$23.00

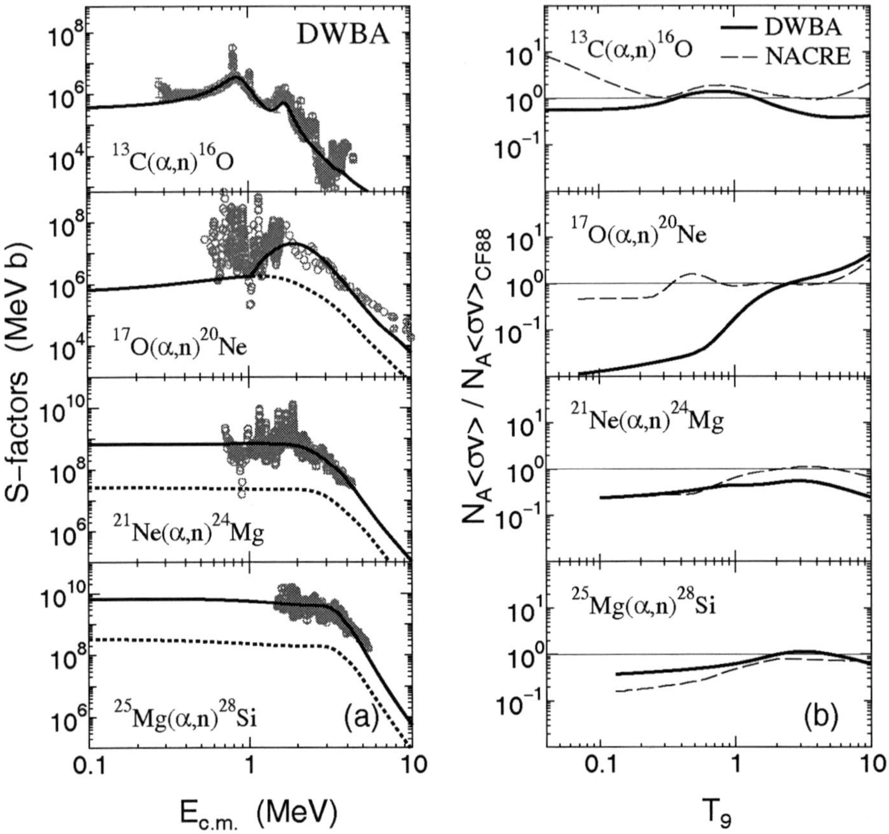


**FIGURE 1.** (a) The astrophysical S-factors of the $^{13}C(\alpha,n)^{16}O$, $^{17}O(\alpha,n)^{20}Ne$, $^{21}Ne(\alpha,n)^{24}Mg$ and $^{25}Mg(\alpha,n)^{28}Si$ reactions. The solid curves are the results obtained from DWBA. The dotted curves are the component of the ground state in the residual nuclei. The experimental data are taken from the NACRE compilation (references therein [2]). (b) The resulting reaction rates are compared with those in the CF88 compilation (the solid curves). The differences between NACRE and CF88 are shown by the dashed curves.

The astrophysical S-factors of the $^{13}C(\alpha,n)^{16}O$, $^{17}O(\alpha,n)^{20}Ne$, $^{21}Ne(\alpha,n)^{24}Mg$ and $^{25}Mg(\alpha,n)^{28}Si$ reactions are shown in Fig. 1 (a). The gross structures of the experimental data are reproduced with the DWBA calculations, shown by the solid curves. The first excited states of $^{20}Ne$, $^{24}Mg$ and $^{28}Si$ are taken into account as the final state of the residual nuclei in the respective reactions, i.e. the solid curves represent the sum of the S-factors for the $A(\alpha,n)B(\text{g.s.})$ and $A(\alpha,n)B(2^+)$ reactions. The dotted curves represent the components of the ground-state residual nuclei.

From Fig. 1(a), the astrophysical S-factors of the $^{21}Ne(\alpha,n)^{24}Mg$ and $^{25}Mg(\alpha,n)^{28}Si$ reactions are found to be dominated by the component of the $2^+$ state. It should be noted that the s-wave neutrons are emitted at low energies when the residual nuclei are excited to the $2^+$ state.

In the $^{17}O(\alpha,n)^{20}Ne$ reaction, the residual nuclei remain in the ground state at low energies since the $^{17}O(\alpha,n)^{20}Ne(2^+)$ channel opens at $E_{c.m.} = 1.05$ MeV. In other words, the astrophysical S-factors are expected to be hindered below this energy.

The gross structure in the $^{13}C(\alpha,n)^{16}O$ reaction including the resonant features around $E_{c.m.} \approx 0.8$ MeV and 1.7 MeV is found to be reproduced with DWBA. The $J^\pi$ value of the resonance around $E_{c.m.} \approx 0.8$ MeV is, however, different from $J^\pi = 3/2^+$ assigned from elastic neutron scattering on $^{16}O$. If the s-wave resonance of $\alpha$+$^{13}C$ near the $\alpha$-particle threshold in $^{17}O$ is assumed, $J^\pi = 1/2^-$ would result. The data reproduction of our calculation assuming the s-wave resonance of $\alpha$+$^{13}C$ may suggest that the $J^\pi$ value of the resonance could be $J^\pi = 1/2^-$. No contribution from the transition to the excited state of $^{16}O$ can be seen below $E_{c.m.} \leq 3.83$ MeV since the excitation energy of the first excited state in $^{16}O$ is 6.05 MeV and the Q-value of the $^{13}C(\alpha,n)^{16}O$ reaction is 2.22 MeV. The contribution from the $^{13}C(\alpha,n)^{16}O^*$ reactions is negligible at the considered energies of astrophysical interest. Note that the $\alpha$+$^{13}C^*$ channels are also closed at low energies.

The derived thermonuclear reaction rates are compared with those in the CF88 compilation in Fig. 1(b). Our reaction rates are obtained from the model calculations shown by the solid curves in Fig. 1(a). For comparison, the differences between NACRE and CF88 are shown by the dashed curves. The reaction rates of $^{13}C(\alpha,n)^{16}O$, $^{21}Ne(\alpha,n)^{24}Mg$ and $^{25}Mg(\alpha,n)^{28}Si$ obtained from DWBA are different from CF88 by a factor of $1/5 \sim 2$. The large deviation of the reaction rates in the $^{17}O(\alpha,n)^{20}Ne$ reaction originates from the difference in the astrophysical S-factors below $E_{c.m.} < 1$ MeV.

An extensive study of the astrophysical S-factors is underway, and will be discussed somewhere else.

## ACKNOWLEDGEMENTS

This work has been supported in part by the Interuniversity Attraction Pole IAP 5/07 of the Belgian Federal Science Policy and by the Konan University – Université Libre de Bruxelles convention "Construction of an Extended Nuclear Database for Astrophysics". The author acknowledges the contributions from M. Arnould, K. Takahashi and H. Utsunomiya.

## REFERENCES

1. G. R. Caughlan, W. A. Fowler, Atom. Data and Nucl. Data Tables **40**, 283 (1988).
2. C. Angulo, M. Arnould, et al., Nucl. Phys. **A 656**, 3 (1999).
3. M. Katsuma, Proc. Tours 2006, AIP Conf. Proc. **891**, 355 (2007); M. Arnould, M. Katsuma, Proc. Int. Conf. Nucl. Data for Science and Technology 2007 (ND2007), Nice, France, in press.
4. G. R. Satchler, *Direct Nuclear Reactions*, International series of monographs on physics, Oxford University Press, 1983.

# Spectroscopy of odd-proton nuclei in the region of $^{254}$No


S.Ketelhut*, P.T.Greenlees*, A.Chatillon†, E.Bouchez†, P.A.Butler**, E.Clément†, O.Dorvaux‡, S.Eeckhaudt*, B.J.P.Gall‡, A.Görgen†, T.Grahn*,**, R.-D.Herzberg**, F.Heßberger§, A.Hürstel†, G.D.Jones**, P.Jones*, R.Julin*, S.Juutinen*, H.Kettunen*, F.Khalfallah‡, W.Korten†, Y.Le Coz†, M.Leino*, A.-P.Leppänen*, S.Moon**, P.Nieminen*, M.Nyman*, J.Pakarinen*,**, J.Perkowski*, P.Rahkila*, M.Rousseau‡, J.Sarén*, C.Scholey*, Ch.Theisen†, J.Uusitalo*, J.N.Wilson†, P.Bonche¶ and P.-H.Heenen‖

**Department of Physics, University of Jyväskylä, P.O.Box 35, Jyväskylä, Finland*
*†CEA-SACLAY, DSM/DAPNIA/SPhN, F-91191 Gif-sur-Yvette Cedex, France*
***Oliver Lodge Laboratory, University of Liverpool, L69 7ZE, United Kingdom*
*‡Institut Pluridisciplinaire Hubert Curien, F-67037 Strasbourg, France*
*§GSI KP2/SHIP, D-64291 Darmstadt, Germany*
*¶CEA-SACLAY, DSM/DAPNIA/SPhT, F-91191 Gif-sur-Yvette Cedex, France*
*‖Service de Physique Nucléaire Théorique, Université Libre de Bruxelles, B-1050 Bruxelles, Belgium*



**Abstract.** Two rotational bands have been found in the transfermium nuclei $^{251}$Md and $^{255}$Lr, the latter being the heaviest nucleus so far studied in-beam. Both are assigned to a $[521]1/2^-$ Nilsson state by comparison to theory. The experiments have been carried out in the Accelerator Laboratory of the University of Jyväskylä (JYFL), where the array of germanium detectors JUROGAM was used in conjunction with the recoil-seperator RITU and the focal-plane setup GREAT for gamma-spectroscopic studies.



**PACS:** 21.10.Re, 23.20.Lv, 25.70.Jj, 27.90.+b


## INTRODUCTION

The coupling of efficient recoil-separators with arrays of germanium detectors has triggered intensive gamma-ray spectroscopic studies of the transfermium region, starting with the first experiments on $^{254}$No at Argonne in 1999 [1]. Nuclei in this region provide experimental information which can be used to test models at the extremes of neutron and proton number. Of special interest are predictions of the next doubly-magic nuclei beyond $^{208}$Pb, which vary depending on the theoretical approach taken [2]. At first, mainly even-even nuclei in the $^{254}$No region were investigated, leading to the discovery of rotational bands in $^{252,254}$No and $^{250}$Fm [1, 3, 4]. Lately, K-isomeric states have been observed in $^{254}$No and $^{250}$Fm [5, 6]. Only scarce experimental information is available on the odd-mass nuclei, as the decay intensity can be fragmented over several bands. The only other odd-mass transfermium nucleus which has been studied using prompt gamma-ray spectroscopy is $^{253}$No [7].


CP1012, *Frontiers in Nuclear Structure, Astrophysics, and Reactions: FINUSTAR 2*
edited by P. Demetriou, R. Julin, and S. V. Harissopulos


© 2008 American Institute of Physics 0-7354-0532-5/08/$23.00

**FIGURE 1.** Experimental ground and first excited states of $^{255}$Lr and $^{251}$Md [10].

## EXPERIMENTAL DETAILS AND ANALYSIS

Both the $^{251}$Md and $^{255}$Lr experiments were carried out at the Accelerator Laboratory of the University of Jyväskylä. The prompt gamma rays were detected by JUROGAM, an array of 43 Compton-suppressed germanium detectors. The gas-filled recoil-separator RITU [8] was used to separate fission fragments and beam-like particles from the fusion-evaporation residues. The residues were guided to the focal-plane setup GREAT [9], where they were implated and subsequent decays measured. A rotational sequence of $\gamma$ transitions at energies of 195, 245, 292, 336, 377, 415, and 450 keV has been found in the analysis of $^{251}$Md [11]. Earlier studies of the $\alpha$-decay properties of the two nuclei determined the 7/2$^-$[514] and 1/2$^-$[521] states to be the ground and first excited states of $^{251}$Md with the latter having an excitation energy of 55 keV (see Fig. 1). As only one signature partner is seen in this band, it can thus be associated to the 1/2$^-$[521] state with a decoupling parameter close to one (see [11] for more details). The recoil-tagged gamma-ray spectrum of $^{255}$Lr (Fig. 2(a)) reveals at least one rotational band, marked by dashed lines, and confirmed by the sum-of-gates spectrum (Fig. 2(c)). It is unambigiously assigned to $^{255}$Lr by a further gate on its $\alpha$ decay (Fig. 2(b)). Theory predicts the three Nilsson configurations 1/2$^-$[521], 7/2$^-$[514] and 7/2$^+$[633] to form the lowest-lying states in $^{255}$Lr [12], as partly confirmed by the experimental level scheme (Fig. 1). As the energies of the $^{255}$Lr band are very similar to the ones of $^{251}$Md, and, again, no signature partner is seen, the same 1/2$^-$[521] band-head is assigned. There is evidence for two more bands in the spectrum, indicated by arrows. Calculations show that a 7/2$^+$ band would decay mainly via M1 transitions, which are highly converted in very heavy nuclei, while a 7/2$^-$ band would decay via the two E2 signature partners of the band [11]. Thus, the two bands are suggested to be based on the 7/2$^-$[514] configuration.

The kinematic moments of inertia of the 1/2$^-$[521] bands are compared to those of even-even nuclei of this region in Fig. 2. The $^{251}$Md moment of inertia is larger than the ones of $^{250}$Fm, $^{252}$No and $^{254}$No at low frequencies, as is typical for odd-mass nuclei, and has the same upbend as the N=150 even-even nuclei at higher frequencies. The $^{255}$Lr moment of inertia behaves similarly, though more data is desirable to extend the band.

## CONCLUSIONS AND OUTLOOK

Two new bands have been found for $^{251}$Md and $^{255}$Lr, for which the Nilsson level of the band could be assigned. There are indications for more bands in the spectra, for

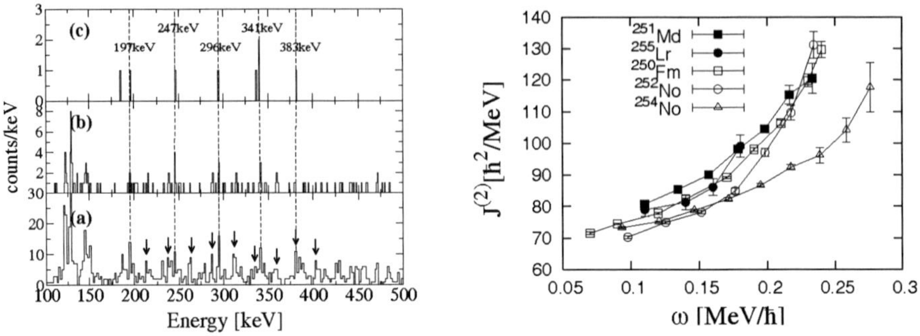


**FIGURE 2.** Left: (a) Recoil-gated spectrum, (b) alpha-tagged spectrum, and (c) sum-of-gates spectrum of $^{255}$Lr with gates on the transitions marked by dashed lines. Right: Moments of inertia for the observed bands compared to bands in even-even nuclei of the same region.

which improved data will be obtained with a new set-up at JYFL. Higher yields at target position can be reached with the new JUROGAM2 array combined with digital electronics, while the combination of an electron-spectrometer with the JUROGAM array in SAGE will enable the investigation of strongly-converted transitions in the rotational bands.

## ACKNOWLEDGMENTS

This work has been supported by the EU 6th Framework programme "Integrating Infrastructure Initiative - Transnational Access", Contract Number: 506065 (EURONS), the Academy of Finland under the Finnish Centre of Excellence Programme 2006-2011 (Nuclear and Accelerator Based Physics Programme at JYFL), EXOTAG (contract HPRI-1999-CT-50017), the European Community Marie Curie Fellowship, and the UK Engineering and Physical Research Council. We also thank the UK/France (EPSRC/IN2P3) Loan Pool and EUROBALL Owners Comittee for the EUROGAM detectors of JUROGAM.

## REFERENCES

1. P. Reiter *et al.*, *Phys. Rev. Lett.* **82**, 509 (1999).
2. M. Bender *et al.*, *Phys. Rev. C* **60**, 034304 (1999).
3. R.-D. Herzberg *et al.*, *Phys. Rev. C* **65**, 014303 (2001).
4. J.E. Bastin *et al.*, *Phys. Rev. C* **73**, 024308 (2006).
5. R.-D. Herzberg *et al.*, *Nature* **442**, 896 (2006).
6. P.T. Greenlees, *to be published*
7. P. Reiter *et al.*, *Phys. Rev. Lett.* **95**, 032501 (2005).
8. M. Leino *et al.*, *Nucl. Instr. and Meth. B* **99**, 653 (1995).
9. R.D. Page *et al.*, *Nucl. Instr. and Meth. B* **204**, 634 (2003).
10. A. Chatillon *et al.*, *Eur. Phys. J. A* **30**, 397 (2006).
11. A. Chatillon *et al.*, *Phys. Rev. Lett* **98**, 132503 (2007).
12. M. Bender *et al.*, *Nuc. Phys.* **A723**, 354 (2003).

# Giant Dipole Radiation And Evidence Of Isospin Mixing Increase With Atomic Number In Hot Nuclei

M. Kicińska-Habior[1], E. Wójcik-Gront[1], M. Kowalczyk[1,2], M. Kisieliński[2,3]

*1 Institute of Experimental Physics, Warsaw University, Hoża 69, 00-681 Warsaw, Poland*
*2 Heavy Ion Laboratory, Warsaw University, Pasteura 5A, 02-093 Warsaw, Poland*
*3 The Andrzej Sołtan Institute for Nuclear Studies (IPJ), 05-400 Świerk/Otwock, Poland*

**Abstract.** We present evidence that the isospin mixing studied in a consistent way in highly excited $^{32}S$, $^{36}Ar$, $^{44}Ti$, and $^{60}Zn$ nuclei at similar excitation energies, increases with increasing atomic number of self-conjugate nuclei. The observed increase is similar to that predicted for ground-state nuclei.

**Keywords:** Isospin symmetry, Isospin mixing, Giant Dipole Resonance,
**PACS:** 21.10.Hw, 24.30.Cz, 24.60.Dr, 25.70.Gh

During recent years, isospin symmetry, its breaking and restoration, have been studied with increasing interest, especially in connection with developing radioactive beam facilities. Several theoretical works discuss the dependence of the isospin mixing on the atomic number for nuclei in the ground states [1-4]. Estimates in a spherical hydrodynamical model have suggested an increase of the mixing in the $N = Z$ nuclei as $Z^{8/3}$ [1], when the recent calculations performed within the deformed Hartree-Fock method showed that the isospin mixing probability is proportional to about $Z^2$ [3]. To our knowledge, no such calculations have been done for highly excited nuclei in which the restoration of isospin symmetry [5-8] may be important. Until now, isospin mixing probability has been measured only for few highly excited nuclei [5, 6, 9] and in independent experiments in which systematic uncertainties have been estimated to be 50% or more. Therefore, it would be difficult to compare absolute magnitudes of the isospin mixing probabilities obtained in these studies.

In this work, we present evidence that isospin mixing studied in a consistent way in highly excited $^{32}S$, $^{36}Ar$, $^{44}Ti$, and $^{60}Zn$ nuclei, at similar excitation energies, increases with the increasing atomic number of the self-conjugate nuclei. Our result is based on inclusive measurements of the statistical decay of the Giant Dipole Resonance (GDR) built on excited states populated in heavy-ion fusion reactions.

We used experimental method [5, 9] which is based on the fact that the E1 decays between states with isospin $T = 0$ are isospin forbidden due to the isovector character of the electric dipole radiation. In this method, self-conjugate compound nuclei are formed in fusion reactions by the entrance channel with the isospin $T = 0$, and for comparison, the same nuclei or, when it is not possible, neighboring compound nuclei at similar excitation energy by the entrance channel with $T \neq 0$. In the statistical decay

CP1012, *Frontiers in Nuclear Structure, Astrophysics, and Reactions: FINUSTAR 2*
edited by P. Demetriou, R. Julin, and S. V. Harissopulos
© 2008 American Institute of Physics 0-7354-0532-5/08/$23.00

of self-conjugate nuclei populated by entrance channel with T = 0, the yield of high-energy $\gamma$-rays heavily depends on the isospin mixing probability. In case of no mixing, the yield arises from allowed GDR $\gamma$-decays populating T = 1 final states in the N = Z compound nucleus, and $\gamma$-decays in daughter nuclei formed by particle emission. This yield is suppressed by a factor of 2-3 compared to the yield from the $T \neq 0$ entrance channels, and in nuclei with $N \neq Z$. However, when isospin mixing occurs, the T = 1 states in N = Z compound nuclei formed by the T = 0 channel are also populated and decay to T = 0 states. Thus, in case of none zero mixing, the $\gamma$-ray yield is larger and depends on the amount of isospin mixing.

We formed nuclei $^{32}$S, $^{36}$Ar, $^{44}$Ti, and $^{60}$Zn with Z increasing from 16 to 30, at similar excitation energies around 50 MeV by the entrance channels with T = 0. Neighboring compound nuclei $^{31}$P, $^{37}$Ar, $^{45}$Ti, and $^{61}$Zn, at similar excitation energy, but with $T \neq 0$ were also populated. To study the reactions $^{20}$Ne+$^{12}$C$\rightarrow$$^{32}$S, $^{12}$C+$^{24}$Mg$\rightarrow$$^{36}$Ar, $^{20}$Ne+$^{24}$Mg$\rightarrow$$^{44}$Ti and $^{36}$Ar+$^{24}$Mg$\rightarrow$$^{60}$Zn, as well as $^{19}$F+$^{12}$C$\rightarrow$$^{31}$P, $^{12}$C+$^{25}$Mg$\rightarrow$$^{37}$Ar, $^{20}$Ne+$^{25}$Mg$\rightarrow$$^{45}$Ti and $^{36}$Ar+$^{25}$Mg$\rightarrow$$^{61}$Zn, we performed experiments using beams from the Warsaw Cyclotron and self-supporting isotopic enriched targets. Gamma rays from the decay of the studied compound nuclei were detected with the multi-detector JANOSIK set-up. The $\gamma$-ray yield was measured by the spectrometer consisting of 25.4 cm $\times$ 29 cm NaI(Tl) crystal surrounded by an active plastic scintillator shield, and by passive $^{6}$LiH and Pb shields. The measured high-energy $\gamma$-ray spectra from the decay of four self-conjugate nuclei $^{32}$S, $^{36}$Ar, $^{44}$Ti, and $^{60}$Zn when compared with the spectra from the appropriate $N \neq Z$ partner, clearly showed that the GDR transitions in N = Z nuclei are suppressed.

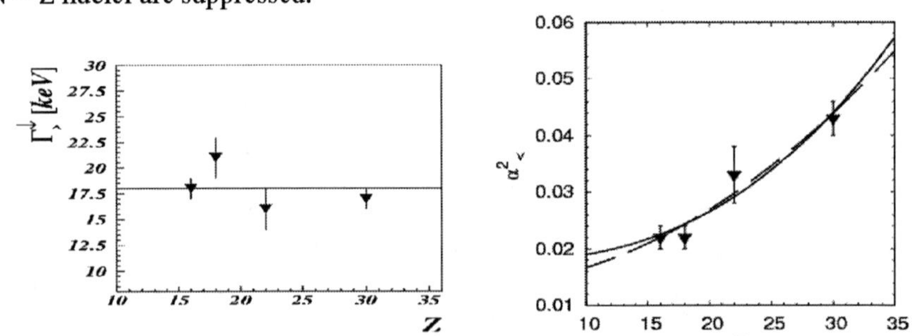


**FIGURE 1.** Extracted isospin-violating spreading width $\Gamma^{\downarrow}_{>}$ (left) and isospin mixing probability $\alpha^2_<$ (right) in highly excited self-conjugate nuclei at $E_x \approx 50$ MeV as a function of atomic number. The curves (in the right box) show fitted functions increasing as $Z^{8/3}$ (dashes) and as $Z^2$ (dots).

We performed a consistent analysis to compare the $\gamma$-ray cross sections measured for the reactions studied with the spectra calculated within the statistical model. We used a modified version of the code CASCADE [10] which included the effect of isospin mixing, parameterized by the isospin-violating spreading width $\Gamma^{\downarrow}_{>}$ according to Ref. [5]. It was assumed that the $\Gamma^{\downarrow}_{>}$ is the same in neighboring nuclei at a given excitation energy. The isospin mixing probability corresponds to the isospin-violating spreading width $\Gamma^{\downarrow}_{>}$ [5, 7]: $\alpha^2_< = (\Gamma^{\downarrow}_< / \Gamma_<) / (1 + \Gamma^{\downarrow}_< / \Gamma_< + \Gamma^{\downarrow}_> / \Gamma_>)$.

Since $\gamma$-ray yield in nuclei with N ≠ Z does not depend heavily on isospin mixing, it allowed to determine the GDR parameters in those nuclei in a way less dependent on the isospin mixing. Thus, each measured spectrum of high-energy $\gamma$-rays from the decay of a given N ≠ Z compound nucleus was fitted with CASCADE calculations using a single Lorentzian GDR strength function and assuming the absence of isospin mixing ($\Gamma^{\downarrow}_{>} = 0$). The extracted GDR parameters (strength $S_{GDR}$, resonance energy $E_{GDR}$ and width $\Gamma_{GDR}$) were then used as constants in the calculations for neighboring N = Z compound nuclei, independently for several values of an isospin-violating spreading width $\Gamma^{\downarrow}_{>}$. The best value of $\Gamma^{\downarrow}_{>}$ and the corresponding isospin mixing probability $\alpha^2_{<}$, for which the measured spectrum is well reproduced, were determined this way. In the next steps, the fitting was repeated for both, the N ≠ Z and N = Z compound nuclei, with the best $\Gamma^{\downarrow}_{>}$ values to extract the best values of the GDR parameters. We found that the GDR parameters for a given N ≠ Z compound nucleus are almost the same for different small values of $\alpha^2_{<}$, thus proving weak dependence of the GDR parameters on isospin mixing in N ≠ Z nuclei. To increase sensitivity to the isospin mixing, we also analyzed the ratios of $\gamma$-ray cross-sections for the reactions forming N = Z and N ≠ Z compound nuclei for the measured and calculated yields.

In conclusion, we have found that the isospin-violating spreading width $\Gamma^{\downarrow}_{>}$ does not depend heavily on the mass and atomic number of the nuclei studied (Fig. 1, left). It was shown for the first time that the extracted isospin mixing probability, $\alpha^2_{<}$, for highly excited states at similar excitation clearly increases with the increasing Z of the self-conjugate nuclei (Fig. 1, right). The observed increase in the studied range of Z is similar to that calculated for the ground-state nuclei in Ref. [1, 3].

## ACKNOWLEDGMENTS

This work was partly supported by the Polish State Committee for Scientific Research (KBN Grant No. 1 P03B 160 29).

## REFERENCES

1. A. Bohr and B.R. Mottelson, *Nuclear Structure,* Vol. I, p. 171, Benjamin, New York (1969).
2. I. Hamamoto and H. Sagawa, Phys. Rev. **C 48**, R960 (1993).
3. J. Dobaczewski, I. Hamamoto, Phys. Lett. **B345**, 181 (1995).
4. G. Colo, M. A. Nagarajan, P. Van Isacker, A. Vitturi, Phys. Rev. **C 52**, R1175 (1995).
5. J. A. Behr, K.A. Snover, C.A. Gossett, M. Kicińska-Habior, J.H. Gundlach, Z.M. Drebi, M.S. Kaplan, D.P. Wells, Phys. Rev. Lett.**70**, 3201 (1993).
6. K.A. Snover, *Nucl. Phys.* **A 553**, 153c (1993).
7. H. L. Harney, A. Richter, and H. A. Weidenmüller, Rev. Mod.Phys. **58**, 607 (1986).
8. H. Sagawa, P. F. Bortignon, G. Colo, Phys. Lett. **B 444**, 1 (1998).
9. M. N. Harakeh, D.H. Dowell, G. Feldman, E.F. Garman, R. Loveman, J.L. Osborne, K.A. Snover, Phys. Lett. **B 176**, 297 (1986).
10. F. Pühlhofer, Nucl. Phys. **A280**, 267 (1977); modified by M. N. Harakeh and J. A. Behr.

# Development of a Spectrometer for Multiple Prompt Gamma-Ray Measurement to Identify Nuclear Levels

Tadahiro Kin, Masumi Oshima, Kazuyoshi Furutaka, Mitsuo Koizumi, Yosuke Toh and Atsushi Kimura

*Japan Atomic Energy Agency,*
*2-4 Shirakata-Shirane, Tokai-mura, Naka-gun, Ibaraki-ken, 319-1195, Japan.*

**Abstract.** We developed a spectrometer for multiple prompt gamma-ray measurements to identify nuclear levels to determine neutron capture cross sections. From a test of finding candidates of $^{15}N$ levels with a developing method, we found performance of the spectrometer is sufficient.

**Keywords:** Multiple Prompt Gammas, JRR-3, Nuclear Level, Ge Detectors' Array.
**PACS:** 21.10.-k, 28.20.Ka, 28.20.Np, 28.41.Rc

## INTRODUCTION

For development of innovative reactors, such as fast breeder reactors or accelerator driven systems, nuclear data of minor actinides (MA) and long-lived fission products (LLFP) are used for simulation of burning rate of fuel and transmutation ratio to reduce the radio-toxicity of the residual waste. However, these data are insufficient at present in quantity and quality. To overcome/improve such a situation, the present authors are measuring capture cross sections of neutrons by MA and LLFP. One way to obtain neutron capture cross sections is to sum partial cross sections of all ground-state transitions[1]. This is a feasible and efficient method for the cross sections compared with other methods[2]. For accurate measurements, identification of all ground-state-transition gamma-ray is required. However, such information of nuclear structure is insufficient in almost all nuclides at present. In order to identify ground-state transitions as much as possible, a gamma-ray spectrometer was developed. It was installed at a cold neutron beam line (C2-3-2) of JRR-3 of JAEA.

## THE SPECTROMETER FOR PROMPT GAMMA-RAY

The spectrometer should be able to detect all prompt gamma rays emitted following a capture event with low crosstalk events effectively. Therefore, a product of detection efficiency and signal-to-background ratio represents performance. An arrangement of the spectrometer was chosen to maximize the product with existing 8 clover-shaped

CP1012, *Frontiers in Nuclear Structure, Astrophysics, and Reactions: FINUSTAR 2*
edited by P. Demetriou, R. Julin, and S. V. Harissopulos
© 2008 American Institute of Physics 0-7354-0532-5/08/$23.00

Ge detectors. BGO Compton-suppressors were also equipped to reduce crosstalk events. A schematic view of the spectrometer is shown in Fig. 1.

We measured detection efficiency and signal-to-background ratio. The detection efficiency of the spectrometer obtained with $^{60}Co$ was determined to 4.41% and the energy resolution was between 2.5- and 3.0-keV at 1.33 MeV. In a measurement of $^{14}N(n_{th},\gamma)^{15}N$ reactions with a 498-mg melamine ($C_3H_6N_6$) sample, signal-to-background ratio of the 5269-keV prompt gamma-ray was around 1.7.

## Performance of The Spectrometer for Nuclear Level Identification

To test the spectrometer, we sought candidates of $^{15}N$ nuclear levels with measured prompt gamma rays from $^{14}N(n_{th},\gamma)^{15}N$ reactions. We are now developing a new method to identify nuclear levels from measured multiple-prompt-gamma-ray events which include all gamma rays emitted following a capture event. This method was used for a test. A brief overview of the method is described below.

When a nucleus captures a neutron, a compound nucleus is excited in a capture state. It is promptly de-excited emitting cascading gamma rays. Because prompt gamma rays are emitted almost at the same time, the order of the gamma rays cannot be determined experimentally. However, although paths of these cascades differ from event to evens, some of paths go through the same exited state. Therefore, frequency of observing level candidates, which is obtained to consider all sequences of gamma rays, through true excited states of a studying nuclide would be higher than other candidates. The new method analyzes distribution of the frequency and figures out nuclear levels automatically as follows.

At first, a frequency distribution as a function of level energy is made by considering all order of the cascading gamma rays and counting all candidates of levels for measured capture events. The first frequency distribution is thought as probability of existing levels for next step. Next, new frequency distribution is made. However, in this step by using the given probability of existing levels, only the most probable order of cascading gamma rays is counted in event by event. The new distribution is used in same way as second step and frequency of level energies that have higher probability is thought to be increased through iteration of these steps. However, frequency of a true and reverse-ordered path is almost same. Therefore, the frequency distribution is symmetric to half energy of capture state. To distinguish between them, if one path of candidates includes at least one of known levels, we choose the path. After choosing the most probable paths of all capture events, new frequency distribution is made. This new distribution is used for choosing candidates again. We make new distribution again and again until it is almost as same as old one. In the last distribution, peaks of the distribution are considered to indicate positions of nuclear levels.

Figure 3 shows an example of results of finding nuclear-level candidates of $^{15}N$ for the test of the spectrometer. In the analysis, three low-lying levels (5270-, 5298- and 6323-keV) were used as the known levels. For convergence of frequency distribution, four times of iteration were required. At almost all known-level energies[3], peaks were observed less than 9 MeV. Therefore, it is demonstrated that the spectrometer has enough performance for identifying nuclear levels based on the method described above.

FIGURE 1. A schematic view of the spectrometer.

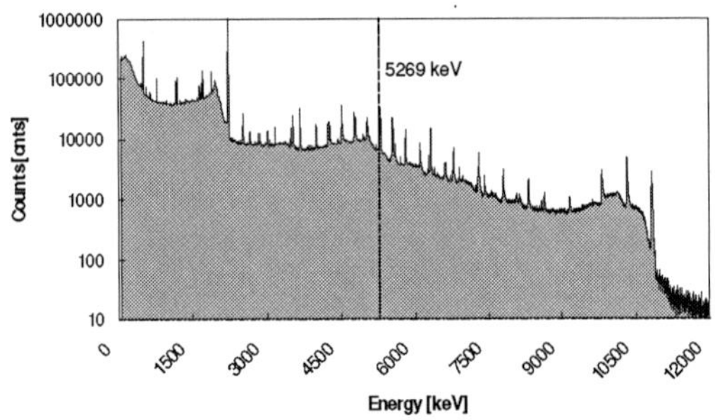


FIGURE 2. An example of prompt gammas' spectra.

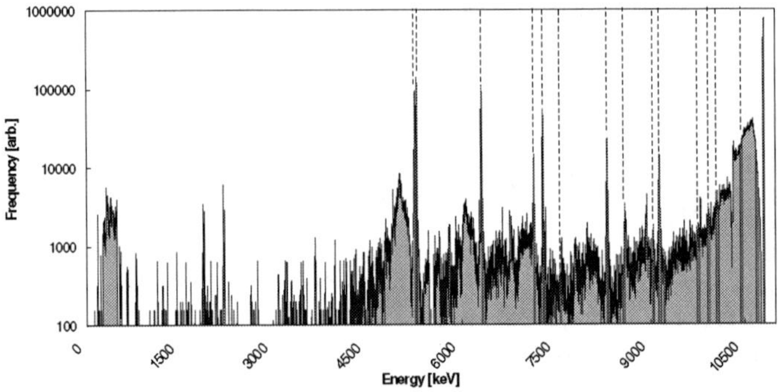


FIGURE 3. A frequency distribution of level energies for $^{15}N$. Broken lines are drawn at all known-level energies.

# CONCLUSION

A spectrometer for identification of nuclear levels by measurements of prompt gamma-ray was constructed. By the developing method for identification of levels, we found the spectrometer has enough potential to identify nuclear levels.

# ACKNOWLEDGMENTS

This study was the result of "Study on nuclear data by using a high intensity pulsed neutron source for advanced nuclear system" entrusted by the Ministry of Education, Culture, Sports, Science and Technology of Japan (MEXT).

# REFERENCES

1. M. Oshima, H. Harada *et.al.*, *Proceedings of The 2007 International Conference on Nuclear Data for Science and Technology,* To be published.
2. K. Furutaka, H. Harada and S. Raman, *J. of Nucl. Sci. and Tech.* **41**[11], 1033-1046 (2004).
3. ENSDF, http://www.nndc.bnl.gov/nndc/ensdf/.

# $^{128}$Xe Lifetime Measurement Using the Coulex-Plunger Technique in Inverse Kinematics

T. Konstantinopoulos[a], A. Lagoyannis[a], S. Harissopulos[a], A. Dewald[b], W. Rother[b], G. Ilie[b], P. Jones[c], P. Rakhila[c], P. Greenlees[c], T. Grahn[c], R. Julin[c] and D. L. Balabanski[d]

[a]*Institute of Nuclear Physics, NCSR "Demokritos", 15310 Aghia Paraskevi, Athens, Greece*
[b] *Institut für Kernphysik, Universität zu Köln, Zülpicherstrasse 77, 50937 Köln, Germany*
[c]*Department of Physics, University of Jyväskylä, P.O. Box 35, 40014 Jyväskylä, Finland*
[d]*INRNE, Sofia, Bulgaria*

**Abstract.** The lifetimes of the lowest collective yrast and non-yrast states in $^{128}$Xe were measured in a Coulomb excitation experiment using the recoil distance method (RDM) in inverse kinematics. Hereby, the Cologne plunger apparatus was employed together with the JUROGAM spectrometer. Excited states in $^{128}$Xe were populated using a $^{128}$Xe beam impinging on a $^{nat}$Fe target with E($^{128}$Xe)≈525 MeV. Recoils were detected by means of an array of solar cells placed at forward angles. Recoil-gated γ-spectra were measured at different plunger distances.

**Keywords:** $^{128}$Xe, Coulomb excitation, inverse kinematics, lifetimes, plunger
**PACS:** 21.10.Tg, 27.60.+j

## MOTIVATION

The E(5) and X(5) critical point symmetries introduced by F. Iachello [1] describe systems that undergo shape-phase transitions. To date, X(5) symmetry is assumed to be well established [2] whereas this does not fully apply for the case of E(5) symmetry. Under these conditions, the confirmation of the E(5) symmetry in nuclear structure is still an open question motivating further experimental investigation. Among the E(5) candidates proposed, $^{128}$Xe and $^{130}$Xe have been underlined as two of the most promising ones. Their excitation spectra are indeed in good agreement with those predicted by the E(5) symmetry. A major problem in determining the lifetimes of these nuclei arises from the fact that they cannot be populated in a "standard" plunger experiment, i.e. in a heavy-ion induced fusion-evaporation reaction. In addition, as they are in gaseous form they cannot be used as targets. A solution to this problem is Xe-induced Coulomb excitation. In this case, however, the analysis of the measured spectra is often not transparent. The combination of Coulomb excitation with the Recoil-Distance Doppler Shift method, known as Coulex-Plunger technique, has proved to be [3] a powerful experimental approach to determine lifetimes in the pico-second regime, enabling lifetime measurements of yrast as well as non-yrast collective states. The reduced transition probabilities of these states can provide a clear signature of the critical-point symmetry in

CP1012, *Frontiers in Nuclear Structure, Astrophysics, and Reactions: FINUSTAR 2*
edited by P. Demetriou, R. Julin, and S. V. Harissopulos
© 2008 American Institute of Physics 0-7354-0532-5/08/$23.00

a nucleus and as such they are of key importance in nuclear structure studies. Applying the Coulex-Plunger technique in inverse kinematics is, from an experimental point-of-view, a very challenging task not only because of various experimental problems but also because of its possible applicability to lifetime measurements with radioactive beams (RIB). On top of confirming the E(5) critical-point symmetry in $^{128}$Xe and $^{130}$Xe, the increasing interest in RIB worldwide was a strong motivation to perform the present feasibility study of the Coulex-Plunger technique in inverse kinematics.

## EXPERIMENTAL SETUP AND MEASUREMENTS

The experimental setup is shown in the left panel of Fig. 1. The $^{128}$Xe beam delivered by the Jyväskylä cyclotron with energy of 525 MeV passed through a 2.1 mg/cm$^2$ Fe foil that served as target (T). The coulomb-excited $^{128}$Xe nuclei were subsequently "retarded" by a 4 mg/cm$^2$-thick Nb foil (R). The Fe recoils were detected by an array of solar cells arranged as shown in the Fig. 1, at angles between 10° and ≈40° around the beam axis. A screening Au foil (S) was placed in front of the solar cells. It was 20.4 mg/cm$^2$ thick to stop the Xe and Nb recoils, thus allowing only the Fe nuclei to be detected by the solar cells. The latter recoils served as trigger for Coulomb excited events occurring in the target. This way, particle-gamma coincidences between the Fe recoils and the $\gamma$ rays depopulating Coulomb-excited states of $^{128}$Xe were recorded at 13 plunger distances ranging from 3 to 300 μm. A typical plunger spectrum is shown in Fig. 2.

From the recoil-gated $\gamma$-ray spectra, decay curves were derived. As an example, the decay curves of the $\gamma$ transitions de-exciting the $2^+_2$ state are plotted in Fig. 3. The analysis of the data will be complemented with the extraction of recoil-gated $\gamma$-$\gamma$ coincidences to apply the DDCM [4] method that allows determining in a very transparent way the lifetimes of the nucleus of interest. This work is in progress.

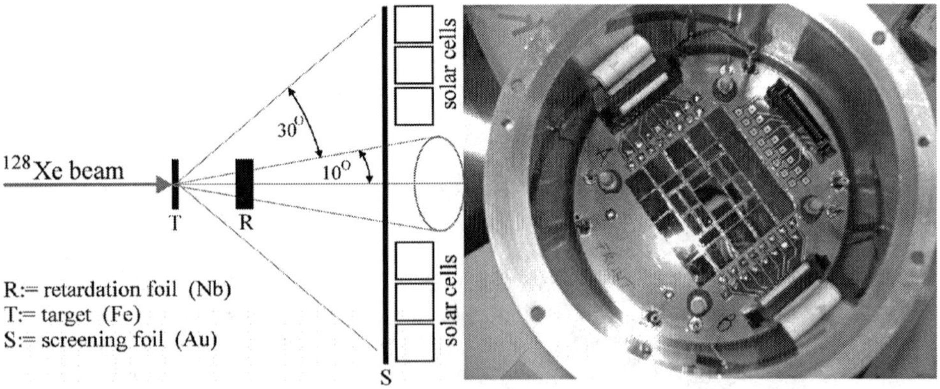


**FIGURE 1.** *Left*: Sketch of the experimental setup. *Right*: The solar-cell array used to detect the Fe recoils.

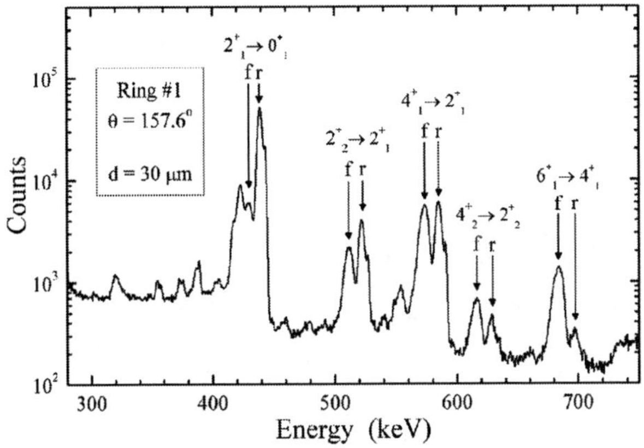


**FIGURE 2**. Particle-gated γ-spectrum measured with the JUROGAM-detector ring at a backward angle (157.6°) at a distance of 30 μm between the target and the retardation foil. The labeled peaks belong to $^{128}$Xe. The fully-shifted and reduced-shifted components are indicated by f and r, respectively.

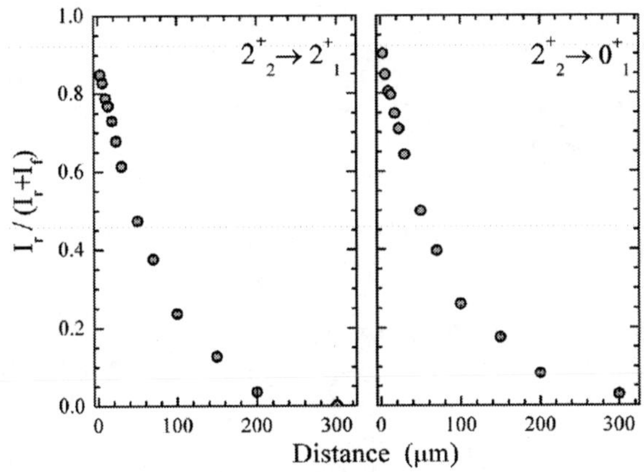


**FIGURE 2**. Decay curves of the γ transitions depopulating the $2^+_2$ excited state. They were obtained from the intensity ratio of the corresponding reduced-shifted component ($I_r$) to the sum of the reduced- ($I_r$) and fully-shifted ($I_f$) components.

# REFERENCES


1. F. Iachello, Phys. Rev. Lett. **85,** 3580 (2000), and Phys. Rev. Lett. **87**, 052502 (2001).
2. A. Dewald *et al.*, J. Phys. G: Nucl. Part. Phys. **31,** S1427 (2005)
3. R. Krücken *et al.*, Phys. Rev. Lett. **88**, 232501 (2002)
4. A. Dewald *et al.*, Z. Phys. **A 334**, 163-175 (1989).

# Puzzling $\gamma$-ray strength functions in $^{44,45}$Sc and $^{50,51}$V

A.C. Larsen*, M. Guttormsen*, R. Chankova†,**, F. Ingebretsen*, T. Lönnroth‡, S. Messelt*, J. Rekstad*, A. Schiller§, S. Siem*, N.U.H. Syed* and A. Voinov§,¶

*Department of Physics, University of Oslo, P.O.Box 1048 Blindern, N-0316 Oslo, Norway
†North Carolina State University, Raleigh, NC 27695, USA
**Triangle Universities Nuclear Laboratory, Durham, NC 27708 USA
‡Department of Physics, Åbo Akademi University, FIN-20500 Åbo, Finland
§Department of Physics and Astronomy, Ohio University, Athens, Ohio 45701, USA
¶Frank Laboratory of Neutron Physics, Joint Institute for Nuclear Research, 141980 Dubna, Moscow region, Russia

**Abstract.** The nuclear physics group at the Oslo Cyclotron Laboratory (OCL) has developed a method to extract nuclear level density and $\gamma$-ray strength function simultaneously from primary $\gamma$-ray spectra. The resulting $\gamma$-ray strength functions of $^{44,45}$Sc and $^{50,51}$V show an unexpected, large enhancement at low $\gamma$-ray energies, indicating an increased probability of emitting low-energy $\gamma$ rays in the quasi-continuum.

**Keywords:** Level density, $\gamma$-ray strength function
**PACS:** 21.10.Ma, 24.10.Pa, 25.55.Ci, 25.55.Hp, 27.40.+z

## INTRODUCTION

The nuclear level density and the $\gamma$-ray strength function give a measure of the gross properties of the nucleus. These average quantities are indispensable in practical applications of nuclear physics. The Oslo Cyclotron group has developed the so-called Oslo method, which enables extraction of both level density and $\gamma$-ray strength function from the distribution of primary $\gamma$ rays at various initial excitation energies. The method will be briefly discussed, and the results for the nuclei $^{44,45}$Sc and $^{50,51}$V will be presented.

## THE OSLO METHOD

The experiments were performed at the OCL (`http://ocl.uio.no`), where the cyclotron provided a beam of $^3$He-ions with energy 38 MeV and 30 MeV on the $^{45}$Sc and $^{51}$V targets, respectively. Particle-$\gamma$ coincidences were measured with the multi-detector array CACTUS [1], which consists of 28 collimated NaI(Tl) $\gamma$-ray detectors and eight Si $\Delta E - E$ particle telescopes. Data from the following reactions were studied:

- The inelastic scattering reactions $^{45}$Sc($^3$He,$^3$He$'\gamma$)$^{45}$Sc and $^{51}$V($^3$He,$^3$He$'\gamma$)$^{51}$V.
- The pickup reactions $^{45}$Sc($^3$He,$\alpha\gamma$)$^{44}$Sc and $^{51}$V($^3$He,$\alpha\gamma$)$^{50}$V.

CP1012, *Frontiers in Nuclear Structure, Astrophysics, and Reactions: FINUSTAR 2*
edited by P. Demetriou, R. Julin, and S. V. Harissopulos
© 2008 American Institute of Physics 0-7354-0532-5/08/$23.00

The Oslo method is designed to disentangle the level density and the transmission coefficient for $\gamma$ emission from particle-$\gamma$ coincidence spectra, see Refs. [2, 3, 4, 5]. The matrix of first-generation (primary) $\gamma$-ray spectra is factorized as a product of the level density at the final excitation energy $\rho(E-E_\gamma)$ and the $\gamma$-ray transmission coefficient $\mathscr{T}(E_\gamma)$ [6, 7, 8]:

$$P(E,E_\gamma) \propto \rho(E-E_\gamma)\mathscr{T}(E_\gamma). \tag{1}$$

A globalized fitting to the data points of the experimental $P(E,E_\gamma)$ matrix determines the functional form of $\rho$ and $\mathscr{T}$ [5]; however, to get the slope and the absolute normalization of the level density and the transmission coefficient, auxiliary data are needed. For more details, see also Refs. [9, 10].

## LEVEL DENSITIES AND $\gamma$-RAY STRENGTH FUNCTIONS

The level densities of $^{44,45}$Sc [10] and $^{50,51}$V [9] are displayed in Fig. 1. The general trend for all nuclei is an approximately exponential increase in the level density as a function of excitation energy. The level density of $^{51}$V differs from the others in the respect that it shows large bumps and structures up to $\sim$ 5 MeV of excitation energy. These features are interpreted as effects of the $N = 28$ shell gap, which also lowers the number of available levels in this nucleus considerably at low excitation energy compared to $^{50}$V.

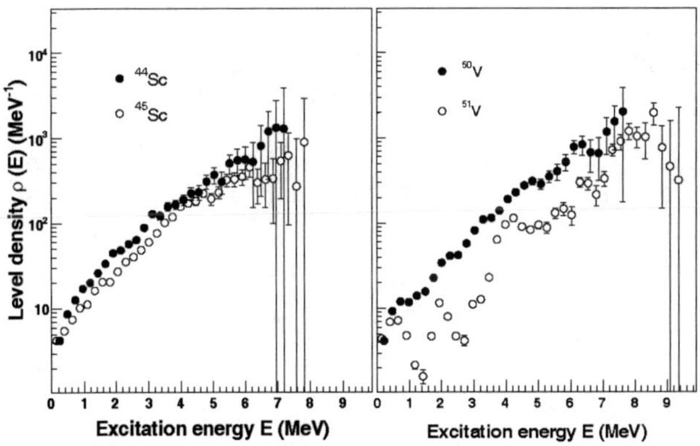


**FIGURE 1.** Level densities of $^{44,45}$Sc and $^{50,51}$V

The $\gamma$-ray strength functions of $^{44,45}$Sc [10] and $^{50,51}$V [9] are shown in Fig. 2. Several interesting features can be seen. In general, for $E_\gamma \geq 3.5$ MeV, the data show that all the $\gamma$-ray strength functions are increasing with $\gamma$ energy. For $\gamma$ energies below $\sim$ 3 MeV, the $\gamma$-ray strength functions have an increase of a factor of $\sim 2-3$ relative to their minimum. This enhancement at low $\gamma$ energies has previously been seen in $^{56,57}$Fe [11] and $^{93-98}$Mo [12]. None of the standard models can account for this behaviour, and it is, as of today, without any established physical explanation.

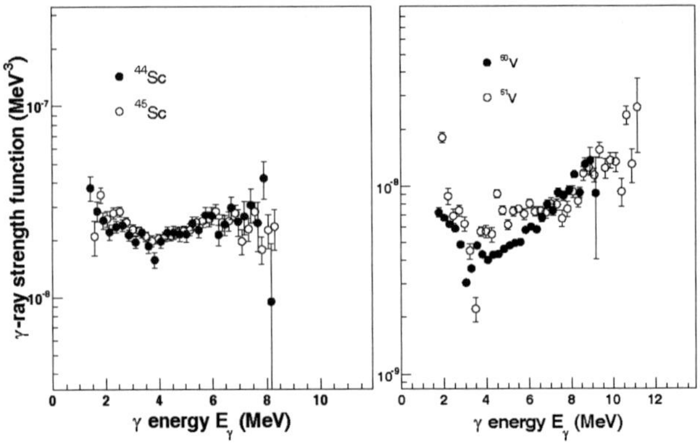


**FIGURE 2.** Gamma-ray strength functions of $^{44,45}$Sc and $^{50,51}$V

## SUMMARY AND CONCLUSIONS

The experimental level densities and $\gamma$-ray strength functions of $^{44,45}$Sc and $^{50,51}$V extracted from primary $\gamma$ spectra with the Oslo method have been presented. The $\gamma$-ray strength functions of all nuclei shown here display an increase with increasing $\gamma$ energy for $E_\gamma \geq 3$ MeV, but also a big enhancement of strength at low $\gamma$ energies. This fascinating behaviour cannot be explained by any known models.

## ACKNOWLEDGMENTS

Financial support from the Norwegian Research Council (NFR) is gratefully acknowledged. A. Schiller acknowledges support from the U. S. National Science Foundation, grant number PHY-06-06007.

## REFERENCES


1. M. Guttormsen *et al.*, Phys. Scr. **T 32**, 54 (1990).
2. M. Guttormsen *et al.*, Nucl. Instrum. Methods Phys. Res. A **374**, 371 (1996).
3. M. Guttormsen, T. Ramsøy, and J. Rekstad, Nucl. Instrum. Methods Phys. Res. A **255**, 518 (1987).
4. L. Henden, L. Bergholt, M. Guttormsen, J. Rekstad, and T.S. Tveter, Nucl. Phys. **A589**, 249 (1995).
5. A. Schiller *et al.*, Nucl. Instrum. Methods Phys. Res. A **447**, 498 (2000).
6. A. Bohr and B. Mottelson, *Nuclear Structure*, (Benjamin, New York, 1969), Vol. I, p. 169, p. 184-185.
7. D. M. Brink, Ph.D. thesis, Oxford University, 1955.
8. P. Axel, Phys. Rev. **126**, 671 (1962).
9. A. C. Larsen *et al.*, Phys. Rev. C **73**, 064301 (2006).
10. A.C. Larsen *et al.*, http://arxiv.org/abs/0706.0533v1, accepted for publication in Phys. Rev. C.
11. A. Voinov *et al.*, Phys. Rev. Lett **93**, 142504 (2004).
12. M. Guttormsen *et al.*, Phys. Rev. C **71**, 044307 (2005).

# DSAM Lifetime Studies for Gd – Nd nuclei with EUROBALL and AFRODITE

E.O. Lieder*,†, R.M. Lieder*,†, A.A. Pasternak**,†, B.G. Carlsson‡, I. Ragnarsson‡, R.A. Bark*, E. Gueorguieva*, J.J. Lawrie*, S.M. Mullins*, P. Papka*, Y. Kheswa*, J.F. Sharpey-Schafer§, W. Gast† and G. Duchêne¶

*iThemba Laboratory for Accelerator Based Sciences, Somerset West 7129, South Africa
†Institut für Kernphysik, Forschungszentrum Jülich, D-52425 Jülich, Germany
**A.F. Ioffe Physical Technical Institute RAS, RU-194021 St. Petersburg, Russia
‡Division of Mathematical Physiscs, Lund Institute of Technology, SE-221 00 Lund, Sweden
§Dept. of Physics, University of the Western Cape, Bellville 7535, South Africa
¶Institut de Recherches Subatomique IReS, F-67037 Strasbourg, France

**Abstract.** Lifetimes of high-spin states have been measured for $^{142}$Gd and $^{134}$Nd with EUROBALL IV and AFRODITE, respectively, using DSAM. From calculations in the cranked Nilsson-Strutinsky model it was concluded that at the high-spin states of the $(+,0)_1$ band, $^{142}$Gd represents a triaxial nucleus rotating around the longest principal axis.

**Keywords:** High-spin states, DSAM lifetimes, Cranked Nilsson-Strutinsky model calculations
**PACS:** 21.10.-k, 21.10.Tg, 27.60.+j

## LIFETIME MEASUREMENT FOR $^{142}$Gd

With EUROBALL IV at IReS Strasbourg a lifetime experiment using the Doppler-shift attenuation method (DSAM) has been carried out [1, 2, 3, 4], which allowed to determine the $B(E2)$ values of quadrupole bands in $^{142}$Gd. This nucleus was produced in a $^{114}$Sn($^{32}$S,2p2n) reaction at a beam energy of 160 MeV. As target a selfsupporting metallic $^{114}$Sn foil of 8 mg/cm$^2$ thickness with an enrichment of 71.1% was used. The lifetime analysis is described in detail in ref. [3]. A partial level scheme of $^{142}$Gd is shown in fig. 1. The resulting $B(E2)$ values are plotted in fig. 2.

For the interpretation of the $(+,0)_1$ band in $^{142}$Gd, calculations in the cranked Nilsson-Strutinsky model have been carried out [4]. Triaxial shapes with well-developed potential energy minima were obtained. For angular momenta around $I = 16$, three minima were found for $\gamma$ values of $\approx 20°$, $\approx -30°$ and $\approx -75°$ corresponding to rotations around each principal axis, but at $I = 26$ only the minimum at $\gamma = -75°$, corresponding to a rotation around the longest principal axis (the axis with the smallest $\mathcal{J}_{rig}$) remains. The $(+,0)_1$ band has from $I = 26$ a $\pi h^4_{11/2} \otimes \nu h^{-2}_{11/2}$ configuration. The present lifetime results support the conclusion that a rotation around the longest principal axis occurs as shown in fig. 2, since for the $26^+ \rightarrow 24^+$ transition the observed $B(E2)$ value coincides with that calculated for the only possible rotation axis at $\gamma = -75°$. This is to our knowledge the first case when experimental features support this suggestion.

CP1012, *Frontiers in Nuclear Structure, Astrophysics, and Reactions: FINUSTAR 2*
edited by P. Demetriou, R. Julin, and S. V. Harissopulos
© 2008 American Institute of Physics 0-7354-0532-5/08/$23.00

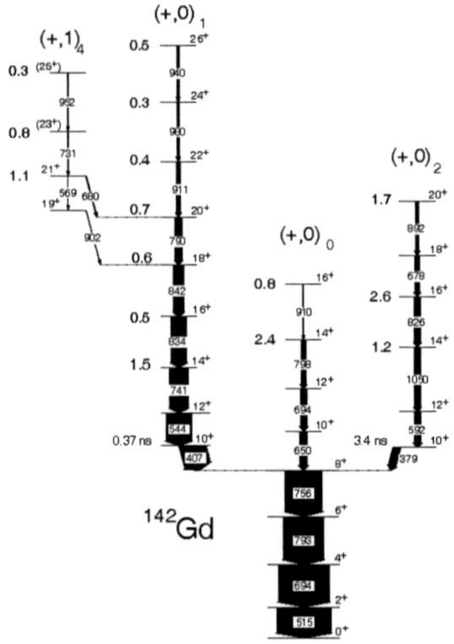


**FIGURE 1.** Partial level scheme of $^{142}$Gd. The lifetimes observed for high-spin states are given in ps.

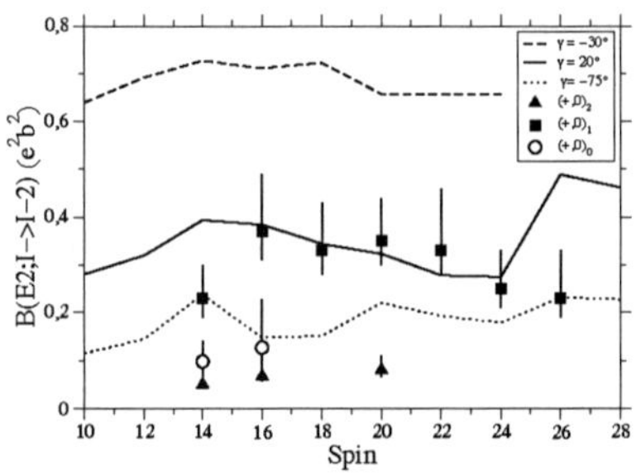


**FIGURE 2.** Comparison of experimental $B(E2)$ values for quadrupole bands in $^{142}$Gd with results of cranked Nilsson-Strutinsky model calculations.

## LIFETIME MEASUREMENT FOR $^{134}$Nd

To investigate lifetimes of quadrupole bands in $^{134}$Nd a DSAM experiment was carried out with the $\gamma$-detector array AFRODITE at iThemba LABS, South Africa. This nucleus was produced in a $^{114}$Cd($^{28}$Si,$\alpha$4n) reaction at a beam energy of 155 MeV. As target a selfsupporting metallic $^{114}$Cd foil of 13 mg/cm$^2$ thickness with an enrichment of 99.1% was used. Four CLOVER detectors each were mounted under 45° and 135°, respectively, in the AFRODITE array.

From this experiment lifetimes can be deduced for several bands in $^{134}$Nd. A partial level scheme of $^{134}$Nd is shown in fig. 3 [5]. Lifetime results have been obtained for the $(+,0)_0$ band from $I = 12$ to $I = 22$. Previously, lifetimes have been measured for $^{134}$Nd using RDM – for the $(+,0)_0$ band up to $I = 12$ [6]. For this spin the obtained lifetime agrees with the present value within uncertainties. A $\nu h_{11/2}$ configuration has

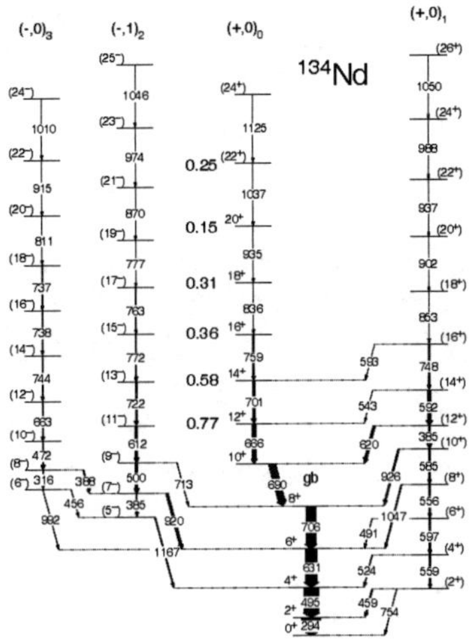


**FIGURE 3.** Partial level scheme of $^{134}$Nd. The lifetimes observed for high-spin states are given in ps.

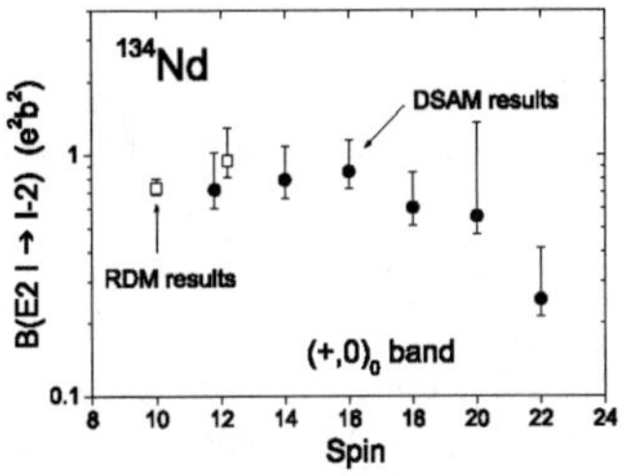


**FIGURE 4.** Experimental $B(E2)$ values for the $(+,0)_0$ band in $^{134}$Nd.

previously been assigned to the $(+,0)_0$ band at $I = 10$ [5]. A comparison of the deduced $B(E2)$ values, shown in fig. 4, with theoretical calculations may elucidate further the configuration assignment and allow to determine the nuclear shape of the $(+,0)_0$ band in $^{134}$Nd.

## REFERENCES

1. E.O. Podsvirova et al., Eur. Phys. J. **A 21**, 1 (2004).
2. A.A. Pasternak et al., Eur. Phys. J. **A 23**, 191 (2005).
3. E.O. Lieder et al., submitted to Eur. Phys. J. **A**, (2007).
4. B.G. Carlsson et al., submitted to Phys. Rev. Lett., (2007).
5. C.M. Petrache et al., Phys. Lett. **B 335**, 307 (1994).
6. T. Klemme et al., Phys. Rev. **C 60**, 034301 (1999).

# Fusion-Fission In The $^{86}$Kr + $^{238}$U Reaction

M. Lipoglavšek[1], R. A. Bark[2], E. A. Gueorguieva[2], J. J. Lawrie[2], E. Lieder[2], R. Lieder[2], E. Lindbo Hansen[1], S. M. Mullins[2], S. S. Ntshangase[2], P. Papka[2], T. Petrovič[1] and M. Vencelj[1]

[1] *Jožef Stefan Institute, Jamova 39, 1000 Ljubljana, Slovenia*
[2] *iThemba LABS, 7130 Somerset West, South Africa*

**Abstract.** The $^{86}$Kr+$^{238}$U reaction has been studied at krypton beam energies about 30 MeV above the Coulomb barrier. Reaction products were detected by an array of 32 photovoltaic cells coupled to the AFRODITE γ-ray detector array at iThemba LABS. A symmetric fission component has been observed at about 600 MeV total kinetic energy. This could possibly be due to fusion-fission with a cross section of 35±20 mb.

**Keywords:** fusion-fission reactions, in-beam spectroscopy, AFRODITE array
**PACS:** 25.70.Jj, 25.85.Ge, 29.30.Ep, 29.30.Kv

Studying nuclei far from the valley of β-stability is now a well established field of nuclear physics. While the proton dripline has been studied in considerable detail due to its accessibility by fusion-evaporation reactions, the other two extremes, namely the superheavy one and the neutron dripline, remain elusive. We performed an experiment that aimed at both of these in April 2007 at iThemba LABS. The aim of the experiment was twofold. Firstly, we wanted to identify fusion-fission in the reaction of $^{86}$Kr with $^{238}$U for its potential to produce superheavy nuclei. Fusion-fission would signal compound nucleus creation, which is, besides the survival probability, the main factor determining the cross section [1]. Secondly, we wanted to study excited states of fission fragments. Since both beam and target nuclei are neutron rich, studying fission of their compound could possibly be a novel method to reach neutron rich nuclei that are inaccessible with conventional methods.

The separated sector cyclotron at iThemba LABS delivered krypton beams with energies of $E_{lab}$=484 and 497 MeV, which brought the compound nucleus $^{324}$128 to a calculated excitation energy of about $E_x$=25 and 35 MeV [2], respectively in the middle of the target. The targets consisted of 400 μg/cm$^2$ natural $UF_4$ on a 40 μg/cm$^2$ carbon backing added for stability. The carbon side was turned against the oncoming beam. The experimental setup consisted of the AFRODITE [3] detector array of eight Ge clover detectors coupled to an array of 32 Si photovoltaic cells used to detect charged particles. Such a setup was used to detect fission fragments in coincidence with X and γ rays. Four clover detectors were placed at an angle of 135° with respect to the beam and four at 90°. The role of the photovoltaic cells was to allow the reconstruction of reaction kinematics and Doppler correction of γ-ray energies. Their size was 2.5 by 0.5 cm. They were grouped in four sets of eight detectors. Each set was placed perpendicularly to each other at a distance of 7.1 cm from the target. Four most forward detectors were placed at a polar angle of 31° and four backward most

CP1012, *Frontiers in Nuclear Structure, Astrophysics, and Reactions: FINUSTAR 2*
edited by P. Demetriou, R. Julin, and S. V. Harissopulos
© 2008 American Institute of Physics 0-7354-0532-5/08/$23.00

detectors at 61°. The angles were chosen such as to minimize the Rutherford scattering count rate, which was the main background in the experiment. Each Si detector had an opening angle $\Delta\theta$ of about 4°. The total solid angle they covered was about 5% which also coincides with the efficiency for fission fragment detection. Their summed energy resolution was about 30 MeV which resulted in mass resolution of 15 mass units. The hardware trigger required two signals from Si detectors at opposite azimuthal angles. About $4\cdot10^7$ particle-particle coincidences were collected at the lower beam energy and about $1.5\cdot10^8$ at the higher one.

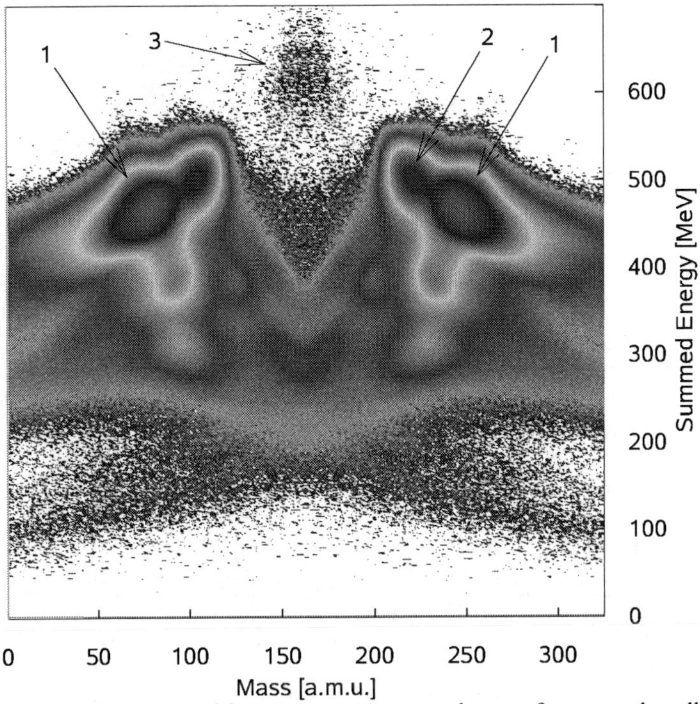


**FIGURE 1.** Reconstructed fragment mass vs. summed energy from two solar cells.

Figure 1 shows a three dimensional plot of the number of counts depending on reconstructed mass and total kinetic energy detected in the two Si detectors. The counts in grey are in logarithmic scale. Part of the data collected at the higher beam energy is shown. The fragment masses $M_{1,2}$ were calculated assuming a binary reaction:

$$M_{1,2} = 324\ \text{a.m.u.}\frac{E_{2,1}\sin^2\theta_{2,1}}{E_1\sin^2\theta_1 + E_2\sin^2\theta_2},$$

where $E_1$, $E_2$, $\theta_1$ and $\theta_2$ are the measured energies and angles. The largest two peaks marked with 1 in the figure are due to elastically scattered $^{86}$Kr and $^{238}$U nuclei. The analysis was severely hampered by about 0.5% tungsten target impurity. Tungsten was identified by $\gamma$ rays associated with events marked with 2 in Figure 1. Despite that, a number of events are observed with ½ of summed mass at about 600 MeV total kinetic energy (marked with 3 in Figure 1). These events were only observed with the higher beam energy. Their angular distribution (Figure 2) is consistent with expected fusion-

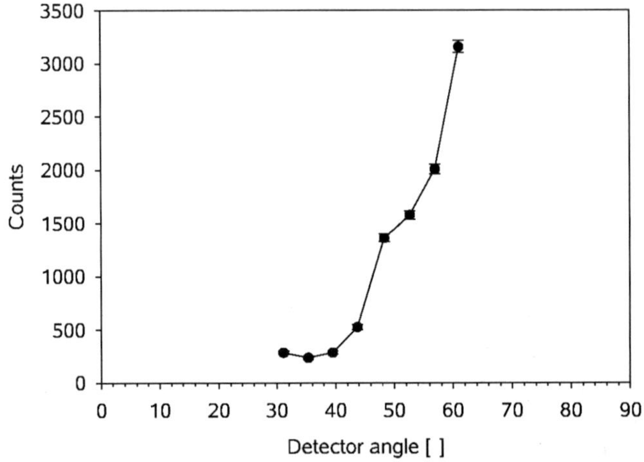

**FIGURE 2.** Angular distribution of fragments inside the symmetric fission blob.

fission distribution. However, the summed energy is too large, since the fusion-fission Q-value is only 90 MeV. The only way of gaining more than 100 MeV in the observed reaction would be by Coulomb induced fission of $^{238}U$. In this case, the angular distribution is not expected to be backward peaked nor the mass distribution to be symmetric. Unfortunately, the low statistics of the symmetric fission component did not allow for identification of individual fragments by X and $\gamma$ rays, which would reveal its origin. The cross section of the symmetric fission component $\sigma_{A/2\pm15}$=35±20 mb calculated by normalization to Rutherford scattering cross section. This compares well with a thick target measurement of the $^{84}Kr + {}^{238}U$ reaction [4], although our mass distribution seems to be narrower. The latter could be explained by lower beam energy and thinner target.

Although we were not able to identify individual fission fragments, we have shown that our experimental technique works very well and would be suitable for a more efficient detector array. If properly identified, the fusion-fission cross section would be comparable to reactions used for production of superheavy nuclei with a $^{48}Ca$ beam [1].

# REFERENCES


1. Y. Oganessian, J. Phys. G **34,** R165 (2007).
2. P. Möller *et al.,* At. Data Nucl. Data Tables **59,** 185 (1995).
3. M. Lipoglavšek *et al.,* Nucl. Instr. Meth. **A557,** 523 (2006).
4. J. V. Kratz *et al.,* Phys. Rev. Lett. **33,** 502 (1974).

# Study of dipole excitations and the single particle structure of neutron rich Ni isotopes

K. Mahata[*], S. Paschalis[†], P. Adrich[**], F. Aksouh[*], T. Aumann[*], M. Babilon[‡], K-H. Behr[*], J. Benlliure[§], T. Berg[¶], M. Boehmer[||], K. Boretzky[*], A. Brünle[*], E. Casarejos[§], M. Chartier[†], A. Chatillon[*], D. Cortina-Gil[§], U. Datta Pramanik[††], L. Deveaux[‡‡], M. Elvers[‡], H. Emling[*], B. Fernandez-Dominguez[†], M. Gorska[*], W. Hüller[*], G. Ickert[*], H. Johansson[*], A. Junghans[§§], C. Karagiannis[*], L. Kern[‡], O. Kiselev[*,¶], A. Klimkiewicz[*], N. Kurz[*], M. Labiche[¶¶], T. Le Bleis[*,***], R. Lemmon[†††], K. Lindenberg[‡], Y. Litvinov[*], P. Maierbeck[||], S. Müller[‡], T. Nilsson[‡‡‡,‡], C. Nociforo[*], R. Palit[§§§], W. Prokopowicz[**], D. Rossi[¶], H. Simon[*], K. Sümmerer[*], A. Wagner[§§], W. Walus[**], H. Weick[*] and M. Winkler[*]

[*]*Gesellschaft für Schwerionenforschung (GSI), D-64291 Darmstadt, Germany*
[†]*Department of Physics, University of Liverpool, Oliver Loge Laboratory, L69 7ZE, United Kingdom*
[**]*Instytut Fizyki, Uniwersytet Jagielloński, PL-30-059 Kraków, Poland*
[‡]*Institut für Kernphysik, Technische Universität, D-64289 Darmstadt, Germany*
[§]*Universidad de Santiago de Compostela, 15706, Santiago de Compostela, Spain*
[¶]*Institut für Kernchemie, Johannes Gutenberg-Universität, D-55099 Mainz, Germany*
[||]*Physik-Department der Technischen Universität München, D-85748 Garching, Germany*
[††]*Saha Institute of Nuclear Physics, Kolkata 700064, India*
[‡‡]*IPN Orsay, IN2P3/CNRS, F-91406 Orsay, France*
[§§]*Institut für Strahlenphysik, Forschungszentrum, Dresden-Rossendorf, D-01314 Dresden, Germany*
[¶¶]*The Institute of Physical Research, University of Paisley, Paisley, United Kingdom*
[***]*Institut Pluridisciplinaire Hubert Curien IN2P3-CNRS/Université Louis Pasteur, F-67037 Strasbourg Cedex 2, France*
[†††]*STFC, Daresbury Laboratory, Daresbury, Warrington, WA4 4AD, United Kingdom*
[‡‡‡]*Fundamental Fysik, Chalmers Tekniska Högskola, S-412 96 Göteborg, Sweden*
[§§§]*Department of Nuclear and Atomic Physics, Tata Institute of Fundamental Research, Mumbai 400005, India*

**Abstract.** An experiment was performed using the FRS-LAND setup at GSI to study the dipole strength distributions above neutron separation threshold for neutron-rich Ni isotopes. Measurements, using the same experimental setup, were also carried out to extract single particle occupancies via knockout reactions to investigate the structure and magicity of the neutron-rich Ni isotopes. The status of the data analysis and preliminary results are presented.

**Keywords:** Coulomb dissociation, Dipole strength distribution, Knockout reaction, Spectroscopic factor
**PACS:** 24.30.Cz, 25.60.-t, 25.70.De, 21.10.Jx, 25.60.Gc

CP1012, *Frontiers in Nuclear Structure, Astrophysics, and Reactions: FINUSTAR 2*
edited by P. Demetriou, R. Julin, and S. V. Harissopulos
© 2008 American Institute of Physics 0-7354-0532-5/08/$23.00

## INTRODUCTION

In previous experiments at GSI, the LAND collaboration has observed a sizeable fraction of low-lying E1 strength (below the giant dipole resonance region) in $^{20,22}$O and $^{130,132}$Sn nuclei [1]. In general, microscopic calculations conclude that in lighter neutron-rich nuclei the low-lying dipole strength arises from single neutron excitation. In contrast, the low-lying dipole states in medium heavy nuclei such as $^{130,132}$Sn display to some extent the feature of a pygmy dipole resonance [2]. More measurements are required to establish the universal collective nature of the low-lying dipole excitation. Determination of the dipole strength functions in exotic nuclei might also help in constraining the isospin and density dependent parts of the effective interaction. With this motivation, an experiment was carried out to study the dipole strength distribution in the continuum by measuring the electromagnetic excitation of the neutron-rich nuclei $^{68,70,72}$Ni. For comparison, electromagnetic excitation of $^{56}$Ni (N=Z) was also measured, thus covering a wide range of isospin from N-Z = 0 to N-Z=16. Another aspect of the experiment is the study of the evolution of shell structure in the Ni isotopes far from stability. Neutron-rich nuclei are often characterized by a very diffuse neutron density distribution at the surface, which would result in a reduced spin-orbit interaction and a more harmonic oscillator like mean field. For the Ni isotopes, this should lead to a shell gap at N=40 and a reduction of that at N=50. Measurements using the same experimental setup were also carried out to extract single particle occupancies via the knockout reaction to study the structure and magicity of the neutron-rich Ni isotopes.

## EXPERIMENTAL DETAILS

The experiments were performed using the FRagment Separator (FRS) and the LAND setup at GSI. Radioactive ions provided as mixed secondary beam of $\sim$500 MeVA by the FRS were unambiguously identified and momentum analyzed on an event-by-event basis. The heavy fragments emerging after dissociation in the Pb target were again identified and momentum analyzed using position sensitive pin diodes, the large acceptance dipole magnet ALADIN, scintillating fiber arrays and a time-of-flight wall consisting of organic scintillators. Gamma-rays and neutrons were detected using an array of CsI(Tl) detectors surrounding the target and the Large Area Neutron Detector (LAND) situated $\sim$15 m downstream, respectively. Additional measurements were performed with a $^{12}$C target and without target to estimate the nuclear contribution and the contributions coming from the other materials present in the beam line, respectively. Finally, a $CH_2$ target was used for the measurement of proton induced knockout reactions. The fast protons from the knockout reaction were detected by plastic scintillators surrounding the beam pipe in the forward direction and by the CsI array.

## DATA ANALYSIS AND RESULTS

All the detectors used for identification and momentum determination of the incoming beam as well as the outgoing heavy fragments have been calibrated.

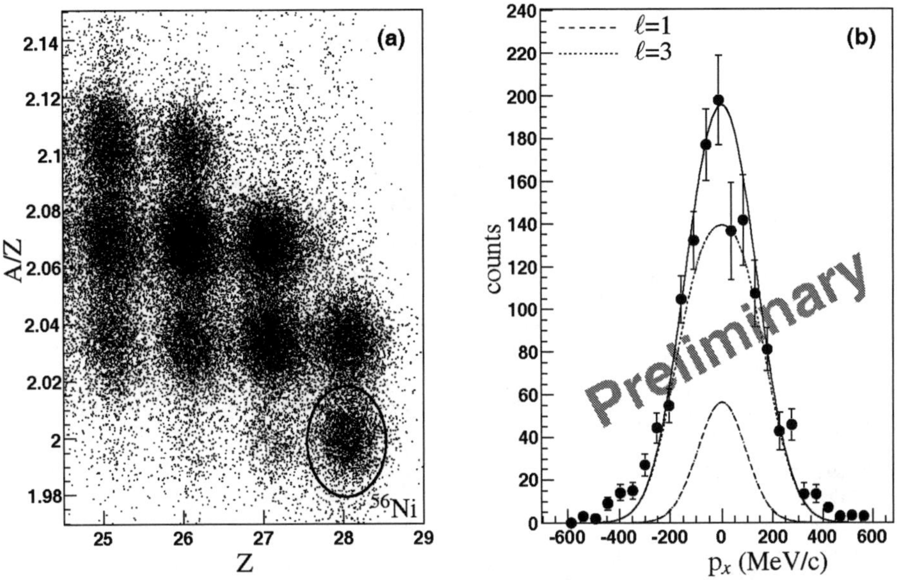


**FIGURE 1.** Identification plot of the outgoing heavy reaction fragments from 500 MeVA $^{57}$Ni incident beam on the $CH_2$ target (left), and the inclusive transverse momentum distribution of the one-neutron removal reaction (right).

Calibrations of the LAND detector, the CsI array and the plastic scintillators surrounding the target are in progress. Fig. 1(a) shows the Z vs. A/Z plot for the outgoing heavy fragments obtained by tracking through the magnetic field of ALADIN and the energy loss measurement. The transverse momentum distribution of one-neutron removal from $^{57}$Ni on the $CH_2$ target is shown in Fig. 1(b). The theoretical predictions corresponding to $\ell = 1$ and $\ell = 3$ folded with the experimental resolution are also shown. The experimental resolution is sufficient to separate contributions coming from different angular momenta. Further analysis of the CsI array, the plastic scintillators surrounding the target and the LAND detector will allow us to isolate contributions from different final states, identify the quasi-free scattering process and reconstruct the invariant mass required for studying dipole strength distributions.

## REFERENCES


1. T. Aumann, *Eur. Phys. J* **A 26**, 441 (2005), and reference cited there.
2. P. Adrich, A. Klimkiewicz, M. Fallot, K. Boretzky, T. Aumann, D. Cortina-Gil, U. D. Pramanik, T. W. Elze, H. Emling, H. Geissel, M. Hellström, K. L. Jones, J. V. Kratz, R. Kulessa, Y. Leifels, C. Nociforo, R. Palit, H. Simon, G. Surówka, K. Sümmerer, and W. Waluś, *Phys. Rev. Lett.* **95**, 132501 (2005).

# Neutron-Rich Hypernuclei Are Coming. What They Tell Us ?

L. Majling*, J. Lukstins†, A.N. Parfenov†, M. Solar** and B. Sopko**

*Nuclear Physics Institute, 25068 Řež, Czech Republic
†Joint Institute for Nuclear Research, 141980 Dubna, Russia
**Czech Technical University, Technická 4, 16607 Prague, Czech Republic

**Abstract.** We continue the discussion of our suggestion presented at FINUSTAR 1: to explore the potential of hypernuclear spectroscopy for investigation the structure of light nuclei beyond neutron dripline.

**Keywords:** hypernuclei
**PACS:** 21.80-a

## HYPERNUCLEUS $^6_\Lambda$H COULD BE IDENTIFIED

The hypernuclei $^6_\Lambda$H and $^7_\Lambda$He could be obtained in the nearest future [1, 2, 3]. So, it would be possible to compare energy spectra of the chain of four (hyper) nuclei with two neutron "halo" and different $s$-shell core: $^5$H, $^6_\Lambda$H, $^6$He and $^7_\Lambda$He, [4, 5].

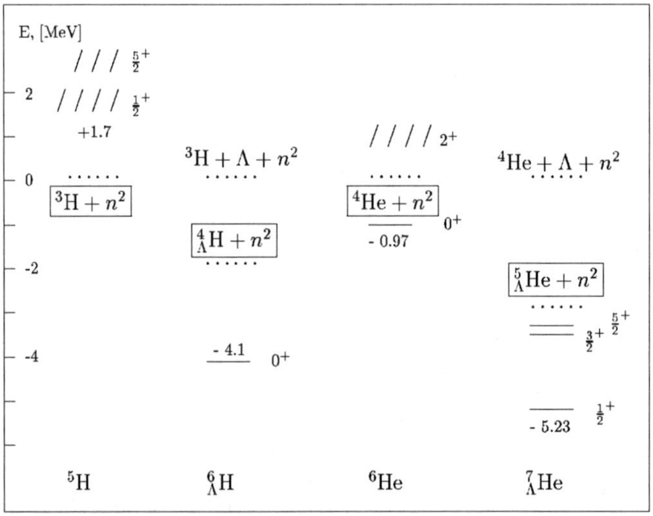


**FIGURE 1.** Spectra of four (hyper) nuclei with four neutrons and different "core"

Such a unique comparison could shed light on a limit of three-body model (core+n+n) [6, 7] as well as on the role of continuum.

CP1012, *Frontiers in Nuclear Structure, Astrophysics, and Reactions: FINUSTAR 2*
edited by P. Demetriou, R. Julin, and S. V. Harissopulos
© 2008 American Institute of Physics 0-7354-0532-5/08/$23.00

A proper shell model description of loosely bound nuclei should take into account the coupling of the discrete bound states with the continuum of scattering states. The tool of choice is Gamow shell model [8] and the continuum shell model [9] . It is obvious that experimental data on ${}^{6}_{\Lambda}$H will be instrumental for further development of such models.

## OUTLOOK: ${}^{8}_{\Lambda}$H

The hypernucleus ${}^{8}_{\Lambda}$H might be stable also [10]. We expect that in ${}^{6}_{\Lambda}$H and ${}^{8}_{\Lambda}$H there is a low-lying state $1^+$. So we obtain a chain of three hypernuclei ${}^{4}_{\Lambda}$H, ${}^{6}_{\Lambda}$H and ${}^{8}_{\Lambda}$H (N/Z = 2, 4 and 6 respectively) with a similar and very simple ground state doublet.

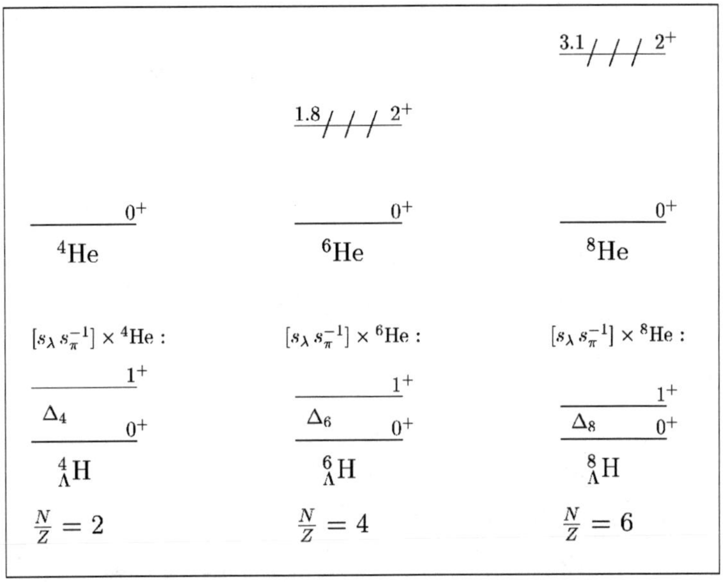


**FIGURE 2.** Spectra of the low lying levels in ${}^{4}$He, ${}^{6}$He and ${}^{8}$He nuclei together with doublet splitting in ${}^{4}_{\Lambda}$H, ${}^{6}_{\Lambda}$H and ${}^{8}_{\Lambda}$H hypernuclei

Hypernuclear community has accumulated a good experience how to extract the spin-dependent part of phenomenological $\Lambda N$ interaction

$$V_{\Lambda N}(r) = V_0(r) + \underbrace{V_\sigma(r)\,\mathbf{s}_N\,\mathbf{s}_\Lambda}_{\text{“}\Delta\text{"}} + \underbrace{V_\Lambda(r)\,\mathbf{l}_{\Lambda N}\cdot\mathbf{s}_\Lambda}_{\text{“}S_\Lambda\text{"}} + \underbrace{V_N(r)\,\mathbf{l}_{\Lambda N}\cdot\mathbf{s}_N}_{\text{“}S_N\text{"}} + \underbrace{V_T(r)\,S_{12}}_{\text{“}T\text{"}}$$

from doublet splittings [11, 12].

We look ahead at the calculations that remain to be done in order to make a link between the energy of the first excited state $1^+$ in ${}^{4}_{\Lambda}$H, ${}^{6}_{\Lambda}$H and ${}^{8}_{\Lambda}$H hypernuclei and the spin dependent part of baryon - baryon interaction in neutron rich systems. The understanding of the structure of isospin asymmetric systems plays a key role in description of systems as diverse as the neutron rich nuclei and neutron stars.

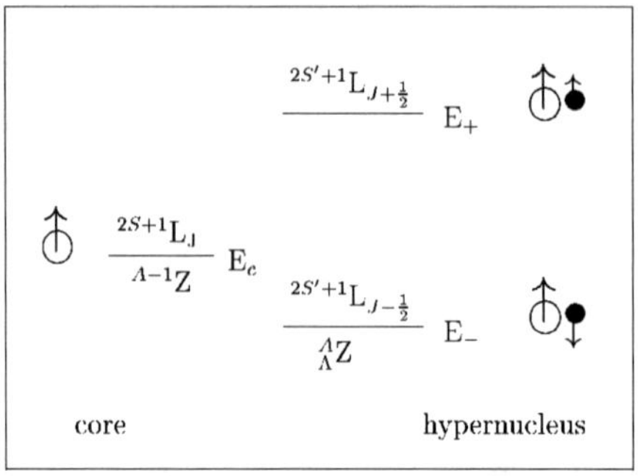


**FIGURE 3.** Schematic picture of low-lying levels

**TABLE 1.** Contributions from the spin-dependent components to the $(1^+, 0^+)$ doublet splitting

| **core** | **doublet** | **interaction:** | $\Delta$ | $S_\Lambda$ | $T$ |
|---|---|---|---|---|---|
| $^2S_{\frac{1}{2}}$ | $(^3S_1, ^1S_0)$ | | 1 | | |
| $^5D_{\frac{1}{2}}$ | $(^5D_1, ^5D_0)$ | | | $-\frac{1}{16}$ | $-\frac{21}{40}$ |
| $^3P_{\frac{1}{2}}$ | $(^3P_1, ^3P_0)$ | | | $-\frac{7}{8}$ | $\frac{3}{2}$ |

## ACKNOWLEDGMENTS

This work was suppotred by Grant Agency of Czech Republic, grant No 2002/02/0930.

## REFERENCES

1. M. Agnello *et al.* (FINUDA Collaboration), Phys. Lett. **B 640**, 145 (2006).
2. S.V. Afanasiev *et al.*, EPJ A (2007), in press (Proceedings HYP06).
3. L. Yuan *et al.*, Phys. Rev. **C 73**, 044607 (2006).
4. L. Majling *et al.*, *in* FINUSTAR, (Eds. S.V. Harissopulos, P. Demetriou, R. Julin), AIP Conference Proceedings, vol. 831, p. 493 (2006).
5. L. Majling, and Š. Gmuca, Phys. At. Nuclei, **70**, 1611 (2007).
6. M.V. Zhukov, B.V. Danilin, D.V. Fedorov, J.M. Bang, I.J. Thompson, and J.S. Vaagen, Phys. Reports **231**, 151 (1993).
7. A.S. Jensen, K. Riisager, D.V. Fedorov, and E. Garrido, Rev. Mod. Phys. **76**, 215 (2004).
8. N. Michel, W. Nazarewicz, J. Okolowicz, and M. Ploszajczak, Nucl. Phys. **A 752**, 335 (2005).
9. A. Volya, and V. Zelevinsky, Phys. Rev. **C 74**, 064314 (2006).
10. R.H. Dalitz, Nucl. Phys. **A 754**, 14c (2005).
11. O. Hashimoto, and H. Tamura, Progr. Part. Nucl. Phys. **57**, 564 (2006).
12. D.J. Millener, *in* Topics in Strangeness Nuclear Physics, (Eds. P. Bydžovský, A. Gal, J. Mareš) Lecture Notes in Physics, **724**, Springer, 2007.

# Decay of relativistic hypernuclei.

Olga Majlingova

*Czech Technical University, Karlovo nam.13, 12000 Praha 2,Czech Rep.*

**Abstract.** The contribution is focused on the analysis of the hypernuclei decay. Hypernuclei, nuclei composed of nucleons and hyperon, enable us to more precise study baryon-baryon interaction, both weak and strong. Several experiments for study new hypernuclear objects are presently taking data or are planned in several laboratories in Italy, Germany, Russia, Japan and USA. The aim of the contribution is the introduction the catalogue of all possible decays of light hypernuclei (A$\leq$12). Created catalogue could be exploited for planning next experiments.

**Keywords:** hypernuclei, decay
**PACS:** 21.80-a

## HYPERNUCLEI

Hypernucleus is a bound system of neutrons, protons and one or more hyperons ($\Lambda, \Sigma, \Xi$). In this contribution only $\Lambda$-hypernuclei are discussed. Notation $^A_\Lambda$Z is used for a hypernucleus with Z protons, A-Z-1 neutrons and a hyperon $\Lambda$. Hypernuclei are a natural bridge connecting nuclear physics of low energies (nuclear spectroscopy) with physics of high energies (physics of elementary particles): hypernuclei are produced in the reactions of high energies, above 1 GeV, by interaction between projectile and one nucleon of target nucleus; the eventual bound state could occur between the hyperon and one fragment of the nucleus. These particular bound states are called hyperfragments. The structure of this many particle system is closely connected to the structure of the host nuclei ($^{A-1}$Z). First hypernuclei were detected by their weak decay in photoemulsions. In these experiments the most important requirement was to have incident particles with energy enough to produce $\Lambda$ hyperon. It was observed that $\Lambda$ particle can stabilize nucleon core. Binding energy of the $\Lambda$ hyperon $B_\Lambda \approx$ A MeV. Later, attention was switched to the quality of the production processes.

For 30 years energy spectra of hypernuclei produced in reactions $^A Z(K^-,\pi^-)^A_\Lambda Z$ and $^A Z(\pi^+,K^+)^A_\Lambda Z$ have been systematically studying. Some time ago, first results obtained from the reaction $^A Z(e,e'K^+)^A_\Lambda(Z-1)$ were published. Recently, Tamura has constructed a large acceptance Ge detector array "Hyperball" dedicated to hypernuclear $\gamma$ ray spectroscopy [1]. With obtained knowledge about hypernuclear structure it was enabled to determine spin dependency in $\Lambda$N interaction and to optimize models of baryon-baryon interaction. In last years hypernuclear physics is used also in astrophysics as a key ingredient for the composition of dense neutron star matter and for cooling history of neutron stars [2].

Hypernuclear physics is currently attracting renewal attention. Several experiments are presently taking data (FINUDA at Frascati, Italy, CEBAF at TJLab, USA) or are planned for both, the near and the more remote future. FINUDA at DA$\Phi$NE is exploring

CP1012, *Frontiers in Nuclear Structure, Astrophysics, and Reactions: FINUSTAR 2*
edited by P. Demetriou, R. Julin, and S. V. Harissopulos
© 2008 American Institute of Physics 0-7354-0532-5/08/$23.00

the production single hypernuclei with tagged stopped kaons. Extended program for experiments with relativistic hypernuclei was offered for Nuclotron at Dubna [3]. The list of "first day" experiments contains not only strictly hypernuclear ones but also the experiments corresponding to the actual problems in nuclear physics, e.g. study of the neutron rich nuclei structure. The production of these hypernuclei in the reactions $(K^-,\pi^+)$ and $(\pi^-,K^+)$ proved to be very difficult. Meaningful alternative could be the uniqueness identification of hyperfragments. Planned new experiments will use the new types of production reactions: neutron rich hypernuclei production (FINUDA, J-PARC, Nuclotron), proton rich hypernuclei production (HypHI). Currently, new experiments on existing facilities are planned as well as new projects for the new accelerators are prepared, so the chart of hypernuclei will quantitatively increase.

New facilities open the door for the study of the qualitatively new effects, concerning with weak decays:

- "rare" weak decay modes are studied in FINUDA,
- identification of ${}^{6}_{\Lambda}$H at Nuclotron is prepared,
- exclusive channels of non mesonic weak decay at Nuclotron are planned.

This was the motivation to create the complete catalogue of possible weak decay channels of light hypernuclei ($A \leq 12$) [4]. This catalogue could be exploited for planning next experiments, for selection of the proper nuclei as well as for kinematic analysis of (multi-particle) decays. We propose to deal not only with two particle decay modes. For three and four particle decays technique of Dalitz plot is used [5]. In next section some examples from the catalogue are presented.

## HYPERNUCLEI DECAY

Main features of the hypernuclear weak decay were mentioned in [5]. Reliable model for three and four particle decays is still open problem in subnuclear physics. The model should respect both mesonic and non-mesonic decay modes, weak and strong fragmentation with charged and neutral ($\pi^0$,n) final particles.

**Rare Decays**

Operating detector FINUDA allows the registration very weak, "rare" two-body decay modes. Current experiments look for ${}^{4}_{\Lambda}\mathrm{H} \rightarrow d+d$ or ${}^{4}_{\Lambda}\mathrm{H} \rightarrow t+p$. With no doubt, we can find more such cases of appropriate decays. Some examples are given in table1.

**TABLE 1.** Examples of "Rare" Decay Modes

| | ${}^{10}_{\Lambda}\mathbf{B}$ | ${}^{10}_{\Lambda}\mathbf{Be}$ | ${}^{9}_{\Lambda}\mathbf{Be}$ | ${}^{7}_{\Lambda}\mathbf{Li}$ |
|---|---|---|---|---|
| *two particles* | | ${}^{7}\mathbf{Li}+\mathbf{t}$<br>${}^{6}\mathbf{He}+\alpha$ | ${}^{7}\mathbf{Li}+\mathbf{d}$<br>${}^{6}\mathbf{Li}+\mathbf{t}$<br>${}^{6}\mathbf{He}+{}^{3}\mathbf{He}$ | $\alpha+\mathbf{t}$ |
| *three particles* | $\alpha+\alpha+\mathbf{d}$ | ${}^{6}\mathbf{He}+\mathbf{d}^2$<br>${}^{4}\mathbf{He}+\mathbf{t}^2$ | ${}^{3}\mathbf{He}+\mathbf{t}^2$ | $\mathbf{t}^2+\mathbf{p}$ |
| *four particles* | | $\mathbf{t}^2+\mathbf{d}^2$ | | |

Special attention should be paid for the Dubna's methods of relativistic hypernuclei production and focused on their decay. Decays were used especially for identifications of hyperfragments as well as for defining of some partial decay modes. Envisaged experiment in Dubna enables development in this way.

**Hydrogen Hypernuclei Decay**

From the table 2, fragment of table of the pionic hydrogen hypernuclei decay given in [5], it is clear that for example ${}^{6}_{\Lambda}$H or ${}^{8}_{\Lambda}$H can be observed through two-particle decay technique. In this way is prepared experiment for identification ${}^{6}_{\Lambda}$H hypernuclei at Nuclotron.

**TABLE 2.** $\pi^-$ Decay Channels ${}^{6}_{\Lambda}$**H** and ${}^{8}_{\Lambda}$**H**

| | ${}^{6}_{\Lambda}$**H** | ${}^{8}_{\Lambda}$**H** |
|---|---|---|
| *two particles* | ${}^{6}$**He** $+\pi^-$ | ${}^{8}$**He** $+\pi^-$ |
| *three particles* | **tt** $+\pi^-$ | |

**Non Mesonic Weak Decays**

Recently was proposed [6] to use the unique feature of the ${}^{9}$Be nucleus, namely that after removing a neutron from its ground state several groups of alpha-particles appear from different excited states of a residual nucleus ${}^{8}$Be. Due to their cluster structure -$\alpha\alpha N\Lambda$ - it may be possible to measure in the ${}^{10}_{\Lambda}$Be and ${}^{10}_{\Lambda}$B hypernuclei the exclusive "alpha-decay widths" $\Gamma^{\tau}_{\alpha\alpha i}$ , corresponding to states of the residual nucleus ${}^{8}$Be$^*(E_i; J_i^{\pi}; T_i)$ which decays through the $\alpha\alpha$-channel [6]. There is also possibility to explore the production of hypernuclei in relativistic heavy ion collisions [3]. According to the catalogue, non mesonic decay can be studied observing ${}^{9}_{\Lambda}$Be or ${}^{10}_{\Lambda}$Li.

# CONCLUSION

We expect confirmation of the existence of the ${}^{6}_{\Lambda}$H nucleus identified as 2-particle decay mode. Methods and techniques for three particle decay will give the possibility to identify more hypernuclei.

# ACKNOWLEDGMENTS

I am grateful to L. Majling for many constructive discussions. This work was supported by grant CTU 0710512.

# REFERENCES

1. O.Hashimoto and H.Tamura, Progr. Part. Nucl. Phys. **57**, 564 (2006).
2. J. Schaffner-Bielich, in *P*roc. HYP06, Mainz, Eur. Phys. J. A (2007), in press.
3. S.V. Afanasiev et al., in *P*roc. HYP06, Mainz, Eur. Phys. J A (2007), in press.
4. L .Majling and O. Majlingova,"Catalogue of Hypernuclear Decay Modes", tbp.
5. O. Majlingova and V. Sopko, AIP Conference Proceedings (Eds. S.V. Harissopulos, P. Demetriou, R. Julin), vol. 831, p. 496 (2006).
6. Y. Batusov et al., Physics Part. and Nuclei **36**, 169 (2005).

# U, Np, Pu & Am Prompt Fission Neutron Spectra

Vladimir M. Maslov

*Joint Institute of Nuclear and Energy Research, 220109, Minsk, Belarus*

**Abstract.** Prompt fission neutron spectra (PFNS) components due to soft and hard pre-fission neutrons are revealed in PFNS data of $^{232}$Th(n,F), $^{238}$U(n,F), $^{235}$U(n,F) and $^{239}$Pu(n,F) reactions for $E_n \leq 20$ MeV. Average energies $\langle E \rangle$ of these PFNS are systematically shifted to higher values, so that Th fission fragments look least heated, while those of Pu - most heated. The average energy $\langle E \rangle$ is correlated with the emissive fission chances contributions to the observed fission cross sections. The predicted contribution of (n, $x$nf) neutrons is most pronounced in case of $^{232}$Th(n,F) reaction. The approach, based on the consistent description of $^{237}$Np(n,F), $^{237}$Np(n,2n)$^{236s}$Np and $^{241}$Am(n,F), $^{241}$Am(n,2n) is used to predict the PFNS of the $^{237}$Np(n,F) and $^{241}$Am(n,F) reactions.

**Keywords:** Prompt Fission Neutron Spectra
**PACS:** 24.75.+1;25.40.h;25.85.Ec

## INTRODUCTION

Analysis of measured prompt fission neutron spectra (PFNS) for neutron-induced fission of $^{238}$U(n,F) shows that a number of data peculiarities could be precisely correlated with the (n, $x$nf) pre-fission neutron spectra [1]. Predicted partial contributions of the $^{238}$U(n, $x$nf) fission reactions to the observed $^{238}$U(n,F) cross section allow a reproduction of the measured PFNS and their average energy $\langle E \rangle$ variation over the incident neutron energy range of $E_n = 5-20$ MeV. These predicted $^{238}$U(n, $x$nf) partial contributions are supported by the surrogate fission data [2, 3] of $^{237}$U(n, f) reaction cross section [4]. The spectra of (n,$x$nf) neutrons are rather soft, as compared with the spectra of neutrons, emitted by the primary fragments after scission of the $^{238}$U and $^{237}$U nuclides. That leads to the lowering of the $\langle E \rangle$ of the PFNS of $^{238}$U(n, F) in the vicinity of $^{238}$U(n, $x$nf) reaction thresholds, observed in [5]. Correlation of this effect with the (n, $x$nf) fission chances is established for the target nuclides with various fissilities [1, 6, 7, 8, 9, 10]. That approach could be used to predict the PFNS of $^{237}$Np(n,F) and $^{241}$Am(n,F) reactions, based on the consistent description of (n,F) and (n,2n) data.

## PROMPT FISSION NEUTRON SPECTRA

Correct estimates of the exclusive pre-fission (n,$x$nf) spectra, alongside with simple modelling of spectra of neutrons, emitted from the fission fragments, allows a reproduction of the PFNS of neutron-induced fission of $^{238}$U(n,F) up to $E_n \sim 20$ MeV. Modelling of the spectra of neutrons, emitted from the fission fragments [11], appears to be rather crude, however it nicely reproduces measured PFNS for $^{238}$U(n, F) [1] and might be used for PFNS prediction for the target nuclides with various fissilities. Increase of the contributions of $^{235}$U(n,nf) or $^{235}$U(n,2nf) to the $^{235}$U observed fission cross section [12, 13, 14],

CP1012, *Frontiers in Nuclear Structure, Astrophysics, and Reactions: FINUSTAR 2*
edited by P. Demetriou, R. Julin, and S. V. Harissopulos
© 2008 American Institute of Physics 0-7354-0532-5/08/$23.00

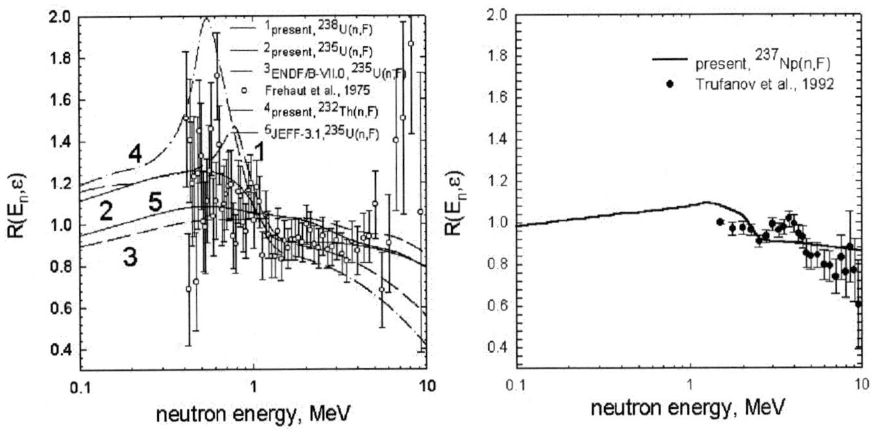


**FIGURE 1.** PFNS, $E_n$=7 MeV (7.8 MeV for $^{237}$Np(n,F)) relative to maxwellian $\langle E \rangle$=2.136 MeV.

actually adds more soft neutrons to the calculated PFNS spectra, but never fills the gap between PFNS data [15] at $E_n$=7 MeV and model calculations [12, 13, 14] (see fig. 1). The effect of (n,nf) neutrons is the strongest for $^{232}$Th(n,F), in case of $^{237}$Np(n,F) reaction [16] the effect of (n,nf) neutrons is shown on the right panel of fig. 1.

Fig. 2 shows the comparison of the present $\langle E \rangle$ of PFNS for $^{232}$Th, $^{238}$U, $^{235}$U and $^{239}$Pu target nuclides, with calculations by Madland [14] for $^{238}$U(n, F), $^{235}$U(n, F) and $^{239}$Pu(n, F). The pre-fission neutrons influence is the strongest in case of $^{232}$Th and weakest in case of $^{239}$Pu. That peculiarity is not due to the fact, that $^{239}$Pu(n, F) fission fragments are most heated, but with the highest contribution of the first chance fission to the $^{239}$Pu(n, F). That is a convincing illustration of the importance of the precise description of the observed PFNS for the investigation of sharing of the excitation energy between the fission fragments and pre-fission neutrons. The right panel of the fig. 2 shows the appreciable influence of the (n,$x$nf) neutrons for the $^{237}$Np(n,F) reaction, which is ignored in nuclear data libraries, used in applied calculations [12, 17].

## CONCLUSIONS

In our calculations there is strict correlation of the "dips" in PFNS average energies with emissive fission contributions to the observed fission cross sections. The pre-fission neutrons influence is the strongest in case of $^{232}$Th and weakest in case of $^{239}$Pu. That peculiarity is not due to the fact, that $^{239}$Pu(n,F) fission fragments are most heated, but with the highest contribution of the first chance fission to the $^{239}$Pu(n,F) fission cross section. That is a convincing illustration of the importance of the precise description of the observed PFNS for the investigation of sharing of the excitation energy between the fission fragments and pre-fission neutrons.

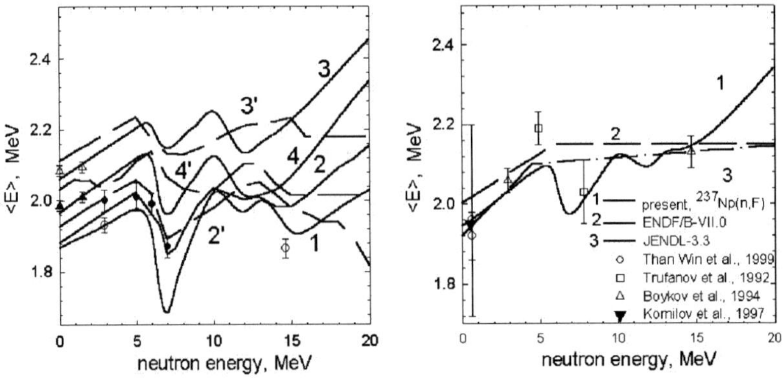


**FIGURE 2.** ⟨E⟩ of PFNS for $^{232}$Th(n,F)(1), $^{238}$U(n,F)(2,2'), $^{235}$U(n,F)(4,4') and $^{239}$Pu(n,F)(3,3'), solid lines - present, dashed lines 2', 4', 3'- [12, 14]; △ - $^{239}$Pu(n,F) [19, 20], ▲ - $^{235}$U(n,F) [19, 20]; • - $^{238}$U(n,F) [18, 21, 22, 23]; ○ - $^{232}$Th(n,F) [18, 21]; right panel - $^{237}$Np(n,F) [16, 22, 24, 25].

# REFERENCES


1. V.M. Maslov et al., Phys. Rev. C **69**, 034607 (2004).
2. C. Plettner et al., Phys. Rev. C **71**, 051602(R) (2005).
3. J.T. Burke et al., Phys. Rev. C **73**, 054604 (2006).
4. V.M. Maslov, Phys. Rev. C **72**, 044607 (2005).
5. T. Ethvignot et al., Phys. Rev. Lett., **94**, 052701 (2005).
6. V.M. Maslov et al., in *Proc. International Conference on Nuclear Data for Science and Technol., Santa Fe, 2004*, (AIP,2004) p. 191.
7. V.M. Maslov et al., Nucl. Phys. A **760**, 274 (2005).
8. V.M. Maslov, VANT, Ser. Fission Reactors, **2**, 33 (2006).
9. V.M. Maslov, Atomnaya Energiya, in press (2007).
10. V.M. Maslov, Phys. Atom. Nucl., in press (2007).
11. N.V. Kornilov et al., Phys. Atom. Nucl. **62**, 173 (1999).
12. M.B. Chadwick et al., Nucl. Data Sheets **107**, 2931 (2006).
13. D.C. Madland and J.R. Nix, Nucl. Sci. Eng. **81**, 213 (1982).
14. D.G. Madland, Nucl. Phys. A **772**, 113 (2006).
15. J. Frehaut et al., in *3d USSR' Conf. on Neutron Physics, Kiev, 1975*. (Moscow, 1975) vol. 5, p. 349.
16. A.M. Trufanov et al., Phys. Atom. Nucl. **55**, 289 (1992).
17. The JEFF-3.1 Nuclear Data Library, Edited by A. Koning et al., NEA No. 6190, OECD.
18. G.N. Lovchikova et al., in: *Proceedings of the XIV International Workshop on Nuclear Fission Physics, Obninsk, 2000* (Obninsk, 2000), p. 72.
19. B.I. Starostov et al., VANT, Ser. Nuclear Constants, **3**, 16 (1985).
20. G.N. Lovchikova et al., Phys. Atom. Nucl. **62**, 1551 (1999).
21. G.S. Boykov et al., Phys. Atom. Nucl. **53**, 628 (1991).
22. G.S. Boykov et al., Ann. Nucl. Energy, **21**, 585 (1994).
23. G.N. Lovchikova et al., Phys. Atom. Nucl. **67**, 1246 (2004).
24. Than Win et al., J. Nucl. Sci. Techn. **36**, 486 (1999).
25. N.V. Kornilov et al., in *Proc. International Conference on Nuclear Data for Science and Technol., Trieste, 1997*, (IPS,1998) p. 577.

# Reaction dynamics with light weakly bound Radioactive Ion Beams at near-barrier energies

M. Mazzocco*, T. Glodariu†, B. Martin**, D. Pierroutsakou**, C. Signorini*, R. Bonetti‡, A. De Rosa**, F. Farinon§, A. Guglielmetti‡, G. Inglima**, M. La Commara**, C. Mazzocchi¶, M. Romoli**, M. Sandoli**, L. Stroe†, F. Soramel‖ and E. Vardaci**

*Dipartimento di Fisica and INFN - Sezione di Padova, via Marzolo 8, I-35131 Padova, Italy
†NIPNE, 407 Atomistilor st., 077125, Magurele, Romania
**Dipartimento di Scienze Fisiche and INFN - Sezione di Napoli, via Cinthia, I-80126 Napoli, Italy
‡Istituto di Fisica Generale ed Applicata and INFN - Sezione di Milano, via Celoria 16, I-20133 Milano, Italy
§GSI, Planckstrasse 1, D-64291 Darmstadt, Germany
¶Istituto di Fisica Generale ed Applicata, via Celoria 16, I-20133 Milano, Italy
‖Dipartimento di Fisica and INFN - Sezione di Udine, via delle Scienze 208, I-33100 Padova, Italy

**Abstract.** In the present contribution, we will briefly present an overview of the last experimental results obtained by the Italian EXOTIC collaboration on the study of the reaction mechanisms around the Coulomb barrier with light weakly bound Radioactive Ion Beams. Several experiments have been performed by our group in different laboratories: at ANL (USA) and, more recently, at LNL (Italy) to study the system $^{17}$F + $^{208}$Pb and at RIKEN (Japan) for the system $^{11}$Be + $^{209}$Bi. A comparative discussion with the results obtained for other reactions induced by light weakly- and tightly-bound projectiles will be finally given.

**Keywords:** Elastic scattering, Interaction and reaction cross section
**PACS:** 25.60.Bx, 25.60.Dz

The interaction of light weakly bound nuclei with heavy target in the energy range around the Coulomb barrier is presently one of the most challenging topics in Nuclear Physics. The exotic features of these projectiles, namely the halo structure, the neutron skin structure and the small binding energy, give rise to a scenario completely different from that reported for reactions induced by stable nuclei. Very loosely bound projectiles can easily break into two or more fragments while they are approaching the target nucleus. Different theoretical predictions exist on the role played by the breakup channel in the reaction dynamics at near-barrier energies. If this channel is considered as an additional reaction mechanism, the fusion cross section is expected to be enhanced with respects to what foreseen by the one-dimension barrier penetration model. On the other side, if the breakup process is considered as a flux removed from the entrance reaction channel, the fusion process is expected to be hindered. Several review have been recently written on this topic [1, 2, 3].

The experimental data collected at near-barrier energies for the $^{6}$He ($S_{2n}$ = 0.975 MeV) halo nuclei interacting with $^{238}$U [4], $^{64}$Zn [5] and $^{63}$Cu [6] targets showed only moderate enhancements of the fusion cross sections with respect to the reactions induced by the stable tightly bound $^{4}$He nucleus. On the contrary, the reaction cross sections

CP1012, *Frontiers in Nuclear Structure, Astrophysics, and Reactions: FINUSTAR 2*
edited by P. Demetriou, R. Julin, and S. V. Harissopulos
© 2008 American Institute of Physics 0-7354-0532-5/08/$23.00

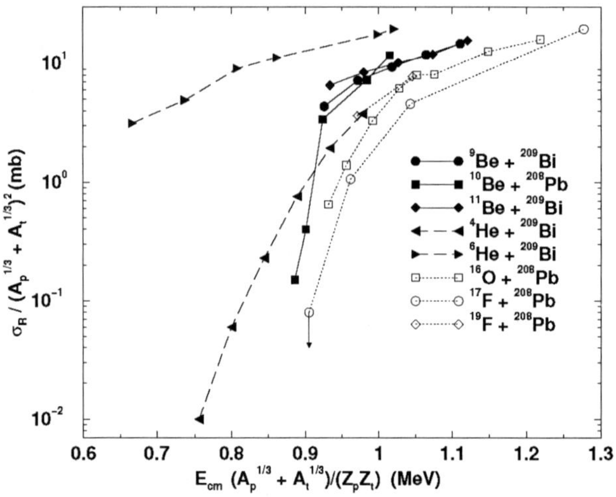


**FIGURE 1.** Reduced reaction cross sections for systems consisting of light projectiles interacting with $^{208}$Pb or $^{209}$Bi targets. The reduction has been carried out according to the procedure described in [15].

turned out to be largely increased with respect to the $^{4}$He reference case. Detailed investigations of the reaction mechanisms revealed that breakup and nucleon transfer channels were the major responsible of such a large enhancement. Data collected with higher statistical accuracy for the systems involving the most weakly bound stable projectiles, namely $^{6}$Li ($S_d$ = 1.475 MeV) + $^{208}$Pb [7] and $^{9}$Be ($S_n$ = 1.675 MeV) + $^{209}$Bi [8], confirmed the relevance of breakup and transfer reaction channels in the energy range around the Coulomb barrier.

Within this framework our collaboration has undertaken the study of the systems $^{17}$F ($S_p$ = 0.601 MeV) + $^{208}$Pb and $^{11}$Be ($S_n$ = 0.504 MeV) + $^{209}$Bi. We mainly performed elastic scattering experiments, since they constitute basic measurements: $(i)$ to determine the interaction potential between the colliding nuclei; $(ii)$ to investigate the influence of the projectile low binding energy and halo structure on the reaction dynamics and $(iii)$ to extract the reaction cross section.

The first system was studied in two experiments. The former was performed at the Argonne National Laboratory (ANL) [9] in 2002 with a $^{17}$F secondary beam of 90.4 MeV ($E_{CM}$ / $V_b$ = 1.04) and an intensity of $\sim$ 1.0 $\times 10^5$ pps. The latter was carried out at the Laboratori Nazionali di Legnaro (LNL) in 2006. The scattering process of $^{17}$F ions from a $^{208}$Pb target was studied at sub-barrier energies ($E_{CM}$ / $V_b$ = 0.98) and the secondary beam had an intensity of $\sim$ 3.6 $\times 10^6$ pps. The reaction $^{11}$Be + $^{209}$Bi was studied at RIKEN in 2004 [12, 13]. In this case the experimental data spanned the energy range from 40 MeV ($E_{CM}$ / $V_b$ = 1.01) to 48 MeV ($E_{CM}$ / $V_b$ = 1.21) and the beam intensity was $\sim 10^5$ (integrated over the whole energy range). The large area and high granularity EXODET array [14] was employed for the detection of the light scattered particles in all the mentioned experiments.

The reaction cross sections deduced from the optical model analysis of the scattering angular distributions are displayed in Fig. 1 together with those obtained for other light projectiles interacting with high-Z spherical targets, $^{208}$Pb or $^{209}$Bi. The data in Fig. 1 have been plotted according to the reduction procedure described in [15]. This reduction accounts for the different Coulomb barrier heights and the geometrical size of the colliding nuclei, without washing out the effects related to the halo structure.

Fig. 1 shows that the reduced cross section for the $^{17}$F weakly bound nucleus does not differ significantly from that measured for the stable doubly magic $^{16}$O [10]. This is quite surprising, in particular considering the very small $^{17}$F binding energy and the proton halo properties of its first excited state (106 keV below the breakup threshold). We also notice that the reaction cross section for the system $^{19}$F + $^{208}$Pb [11] is larger than that we measured for the reaction induced by the radioactive projectile. We conclude therefore that in this energy range it seems to be easier to excite collective structures (i.e. the $^{19}$F ground-state rotational band) than single-particle states, even in very loosely bound nuclei as $^{19}$F.

On the other side, within the experimental uncertainties, no differences can be observed between the reduced cross sections for the systems $^{9,11}$Be + $^{209}$Bi [13] and $^{10}$Be + $^{208}$Pb [16] in the energy range above the Coulomb barrier (see Fig. 1). This result is also rather unexpected, in particular considering the very small $^{11}$Be binding energy and its well known halo structure. However, one has also to consider that the reaction cross section for $^{9}$Be could be enhanced because of the excitation of its ground state rotational band, while the reaction induced by the well-bound $^{10}$Be ($S_n = 6.81$ MeV) nucleus has a $2n$-transfer channel with positive Q-value, which can increase the reaction probability.

In conclusion, our results indicate that the breakup related effects on the reaction cross section cannot be predicted considering only the projectile halo structure and/or its small binding energy. Couplings to direct reaction channels (as observed for $^{6}$He and $^{10}$Be) as well as specific collective structures (e.g. for $^{9}$Be and $^{19}$F) or single-particle properties (e.g. for $^{17}$F) of the colliding nuclei could dramatically influence the reaction dynamics around the barrier. Of course, new experimental data with higher statistical accuracy and able to unambiguously distinguish between different inelastic channels are welcome to solve the reaction mechanism *puzzle* at near-barrier energies.

## REFERENCES


1. J.F. Liang and C. Signorini, *Int. J. Mod. Phys. E* **14**, 1121 (2005).
2. L.F. Canto, P.R.S. Gomes, R. Donangelo and M.S. Hussein, *Phys. Rep.* **424**, 1 (2006).
3. N. Keeley, R. Raabe, N. Alamanos and J.L. Sida, *Prog. Part. Nucl. Phys.* (in press).
4. R. Raabe et al., *Nature* **431**, 823 (2004)
5. A. Di Pietro et al., *Phys. Rev. C* **69**, 044613 (2004).
6. A. Navin et al., *Phys. Rev. C* **70**, 044601 (2004)
7. C. Signorini et al., *Phys. Rev. C* **67**, 044607 (2003).
8. C. Signorini et al., *Proceedings of the International Conference BO2000*, p. 413.
9. M. Romoli et al., *Phys. Rev. C* **69**, 064614 (2004).
10. I.J. Thompson et al., *Nucl. Phys. A* **505**, 84 (1989).
11. C.J. Lin et al., *Phys. Rev. C* **63**, 064606 (2001).
12. M. Mazzocco et al, *Eur. Phys. J. A* **28**, 295 (2006).
13. M. Mazzocco et al., *Eur. Phys. J. ST* (in press).
14. M. Romoli et al., *IEEE Trans. Nucl. Sci.* **52**, 1860 (2005)
15. P.R.S. Gomes et al., *Phys. Rev. C* **71**, (2005).
16. J.J. Kolata et al., *Phys. Rev. C* **69**, 047601 (2004).

# A DIAMANT Wedding For AFRODITE : Probing Structure and Characterizing Reaction Properties Via Charged-Particle-$\gamma$ Correlations

S.M. Mullins[1], B.M. Nyakó[2], J. Timár[2], G. Berek[2], J. Gál[2], G. Kalinka[2], J. Molnár[2], S.H.T. Murray[1], R.A. Bark[1], E. Gueorguieva[1], K. Juhász[3], A. Krasznahorkay[2], J.J. Lawrie[1], E.O. Lieder[1], R.M. Lieder[1], M. Lipoglavšek[4], S.S. Ntshangase[1,5], P. Papka[1], J.N. Scheurer[6], J.F. Sharpey-Schafer[7,8], O. Shirinda[7], L. Zolnai[2]

[1]*iThemba LABS, P.O.Box 722, Somerset West, 7129, South Africa*
[2]*Institute of Nuclear Research (ATOMKI), P.O.Box 51, H-4001 Debrecen, Hungary*
[3]*Dept. of Information Technology, University of Debrecen, H-4032, Debrecen, Hungary*
[4]*Josef Stefan Institute, 1000 Ljubljana, Slovenia*
[5]*Dept. of Physics, University of Cape Town, Cape Town 7701, South Africa*
[6]*CNRS-IN2P3-Université de Bordeaux I, 33175 Gradignan Cedex, France*
[7]*Dept. of Physics, University of the Western Cape, Bellville 7535, South Africa*
[8]*Dept. of Physics & Engineering, University of Zululand, Kwazulu-Natal, South Africa*

**Abstract.** The DIAMANT-AFRODITE combination has been used to investigate incomplete fusions reactions via the $^{13}$C + $^{170}$Er entrance channel. The intensity of $^{176}$Hf (populated via the $\alpha$ 3n exit channel) is ~8% of $^{178}$W (populated via 5n evaporation) which is ~8 times stronger than that expected from complete fusion. Moreover, 2$\alpha$xn exit channels leading to Yb nuclei are observed with intensities that are ~30-to-40% of $^{176}$Hf, for which no yield is expected from complete fusion. A comparison of the intensities from the two-$\alpha$- and one-$\alpha$-gated data is consistent with fragmentation of the $^{13}$C beam into ($\alpha$-$\alpha$-$\alpha$-$n$) which suggests that the population of Yb nuclei results from fusion (or "massive transfer") of one the break-up $\alpha$-particles. A campaign of measurements is scheduled for late 2007 with further investigations planned for 2008, including the continuation of the study of superdeformation in $^{32}$S.

**Keywords:** HpGe, CsI detectors, $\alpha$-$\gamma$ coincidences, incomplete fusion, superdeformation.
**PACS:** 27.70.+t, 27.70.+q, 21.10.-k, 23.20.Lv, 29.30.Kv, 29.30.Ep

An ongoing bilateral agreement between the governments of South Africa and Hungary has enabled the coupling of the DIAMANT CsI charged-particle detector [1] with the AFRODITE gamma-ray spectrometer [2] at iThemba LABS. The first measurement utilized the 24-detector "Chessboard" from ATOMKI (see left panel of figure 1). This was placed at forward angles in the AFRODITE target chamber in order to record $\alpha$-$\gamma$-$\gamma$ coincidences following the bombardment of a 4mg/cm$^2$ $^{170}$Er target with a $^{13}$C beam at energies of 80 and 70 MeV for one weekend each, respectively. Single-$\alpha$-gated and two-$\alpha$-gated $E_\gamma$-$E_\gamma$ correlation matrices were generated, the total projections of which are shown in the right panel of figure 1 for

CP1012, *Frontiers in Nuclear Structure, Astrophysics, and Reactions: FINUSTAR 2*
edited by P. Demetriou, R. Julin, and S. V. Harissopulos
© 2008 American Institute of Physics 0-7354-0532-5/08/$23.00

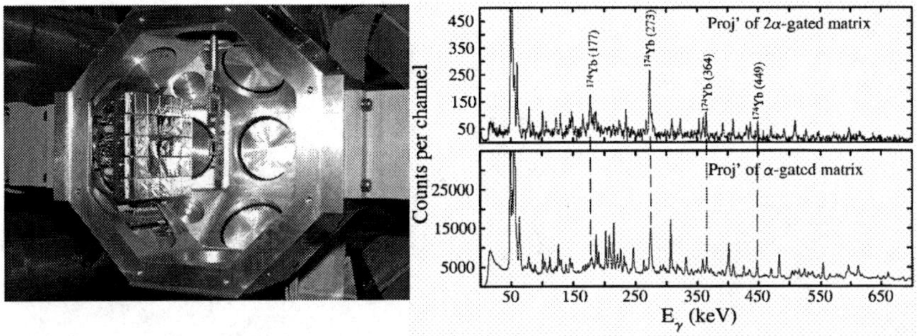


**FIGURE 1.** *Left* : The "Chessboard" inside AFRODITE. *Right* : Total projections of $E_\gamma$-$E_\gamma$ matrices selected by one α (lower panel) and two α particles (upper panel) from the 80 MeV $^{13}$C + $^{170}$Er data.

the 80 MeV data. The comparison of intensities of transitions in the ground-state bands of $^{176}$Hf and $^{178}$W extracted from an $E_\gamma$-$E_\gamma$ matrix with no α selection shows that the former is populated with an intensity that is ~8% of the latter, whereas evaporation codes such as PACE4 [3] predict no more that ~1% of compound nucleus events decay via α*xn* channels. Furthermore, the single-α-gated projection contains transitions not only from $^{176}$Hf and $^{175}$Hf, but also from $^{172}$Yb, $^{173}$Yb and $^{174}$Yb that are the result of 2α*xn* exit channels for which absolutely no measurable yield is expected from fusion-evaporation processes. The Yb nuclei are observed with ~30-to-40% of the intensity of $^{176}$Hf in the single-α-gated matrix. This raises the question whether the $^{13}$C beam predominantly breaks into α + $^{9}$Be components in the incomplete fusion process, so that fusion-evaporation instigated by the $^{9}$Be or α fragment populates Hf or Yb nuclei, respectively. This would imply there was no separation between an α particle or a $^{9}$Be fragment in the CsI detectors. Comparison of the intensities of $^{176}$Hf ground-state band transitions in a matrix generated purely from the $E_\gamma$-$E_\gamma$ coincidence data with those in the single-α-gated matrix indicate that the efficiency for the detection of a single α particle was ~15%. Hence, the efficiency for the detection of two α particles should be $(\sim 0.15)^2 \times (23/24) \approx 2\%$. The ratio of the intensities for the $^{174}$Yb ground-state band taken from the single-α and two-α-gated matrices is ~1.7%. This suggests that the predominant mode of population of the Yb nuclei is via complete break-up of the $^{13}$C beam into (α-α-α-*n*) with fusion-evaporation initiated by massive transfer of one of the α particles. A more detailed understanding of the break-up/massive transfer/incomplete fusion process would benefit from a measurement with the full DIAMANT array in order to have greater angular coverage and granularity.

In figure 2 is shown a potential energy surface for $^{32}$S with a deep minimum which corresponds to a superdeformed (SD) shape [4]. The lowest two states that belong to the SD band are in fact known (see [5]) and the 4.2 MeV γ-ray transition that de-excites the $2^+$ member to the ground state has been previously identified [6]. This prompted us to investigate $^{32}$S to identify the higher spin members of the SD band, making use of the high intrinsic efficiency of the AFRODITE clovers detectors.

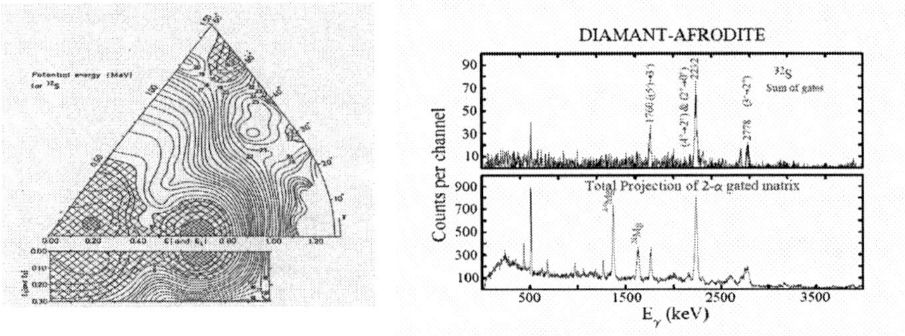


**FIGURE 2.** *Left* : Potential Energy Surface for $^{32}S$ which shows the superdeformed minimum [4]. *Right* : (lower) Total projection of a two-α-gated matrix ; (upper) Sum-of-gates coincidences pectrum for $^{32}S$.

States in $^{32}S$ were populated via the $^{24}Mg(^{16}O,2\alpha)^{32}S$ reaction at a beam energy of 50 MeV. The total projection spectrum of an $E_\gamma$-$E_\gamma$ matrix selected by 2α particles is shown in the lower panel in the right of figure 2. Transitions belonging to $^{32}S$ can be seen, as well as some from $^{24}Mg$ which resulted from $^{16}O(^{16}O,2\alpha)^{24}Mg$ reactions due to target oxidation. However, there is no sign of the 4.2 MeV transition which depopulates the $2^+$ SD state, whereas members of the $^{38}Ar$ SD band [7] were populated via the 2p exit channel, as could be seen in a 2p-gated $E_\gamma$-$E_\gamma$ matrix.

Further measurements will take place in October and November of 2007 with the full implementation of DIAMANT. These will focus on the study of the neutron-rich nucleus $^{32}Si$ and the investigation of the role of incomplete fusion in the population of $^{142}Eu$ via the $^{51}V + ^{97}Mo \rightarrow \alpha 2n$ or 2p4n reactions. Further measurements, including a revisit to $^{32}S$ with a different reaction, are planned to take place during the course of 2008 with DIAMANT coupled to AFRODITE enhanced with additional clover detectors in order to be able to collect charged-particle-triple-γ coincidence data.

## ACKNOWLEDGMENTS

This work is supported under a governmental bilateral agreement via grants from the South African National Research Foundation (GUN 2067469 and GUN 61851), and the Hungarian NKTH and KPI (TéT DAK-5/03 and *ZA*-7/2006), respectively.

## REFERENCES

1. J. N. Scheurer et al., Nucl. Instr. Meth A **385**, 501, (1997).
2. R.T. Newman *et al, Balkan. Phys. Lett.* Special Issue, 182 (1998).
3. www.nscl.msu.edu/lise
4. G. Leander and S. E. Larsson, Nucl.Phys. **A239**, 93 (1975).
5. R. K. Sheline, I. Ragnarsson and S. G. Nilsson, Phys. Lett. **41B**, 115 (1972).
6. P. E. Carr, D. C. Bailey, L. L. Green, A. N. James, J. F. Sharpey-Schafer and D. A. Viggars, J. Phys. A: Math. Nucl. Gen. **6**, 705 (1973).
7. D. Rudolph et al., Phys. Rev **C 65**, 034305 (2002).

# Alpha particle condensation in $^{12}$C and nuclear rainbow scattering

S. Ohkubo* and Y. Hirabayashi†

* *Department of Applied Science and Environment, Kochi Women's University, Kochi 780-8515, Japan*
† *Information Initiative Center, Hokkaido University, Sapporo 060-0811, Japan*

**Abstract.** It is shown that the large radius of the Hoyle state of $^{12}$C with a dilute density distribution in an $\alpha$ particle condensate can be clearly seen in the shift of the rainbow angle (therefore the Airy minimum) to a larger angle in $\alpha$+$^{12}$C rainbow scattering at the high energy region and prerainbow oscillations in $^3$He+$^{12}$C scattering at the lower energy region.

**Keywords:** Bose Einstein condensation, alpha particle, cluster, $^{12}$C, nuclear rainbow
**PACS:** 03.75.Nt,21.60.Gx,25.55.Ci,27.20.+n

Much attention has been paid to $\alpha$ particle condensation in nuclei. $^{12}$C is a key nucleus in the nucleosynthesis. The cluster structure of $^{12}$C has been thoroughly investigated in the cluster model by Uegaki *et al.* of Hokkaido group [1] and Kamimura *et al.* of Kyushu group [2], which shows that the $0_2^+$ (7.65 MeV) state of $^{12}$C, which is 0.39 MeV above the $\alpha$ threshold, has a well-developed $\alpha$ cluster structure with the $^8$Be$\otimes\alpha$ configuration. Tohsaki *et al.* [3] conjectured that this Hoyle state may be a BEC state of three $\alpha$ particles with a dilute density distribution obeying Bose statistics. For an $\alpha$ particle condensate, a physical quantity peculiar to condensation would appear. Because of the dilute property, one of the characteristic features of the $\alpha$ particle condensate is that its radius is very large compared with a normal state. However, it is difficult to measure the radius of the excited Hoyle state and no experiment has been reported. Therefore it is important to know whether the radius of the Hoyle state is very large.

We show that nuclear rainbow scattering [4] is very useful in confirming the large radius of the Hoyle state. The large radius means that the potential for the Hoyle state is extended compared with the normal ground state. The nucleus acts as a lens. The huge radius of the Hoyle state means that the refraction of an incident particle is much stronger than for the normal ground state.

On the other hand, it has been shown that inelastic Airy structure can be understood in a way similar to elastic scattering [5]. This suggests that inelastic rainbow and prerainbow oscillation may also be useful for the study of the nuclear properties of the excited states of the target nucleus because the internal region of the interaction potential can be well determined even for inelastic channels.

We analyze $\alpha$+$^{12}$C scattering with a double folding model potential in the coupled channel calculations. The diagonal and coupling potentials for the $\alpha$+$^{12}$C system are calculated from the transition nucleon density of $^{12}$C derived from the precise microscopic cluster model wave functions [2]. The wave function for the Hoyle state of $^{12}$C [2] used is almost completely in agreement with the BEC wave function. In the coupled

CP1012, *Frontiers in Nuclear Structure, Astrophysics, and Reactions: FINUSTAR 2*
edited by P. Demetriou, R. Julin, and S. V. Harissopulos
© 2008 American Institute of Physics 0-7354-0532-5/08/$23.00

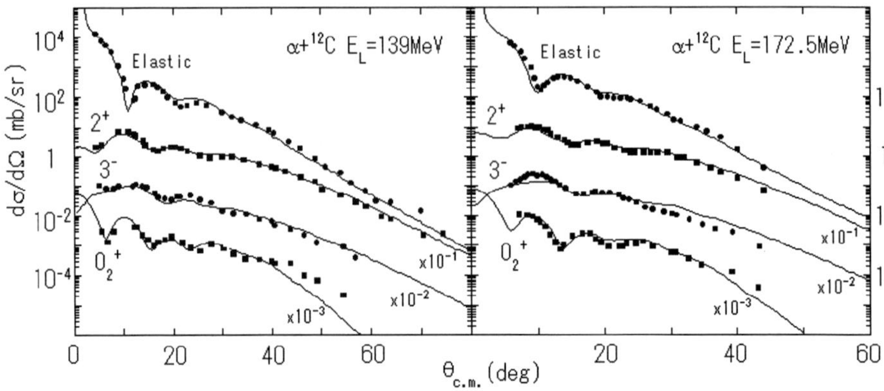


**FIGURE 1.** Calculated angular distributions (solid lines) in $\alpha$+$^{12}$C rainbow scattering at 139 and 172.5 MeV are compared with the experimental data (points).

channel calculations the ground state $0_1^+$, $2_1^+$ (4.43 MeV), $3_1^-$ (9.63 MeV) and $0_2^+$ (7.65 MeV) states are simultaneously taken into account. In the calculation the normalization factor $N_R$ for the real part of the potential is introduced. Imaginary potentials are introduced phenomenologically for each channel.

In Fig. 1 the calculated angular distributions in $\alpha$+$^{12}$C rainbow scattering reproduce the experimental data well. We note that the rainbow angle (therefore the Airy minimum) for the Hoyle state is shifted considerably to a larger angle compared with that for the ground state in agreement with the dilute density distribution due to the $\alpha$ particle condensation [6].

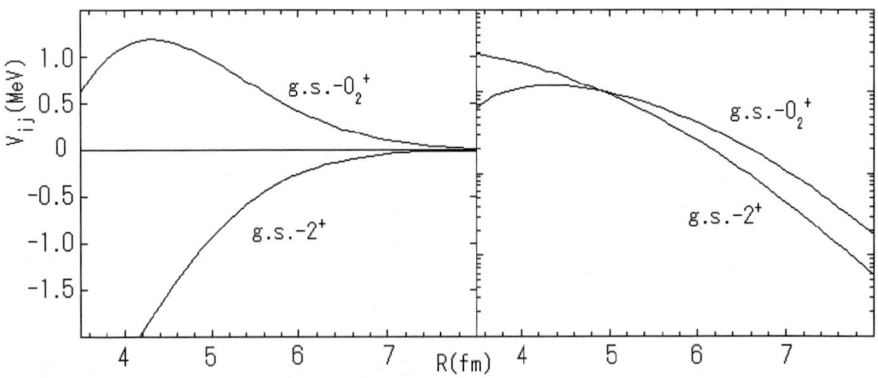


**FIGURE 2.** Coupling potentials for the g.s.-$2_1^+$ and g.s.- $0_2^+$ channels in $\alpha$+$^{12}$C scattering at $E_L$=139 MeV are shown on a linear scale (left) and their absolute values are shown on a logarithmic scale (right) to emphasize the difference at the surface region.

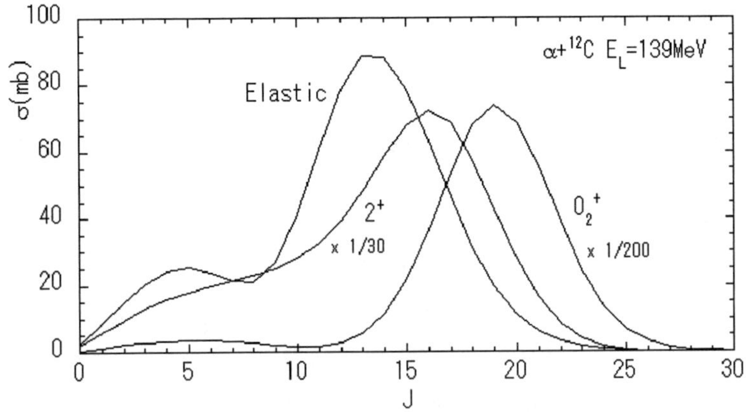


**FIGURE 3.** Calculated partial cross sections for elastic $\alpha+^{12}$C scattering and inelastic scattering to the $2_1^+$ and $0_2^+$ state at $E_L$=139 MeV are shown as a function of total angular momentum.

In Fig. 2. the coupling potentials for the g.s.-$2_1^+$ and g.s.- $0_2^+$ channels are compared. It is clear that the coupling potential for the g.s.-$0_2^+$ channel is considerably extended to the outer region and the inelastic scattering to the Hoyle state is enhanced in this surface region. In Fig. 3 we see that the inelastic scattering to the Hoyle state occurs at very large angular momenta (large radius) compared with the normal g.s. and $2_1^+$ states. These facts account for the shift of the rainbow angle (Airy minimum) of the Hoyle state to a larger angle and for the idea that the Hoyle state has a large radius compared with the normal ground state.

The above finding is also confirmed in the prerainbow oscillations at the lower energy region in $^3$He+$^{12}$C scattering [7] and details will be reported elsewhere.

## ACKNOWLEDGMENTS

One of the authors (S.O.) has been supported by a Grant-in-aid for Scientific Research of the Japan Society for Promotion of Science (No. 16540265).

## REFERENCES


1. E. Uegaki , S. Okabe, Y. Abe, and H. Tanaka, *Prog. Theor. Phys.* **57**, 1262 (1977); E. Uegaki, Y. Abe, S. Okabe, and H. Tanaka, *Prog. Theor. Phys.* **59**, 1031 (1978); **62**, 1621 (1979).
2. M. Kamimura, *Nucl. Phys.* **A351**, 456 (1981); Y. Fukushima and M. Kamimura, *Proceedings of the International Conference on Nuclear Structure, Tokyo, 1977*, edited by T. Marumori *J. Phys. Soc. Japan Suppl.* **44**,225 (1978).
3. A. Tohsaki , H. Horiuchi, P. Schuck, and G. Röpke, *Phys. Rev. Lett.* **87**, 192501 (2001).
4. D. T. Khoa, W. von Oertzen, H. G. Bohlen, and S. Ohkubo, *J. Phys.* **G 34**, R111 (2007).
5. F. Michel and S. Ohkubo, *Phys. Rev.* **C 70**, 044609 (2004).
6. S. Ohkubo and Y. Hirabayashi, *Phys. Rev.* **C 70**, 041602(R) (2004).
7. S. Ohkubo and Y. Hirabayashi, *Phys. Rev.* **C 75**, 044609 (2007).

# Electron screening and stellar rates in the $^3$He ($^3$He, 2p) $^4$He and $^3$He (d, p) $^4$He fusion reactions

L. Messahel, S. Ouichaoui, A. Belhout, M. Fouka and A. Trabelsi

*U.S.T.H.B., Faculté de Physique, B.P. 32, El - Alia, 16111 Bab Ezzouar, Algiers, Algeria.*

**Abstract.** The astrophysical S(E) factor experimental data available over the energy region *E* (C.M.) < 1.0 MeV for the $^3He(^3He,2p)^4He$ and $^3He(d,p)^4He$ fusion reactions are analyzed using a polynomial expression and the R-Matrix formalism, respectively. The reaction thermonuclear rates for bare nuclei are determined and compared to previous ones after a precise assessment of the electron screening factors. New level parameter values are deduced for the $^5$Li nucleus.

**Keywords:** resonance fusion reaction, R-Matrix theory, stellar nucleosynthesis, reaction rate, electronic screening potential
**PACS:** 24.30.-v, 25.60.Pj, 26.20.+f, 26.10.+a.

The reactions $^3$He ($^3$He, 2p)$^4$He and $^3$He(d, p)$^4$He are of considerable importance in nuclear astrophysics. The former contributes to 86% in the stellar p-p chain and was suspected as a possible solution to the solar neutrino puzzle, while the latter is involved in the primordial nucleosynthesis generating light elements D, $^3$He, $^4$He and $^7$Li, and could be used for clean energy production in fusion reactors. However, both two reactions are strongly affected by the electron screening effect which requires be accurately evaluated. This work consists in a re-analysis of the published experimental cross sections of the two reactions at astrophysical energies, E < 1MeV, in terms of the astrophysical S(E) factor. Due to inconsistencies in previous analyses and to the large uncertainties in the published values of the $U_e$ electron screening potential, the most precise determination of the latter is aimed at herein. The most recent astrophysical S(E) factor experimental data for both two reactions have been considered for evaluating the cross section enhancement factor, fitted by the exponential function:

$$f_e(E) \approx \exp(\pi\eta\, U_e / E)\ , \tag{1}$$

where $\eta$ is the Sommerfeld parameter.

For the $^3He(^3He,2p)^4He$ reaction, the S(E) data from Refs.[1-4] have been considered in a linear polynomial fit assuming the contributions of only s and p-waves to which the total astrophysical *S*(*E*) factor has then been decomposed. Over the energy range $E$ = 16.5 - 342.5 keV where experimental data are considered (for stellar temperatures ≤ 0. ), the linear expression:

$$S(E) = 5.29 - 2.84\,\mathrm{E}\ , \tag{2}$$

with *E* being expressed in MeV units and the zero-energy value given by

$$S(0) = 5.296 \pm 0.6784\ \mathrm{MeV\,b}\ , \tag{3}$$

CP1012, *Frontiers in Nuclear Structure, Astrophysics, and Reactions: FINUSTAR 2*
edited by P. Demetriou, R. Julin, and S. V. Harissopulos
© 2008 American Institute of Physics 0-7354-0532-5/08/$23.00

is obtained for bare nuclei.

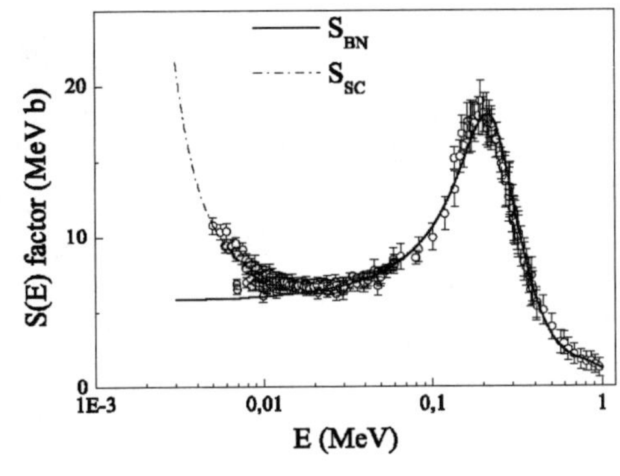


**FIGURE 1.** $^3$He(d,p)$^4$He S($E$) data and R-Matrix fits for screened and bare nuclei.

The calculated reaction rate proves to be consistent with the NACRE compilation [5] and differs, as expected, from the superseded CF88 rate [6]. The extrapolation has been performed by excluding the data points within the energy range E ≤ 40 keV affected by the electron screening, assessed by considering all the experimental data. Fitting Eq. (1) to the $f_e(E)$ data leads to a value $U_e$ = 327 ± 42 eV, in very good agreement with the theoretical one of 328 eV predicted by Bencze's model [7]. The resulting fits to experimental data confirm that the abrupt enhancement of the cross section in the Gamow peak region definitely reflects only the electron screening effect without presence of any hypothetical resonance.

The S(E) data reported in Refs. [1, 8-10] for the $^3$He(d,p)$^4$He reaction have been analyzed using the R-matrix theory, assuming a resonant interaction mechanism. The angular distribution data from Ref. [10] and $^3$He(d,d)$^3$He elastic scattering excitation functions [11] have been considered. Satisfactory simultaneous fits to all experimental data sets have been obtained. Fig.1 reports the S(E) data and corresponding best theoretical fits for screened and bare nuclei, obtained by assuming the g.s. and four excited states of the $^5$Li compound nucleus, whose parameters are reported in Table 1.

An extrapolated S-factor value for bare nuclei, $S_{BN}(0)$ = 5.69 ± 0.81 MeV b, is derived. The calculated reaction rate is found to be in good agreement with the NACRE [5] and LANL01 [12] rates but in net disagreement with the CF88 [6] one, as shown by Fig. 2.

Our fit to the electron screening factor data for this reaction using Eq. (1) leads to a value, $U_e$= 199 ± 23 eV, of the electron screening potential, which is in good agreement with that reported in Ref. [13] within the adiabatic approximation.

**TABLE 1.** Level parameters of $^5$Li deduced by R-Matrix analysis of the $^3$He(d,p)$^4$He reaction data.

| Level number | $J^\pi$ | $E_x(MeV)$ | $E_R(MeV$ | $\Gamma_{cm}(MeV)$ | $\Gamma_p/\Gamma_{to}$ | $\gamma_d^2(MeV)$ |
|---|---|---|---|---|---|---|
| 0 | $\frac{3}{2}^-$ | *g.s.* | – | – | – | – |
| 1 | $\frac{1}{2}^-$ | 1.49 | –14.9 | 1.247 | 1 | – |
| 2 | $\frac{3}{2}^+$ | 16.63 | 0.238 | – | –0.852 | 0.234 |
| 3 | $\frac{1}{2}^+$ | 16.92 | 0.533 | 1.044 | –0.053 | 2.550 |
| 4 | $\frac{3}{2}^-$ | 17.41 | 1.022 | 1.208 | –0.206 | 4.697 |

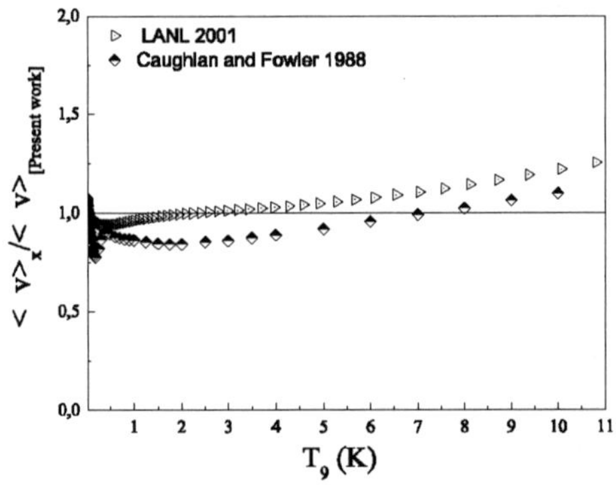


**FIGURE 2.** $^3$He(d,p)$^4$He reaction rate compared to those from Refs. [5, 12].

## REFERENCES


1. A. Krauss et al., Nucl. Phys. A **467**, 273-290 (1987).
2. M. Junker et al., Nucl. Phys. B **70**, 382-385 (1999).
3. C. Arpesella et al., Phys. Lett. B **389**, 452-456 (1996).
4. Nobuyuki Kudomi et al., Phys. Rev. C **69**, 015802 (2004).
5. C. Angulo et al. Nuclear Physics. A **656**, 3-183 (1999).
6. G. R. Caughlan and W. A. Fowler, At. Data Nucl. Data Tables **40**, 283 (1988).
7. G. Bencze. Nucl. Phys. **A492**, 459 (1989).
8. W. Möller and F. Besenbacher, Nucl. Instr. Meth **168**, 111 (1980).
9 S. Engstler et al., Phys. Lett. **B 202**, 179 (1988).
10. W. H. Geist, Phys. Rev. C **60**, 054003 (2003).
11. L. Kraus, Ph. D. thesis, Strasbourg (1971).
12. Los Alamos National Laboratory, http ://t2.lanl.gov/.
13. F. C. Barker, Nuclear Physics A**707**, 277-300 (2002).

# Evidence for a strongly-coupled band in very neutron-deficient odd-mass nucleus $^{185}$Pb

J. Pakarinen*, A.N. Andreyev†, S. Antalic**, L. Bianco*, I.G. Darby*, T. Grahn*, P.T. Greenlees‡, R. Julin‡, S. Juutinen‡, H. Kettunen‡, M. Leino‡, A.-P. Leppänen‡, P. Nieminen‡, D.G. Jenkins§, P. Jones‡, P. Joshi§, M. Nyman‡, R.D. Page*, J. Perkowski‡, P.M. Raddon§, P. Rahkila‡, D. Rostron*, J. Saren‡, C. Scholey‡, B. Streicher**, J. Sorri‡, J. Uusitalo‡, K. Van de Vel‡, M. Venhart‡, D. Wiseman* and R. Wadsworth§

*Department of Physics, Oliver Lodge Laboratory, University of Liverpool, Oxford Street, Liverpool, L69 7ZE, United Kingdom

†TRIUMF, 4004 Wesbrook Mall, Vancouver, British Columbia, Canada V6T 2A3

**Department of Nuclear Physics and Biophysics, Comenius University, Bratislava, SK-84248, Slovakia

‡Department of Physics, University of Jyväskylä, P.O.Box 35, 40014 Jyväskylä, Finland

§Department of Physics, University of York, Heslington, York, YO10 5DD, United Kingdom

**Abstract.** For the first time, excitated states in very neutron-deficient odd-mass nucleus $^{185}$Pb have been observed in an in-beam $\gamma$-ray measurement. The present data provide evidence for a strong-coupling of an odd-neutron hole to the even-mass prolate core in the light Pb isotopes.

**Keywords:** Shape coexistence, $\gamma$-ray spectroscopy, recoil-decay tagging

**PACS:** 27.70.+q, 21.10.Re, 23.20.Lv, 25.70.Gh

## INTRODUCTION

In the last couple of decade, lots of experimental and theoretical effort has been put into investigations of neutron-deficient Pb nuclei [1, 2, 3, 4, 5] providing a unique laboratory to study competing structures and coexisting shapes in atomic nuclei. However, most of the experimental data are available only for even-mass nuclei, whereas data on odd-mass nuclei is eminently lacking (see Refs. [6, 7] and references therein). In the present work, the excited states in odd-mass $^{185}$Pb nucleus, situated in the heart of triple-shape coexistence nuclei, have been studied.

## EXPERIMENTAL DETAILS AND RESULTS

The present work incorporates data from two different in-beam recoil-decay tagging (RDT) experiments carried out at the Accelerator Laboratory of Jyväskylä. The first one was optimized for production of $^{184}$Pb nuclei by bombarding a $^{104}$Pd target with $^{83}$Kr ions at a beam energy of 362 MeV. The second experiment was dedicated to study $^{185}$Pb nuclei via the $^{106}$Pd($^{82}$Kr,3$n$)$^{185}$Pb reaction at a beam energy of 367 MeV. In Fig. 1(a), the $\alpha$-particle energy spectrum correlated with implanted recoils is shown. During 300

CP1012, *Frontiers in Nuclear Structure, Astrophysics, and Reactions: FINUSTAR 2*
edited by P. Demetriou, R. Julin, and S. V. Harissopulos
© 2008 American Institute of Physics 0-7354-0532-5/08/$23.00

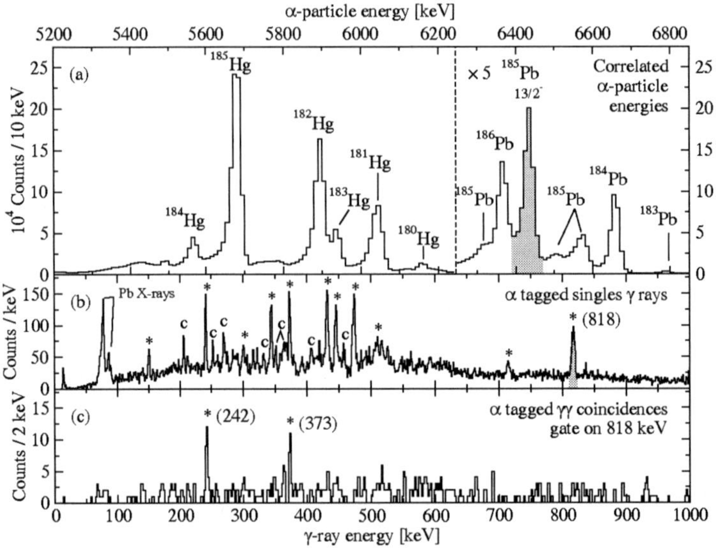


**FIGURE 1.** (a) Energy spectrum of $\alpha$ particles measured in the DSSSDs of the GREAT spectrometer. (b) Singles $\gamma$-ray energy spectrum gated with recoils and tagged with $^{185}$Pb $\alpha$ decays. (c) Recoil-gated, $\alpha$-tagged $\gamma\gamma$-coincidence spectrum with a gate on the 818 keV transition. Shaded areas correspond to the energy gates used for tagging or gating. The transition associated with $^{185}$Pb and the contaminant transitions are marked with * and c, respectively.

hours of effective beam time, a total of $10^4$ events associated with the $\alpha$ decay of the 13/2$^+$ isomeric state in $^{185}$Pb were recorded. Due to overlapping $\alpha$-decay energies it was difficult to employ the decay of the 3/2$^-$ state for RDT purposes.

Prompt $\gamma$ rays were measured using the JUROGAM Ge-detector array. Fusion-evaporation residues were separated from the primary beam and transported to the focal plane using the gas-filled recoil-separator RITU [8], where they were implanted into the Double-Sided Silicon Strip Detector (DSSSD) of the GREAT spectrometer [9]. Detector signals were recorded using the triggerless Total Data Readout data acquisition system [10], where they were time stamped with a precision of 10 ns to allow accurate temporal correlations between prompt $\gamma$ rays, recoil implants and their subsequent $\alpha$ decays. Prompt $\gamma$ rays detected in coincidence with an implanted recoil and followed by a subsequent $^{185}$Pb $\alpha$-decay in the same DSSSD pixel within 12 s were selected in data analysis. In Fig. 1(b), recoil-gated, $\alpha$-tagged singles $\gamma$-ray spectrum is shown. It mainly consists of transitions associated with $^{185}$Pb nuclei, but shows also contaminant transitions arising from random correlations with $^{185}$Tl, $^{184}$Hg and $^{182}$Hg nuclei produced with much larger cross sections. An example of recoil-gated, $\alpha$-tagged $\gamma\gamma$ coincidence spectrum with a gate on the 818 keV transition is shown in Fig. 1(c). Based on similar conditions, the level scheme shown in Fig. 2 was deduced. All spin and parity assignments are tentative at the present stage of the analysis.

The level structure suggests the odd-neutron hole couples strongly to the neighbouring even-mass prolate core. Similar band, associated with strong coupling, has been

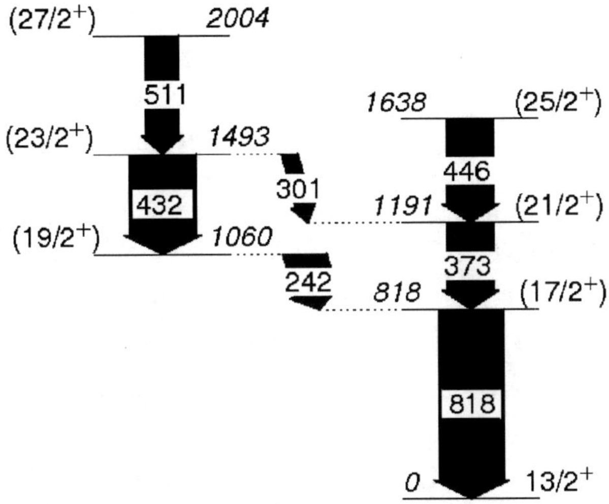


**FIGURE 2.** Proposed level scheme of $^{185}$Pb deduced from the present data. The widths of the arrows are proportional to the measured $\gamma$-ray intensities. The level energies are relative to the 13/2$^+$ isomeric state.

observed in its Hg isotone [11]. This would mean a change in the coupling scheme compared to heavier Pb isotopes, where the odd-neutron hole is proposed to be weakly coupled to nearly spherical even-mass core.

## ACKNOWLEDGMENTS

This work has been supported by the EU 6th Framework programme "I3 - Transnational Access", Contract: 506065 (EURONS) and by the Academy of Finland under the Finnish Centre of Excellence Programme 2006-2011 (Nuclear and Accelerator Based Physics Programme at JYFL). We also thank the UK/France (EPSRC/IN2P3) Loan Pool and EUROBALL Owners Committee for the EUROGAM detectors of JUROGAM.

## REFERENCES

1. J.L. Wood, K. Heyde, W. Nazarewicz, M. Huyse, P. Van Duppen, Phys. Rep. **215**, 101 (1992).
2. R. Julin, K. Helariutta, M. Muikku, J. Phys. G: Nucl. Part. Phys. **27**, R109 (2001).
3. W. Nazarewicz, Phys. Lett. B **305**, 195 (1993).
4. A.N. Andreyev *et al.*, Nature **405**, 430 (2000).
5. K. Heyde, P. Van Isacker, M. Waroquier, J.L. Wood, R.A. Meyer, Phys. Rep. **102**, 291 (1983).
6. J. Pakarinen *et al.*, Phys. Rev. C **75**, 014302 (2007).
7. A.M. Baxter *et al.*, Phys. Rev. C **58**, 2671 (1998).
8. M. Leino *et al.*, Nucl. Instr. and Meth. B **99**, 653 (1995).
9. R.D. Page *et al.*, Nucl. Instr. and Meth. B **204**, 634 (2003).
10. I.H. Lazarus *et al.*, IEEE Trans. on Nucl. Sci. **48**, 12 (2001).
11. K.S. Bindra *et al.*, Phys. Lett. B **318**, 41 (1993).

# One Nucleon Transfer Reactions Around $^{68}$Ni at REX-ISOLDE

N. Patronis*, R. Raabe*, V. Bildstein†, N .Bree*, R.Gernhäuser†, M. Huyse*, Th. Kröll†, R. Krücken†, M. Mahgoub†, P. Maierbeck†, I. Stefanescu*, J. Van De Walle* and P. Van Duppen*

**Instituut voor Kern- en Stralingsfysica, K. U. Leuven, Celestijnenlaan 200D, B-3001 Leuven, Belgium*

*†Physik- Department E12, Technische Universität München, Garching, Germany*

**Abstract.** The newly built position sensitive Si detectors array of nearly $4\pi$ angular coverage which is going to be installed at the REX-ISOLDE facility at CERN is briefly presented. This setup will be combined with the Miniball detectors array, constituting a unique tool for the study of one-nucleon transfer reactions. The experimental study of d($^{66}$Ni,p)$^{67}$Ni reaction will be proposed, as a starting point for a series of experiments aiming to the study of the single particle character of the levels of the odd mass neutron reach unstable Ni isotopes. In this contribution, the feasibility and sensitivity of the experiment is presented.

**Keywords:** direct transfer reactions, spectroscopic factors, radioactive ion beams, shell model
**PACS:** 21.10.-k, 24.50.+g., 25.60.Je, 25.60.-t

## INTRODUCTION

The magic numbers can be considered as the keystones for the modelling the nuclear structure. But recent studies suggest that by changing the number of neutrons with respect to the protons the size of shell gaps alters [1, 2]. Weakening of the spin-orbit force caused by the diffuse neutron matter [3], or the effect of tensor monopole neutron-proton interaction [4], are two possible causes for this effect. One interesting region of the nuclear chart is situated between $^{68}$Ni and $^{78}$Ni because of the closed proton shell (Z=28) and the closed neutron harmonic-oscillator sub-shell (N=40), where the $1g_{9/2}$ unique parity orbital plays a key role.

Up to now the collective properties of the neutron rich unstable isotopes of Ni, Zn and Cu at the N=40/50 mass region have been experimentally studied through the "safe" low energy Coulomb excitation [5, 6], through intermediate energy Coulomb excitation [7, 8] and $\beta$-decay studies [9, 10]. The experimental determination of the single particle character of the ground and first excited states of the odd-Ni isotopes will shed more light on the nuclear structure in this mass region.

## EXPERIMENTAL SETUP

The REX-ISOLDE facility at CERN provides intensive and well-defined beams of unstable nuclei at energies up to 3A MeV. Furthermore, the newly built position sensitive Si detectors array of nearly $4\pi$ angular coverage combined with the Miniball detectors

CP1012, *Frontiers in Nuclear Structure, Astrophysics, and Reactions: FINUSTAR 2*
edited by P. Demetriou, R. Julin, and S. V. Harissopulos
© 2008 American Institute of Physics 0-7354-0532-5/08/$23.00

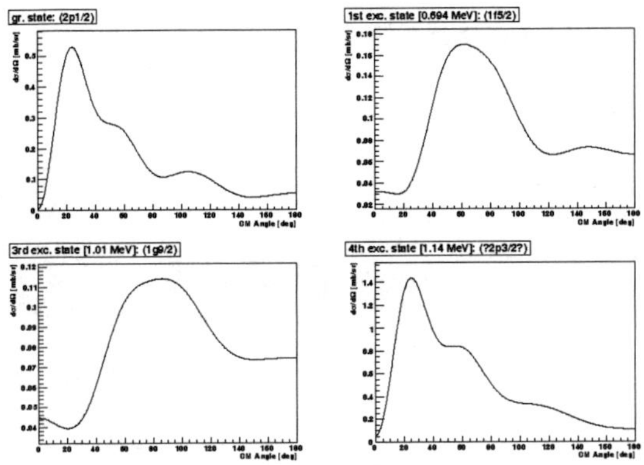


**FIGURE 1.** The differential cross section in CM system as obtained from DWBA calculations applying the code FRESCO [11]. The spin and parity of the forth excited state is assumed.

array, constitute a unique tool for the study of one-nucleon transfer reactions like the (d,p) one-neutron transfer reactions [12].

The position sensitive Si detector array will cover an extended angular range $8° < \theta_{lab} < 75°$ and $105° < \theta_{lab} < 172°$ which will allow the determination of the angular distribution of protons in a wide angular range and correspondingly the unambiguous assignment of spin and parity of the populated levels. The necessary particle identification at the forward angles will be performed by means of the double layer $\Delta$E-E configuration. The barrel array will be installed in a scattering chamber surrounded by the Miniball detector array.

## PHYSICAL CASE

The experimental study of d($^{66}$Ni,p)$^{67}$Ni reaction will be proposed, using the $^{66}$Ni radioactive ion beam from REX-ISOLDE [13], as a starting point for a series of experiments aiming to the study of the single particle character of the levels of the odd mass neutron rich unstable Ni isotopes. The objectives of this work are the unambiguous determination of spin and parities of the ground and first excited states of $^{67}$Ni and of the corresponding relative spectroscopic factors (SF) that will be compared with those from large-scale shell model calculations.

The main issue of the study of the single particle character of the odd-even neutron rich Ni isotopes arises from the fact that some of the states are isomeric, mostly due to the occupation of the $1g_{9/2}$ orbital, a unique parity level in the fp shell. For that reason, the particle gamma coincidence technique can not always be implemented for the identification of the populated energy levels. That means that the overall resolution of the silicon array is of major importance and a thin $CD_2$target has to be used. Extended GEANT4 Monte-Carlo (MC) calculations have been performed where the full experimental setup was included. The simulated angular distribution of the protons has been obtained through DWBA calculations (Fig. 1). As can be seen in Fig. 2, four groups of

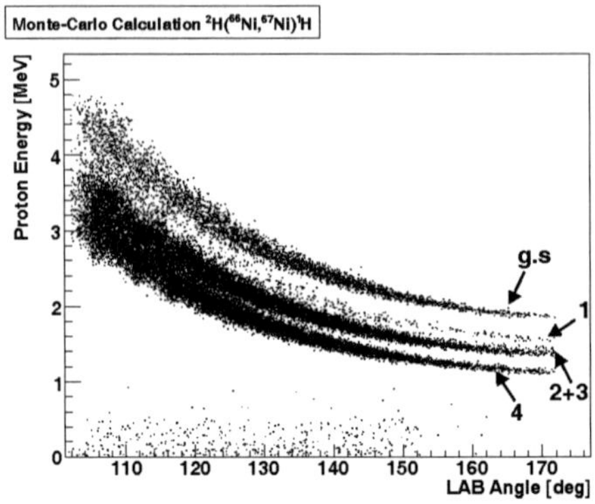


**FIGURE 2.** Results of the Monte-Carlo calculations. The different proton groups assigned to the population of the ground and the first four excited states of $^{67}$Ni are indicated.

protons can be easily identified out of the five levels that were included to the MC calculations. The unresolved proton groups correspond to the second and third excitation levels at 1.0 and 1.14 MeV correspondingly.

## CONCLUSIONS

As a starting point of the study of the single particle character around N=40 the d($^{66}$Ni,p)$^{67}$Ni reaction will be proposed at REX-ISOLDE. The counting rate and the expected resolution of the newly build barrel configuration have been studied in the present work. In most cases, by using thin $CD_2$ target, the assignment of each proton group will be feasible even for the singles proton spectra. The simulations show that the unambiguous determination of the spin and parity as well as the determination of the relative SF 's of the populated levels is possible.

## REFERENCES


1. T. Otsuka *et al.*, Phys. Rev. Lett. **87**, 082502 (2001).
2. D. Warner, Nature (London) **430**, 517 (2004).
3. J. Dobaczewski *et al.*, Phys. Rev. Lett. **72**, 981 (1994).
4. T. Otsuka *et al.*, Phys. Rev. Lett. **95**, 232502 (2005).
5. I. Stefanescu *et al.*, Phys. Rev. Lett. **98**, 122701 (2007).
6. J. Van De Walle *et al.*, accepted to Phys. Rev. Lett.
7. O. Sorlin *et al.*, Phys. Rev. Lett. **88**, 092501 (2002).
8. O. Perru *et al.*, Phys. Rev. Lett. **96**, 232501 (2006).
9. W. F. Mueller *et al.*, Phys. Rev. Lett. **83**, 3613 (1999).
10. S. Franchoo *et al.*, Phys. Rev. Lett. **81**, 3100 (1998).
11. I. J. Thompson, Comp. Phys. Reports **7**, 167 (1988).
12. V. Bildstein *et al.*, Progr. in Part. and Nucl. Physics **59**, 386 (2007).
13. O. Kester *et al.*, Nucl. Instr. and Meth. in Phys. Res. **B204**, 30 (2003).

# On-line separator for $\gamma$-spectroscopic studies

A.G. Popeko*, A.V. Belozerov*, V.I. Chepigin*, O. Dorvaux†, K. Hauschild**, A.P. Kabachenko*, A. Korichi**, A. Lopez-Martens**, O.N. Malyshev*, A.V. Shutov*, A.I. Svirikhin* and A.V. Yeremin*

*FLNR, JINR, 141980, Dubna, Russia
†IPHC, IN2P3-CNRS, F-67037 Strasbourg, France
**CSNSM, IN2P3-CNRS, F-91405 Orsay Campus, France

**Abstract.** We report about R&D of the new separator for spectroscopic studies which we plan to install at the beam of the modernized U400R FLNR cyclotron and which will allow to realize new possibilities of the JINR - IN2P3 collaboration project GABRIELA [1, 2].

**Keywords:** Fusion and fusion-fission reactions, 220≤A, Charged particle spectrometers: electric and magnetic
**PACS:** 25.70.Jj; 27.90.+b; 29.30.Aj

## INTRODUCTION

High intensity of the FLNR U400 cyclotron beams of accelerated iones of rare isotopes, availability of unique radioactive targets like $^{244}$Pu, $^{248}$Cm, and the electrostatic on-line separator VASSILISSA [3] made it possible to perform spectroscopic studies of heavy neutron rich short living nuclei with $T_{1/2} > 1\mu$s, having formation cross sections in complete fusion reactions of higher than 1 nb.

## CHOICE OF A SEPARATOR

The only way to investigate short living isotopes with $T_{1/2} < 1$ s is the use of electromagnetic or electrostatic on-line separators, which can be divided into two groups - devices operating in the gas-filled mode and devices operating in the vacuum mode.

The separation of complete fusion reaction products from full energy beam particles, from scattered beam particles and from unwanted products of incomplete fusion or multinucleon transfer reactions can be based on differences in kinematic characteristics (impulse, angle) and their derivatives - range in some media, magnetic or electric rigidities.

Before starting the detailed design of the separator, the analysis of cinematic and other characteristics of possible nuclear reactions, in which isotope of our interest could be produced has been performed.

In Table 1 some characteristic data (magnetic rigidities of particles moving in vacuum after stripping on carbon foil, in helium, velocity, electric rigidity in vacuum) for "asymmetric" $^{22}$Ne + $^{238}$U $\rightarrow$ $^{260}$No* and "symmetric" $^{136}$Xe + $^{136}$Xe $\rightarrow$ $^{266}$Hs* reactions are presented.

CP1012, *Frontiers in Nuclear Structure, Astrophysics, and Reactions: FINUSTAR 2*
edited by P. Demetriou, R. Julin, and S. V. Harissopulos
© 2008 American Institute of Physics 0-7354-0532-5/08/$23.00

**TABLE 1.** Characteristic data of particles produced in "asymmetric" and "symmetric" reactions

| $^{22}Ne + ^{238}U \rightarrow ^{260}No^*$ | | | | | | |
|---|---|---|---|---|---|---|
| Particle | | Energy (MeV) | $B\rho_{vac}$ (T·m) | $B\rho_{He}$ (T·m) | v (cm/ns) | $E/q_{vac}$ (MV) |
| Projectile | $^{22}Ne$ | 112 | 0.76 | 0.76 | 3.14 | 11.71 |
| Recoil | $^{260}No$ | 9.5 | 0.74 | 1.82 | 0.26 | 0.99 |
| Target-like | $^{238}U$ | 34.8 | 0.72 | 1.89 | 0.53 | 1.93 |

| $^{136}Xe + ^{136}Xe \rightarrow ^{266}Hs^*$ | | | | | | |
|---|---|---|---|---|---|---|
| Particle | | Energy (MeV) | $B\rho_{vac}$ (T·m) | $B\rho_{He}$ (T·m) | v (cm/ns) | $E/q_{vac}$ (MV) |
| Projectile | $^{136}Xe$ | 600 | 1.08 | 1.38 | 3.0 | 15.7 |
| Recoil | $^{266}Hs$ | 300 | 0.94 | 1.39 | 1.46 | 6.85 |
| Target-like | $^{136}Xe$ | 600 | 1.08 | 1.38 | 3.0 | 15.7 |

As we see from the Table 1 it is impossible to separate particles according to their magnetic rigidities in vacuum. In the most difficult cases of symmetric reaction we have to deal with particles with magnetic rigidities up to 2.0 T·m and with electric rigidities up to 7.0 MV.

Gas filled separators are successfully used as well for nuclear spectroscopy as for synthesis and investigation of the heaviest isotopes.

The advantages of gas filled separators are: high efficiency of complete fusion reaction products transportation, simplicity and relatively low costs and the ability to accept high beam currents due to the cooling of the target by the filling gas. But a gas filled separator can be successfully applied for the investigation of reactions with medium mass asymmetry in the entrance channel.

The vacuum mode separators have the transmission which is by few % less than by the gas filled set-ups. They have higher, by factor of 20 – 100, suppression factors of incomplete fusion reaction products independent from the mass asymmetry in the entrance channel.

These are the reasons why we decided to construct the new separator as a device operating in the vacuum mode.

## THE NEW SEPARATOR

For the analysis of different variants of the ion optical scheme of the separator and for the optimization of characteristics of elements and their positions the computer codes COSY [4] and GICOSY [5] are used.

For the more realistic analysis the special computer code VASFIT [6] was employed. The code generates "real" particles, considers the evaporation of neutrons, protons and $\alpha$s from the excited ER, the reaction cinematics, scattering and energy losses of ERs in the target and the ion charge distribution of the ERs. Generated "real" particles will be transported through the system using the first-order optics.

We propose to use the optical scheme MQ-MQ-MQ-ES-DI-DI-ES-MQ-MQ-MQ-DI, where MQ stands for quadrupole lenses, ES for electrostatic sector and DI for dipole magnets. The principal component of the new separator is a system consisting of electrostatic and magnetic dipoles, which accomplish spatial filtering of recoil nuclei, multinucleon transfer reaction products and beam particles dispersing them in their velocity and magnetic rigidity. The input focusing system consists of a triplet of magnetic quadrupoles. The input triplet is located just behind the target and focuses the evaporation residues emerging from the target, forming a quasi parallel beam.

The beam is dumped behind the dipole magnet and can be effectively shielded by concrete walls to reduce background in both directions - to the focal plane detector position and to the target position, where detectors of prompt $\gamma$-quanta can be placed.

After the dispersing system an output triplet of quadrupole lenses is used, which serves for better collecting of ERs onto the focal plane detectors. Fig. 1 shows the layout and simulated using [6] ion trajectories in the proposed separator.

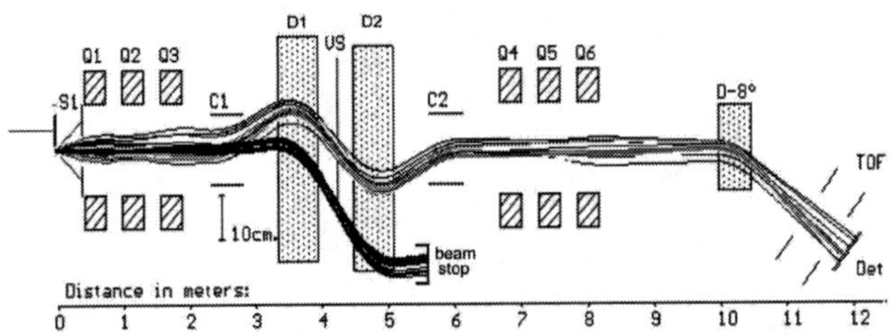


**FIGURE 1.** Layout and simulated ion trajectories in the proposed separator.

For the design of the new separator many developments, made for VASSILISSA - quadrupole lenses, focal plane detectors, data acquisition system can be employed. This will significantly accelerate the R&D process.

## ACKNOWLEDGMENTS

The authors would like to thank Profs. S. Hofmann and M. Leino for interesting discussions.

The work at FLNR has been performed partially under the financial support of the Russian Foundation for Basic Research, contract No 05-02-16198.

## REFERENCES

1. A.G. Popeko, *et. al.*, Phys. of At. Nucl. **69**, 1183 (2006).
2. K. Hauschild, *et al.*, Nucl. Instr. & Meth. **A560**, 388 (2006).
3. A.V. Yeremin, *et al.*, Nucl. Instr. & Meth. **A350**, 608 (1994).
4. COSY INFINITY, Version 8.1, http://cosy.pa.msu.edu/
5. GICOSY Ion optics program, http://www-linux.gsi.de/~weick/gicosy/
6. A.G. Popeko, *et. al.*, Nucl. Instr. & Meth., **A427** 166 (1999).

# Production and Separation of Long-lived Radionuclides for Nuclear Physics Experiments

Dorothea Schumann

*Paul Scherrer Institute Villigen, 5232 Villigen PSI*
*Switzerland*

**Abstract.** Activated components from a high-intensity proton accelerator facility are used as a source for exotic long-lived radionuclides. These isotopes represent valuable material in several science fields like nuclear astrophysics, nuclear structure studies, accelerator mass spectrometry, radioactive beam development, geophysics as well as radiopharmaceutical chemistry. Elaborate radiochemical separation techniques have to be developed to gain suitable amounts of radionuclides.

**Keywords:** Accelerator waste, radiochemical separation, neutron capture
**PACS:** 28.60+s; 28.41Kw; 26.50+x; 25.40Sc

## INTRODUCTION

Long-lived exotic radionuclides such as $^{60}$Fe, $^{26}$Al, $^{44}$Ti or $^{10}$Be are of great interest in several fields of research like radiopharmacy, basic nuclear physics, astrophysics and/or radioactive ion beam techniques. Some examples for the use of such rare isotopes in the research area of nuclear astrophysics are the application of $^{44}$Ti for investigations of core collapse supernovae or studies of several neutron capture reactions on radioactive isotopes like the $^{60}$Fe(n,$\gamma$)$^{61}$Fe reaction at stellar energies. The direct production of these nuclides in sufficient amounts is, in some cases, nearly impossible, in others very time consuming and extremely expensive.

Until very recently, no concerted effort had been made to reclaim such radionuclides for future use from activated components of particle accelerators. This is due, in large part, to the small amounts of suitably irradiated material. Nowadays, this situation changed drastically with the construction of new large accelerator facilities, spallation neutron sources and radioactive-beam-facilities. An example is the EURISOL project [1].

The Paul Scherrer Institute operates the most powerful spallation neutron source (SINQ) in Europe, driven by a high power proton accelerator, the 590-MeV ring cyclotron with a proton beam current up to 1.8 mA. Previous radiochemical analyses showed that beam dumps, shielding and target materials from these facilities contain long-lived radionuclides in such high amounts, that chemical separation for several applications seems to be attractive [2]. It was found that the isotopes $^{60}$Fe, $^{26}$Al, $^{10}$Be, $^{44}$Ti and probably many others can be separated already now in amounts of $10^{16}$ - $10^{18}$ atoms [3,4].

CP1012, *Frontiers in Nuclear Structure, Astrophysics, and Reactions: FINUSTAR 2*
edited by P. Demetriou, R. Julin, and S. V. Harissopulos
© 2008 American Institute of Physics 0-7354-0532-5/08/$23.00

## AVAILABLE MATERIAL AND CHEMICAL SEPARATION

At the moment, PSI can provide highly-active samples from a proton-irradiated copper beam dump as well as former graphite targets from the myon production station (so-called target E station) with a high content of $^{7}Be$ and/or $^{10}Be$. Additionally, first irradiated lead targets from the spallation neutron source SINQ will be ready for use after the necessary cooling time in the near future. The radiochemical separation procedure depends mainly on the requirements from the user side, e.g. chemical form, allowable impurities, carrier-free of carrier-added form and so on. Depending on the target material, a bright variety of radionuclides has to be expected which influences the separation technique. Therefore, no "universal recipe" can be given, but the separation procedures have to be modified according to the special requirements. Mainly, chemistry in aqueous solution is applied using classical separation methods like precipitation, extraction and ion exchange.

### Copper beam dump

The copper beam dump served as a beam stop for 590-MeV protons in the former pion therapy station (**Bio-Medical-Area**, BMA). The proton- and neutron-induced spallation reactions lead amongst others to long lived residues like $^{60}Fe$, $^{44}Ti$ and $^{26}Al$. Due to the high content of 60Co, the samples show a high dose rate, which requires special radiation safety measures.

The matrix material can easily be dissolved in half-concentrated $HNO_3$. After evaporation and re-dissolving the residue in 7 M HCl, iron can be removed by extraction methods.

$^{44}Ti$ and $^{26}Al$ are separated by precipitation using isotopic or non-isotopic carrier. A final purification is achieved by ion exchange.

By use of the mentioned methods, samples in the range of $10^{14}$ ($^{26}Al$) - $10^{16}$ ($^{60}Fe$, $^{44}Ti$) atoms could be separated. Details of the sample separation can be found in [4].

### Graphite targets

Carbon targets are used for myon production in the target E station and serve as a source for $^{7}Be$ and $^{10}Be$. Usual exposure times are 1-3-years. The beam doses vary from 4 to 11 Ah (590 MeV proton beam). The main additional activation products are $^{3}He$ and $^{14}C$, low-energy $\beta$-emitters, which allows easy handling.

Graphite samples were dissolved in a mixture of concentrated acids. Iron was added as non- isotopic carrier, and beryllium was separated by precipitating the hydroxide. The chemical yield was monitored measuring the $\gamma$-radiation of $^{7}Be$. Remove of iron and final purification was done by ion exchange.

$10^{18}$ atoms of $^{10}Be$ could be separated. Details of the separation technique are described in [3].

### Lead targets

Two proton-irradiated lead targets (EOB 1999 and 2001) from the spallation neutron source at PSI are available for investigations and preparative separations.

Besides the radionuclides, which we can find in the copper beam dump, these material can serve also as a source for residues with higher mass numbers like $^{182}Hf$ and several long-lived rare earth elements. However, only a few analytical studies had been carried out till now in order to get an overview on the radionuclide inventory and, according to this, to develop the necessary chemical separation techniques. Exploitation of the material is only possible after complete analytical characterization of the targets. Moreover, the high dose rate requires a fully-equipped shielded Hot-cell and automated systems for the chemical separation procedures.

## INTERNATIONAL COLLABORATION

An exploratory workshop sponsored by the European Science Foundation (ESF) aimed to search for potential collaboration partners and establish an international network for the exploitation of accelerator waste materials with regard to the use of exotic radionuclides in basic science and for technological and medical applications was initiated by and held at PSI in Nov. 2006 (ERAWAST: **E**xotic **R**adionuclides from **A**ccelerator **WA**ste for **S**cience and **T**echnology [5]). 30 participants from 12 countries decided to create an international collaboration. As one of the ERAWAST workhop results, a working program consisting of both long-term and short (mid) -term goals had been established. Further collaboration in experiments as well as joint

## ACKNOWLEDGMENTS

We thank the "PSI-West Hallendienst" crew as well as the technicians from the PSI "Hot-Labor" for cutting and drilling the sample material, R. Weinreich, S. Teichmann and G. Heidenreich for fruitful discussions and S. Horn for her support concerning the radiochemical separation.

## REFERENCES

1. http://www.ganil.fr/eurisol/
2. Schumann D, Neuhausen J, Weinreich R, Atchison F, Kubik P, Synal H-A, Korschinek G, Faestermann Th, Rugel G, NIM B 2007, in print
3. D. Schumann, J. Neuhausen, S. Horn, P. W. Kubik, I. Günther-Leopold, Radiochemical separation and analytical determination of $^{10}Be$ from proton-irradiated graphite targets Radiochim. Acta 2007, in print
4. D. Schumann, J. Neuhausen, Accelerator Waste as a Source for Exotic Radionuclides, Journal of Physics G 2007, in print
5. http://lch.web.psi.ch/radwaste/workshop/index.html

# Level densities and $\gamma$-strength functions in Sm isotopes

S. Siem*, M. Guttormsen*, R. Chankova†,**, A.C. Larsen*, H.T. Nyhus*, J.Rekstad*, N.U.H. Syed*, A. Schiller‡, A. Voinov‡,§ and S.W. Ødegård*

*Department of Physics, University of Oslo, P.O.Box 1048 Blindern, N-0316 Oslo, Norway
†North Carolina State University, Raleigh, NC 27695, USA
**Triangle Universities Nuclear Laboratory, Durham, NC 27708, USA
‡Department of Physics and Astronomy, Ohio University, Athens, Ohio 45701, USA
§Frank Laboratory of Neutron Physics, Joint Institute of Nuclear Research, 141980 Dubna, Moscow region, Russia

**Abstract.** The level density and radiative strength functions (RSF) for $^{146,147}$Sm have been extracted. As one approaches the closed $N = 82$ neutron shell we see that the structures in the level density become more pronounced due to shell effects. The experimental level densities can be used to explore thermodynamical properties of the nucleus within the microcanonical ensemble.

**Keywords:** Level density, $\gamma$-ray strength function
**PACS:** 21.10.Ma,24.30.Gd,25.55.Hp,27.60.+j

## INTRODUCTION

Nuclear structure changes significantly when approaching closed shells. The change of the nuclear shape from deformed to spherical is expected to influence the level density as well as the radiative strength function (RSF). These quantities, which are important input parameters in large network calculations of stellar evolution, have been extracted using the Oslo method for a wide range of nuclei. The experiments were carried out at the Oslo Cyclotron Laboratory, using the ($^3$He,$\alpha$) and ($^3$He,$^3$He')-reactions with the 45 MeV $^3$He-beam delivered by the MC-35 cyclotron. Charged particles and $\gamma$-rays were recorded with the multi-detector system CACTUS, which includes a NaI $\gamma$-detector array with a total efficiency of $\sim 15$ %. The Oslo group has developed a technique [1] to measure simultaneously the level density and radiative strength function.

## LEVEL DENSITY

The extracted level density of $^{146,147}$Sm (left panel of Fig. 1) show a very pronounced step structure. The difference in the level density between the odd-even and even-even can be described as a simple energy shift due to pairing. In the right panel of Fig. 1, the structures in the level density become more pronounced due to shell effects going from well deformed midshell Dy, via more spherical Sm nuclei approaching the $N = 82$ closed shell, to the doubly magic $^{208}$Pb. In the same way as the lowering of the ground state in even-even nuclei due to pairing leads to an energy shift in the level density, the shell-effects can also lead to an energy shift. In the case of a closed shell the particles

CP1012, *Frontiers in Nuclear Structure, Astrophysics, and Reactions: FINUSTAR 2*
edited by P. Demetriou, R. Julin, and S. V. Harissopulos
© 2008 American Institute of Physics 0-7354-0532-5/08/$23.00

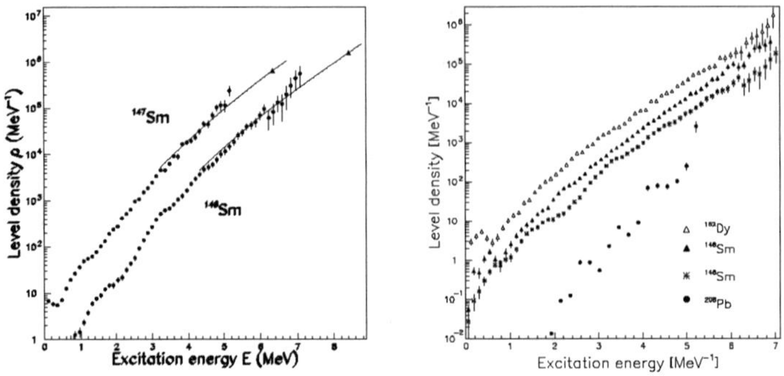


**FIGURE 1.** The level density of $^{146,147}$Sm, left panel. Right panel: level density of $^{162}$Dy (open triangle), $^{148}$Sm (filled triangle), $^{146}$Sm (asterisk) and $^{208}$Pb (filled circle).

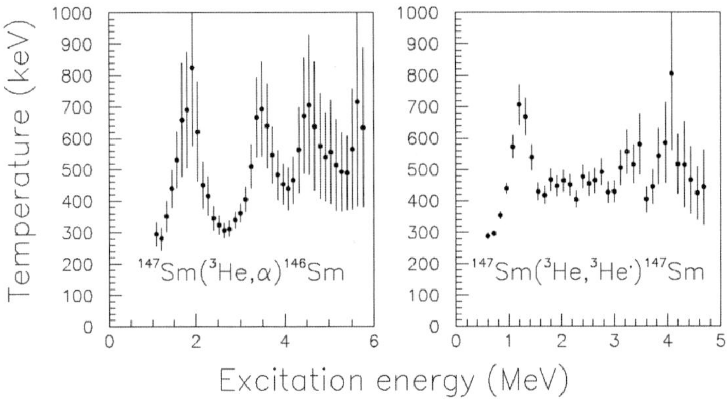


**FIGURE 2.** The temperature curve for $^{146,147}$Sm derived within the microcanonical ensemble.

need enough energy to cross the shell gap before having access to many levels, similar to overcoming the pair-gap in even-even nuclei. From looking at the energy shift as we approach the closed neutron shell of $N = 82$, we can get information on the size of the shell gap. When using the micro-canonical ensemble the partition function is simply the multiplicity of nuclear states $\omega(E)$, which experimentally corresponds to the level density of accessible states. Thus, the entropy is defined as: $S(E) = k_B \ln \omega(E)$, where $k_B$ is the Boltzmann constant. The micro-canonical temperature in units of MeV is given by: $T(E) = k_B \left(\frac{\partial S}{\partial E}\right)^{-1}$. Due to the step structure of the level density, very pronounced peaks in the temperature spectrum of $^{146,147}$Sm are seen in Fig. 2. When particle pairs are broken, new degrees of freedom open up i.e. a more than normal opening of new domains of the phase space, leading to an unusual increase in the entropy and thus the decrease in the temperature. We therefore interpret that the break up of a nucleon pair happens in the region where the micro-canonical temperature curve has a negative slope.

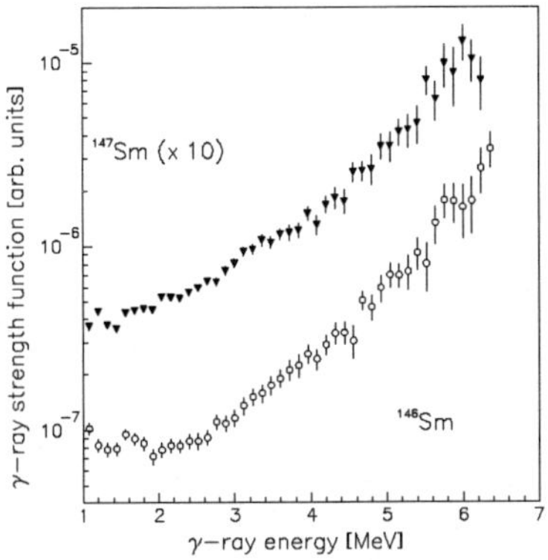


**FIGURE 3.** The radiative strength function for $^{147}$Sm, filled triangles and $^{146}$Sm open circles. The $^{147}$Sm values are multiplied by a factor of 10.

# THE RADIATIVE STRENGTH FUNCTION

A pygmy resonance at around 3 MeV has been observed in the RSF of several rare earth nuclei, being most pronounced in the well deformed $^{161,162}$Dy, $^{166,167}$Er and $^{171,172}$Yb nuclei and less obvious in the more spherical $^{148,149}$Sm nuclei [2]. The pygmy resonance was shown to have M1 multipolarity in $^{172}$Yb [3]. It is interpreted as the scissors mode built on exited states, which is deformation dependent. And indeed when approaching the $N = 82$ closed shell, where the nuclei are less deformed, the pygmy resonance vanishes, as seen in the radiative strength functions of $^{146,147}$Sm (see Fig. 3). Both strength functions flatten out at low $\gamma$-ray energies, but the previously observed enhancement of the total RSF of Fe [4] and Mo [5] isotopes at low $\gamma$-ray energies, is not seen in the Sm nuclei.

# ACKNOWLEDGMENTS

Financial support from the Norwegian Research Council (NFR) is gratefully acknowledged.

# REFERENCES

1. A. Schiller et al., *Nucl. Instrum. Methods* **A447**, 498 (2000).
2. S. Siem et al., *Phys. Rev.* **C65**, 044318 (2002) and refs. therein.
3. A. Schiller et al., *Phys. Lett.* **B 633** (2006) 225.
4. A. Voinov et al., *Phys. Rev. Lett.* **93**, 142504 (2004).
5. M. Guttormsen et al., *Phys. Rev.* **C71** 044307 (2005).

# The Oslo Method and Its Application to Lead Isotopes

N.U.H. Syed*, M. Guttormsen*, F. Ingebretsen*, A.C. Larsen*, S. Siem*, T. Lönnroth†, A. Schiller** and A. Voinov**

*Department of Physics, University of Oslo, Norway
†Åbo Akademi University, FIN-20500 Åbo, Finland
**Department of Physics and Astronomy, Ohio University
Athens, OH, USA

**Abstract.** The Oslo method has been tested on $^{205-208}$Pb nuclei utilizing the neutron pick-up and inelastic scattering reactions for the extraction of the level density and $\gamma$ strength function. The preliminary results agree well with the known characteristics of these nuclei.

**Keywords:** The Oslo Method, level density, $\gamma$-ray strength function
**PACS:** 21.10.Ma, 25.55.Ci, 25.55.Hp, 27.80.+w

## INTRODUCTION

The Oslo Cyclotron (OCL) group has developed a method [1], called "The Oslo method", to extract the nuclear level density and $\gamma$-ray strength function simultaneously from the particle-$\gamma$ coincidences. The method involves three mains steps in the process of extraction, namely, unfolding [2], extraction of first generation $\gamma$-rays [3], and application of Brink-Axel hypothesis [4, 5]. The important requirements are that spin and parity distributions should remain the same for all excitation energy bins and that the nucleus should thermalize at each step within a given $\gamma$ cascade [3].

## EXPERIMENTAL DETAILS

The experiments were conducted at Oslo Cyclotron Laboratory using 38-MeV $^3$He ion beams. The self supporting targets were $^{206}$Pb and $^{208}$Pb having a thickness of 4.71 mg/cm$^2$ and 1.40 mg/cm$^2$, respectively. The experiments ran for $\sim$ 8 days with beam currents $\sim$ 0.6 nA.

The particles are detected by eight collimated Si $\Delta$E-E type detectors, which are 145$\mu$m and 1500$\mu$m thick, respectively and make an angle of 45° with the beam line. For the $\gamma$-rays 28 5"x5" NaI detectors, called CACTUS [6], are used with the total efficiency $\sim$ 15 % of $4\pi$.The CACTUS array surrounds the target and Si particle detectors. The reactions of particular interest are; $^{208}$Pb($^3$He, $^3$He')$^{208}$Pb, $^{208}$Pb($^3$He, $\alpha$)$^{207}$Pb, $^{206}$Pb($^3$He, $^3$He')$^{206}$Pb and $^{206}$Pb($^3$He, $\alpha$)$^{205}$Pb.

CP1012, *Frontiers in Nuclear Structure, Astrophysics, and Reactions: FINUSTAR 2*
edited by P. Demetriou, R. Julin, and S. V. Harissopulos
© 2008 American Institute of Physics 0-7354-0532-5/08/$23.00

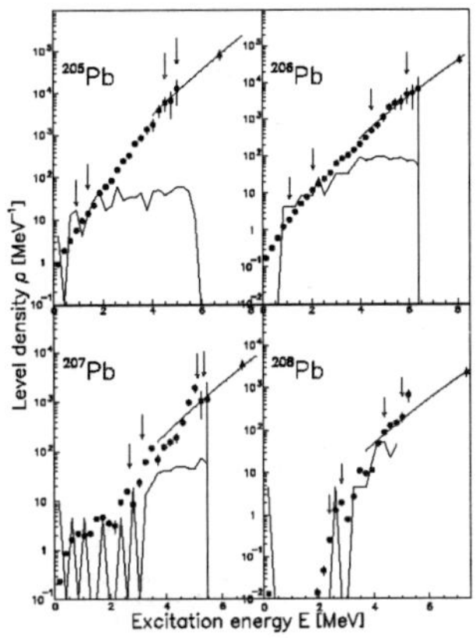


**FIGURE 1.** Normalization of the level density for the $^{205-208}$Pb nuclei (preliminary). The arrows indicate the fitting of data points (closed circles) with known levels (zigzag line). The back-shifted Fermi gas model based level density (solid line) is also drawn for comparison.

## NUCLEAR LEVEL DENSITY AND $\gamma$ STRENGTH FUNCTION

The energy distribution of primary $\gamma$-rays emitted contains the information of level density and $\gamma$-ray strength function. The Brink-Axel hypothesis [4, 5] is employed to factorize the normalized primary $\gamma$-ray matrix $P(E,E_\gamma)$ in the statistical region i.e.,

$$P(E,E_\gamma) \propto \rho(E-E_\gamma)\mathscr{T}(E_\gamma), \tag{1}$$

where the level density $\rho$ and transmission coefficient $\mathscr{T}$ are determined by an iterative procedure [1] through the adjustment of all data points until a global $\chi^2$ minimum with $P(E,E_\gamma)$ is reached. The entries of $P(E,E_\gamma)$ remain invariant under the transformation [1]

$$\tilde{\rho}(E-E_\gamma) = A\exp[\alpha(E-E_\gamma)]\rho(E-E_\gamma), \tag{2}$$

$$\tilde{\mathscr{T}}(E_\gamma) = B\exp(\alpha E_\gamma)\mathscr{T}(E_\gamma). \tag{3}$$

The free parameters $A$, $B$, $\alpha$ are the normalization parameters. The normalization of the level densities of $^{205,206,207,208}$Pb are shown in Fig. 1. The effects of shell closure are apparent by the opening of neutron shell closure at $N$=126 to one, two and three vacancies. The extracted level densities are in good agreement with the spectroscopic findings of levels at excitation energies below $\sim$ 2-4 MeV. The low level density and the presence of structures are the signatures of closed shell nuclei.

The transmission coefficient $\mathscr{T}$ in Eq. (3) is normalized through a procedure described in [7]. The $\gamma$-ray strength function can be deduced using,

$$f(E_\gamma) = \frac{1}{2\pi}\frac{\mathscr{T}(E_\gamma)}{E_\gamma^3}. \tag{4}$$

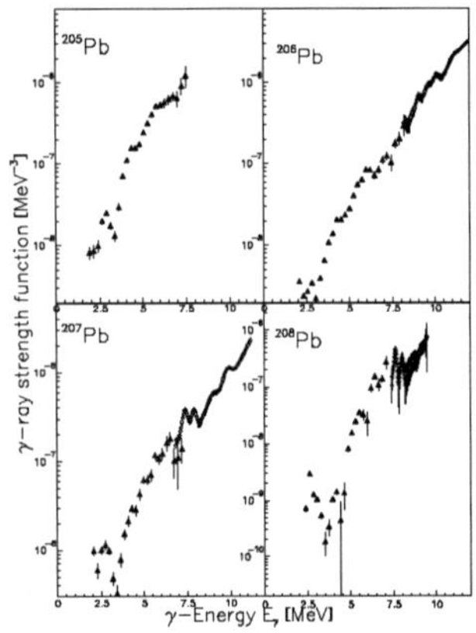


**FIGURE 2.** The preliminary $\gamma$ strength function of $^{205,206,207,208}$Pb. The triangular points are Oslo data, while open circles are the ($\gamma$, n) data given in [8].

The strength functions of the $^{205,206,207,208}$Pb as a function of $\gamma$-ray energy are shown in Fig. 2. The Oslo data for $^{206,207,208}$Pb are compared with the respective ($\gamma$, n) data of [8]. Intermediate resonances have been observed in Oslo data below neutron binding energy ($B_n$) as they were seen above $B_n$ in the ($\gamma$, n) data findings.

## CONCLUSION

The extraction of level density and $\gamma$ strength function in heavy closed shell, Pb nuclei is an ideal case of studying the Oslo method for its applicability on spherical nuclei. The preliminary results show that the method is robust and workable for the closed shell magic nuclei.

## ACKNOWLEDGMENTS


The Norwegian Research Council (NFR) is gratefully acknowledged for its financial support for the present work.


## REFERENCES


1. A. Schiller *et al.*, Nucl. Instr. Methods Phys. Res. **A447**, 498 (2000).
2. M. Guttormsen *et al.*, Nucl. Instr. Methods Phys. Res. **A374**, 371 (1996).
3. M. Guttormsen *et al.*, Nucl. Instr. Methods Phys. Res. **A225**, 518 (1987).
4. P. Axel, Phys. Rev. **126**, 671 (1962).
5. D. M. Brink, Ph.D. thesis, Oxford University, 1955.
6. M. Guttormsen *et al.* Phys. Scr. **T32**, 54 (1990).
7. A. Voinov *et al.*, Phys. Rev. **C63**, 044313 (2001).
8. S.N. Belyaev *et al.*, Yad. Fys. **42**, 1050 (1985).

# Shape coexistence in the neutron-deficient Pt isotopes in the configuration-mixed IBM

Carlos E. Vargas*, Irving O. Morales†, Alejandro Frank†,
Piet Van Isacker** and Cuauhtemoc Campuzano*

*Facultad de Física e Inteligencia Artificial, Universidad Veracruzana, Sebastián Camacho 5, Centro, CP 91000, Xalapa, Ver. Mexico.
†Instituto de Ciencias Nucleares, UNAM, Apdo. Postal 70-543, 04510 México, D.F. México
** GANIL, CEA/DSM–CNRS/IN2P3, BP 55027, F-14076 Caen Cedex 5, France

**Abstract.** The matrix-coherent state approach in the IBM with configuration mixing is used to describe the geometry of neutron-deficient Pt isotopes. Employing a parameter set for all isotopes determined previously, it is found that the lowest minimum goes from spherical to oblate and finally acquires a prolate shape when approaching the mid-shell Pt isotopes.

**Keywords:** Nuclear structure, shape coexistence, phase transitions, configuration mixing
**PACS:** 21.10.Re, 21.60.Ev, 21.60.Fw, 27.70.+q

## INTRODUCTION

The phenomenon of shape coexistence in nuclei has been studied for many decades and is still an active topic of research. For example, advances in experimental techniques have allowed to identify three $0^+$ states in $^{186}$Pb [1], each interpreted in terms of a distinct shape and originating from multiple particle-hole excitations across the $Z = 82$ proton shell gap. Also in the platinum isotopes the mixing between different configurations was claimed to influence low-lying states [2]. Deformed mean-field calculations have indicated the possibility to find close-lying oblate and prolate minima next to the spherical ground-state configuration in several isotopes [3]. Such studies have focussed mainly on the static potential energy surface. On the other hand, shell-model calculations are very difficult to realize due to the large spaces involved [4]. The Interacting Boson Model-1 (IBM-1) [5] adopts a Hilbert space which is highly truncated as compared to the shell model, with a Hamiltonian that allows a transition from U(5) vibrational to SU(3) axially-symmetric rotational nuclei. Moreover, the coherent-state formalism allows a geometric visualization of these algebraic results. In this contribution we use the IBM-1 to study the evolution of the shape and the phenomenon of shape coexistence in the platinum isotopes.

## THE MODEL HAMILTONIAN

The model Hamiltonian and its parameters have been taken without modification from Ref. [6]. The model space employed consists of the sum of two symmetric U(6) repre-

CP1012, *Frontiers in Nuclear Structure, Astrophysics, and Reactions: FINUSTAR 2*
edited by P. Demetriou, R. Julin, and S. V. Harissopulos
© 2008 American Institute of Physics 0-7354-0532-5/08/$23.00

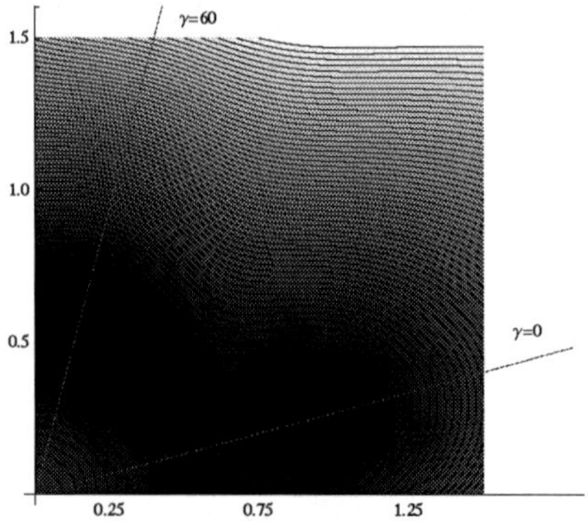


**FIGURE 1.** Potential energy surface $V(\beta,\gamma)$ for $^{186}$Pt.

sentations [N] $\oplus$ [N+2]. The Hamiltonian considered is

$$\hat{H} = \hat{H}_{0p-0h} + \hat{H}_{2p-2h} + \hat{H}_{mix} \ , \tag{1}$$

where the different pieces are the operators $\hat{H}_i$ for each configuration ($i$ = 0p-0h, 2p-2h)

$$\hat{H}_i = \varepsilon_i \hat{n}_d + \kappa_i \hat{Q}_i \cdot \hat{Q}_i + \kappa_i' \hat{L} \cdot \hat{L} \ , \tag{2}$$

and $\hat{H}_{mix}$ represents the mixing between the two configurations. From the coherent-state formalism we obtain the energy surface for the configuration with $N$ bosons,

$$\begin{aligned} E_i(N,\beta,\gamma) &= (\varepsilon_i + 6\kappa_i')\frac{N}{1+\beta^2}\beta^2 + \kappa_i\left[\frac{N}{1+\beta^2}(5+(1+\chi_i^2)\beta^2)\right. \\ &\quad \left. + \frac{N(N-1)}{(1+\beta^2)^2}\left(\frac{2}{7}\chi_i^2\beta^4 - 4\sqrt{\frac{2}{7}}\chi_i\beta^3\cos 3\gamma + 4\beta^2\right)\right] \ , \end{aligned} \tag{3}$$

and the non-diagonal matrix elements which mix both configurations

$$\langle N+2,\beta,\gamma|H_{mix}|N,\beta,\gamma\rangle = \frac{\sqrt{(N+2)(N+1)}}{1+\beta^2}\left(\omega + \frac{\omega}{\sqrt{5}}\beta^2\right) \ . \tag{4}$$

The 0p-0h configuration corresponds to $N$ bosons while the 2p-2h excitations require two extra bosons [7]. This leads to a $2\times2$ potential energy matrix [8]. After numerical diagonalization of this matrix we obtain the eigenpotentials. The full explanation of the method, the results and a detailed analysis of them will be published elsewhere [9].

## SOME RESULTS

The platinum isotopes with mass number between 204 and 194 have an sperical shape, associated with the dominance of the regular configuration. The region between 192 and

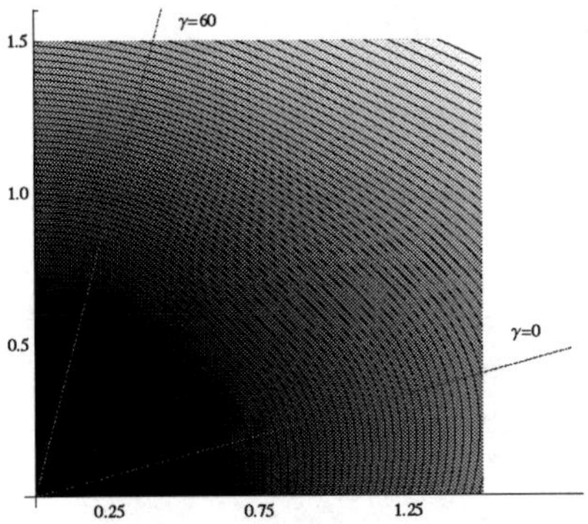


**FIGURE 2.** Potential energy surface $V(\beta,\gamma)$ for $^{192}$Pt with a strong mixing $\omega = 200$ keV.

188 shows an oblate predominance but the potential is almost flat in $\gamma$. If the mixing is strong, these nuclei become $\gamma$ unstable. In $^{186}$Pt and lighter isotopes the potential energy surface obtained implies coexistence, as can be seen from Fig. 1. The $^{190-194}$Pt isotopes have an potential energy surface which is flatter than those close to the semi-magic $^{204}$Pt isotope, resulting in a region of diverse shapes. As an example, Fig. 2 shows the $\gamma$-unstable surface of $^{192}$Pt found with a strong mixing. The phase diagram for all platinum isotopes will be published elsewhere [9].

A persistent problem of the IBM-1 calculations with configuration mixing is that the mixing parameter $\omega$ is ill determined. The measurement of two-proton or two-neutron (one-boson) transfer intensities should give direct information on the actual mixing. Depending on the strength of the mixing, transfer probabilities might show a sudden (phase-transitional?) change. Transfer experiments in this region might thus confirm or reject the predictions of the model.

This work was supported in part by CONACyT (Mexico).

## REFERENCES

1. A.N. Andreyev *et al.*, Nature **405**, 430 (2000).
2. G.D. Dracoulis *et al.*, Phys. Rev. C **44**, 1246(R) (1991); P.M. Davidson *et al.*, Nucl. Phys. A **657**, 219 (1999).
3. F.R. May *et al.*, Phys. Lett. B **68**, 113 (1977).
4. E. Caurier *et al.*, Rev. Mod. Phys. **77**, 427 (2005).
5. F. Iachello and A. Arima, *The Interacting Boson Model* (Cambridge University Press, Cambridge, 1987); A. Frank and P. Van Isacker, *Algebraic Methods in Molecular and Nuclear Structure Physics* (Wiley Interscience, New York, 1994).
6. M.K. Harder, K.T. Tang, and P. Van Isacker, Phys. Lett. B **405**, 25 (1997).
7. P.D. Duval and B.R. Barrett, Nucl. Phys. A **376**, 213 (1982).
8. A. Frank, P. Van Isacker, and C.E. Vargas, Phys. Rev. C **69**, 034323 (2004).
9. I. Morales, A. Frank, P. Van Isacker, and C.E. Vargas, to be published.

# List of Participants


**Andreyev, Andrei**
Insituut voor Kern- en Stralingfysica -IKS
University of Leuven,
Celestijnenlaan 200 D,
B-3001 Leuven, Belgium
Email: andrei.andreyev@fys.kuleuven.be

**Aoyama, Shigeyoshi**
Niigata University,
Nishiku Igarashi 2-8050,
Niigatashi, 950-2015, Japan
Email: aoyama@cc.niigat-u.ac.jp

**Astier, Alain**
CSNSM Orsay, bat. 104,
91405 Orsay Campus,
France
Email: alain.astier@csnsm.in2p3.fr

**Axiotis Michael**
Institute of Nuclear Physics,
NCSR "Demokritos"
153.10 Aghia Paraskevi, Athens, Greece

**Azaiez, Faisal**
IPN Orsay 91406
Orsay Cedex, France
Email: azaiez@ipno.in2p3.fr

**Ball, Gordon C.**
TRIUMF, 4004 Wesbrook Mall
Vancouver BC, V6T 2A3
Canada
Email: ball@triumf.ca

**Beck, Christian**
CNRS-Institut Pluridisciplinaire Hubert Curien,
Univ. Louis Pasteur Strasbourg,
23 rue du Loess, B.P. 28,
F-67037, Strasbourg Cedex 2, France
Email: christian.beck@ires.in2p3.fr

**Becker, Hans-Werner**
DTL and Institut für Experimentalphysik III
Ruhr-Universität Bochum, NB 3/173
Universitätsstr 150,
D-44780 Bochum, Germany
Email: hans-werner.becker@rub.de

**Belgaid, Mohamed**
Faculty of Physics
USTHB University
BP32 El-Ali BEZ Algiers, Algeria
Email: belgaidm@yahoo.com

**Bender, Michael**
Centre d'Etudes Nucleaires de Bordeaux
Gradignan – CENBG,
Chemin du Solarium, Le Haut Vigneau, BP120,
F-33175 Gradignan Cedex, France
Email : bender@cenbg.in2p3.fr

**Bentley, Michael A**
Department of Physics, University of York,
Heslington, York, YO10 5DD, UK
Email: mab503@york.ac.uk

**Benzoni, Giovanna**
INFN-Sezione di Milano,
via Celoria 16
20133 Milano, Italy.
Email: giovanna.benzoni@mi.infn.it

**Betak, Emil**
Institute of Physics SAS,
Dubravska cesta 9,
84511 Bratislava, Slovakia
Email: betak@savba.sk

**Billowes, Jon**
University of Manchester,
Shuster Laboratory
Manchester M13 9PL, UK
Email: jon.billowes@manchester.ac.uk

**Bonaccorso, Angela**
INFN, Sez. di Pisa,
Largo Pontecorvo 3, Edificio C,
56127 Pisa, Italy.
Email: bonac@df.unipi.it

**Boretzky, Konstanze**
GSI,
Planckstr. 1,
64291 Darmstadt, Germany
Email: k.boretzky@gsi.de

**Borge G, Maria José**
Instituto de Estructura de la Materia, CSIC,
Serrano 113bis, 28006-Madrid, Spain
Email: borge@iem.cfmac.csic.es

**Bredeweg, Tod**
Los Alamos National Laboratory,
PO Box 1663, MS J514, Los Alamos,
NM 87545, USA
Email: toddb@lanl.gov

**Brown, James R**
Department of Physics,
University of York,
Heslington, York, YO10 5DD, UK
Email: jb536@york.ac.uk

**Bruce, Alison**
School of Engineering,
University of Brighton,
Brighton, BN2 4GJ, UK
Email: alison.bruce@brighton.ac.uk

**Bürvenich, Thomas J.**
Frankfurt Institute for Advanced Studies,
Johann Wolfgang Goethe University
Ruth-Moufang Str. 1,
60438 Frankfurt am Main, Germany
Email: buervenich@fias.uni-frankfurt.de

**Burjan, Václav**
Nuclear Physics Institute CAS,
Řež near Prague,
253 68 Czech Republic
Email: burjan@ujf.cas.cz

**Capel, Pierre**
PNTPM (C.P. 229),
Université Libre de Bruxelles,
Bvd. du Triomphe, Campus Plaine
B-1050 Brussels, Belgium
Email: pierre.capel@centraliens.net

**Chadwick, Mark**
Los Alamos National Laboratory,
MS F699, Los Alamos,
NM 87544, USA
Email: mbchadwick@lanl.gov

**Chomaz, Philippe**
GANIL, Bd Henri Becquerel,
BP 55027, 14076 Caen, France
Email: chomaz@ganil.fr

**Chowdhury, Partha**
University of Massachusetts Lowell,
Department of Physics,
1 University Avenue, Lowell, MA 01854, USA
Email: partha_chowdhury@uml.ed

**Cullen, Dave M.**
University of Manchester,
Schuster Laboratory,
Manchester M13 9PL, UK
Email: dave.cullen@manchester.ac.uk

**D' Erasmo, Ginevra**
Dipartimento di Fisiac et INFN. Sezione di Bari,
Via Amendola 173,
70126 Bari, Italy
Email: derasmo@ba.infn.it

**Demetriou, Paraskevi**
Institute of Nuclear Physics,
NCSR "Demokritos",
153.10 Aghia Paraskevi, Athens, Greece
Email: vivian@inp.demokritos.gr

**De Angelis, Giacomo**
INFN-Laboratori Nazionali di Legnaro,
viale dell'Universita 2,
Legnaro, Italy
Email: deangelis@lnl.infn.it

**De Leo, Raffaele**
Physics Department,
University of Bari,
Bari, Italy
Email: deleo@ba.infn.it

**Dobes, Jan**
Nuclear Physics Institute ASCR,
250 68 Rez, Czech Republic
Email: dobes@ujf.cas.cz

**Dorvaux, Olivier**
Institut Pluridisciplinaire Hubert Curien,
Department of Subatomic Physics,
23 rue du Loess,
F-67037 Strasbourg, France
Email: olivier.dorvaux@ires.in2p3.fr

**Draayer, Jerry P.**
Dept. of Physics,
Louisiana State University,
Baton Rouge, LA 70808, USA
Email: draayer@sura.org

**Dressler, Rugard**
Laboratory for Radiochemistry and Environmental Chemistry,
Paul Scherrer Institute
CH-5232 Villigen-PSI, Switzerland
Email: rugard.dressler@psi.ch

**Ekström, Andreas**
Lund University,
Department of Physics, Box 118,
SE-221 00 Lund, Sweden
Email: andreas.ekstrom@nuclear.lu.se

**El Masri, Youssef**
Institute of Nuclear Physics,
Catholic University of Louvain,
Chemin du Cyclotron 2,
1348 Louvain-la-Neuve, Belgium
Email: elma@fynu.ucl.ac.be

**Ferreira, Lidia S**
Centro de Fisica Int. Fundamentais e Dep Fisica,
Instituto Superior Tecnico,
Av Rovisco Pais,
1049-001 Lisboa, Portugal
Email: flidia@ist.utl.pt

**Flocard, Hubert**
IN2P3/CNRS, CSNSM Bat 108,
91405 Orsay Campus,
France
Email: flocard@csnsm.in2p3.fr

**Foteinou, Varvara**
Tandem Accelerator Laboratory,
Institute of Nuclear Physics, NCSR
"Demokritos"
153.10 Aghia Paraskevi, Athens, Greece

**Fujimoto, Shin-ichirou**
Kumamoto National College,
2659-2 Suya, Goshi,
Kumamoto 861-1102, Japan
Email: fujimoto@ec.knct.ac.jp

**Gales, Sydney**
GANIL, Bd Henri Becquerel,
BP 55027, 14076 Caen, France
Email: gales@ganil.fr

**Garibaldi, Franco**
INFN-Roma 1-gr. SANITA
viale Regina Elena 299
00161-Rome, Italy
Email: franco.garibaldi@iss.infn.it

**Garrett, Paul**
Department of Physics,
University of Guelph, Guelph,
ON, N1G2W1 Canada
Email: pgarrett@physics.uoguelph.ca

**Geissel, Hans**
GSI,
Planckstr. 1,
64291 Darmstadt, Germany
Email: h.geissel@gsi.de

**Georgiev, Georgi**
CSNSM Orsay, bat. 104
91405 Orsay Campus,
France
Email: georgi.georgiev@csnsm.in2p3.fr

**Goriely, Stephane**
Institut d'Astronomie et d'Astrophysique - IAA,
Universite Libre de Bruxelles, Campus de la
Plaine - CP 226,B-1050 Brussels, Belgium
Email: sgoriely@astro.ulb.ac.be

**Grahn, Tuomas**
Department of Physics, Oliver Lodge Laboratory,
University of Liverpool,
Liverpool L69 7ZE, UK
Email: tg@ns.ph.liv.ac.uk

**Greenlees, Paul**
Department of Physics,
P.O. Box 35 (YFL),
FI-40014 University of Jyväskylä, Finland
Email: ptg@phys.jyu.fi

**Guttormsen, Magne**
Oslo Cyclotron Laboratory,
Department of Physics, University of Oslo,
P.O.Box 1048 Blindern, N-0316 Oslo, Norway
Email: magne.guttormsen@fys.uio.no

**Gyürky, György**
Institute of Nuclear Research (ATOMKI),
Bem ter 18/c,
H-4026 Debrecen, Hungary
Email: gyurky@atomki.hu

**Hanappe, Francis**
PNTPM, Université Libre de Bruxelles - ULB
Avenue F.D.Roosevelt 50,
B-1050, Bruxelles, Belgium
Email: fhanappe@ulb.ac.be

**Harissopulos, Sotirios**
Tandem Accelerator Laboratory,
Institute of Nuclear Physics, NCSR "Demokritos"
153.10 Aghia Paraskevi, Athens, Greece
Email: sharisop@inp.demokritos.gr

**Hilaire, Stephane**
CEA/DAM Ile-de-France,
DPTA/Servie de Physique Nucleaire,
BP12, 91680-Bruyeres-le-Chatel, France
Email: stephane.hilaire@cea.fr

**Hlavac, Stanislav**
Institute of Physics SAS,
Dubrovska cesta 9,
84511 Bratislava, Slovakia
Email: hlavac@savba.sk

**Huiza Pari, Juan Felix**
Universidade de São Paulo,
CRUSP-Cidade Universitária,
São Paulo - Brasil, CEP: 05508-060
Email: jhuiza@dfn.if.usp.br

**Jolie, Jan**
Institut für Kernphysik,
Universität zu Köln,
Zülpicher Str. 77,
D-50937 Köln, Germany
Email: jolie@ikp.uni-koeln.de

**Jones, Pete**
Department of Physics,
P.O. Box 35 (YFL),
FI-40014 University of Jyväskylä, Finland
Email: Pete.Jones@phys.jyu.fi

**Julin, Rauno**
Department of Physics,
P.O. Box 35 (YFL),
FI-40014 University of Jyväskylä, Finland
Email: julin@phys.jyu.fi

**Karabaya, Sofia**
Tandem Accelerator Laboratory,
Institute of Nuclear Physics, NCSR "Demokritos"
153.10 Aghia Paraskevi, Athens, Greece

**Karataglidis, Steven**
Department of Physics and Electronics,
Rhodes University,
P.O. Box 94,Grahamstown, 6140, South Africa
Email: S.Karataglidis@ru.ac.za

**Katsuma, Masahiko**
Institut d'Astronomie et d'Astrophysique - IAA,
Universite Libre de Bruxelles, Campus de la
Plaine - CP 226,B-1050 Brussels, Belgium
Email: mkatsuma@ulb.ac.be

**Ketelhut, Steffen**
Department of Physics,
P.O. Box 35 (YFL),
FI-40014 University of Jyväskylä, Finland
Email: steffen.ketelhut@phys.jyu.fi

**Kicinska-Habior, Marta**
Warsaw University,
ul Hoza 69, 00-681 Warszawa, Poland
Email: marta@fuw.edu.pl

**Kin, Tadahiro**
Japan Atomic Energy Agency,
2-4 Shirakata-Shirane, Tokai-Mura,
Naka-gun, Ibaraki-ken, 319-1195, Japan
Email: kin.tadahiro@jaea.go.jp

**Körner, Gabriele-Elisabeth**
NuPECC,
James-Franck-Str. 1,
D-85748 Garching, Germany
Email: sissy.koerner@ph.tum.de

**Konstantinopoulos, Theodoros**
Tandem Accelerator Laboratory,
Institute of Nuclear Physics, NCSR "Demokritos"
153.10 Aghia Paraskevi, Athens, Greece
Email: theodorekon@inp.demokritos.gr

**Kröll, Thorsten**
Physik-Department E12, TU Muenchen,
James-Franck-Strasse 1,
D-85748 Garching, Germany
Email: thorsten.kroell@ph.tum.de

**Kroha, Vaclav**
Nuclear Physics Institute CAS,
250 68 Řež, Czech Republic
Email: kroha@ujf.cas.cz

**Lagoyannis, Anastasios**
Tandem Accelerator Laboratory,
Institute of Nuclear Physics, NCSR "Demokritos"
153.10 Aghia Paraskevi, Athens, Greece
Email: lagoya@inp.demokritos.gr

**Larsen, Ann-Cecilie**
Oslo Cyclotron Laboratory,
Department of Physics,
University of Oslo,
P.O.Box 1048 Blindern, N-0316 Oslo, Norway
Email: a.c.larsen@fys.uio.no

**Leeb, Helmut**
Technische Universitaet Wien, Atominstitut,
Wiedner Hauptstrasse 8-10,
1040 Vienna, Austria
Email: leeb@kph.tuwien.ac.at

**Lefebvre-Schuhl, Anne**
CSNSM CNRS/IN2P3 -
Bat. 104, Universite Paris-Sud,
F-91405 Orsay-Campus, France
Email: lefebvre@csnsm.in2p3.fr

**Leoni, Silvia**
Dipartimento di Fisica,
Università di Milano and INFN Sez. di Milano,
Via Celoria 16, 20133 Milano, Italy
Email: silvia.leoni@mi.infn.it

**Lieder, Evgenia**
iThemba LABS,
23 Uitsig Dr,
Somerset West 7129, South Africa
Email: lieder@tlabs.ac.za

**Lieder, Reiner M.**
iThemba LABS,
23 Uitsig Dr,
Somerset West 7129, South Africa
Email: lieder@tlabs.ac.za

**Lipoglavsek, Matej**
Jozef Stefan Institute,
Jamova cesta 39, 1000 Ljubljana, Slovenia
Email: matej.lipoglavsek@ijs.si

**Lister, C.J. (Kim)**
Argonne National Laboratory,
9700 South Cass Ave,
Argonne, IL 60439, USA
Email: Lister@anl.gov

**Lo Iudice, Nicola**
Universita di Napoli Federico II,
Dipartimento di Scienze Fisiche
Monte S. Angelo, via Cintia,
80126 Napoli, Italy
Email: loiudice@na.infn.it

**Lubian Rios, Jesus**
Universidade Federal Fluminense,
Avenida Litoranea s/n Gragotá, Niteroi,
Rio de Janeiro, Brazil
Email: lubian@if.uff.br

**Lunardi, Santo**
Dipartimento di Fisica,
Università di Padova,
Via Marzolo 8, 35131 Padova, Italy
Email: lunardi@pd.infn.it

**Mach, Henryk**
Uppsala University,
Uppsala, Sweden
Email: henryk.mach@tsl.uu.se

**Maglione, Enrico**
INFN and Dipartimento di Fisica-
Università di Padova,
Via Marzolo 8, 35131 Padova, Italy
Email: maglione@pd.infn.it

**Mahata, Kripamay**
GSI,
Planck Str. 1,
D-64291 Darmstadt, Germany
Email: k.mahata@gsi.de

**Maierbeck, Peter**
Physik-Department E12, TU Muenchen,
James-Franck-Strasse 1,
D-85748 Garching, Germany
Email: peter.maierbeck@physik.tu-muenchen.de

**Maj, Adam**
IFJ PAN,
ul. Radzikowskiego 152,
31-342 Krakow, Poland
Email: Adam.Maj@ifj.edu.pl

**Majling, Lubomir**
Nuclear Physics Institute CAS,
250 68 Řež, Czech Republic
Email: majling@ujf.cas.cz

**Majlingova, Olga**
Czech Technical University,
Karlovo nam. 13,
120 00 Prague 2, Czech Republic
Email: olga.majlingova@fs.cvut.cz

**Martel, Ismael**
Universidad de Huelva,
Av. Fuerzas Armadas SN,
21071 Huelva, Spain
Email: imartel@uhu.es

**Maslov, Vladimir M**
JINER, 220109, Minsk-Sosny,
Krasina str. 99, Belarus
Email: maslov@bas-net.by

**Mazzocco, Marco**
INFN and Università di Padova,
Via Marzolo 8, 35131 Padova, Italy
Email: marco.mazzocco@pd.infn.it

**Meyer, Bradley S**
Department of Physics and Astronomy,
Clemson University,
Clemson, SC 29634-0978, USA
Email: mbradle@clemson.edu

**Mullins, Simon**
iThemba LABS, P.O.Box 722,
Somerset West 7129, South Africa
Email: smm@tlabs.ac.za

**Neyens, Gerda**
IKS, K.U. Leuven,
Celestijnenlaan 200 D,
B-3001 Leuven, Belgium
Email: gerda.neyens@fys.kuleuven.be

**Ohkubo, Shigeo**
Department of Applied Science,
Kochi Women's University,
Kochi 780-8515, Japan
Email: shigeo@cc.kochi-wu.ac.jp

**Ouichaoui, Saâd**
University of Sciences and Technology
H. Boumediéne (USTHB)
B.P. 32, El Alia, 16111 Bab Ezzouar, Algeria
Email: souichaoui@yahoo.co.uk

**Paar, Nils**
Department of Physics,
University of Zagreb
Ljudevita Gasa 26, 1043 Samobar, Croatia
Email: npaar@phy.hr

**Pakarinen, Janne**
Oliver Lodge Laboratory
University of Liverpool, Oxford Street,
L69 7ZE, UK
Email: jp@ns.ph.liv.ac.uk

**Pantaleo, Ambrogio**
INFN. Sezione di Bari,
Via Amendola 173,
70126 Bari, Italy
Email: pantaleo@ba.infn.it

**Parpottas, Yannis**
Department of Physics,
University of Cyprus,
Nicosia 1678, Cyprus
Email: parpottas@ucy.ac.cy

**Patronis, Nikolas**
IKS, K.U. Leuven, Celestijnenlaan 200 D,
B-3001 Leuven, Belgium
Email: Nikolaos.Patronis@fys.kuleuven.be

**Pedro de Jesus, Adelaide**
Centro de Física Nuclear, University of Lisbon,
Av. Gama Pinto, nº 2,
1649–003 Lisboa, Portugal
Email: apjesus@cii.fc.ul.pt

**Piasecki, Eryk**
Heavy Ion Laboratory, Warsaw University and
Institute for Nuclear Studies,
Hoza 69, 00-681 Warsaw, Poland
Email: Eryk.Piasecki@fuw.edu.pl

**Pizzone, Rosario Gianluca**
INFN - LNS,
via S. Sofia 62,
95123 Catania, Italy
Email: rgpizzone@lns.infn.it

**Popeko, Andrey**
FLNR JINR, Joliot Curie str, 6,
Dubna, Moscow region, 141980,
Russia
Email: popeko@jinr.ru

**Prati, Paolo**
INFN Sez. di Genova,
Via Dodecaneso 33,
16146 Genova, Italy
Email: prati@ge.infn.it

**Prieels, Rene**
Institut de Physique Nucleaire,
Universite catholique de Louvain,
2, chemin du Cyclotron,
B-1348 Louvain-la-Neuve, Belgium
Email: R.Prieels@fynu.ucl.ac.be

**Provatas, Georgios**
Tandem Accelerator Laboratory,
Institute of Nuclear Physics,
NCSR "Demokritos"
153.10 Aghia Paraskevi, Athens, Greece

**Riisager, Karsten**
PH Department, CERN,
CH-1211 Geneve 23,
Switzerland
Email: Karsten.Riisager@cern.ch

**Rosse, Bertrand**
CEA/DAM/DPTA/SPN BP 12,
91680 Bruyeres-le-Chatel, France
Email: bertrand.rosse@cea.fr

**Sahin, Eda**
Istanbul University,
Department of Physics, Beyazit-Istanbul, Turkey
Email:eda.sahin@lnl.infn.it/
edasahin@istanbul.edu.tr

**Sarén, Jan**
Department of Physics,
P.O. Box 35 (YFL),
FI-40014 University of Jyväskylä, Finland
Email: jan.saren@phys.jyu.fi

**Sarriguren, Pedro**
Instituto Estructura de la Materia, CSIC,
Serrano 123,
28006 Madrid, Spain
Email: sarriguren@iem.cfmac.csic.es

**Schumann, Dorothea**
Paul Scherrer Institute,
5232 Villigen-PSI,
Switzerland
Email: dorothea.schumann@psi.ch

**Seweryniak, Dariusz**
Argonne National Laboratory,
9700 South Cass Ave,
Argonne, IL 60439, USA
Email: seweryniak@anl.gov

**Sfienti, Concettina**
INFN Sez. di Catania & Univerita di Catania
Via. S. Sofia, I-95123
Catania, Italy
Email: c.sfienti@gsi.de

**Sharpey-Schafer, John F.**
University of Western Cape, Cape Town,
37 Companje Weg, Somerset West,
ZA 7130, South Africa
Email: jfss@tlabs.ac.za

**Siem, Sunniva**
Oslo Cyclotron Laboratory,
Department of Physics, University of Oslo,
P.O.Box 1048 Blindern, N-0316 Oslo, Norway
Email: sunniva.siem@fys.uio.no

**Sorlin, Olivier**
GANIL, Bd Henri Becquerel,
BP 55027, 14076 Caen, France
Email: sorlin@ganil.fr

**Spillane, Timothy**
University of Connecticut,
2152 Hillside Rd.,
Storrs, CT, USA, 06269
Email: spillane@phys.uconn.edu

**Styczen, Jan**
IFJ PAN, ul. Radzikowskiego 152,
31-342 Krakow, Poland
Email: Jan.Styczen@ifj.edu.pl

**Sugimoto, Satoru**
Department of Physics,
Kyoto University,
Kitashirakawa, Kyoto, 606-8502 Japan
Email: satoru@ruby.scphys.kyoto-u.ac.jp

**Suzuki, Toshio**
Nihon University, Department of Physics,
College of Humanities and Sciences,
Sakurajosui 3-25-40, Setagaya-ku,
Tokyo 156-8550, Japan
Email: suzuki@chs.nihon-u.ac.jp

**Syed, Naeem Ul Hasan**
Oslo Cyclotron Laboratory,
Department of Physics, University of Oslo,
P.O.Box 1048 Blindern, N-0316 Oslo, Norway
Email: n.u.h.syed@fys.uio.no

**Szanto de Toledo, Alejandro**
Universidade de Sao Paulo, Instituto di Fisica,
Rua do Matao, Travessa R, 187
CEP 05508-900 Cidade Universitaria
Sao Paulo, Brazil
Email: szanto@dfn.if.usp.br

**Utsunomiya, Hiroaki**
Konan University, Department of Physics,
8-9-1 Okamoto, Higashinada,
Kobe 658-8501, Japan
Email: hiro@konan-u.ac.jp

**Van de Walle, Jarno**
IKS, K.U. Leuven,
Celestijnenlaan 200 D,
B-3001 Leuven, Belgium
Email: jarno.van.de.walle@cern.ch

**Vargas, Carlos M.**
Facultad de Fisica e Inteligencia Artificial,
Universidad Veracruzana,
Sebastian Camacho 5, Centro, CP 91000,
Xalapa, Ver. Mexico.
Email: cavargas@uv.mx

**Wörtche, Heinrich Johannes**
Kernfysisch Versneller Instituut (KVI),
University of Groningen, Zernikelaan 25,
9747 AA Groningen, The Netherlands
Email: wortche@kvi.nl

**Wuosmaa, Alan H.**
Physics Department,
Western Michigan University,
1903 W. Michigan Ave,
Kalamazoo MI, 49008-5252 USA
Email: alan.wuosmaa@wmich.edu

**Yamada, Taiichi**
Laboratory of Physics,
Kanto Gakuin University,
Mutsuura Higashi 1-50,
Yokohama 236-8501, Japan
Email: yamada@kanto-gakuin.ac.jp

**Zilges, Andreas**
Institut fuer Kernphysik, TU Darmstadt,
Schlossgartenstr. 9,
D-64289 Darmstadt, Germany
Email: zilges@ikp.tu-darmstadt.de

# Author Index

## A


Ackermann, D., 56, 69, 139
Acosta, L., 333
Adrich, P., 250, 389
Akimune, H., 173
Aksenov, N. V., 69
Aksouh, F., 89, 291, 389
Aldrich, P., 347
Algora, A., 139
Alvarez, C. R., 305
Álvarez, M. A. G., 333
Alvarez-Pol, H., 89
Ames, F., 84, 291, 296
Amokrane, A., 344
Amorini, F., 208
Amos, K., 328
Andrés, M. V., 333
Andreyev, A. N., 413
Angulo, C., 333
Antalic, S., 69, 413
Antonov, A. N., 255
Arumugam, P., 341
Astier, A., 134
Aumann, T., 89, 250, 389
Aydin, S., 139
Azaiez, F., 134, 139, 300


## B


Babilon, M., 389
Balabanski, D., 377
Bark, R. A., 19, 383, 386, 404
Battacharyya, S., 300
Bauge, E., 123
Bazin, D., 347
Bazzacco, D., 26
Beck, C., 233
Becker, H.-W., 144, 179
Behr, K., 89
Behr, K.-H., 389
Behrens, T., 84, 291
Belgaid, M., 344
Belhout, A., 410
Belozerov, A. V., 64, 69, 419
Bém, P., 323
Bemmerer, D., 305
Bender, M., 278
Benjamin, E. A., 89
Benlliure, J., 89, 389
Bentley, M. A., 347
Benzoni, G., 11, 300
Berek, G., 404
Berg, T., 389
Běták, E., 313
Bettermann, L., 188
Bianco, L., 413
Bildstein, V., 84, 89, 291, 416
Billowes, J., 213
Blasi, N., 11, 300
Blazhev, A., 291
Boehmer, M., 389
Bogachev, A., 134
Böhmer, M., 89
Bonaccorso, A., 270
Bonche, P., 368
Bonetti, R., 305, 401
Bordeanu, C., 144
Boretzky, K., 89, 250, 389
Borge, M. J. G., 89, 193, 333
Bostan, M., 139
Bouchez, E., 368
Bozhikov, G. A., 69
Bracco, A., 11, 300
Braun, N., 188
Bredeweg, T., 111
Bree, N., 416
Briançon, Ch., 64
Broggini, C., 305
Brown, J. R., 347
Brünle, A., 89, 389
Bürger, A., 89
Burjan, V., 323
Burke, D. G., 188
Burkhard, H. G., 69
Bürvenich, T. J., 74
Buta, A., 134
Butler, P. A., 56, 84, 296, 368
Byrski, T., 139

## C

Caamano, M., 89
Camera, F., 11, 300
Campuzano, C., 431
Capel, P., 165
Cardella, G., 208
Carlsson, B. G., 383
Carpenter, M. P., 79, 265
Casarejos, E., 89, 333, 389
Cederkäll, J., 84, 291, 296
Chankova, R., 380, 425
Chartier, M., 389
Chatillon, A., 56, 89, 368, 389
Chelnokov, M. L., 64
Chepigin, V. I., 64, 69, 419
Cherubini, S., 155, 323
Chiba, S., 51
Chowdhury, P., 265
Chulkov, L. V., 89
Clément, E., 84, 291, 368
Cline, D., 265
Cocolios, T. E., 291
Comas, V. F., 69
Confortola, F., 305
Conti, E., 305
Cook, J. M., 347
Coquard, L., 84
Corradi, L., 139, 300
Corsi, A., 300
Cortina-Gil, D., 250, 389
Corvisiero, P., 305
Costantini, H., 305
Crema, E., 203, 362
Crespi, F. C. L., 11, 300
Crucillà, V., 155
Cullen, D. M., 220
Curien, D., 64, 134, 139

## D

Darby, I. G., 413
Davinson, T., 84, 291, 296
de Angelis, G., 26, 139, 300
De France, G., 139, 300
Delahaye, P., 84, 291
Delaroche, J.-P., 350
Della Vedova, F., 139
Demand, G. A., 188
Demetriou, P., 118, 179
De Napoli, M., 208
De Rosa, A., 401
Désesquelles, P., 64
Deveaux, L., 389
Dewald, A., 3, 260, 377
Diaz-Torres, A., 233
Diget, C. A., 347
Diriken, J., 84
Dmitriev, S. N., 69, 238
Dorvaux, O., 64, 134, 139, 368, 419
Dressler, R., 69
Duchêne, G., 134, 139, 383

## E

Eberth, J., 84, 291, 296
Eeckhaudt, S., 56, 368
Eichler, R., 69
Ekström, A., 84, 291, 296
Elekes, Z., 305
El Masri, Y., 118
Elvers, M., 389
Elze, Th. W., 250
Emling, H., 250, 389
Enders, J., 89
Eppinger, K., 89
Erduran, M. N., 139
Erturk, S., 139
Escrig, D., 333
Espino, J. M., 333

## F

Fabbietti, L., 89
Faestermann, T., 84, 89, 188
Fahlander, C., 296
Fallot, M., 250
Farinon, F., 401
Farnea, E., 139, 300
Faul, T., 139
Fedorov, D. V., 291
Fedosseev, V. N., 291
Fernandez-Dominguez, B., 389
Ferreira, L. S., 341
Finke, F., 84
Finlay, P., 188
Fioretto, E., 139

Formicola, A., 305
Fouka, M., 410
Fraile, L. M., 84, 193, 291
Franchoo, S., 84, 291, 300
Frank, A., 431
Friese, J., 89
Fu, C., 155
Fujimoto, S.-I., 318
Fülöp, Zs., 305
Furutaka, K., 374
Fynbo, H. O. U., 193

## G

Gade, A., 347
Gadea, A., 139, 300
Gäggeler, H. W., 69
Gagliardi, C. A., 323
Gaidarov, M. K., 255
Gál, J., 404
Galaviz, D., 333
Gales, S., 94
Gall, B., 64, 139
Gall, B. J. P., 56, 134, 368
Gan, Z., 69
García-Ramos, J. E., 333
Garrett, P. E., 188
Gascon, M., 89
Gast, W., 26, 383
Geissel, H., 89, 250
Georgiev, G., 84, 291, 296
Gerl, J., 89
Gernhäuser, R., 84, 89, 291, 416
Gervino, G., 305
Giacoppo, F., 208
Gialanella, L., 144
Gil, D. C., 89
Girod, M., 359
Glasmacher, T., 347
Glodariu, T., 401
Goko, S., 173
Goldberg, V., 155
Gomes, P. R. S., 203, 362
Gómez-Camacho, J., 333
Görgen, A., 56, 260, 296, 368
Goriely, S., 38, 123, 173, 350, 353, 356
Gorshkov, V. A., 64, 69
Górska, M., 89, 296, 389
Grahn, T., 56, 220, 260, 368, 377, 413
Gray-Jones, C., 56
Green, K. L., 188
Greene, J. P., 225
Greenlees, P. T., 56, 220, 260, 368, 377, 413
Greiner, W., 74
Gros, S., 79
Gueorguieva, E. A. , 19, 383, 386, 404
Guglielmetti, A., 305, 401
Gulino, M., 155
Gustavino, C., 305
Guttormsen, M., 64, 198, 380, 425, 428
Gyürky, Gy., 305

## H

Haas, F., 139
Habs, D., 84, 291, 296
Haenssler, F., 69
Hanappe, F., 64
Hansen, E. L., 386
Hansen, P. G., 89
Harano, H., 173
Harissopulos, S., 179, 377
Hashimoto, M.-A., 318
Hauschild, K., 64, 419
Hecht, A. A., 79
Heenen, P.-H., 368
Heinz, S., 69
Helariutta, K., 260
Hellström, M., 250
Henderson, D. J., 225
Heredia, J. A., 69
Hertenberger, R., 188
Herzberg, R.-D., 56, 368
Heßberger, F.-P., 56, 368
Heßberger, H. F., 69
Heyde, K., 291
Higashiyama, K., 51
Hilaire, S., 123, 173, 356, 359
Hirabayashi, Y., 407
Hjorth-Jensen, M., 296
Hofmann, S., 69
Honma, M., 51
Hons, Z., 323
Hornillos, M. B. G., 260
Hoteling, N., 79
Huber, G., 291
Huiza, J. F. P., 203, 362

Hüller, W., 389
Hurst, A., 84
Hürstel, A., 368
Huyse, M., 84, 291, 296, 416

## I

Ibrahim, F., 291, 300
Ickert, G., 389
Ikeda, K., 286
Iliadis, C., 160
Ilie, G., 377
Imbriani, G., 305
Ingebretsen, F., 380, 428
Inglima, G., 401
Isaev, S., 118
Itkis, M. G., 69
Ivanov, O., 84, 291, 296
Iwanicki, J., 84, 291, 296

## J

Jakobsson, U., 56
Janssens, R. V. F., 79, 225, 265
Jastrzębski, J., 238
Jenkins, D. G., 413
Jisonna, L., 225
Johansson, H., 389
Jolie, J., 3, 260, 291
Jones, G. D., 56, 368
Jones, K. L., 250
Jones, P. M. , 56, 220, 260, 368, 377, 413
Jonson, B., 89, 193
Joshi, P., 413
Juhász, K., 404
Julin, R., 56, 220, 260, 368, 377, 413
Jungclaus, A., 84
Junghans, A., 389
Junker, M., 305
Juutinen, S., 56, 220, 260, 368, 413

## K

Kabachenko, A. P., 64, 419
Kadem, F., 344
Kaihori, T., 173
Kajino, T., 51
Kalinka, G., 404
Kanungo, R., 89
Karagiannis, C., 389
Karataglidis, S., 328
Katsuma, M., 365
Kau, J., 19
Keeley, N., 233
Kent, P. E., 84
Kern, L., 389
Kester, O., 84, 291, 296
Ketelhut, S., 56, 260, 368
Kettunen, H., 368, 413
Keutgen, Th., 118
Khalfallah, F., 64, 134, 368
Khan, S., 220
Kheswa, Y., 383
Khoo, T.-L., 56, 79, 265
Khouaja, A., 64
Khuyagbaatar, J., 69
Kicińska-Habior, M., 371
Kimura, A., 374
Kin, T., 374
Kindler, B., 69
Kiselev, O., 89, 389
Kishada, A. M., 220
Kisieliński, M., 371
Klimkiewicz, A., 89, 250, 389
Koizumi, M., 374
Kojouharov, I., 69, 89
Komati, F., 19
Kondo, T., 173
Koning, A. J., 123, 173
Konstantinopoulos, T., 179, 377
Korichi, A., 64, 419
Korten, W., 56, 368
Köster, U., 84, 291, 296
Kowalczyk, M., 238, 371
Kozulin, E., 238
Kr, R., 188, 260
Krasznahorkay, A., 404
Kratz, J. V., 250
Krücken, R., 291
Krogulski, T., 238
Kroha, V., 323
Kröll, T., 260, 291
Kröll, Th., 84, 89, 416
Krücken, R., 84, 89, 416
Kulessa, R., 250
Kurtukian, T., 89
Kurz, N., 89, 389

Kuusiniemi, P., 69

## L

Labiche, M., 389
La Cognata, M., 155, 323
La Commara, M., 401
Lagoyannis, A., 179, 377
Lamia, L., 155
Larsen, A. C., 64, 198, 380, 425, 428
Larsson, K., 89
Laube, A., 69
Lauer, M., 291
Lauritsen, T., 79, 265
Lawrie, J. J., 19, 383, 386, 404
Leach, K. G., 188
Lebedev, V. Ya., 69
Le Bleis, T., 89, 389
Le Coz, Y., 368
Lefebvre-Schuhl, A., 150
Leino, M., 56, 69, 220, 260, 368, 413
Lemmon, R., 389
Lemut, A., 305
Lenzi, S., 139
Leoni, S., 11, 300
Leppänen, A. P., 220
Leppänen, A.-P., 368, 413
Leske, J., 84
Liang, C. L., 225
Liang, X., 139
Lieder, E. O., 26, 383, 386, 404
Lieder, R. M., 26, 383, 386, 404
Lighthall, J. C., 225
Limata, B., 305
Lindenberg, K., 389
Lipoglav, M., 386, 404
Lisetskiy, A. F., 291
Lister, C. J., 79, 128, 265
Litvinov, Y., 389
Ljungvall, J., 260
Lo Iudice, N., 183
Lommel, B., 69
Lönnroth, T., 380, 428
Lopez-Martens, A., 64, 419
Lotay, G., 79
Lucas, R., 134
Lui, Y.-W., 173
Lukstins, J., 392
Lunardi, S., 139, 300
Lutter, R., 84, 291

## M

Madurga, M., 193
Maglione, E., 341
Mahata, K., 89, 389
Mahgoub, M., 84, 416
Maier, L., 89
Maierbeck, P., 84, 89, 260, 389, 416
Maine, P., 19
Majling, L., 392
Majlingova, O., 395
Makinaga, A., 173
Malyshev, O. N., 64, 69, 419
Mann, R., 69
Marginean, N., 139
Marketin, T., 46
Marley, S. T., 225
Marsh, B. A., 291, 296
Marta, M., 305
Martel, I., 333
Martin, B., 401
Martin, D., 84
Maslov, V. M., 398
Mason, P. J. R., 139, 220, 300
Matsumoto, T., 173
Mayet, P., 84, 291
Mazzocchi, C., 305, 401
Mazzocco, M., 401
McDaniel, S. M., 347
Melon, B., 260
Menegazzo, R., 305
Messahel, L., 410
Messelt, S., 380
Meyer, B. S., 31
Meyer, M., 134
Million, B., 11, 300
Minkova, A., 19, 64
Mishustin, I. N., 74
Modamio, V., 84
Molnár, J., 404
Montagnoli, G., 139, 300
Montanari, D., 11, 300
Monteiro, D. S., 203, 362
Moon, S., 56, 368
Moore, E. F., 225
Morales, I. O., 431
Morgan, T., 84

Moro, A. M., 333
Moya de Guerra, E., 255
Mrázek, J., 323
Mukha, I., 333
Mukhamedzhanov, A. M., 155, 323
Müller, S., 389
Mullins, S. M., 19, 383, 386, 404
Münch, M., 84
Murray, S. H. T., 19, 404
Mutterer, M., 238
Müzenberg, G., 69

**N**

Napoli, D. R., 139
Nardelli, S., 84
Navin, A., 300
Ncapayi, N. J., 19
Nespolo, M., 139
Nicolini, R., 300
Niedermaier, O., 84, 291
Nieminen, P., 220, 368, 413
Nilsson, T., 89, 193, 291, 389
Nishimura, N., 318
Nishio, K., 69
Nociforo, C., 89, 250, 389
Nomoto, K., 51
Novák, J., 323
Nowacki, F., 139
Ntshangase, S. S., 386, 404
Nyakó, B. M., 404
Nyhaus, H. T., 425
Nyman, G., 89, 193
Nyman, M., 56, 220, 260, 368, 413

**O**

Ødegård, S. W., 425
Oganessian, Yu. Ts., 64, 69
Ogloblin, A., 238
Ohkubo, S., 407
Oliveira, J. R. B., 139
Orlandi, R., 139
Oshima, M., 374
Otsuka, T., 51
Ouichaoui, S., 410

**P**

Paar, N., 46, 250
Page, R. D., 260, 413
Pakarinen, J., 56, 220, 368, 413
Palit, R., 250, 389
Pantea, M., 84, 291
Papadakis, P., 56
Papka, P., 383, 386, 404
Pardo, R. C., 225
Parfenov, A. N., 392
Parpottas, Y., 160
Paschalis, S., 389
Pascual-Izarra, C., 89
Pasini, M., 84
Pasternak, A. A., 26, 383
Patel, N., 225
Patronis, N., 84, 416
Paul, M., 225
Pearson, J. M., 353
Perea, A., 89
Perez, D., 89
Pérez-Bernal, F., 333
Perkowski, J., 368, 413
Perru, O., 291
Peterson, D., 79, 225
Petrovič, T., 386
Petrushkin, O. V., 69
Petts, A., 260
Peura, P., 56
Phillips, A. A., 188
Piasecki, E., 238
Piasecki, K., 238
Pieper, S. C., 225
Pierroutsakou, D., 401
Pietralla, N., 84
Piguet, D., 69
Piqueras, I., 134
Pisko, Š., 323
Pissulla, Th., 260
Pizzone, R. G., 155, 323
Plunkett, A., 323
Pollarolo, G., 139, 300
Popeko, A. G., 64, 69, 419
Porquet, M.-G., 134
Pramanik, U. D., 250, 389
Prati, P., 305
Prezado, Y., 193
Prieels, R., 118
Pritychenko, B., 347

Prochazka, A., 89
Prokopowicz, W., 389
Puglia, S. M. R., 155

## R

Raabe, R., 291, 416
Raciti, G., 208
Raddon, P. M., 413
Ragnarsson, I., 383
Rahkila, P., 56, 220, 260, 368, 413
Raiola, F., 144
Rakhila, P., 377
Rapisarda, E., 208
Rapisarda, G. G., 155
Rasmussen, P., 69
Ratkiewicz, A., 347
Recchia, F., 139
Redon, N., 134
Rehm, K. E., 225
Reiter, P., 84, 291, 296
Rejmund, M., 300
Rekstad, J., 198, 380, 425
Richer, M., 56
Rigby, S. V., 220
Riisager, K., 106, 193
Ring, P., 46
Robin, J., 139
Robinson, A. P., 79
Roca, V., 305
Rodríguez, D., 333
Rodriguez-Tajes, C., 89
Rolfs, C., 144, 305
Romano, M., 144
Romano, S., 155, 323
Romoli, M., 401
Rossi, D., 89, 389
Rostron, D., 56, 413
Rother, W., 377
Rousseau, M., 64, 134, 368
Rowley, N., 64, 238
Rowley, N., H. J. Wörtche, 238
Rudolph, K., 84
Rusek, K., 238, 333

## S

Sagaidak, R. N., 64
Sahin, E., 139
Sánchez-Benítez, A. M., 333
Sandoli, M., 401
Santos, C., 56
Sarén, J., 56, 260, 368, 413
Saro, S., 69
Sarriguren, P., 255
Savard, G., 225
Sawicka, M., 291
Scarlassara, F., 139, 300
Schaffner, H., 89
Scheilke, S., 84
Scheit, H., 84, 291, 296
Scherillo, A., 84
Scheurer, J. N., 404
Schiffer, J. P., 225
Schiller, A., 198, 380, 425, 428
Schmaker, M. A., 188
Scholey, C., 56, 220, 260, 368, 413
Schött, H. J., 69
Schrieder, G., 89, 291
Schumann, D., 422
Schürmann, D., 144
Schwalm, D., 84, 291, 296
Schweitzer, J., 144
Schwertel, S., 89
Segel, R. E., 225
Seidlitz, M., 84
Seliverstov, M. D., 291
Sergi, S., 155
Seweryniak, D., 79, 265
Sfienti, C., 208
Sharo, S., 64
Sharpey-Schafer, J. F., 19, 383, 404
Shirikova, N. Yu., 183
Shirinda, O., 404
Shishkin, S. V., 69
Shorto, J. M. B., 203, 362
Shutov, A. V., 64, 69, 419
Sieber, T., 84, 291
Siem, S., 64, 198, 296, 380, 425, 428
Siemssen, R. H., 225
Signorini, C., 401
Sillanpää, M., 238
Sim, J. M. B., 362
Šimečková, E., 323
Simões, R. F., 203
Simon, H., 89, 250, 389
Singh, R. P., 139
Sitar, B., 89
Siwek, K., 347

Sletten, G., 291
Smirnov, S., 238
Smirnova, N., 291
Solar, M., 392
Somorjai, E., 305
Sopko, B., 392
Soramel, F., 401
Sorri, J., 56, 260, 413
Speidel, K.-H., 84
Spillane, T., 144
Spitaleri, C., 155, 323
Stanoiu, M., 89, 291
Stefan, I., 300
Stefanescu, I., 84, 291, 296, 416
Stefanini, A. M., 139, 300
Stézowski, O., 134
Straniero, O., 305
Streicher, B., 69, 413
Strieder, F., 144, 305
Stroe, L., 401
Stuchlík, Z., 313
Stuttgé, V. I. L., 64
Sugimoto, S., 286
Sullignano, B., 69
Sümmerer, K., 89, 250, 389
Surówka, G., 250
Sushkov, A. V., 183
Suzuki, T., 51
Svensson, C. E., 188
Svirikhin, A. I., 64, 69, 419
Świderski, Ł., 238
Syed, N. U. H., 64, 198, 380, 425, 428
Szilner, S., 139

## T

Tabacaru, G., 155
Tandel, S. K., 56, 265
Tandel, U. S., 265
Tang, X., 225
Taylor, M. J., 347
Tengblad, O., 89, 193, 333
Tereshatov, E. E., 69
Terrasi, F., 305
Theisen, Ch., 56, 64, 368
Thirolf, P., 84
Thomas, J.-C., 291
Timár, J., 404
Tiourin, G., 238
Toh, Y., 374
Toki, H., 286
Toyokawa, H., 173
Trabelsi, A., 410
Trache, L., 155, 323
Trautvetter, H. P., 305
Tribble, R., 155
Tribble, R. E., 323
Trzaska, W. H., 238
Tsertos, H., 160
Tumino, A., 155, 323
Tveten, G. M., 296

## U

Umeda, H., 51
Ur, C., 139
Urbanec, M., 313
Utsunomiya, H., 173
Uusitalo, J., 56, 69, 220, 260, 368, 413

## V

Valiente-Dobón, J. J., 139, 291, 300
Van de Vel, K., 413
Van de Walle, J., 84, 291, 296, 416
Van Duppen, P., 84, 291, 296, 416
Van Isacker, P., 431
Van Mol, J., 118
Vardaci, E., 401
Vargas, C. E., 431
Varley, B. J., 220
Vencelj, M., 386
Venhart, M., 56, 69, 413
Venkova, Ts., 134
Verney, D., 291, 300
Vincour, J., 323
Voinov, A., 198, 380, 425, 428
Vostokin, G. K., 69
Voulot, D., 84, 291, 296
Vretenar, D., 46, 250
Vymers, P., 19

## W

Wadsworth, R., 260, 413
Wagner, A., 389

Walker, J., 84
Walters, W. B., 79
Walus, W., 389
Waluś, W., 250
Wang, X., 79
Warr, N., 84, 291, 296
Wegrzecki, M., 69
Weick, H., 89, 389
Weinzierl, W., 84
Weisshaar, D., 84, 291, 296, 347
Wenander, F., 84, 291, 296
Wiedemann, K. T., 139
Wieland, O., 11, 300
Wilson, J. N., 368
Winkler, M., 389
Winkler, S., 89
Wiringa, R. B., 225
Wirth, H.-F., 188
Wiseman, D., 413
Wójcik-Gront, E., 371
Wolf, B. H., 291
Wong, J., 188
Woods, P. J., 79
Wörtche, H. J., 243
Wu, C.-Y., 265
Wuosmaa, A. H., 225

**Y**

Yamagata, T., 173
Yeremin, A. V., 64, 69, 419
Yoshida, T., 51

**Z**

Zalite, A., 300
Zeng, S., 144
Zhang, Y. H., 139
Zhou, X. H., 139
Zhu, S., 79, 265
Zielińska, M., 260, 291, 296
Zocca, F., 300
Zolnai, L., 404